Lecture Notes in Artificial Intelligence 12469

Subseries of Lecture Notes in Computer Science

More information about this series at http://www.springer.com/series/1244

Lourdes Martínez-Villaseñor ·
Oscar Herrera-Alcántara ·
Hiram Ponce · Félix A. Castro-Espinoza (Eds.)

Advances in Computational Intelligence

19th Mexican International Conference
on Artificial Intelligence, MICAI 2020
Mexico City, Mexico, October 12–17, 2020
Proceedings, Part II

 Springer

Editors
Lourdes Martínez-Villaseñor (ID)
Facultad de Ingeniería
Universidad Panamericana
Mexico City, Mexico

Hiram Ponce (ID)
Facultad de Ingeniería
Universidad Panamericana
Mexico City, Mexico

Oscar Herrera-Alcántara (ID)
Universidad Autónoma Metropolitana
Mexico City, Mexico

Félix A. Castro-Espinoza (ID)
Universidad Autónoma del Estado
de Hidalgo
Hidalgo, Mexico

ISSN 0302-9743   ISSN 1611-3349   (electronic)
Lecture Notes in Artificial Intelligence
ISBN 978-3-030-60886-6   ISBN 978-3-030-60887-3   (eBook)
https://doi.org/10.1007/978-3-030-60887-3

LNCS Sublibrary: SL7 – Artificial Intelligence

This Springer imprint is published by the registered company Springer Nature Switzerland AG
The registered company address is: Gewerbestrasse 11, 6330 Cham, Switzerland

Preface

The Mexican International Conference on Artificial Intelligence (MICAI) is a yearly international conference series that has been organized by the Mexican Society for Artificial Intelligence (SMIA) since 2000. MICAI is a major international artificial intelligence (AI) forum and the main event in the academic life of the country's growing AI community.

MICAI conferences publish high-quality papers in all areas of AI and its applications. The proceedings of the previous MICAI events have been published by Springer in its *Lecture Notes in Artificial Intelligence* (LNAI) series, vol. 1793, 2313, 2972, 3789, 4293, 4827, 5317, 5845, 6437, 6438, 7094, 7095, 7629, 7630, 8265, 8266, 8856, 8857, 9413, 9414, 10061, 10062, 10632, 10633, 11288, 11289, and 11835. Since its foundation in 2000, the conference has been growing in popularity and improving in quality.

The proceedings of MICAI 2020 are published in two volumes. The first volume, *Advances in Soft Computing*, contains 37 papers structured into three sections:

- Machine and Deep Learning
- Evolutionary and Metaheuristic Algorithms
- Soft Computing

The second volume, *Advances in Computational Intelligence*, contains 40 papers structured into three sections:

- Natural Language Processing
- Image Processing and Pattern Recognition
- Intelligent Applications and Robotics

The two-volume set will be of interest for researchers in all fields of AI, students specializing in related topics, and for the public in general interested in recent developments in AI.

The conference received for evaluation 186 submissions from 26 countries: Argentina, Armenia, Austria, Belgium, Brazil, Colombia, Cuba, Czech Republic, Ecuador, Finland, France, India, Ireland, Kazakhstan, Mexico, Nepal, Nigeria, Pakistan, Peru, Portugal, Russia, South Africa, Spain, Switzerland, Ukraine, and USA. From these submissions, 77 papers were selected for publication in these two volumes after a peer-reviewing process carried out by the International Program Committee. The acceptance rate was 41%.

The International Program Committee consisted of 152 experts from 16 countries: Australia, Brazil, Colombia, France, Greece, Ireland, Japan, Kazakhstan, Malaysia, Mexico, Philippines, Portugal, Russia, Spain, the UK, and the USA.

MICAI 2020 was honored by the presence of renowned experts who gave excellent keynote lectures:

- Ljiljana Trajkovic, Simon Fraser University, Canada
- Pedro Larrañaga, Technical University of Madrid, Spain

- Manuel Morales, University of Montreal, Canada
- Soujanya Poria, Singapore University of Technology and Design, Singapore
- Francisco Hiram Calvo Castro, CIC-IPN, Mexico

Four workshops were held jointly with the conference:

- The 13th Workshop on Intelligent Learning Environments (WILE 2020)
- The 13th Workshop of Hybrid Intelligent Systems (HIS 2020)
- The Second Workshop on New Trends in Computational Intelligence and Applications (CIAPP 2020)
- The 6th Workshop on Intelligent Decision Support Systems for Industry (WIDSSI 2020)

The authors of the following papers received the Best Paper Awards based on the paper's overall quality, significance, and originality of the reported results:

- First place: "An NSGA-III-based Multi-Objective Intelligent Autoscaler for Executing Engineering Applications in Cloud Infrastructures," by Virginia Yannibelli, Elina Pacini, David Monge, Cristian Mateos, and Guillermo Rodriguez (Argentina)
- Second place: "Speaker Identification using Entropygrams and Convolutional Neural Networks," by Antonio Camarena-Ibarrola, Karina Figueroa, and Jonathan García (Mexico)
- Third place: "Dissimilarity-Based Correlation of Movements and Events on Circular Scales of Space and Time," by Ildar Batyrshin, Nailya Kubysheva, and Valery Tarassov (Mexico/Russia)

We want to thank all the people involved in the organization of this conference: to the authors of the papers published in these two volumes – it is their research work that gives value proceedings – and to the organizers for their work. We thank to the reviewers for their great effort spent on reviewing the submissions, to the track chairs for their hard work, and to the Program and Organizing Committee members.

We are deeply grateful to the Universidad Panamericana for their warm hospitality of MICAI 2020. We would like to express our gratitude to Dr. Santiago García Álvarez, Rector of the Universidad Panamericana, Campus Mexico, Dr. Alejandro Ordoñez Torres, Director of the School of Engineering of the Universidad Panamericana, Campus Mexico, and Dr. Roberto González Ojeda, Secretary of Research at the School of Engineering.

The entire submission, reviewing, and selection process, as well as preparation of the proceedings, was supported by the EasyChair system (www.easychair.org). Last but not least, we are grateful to Springer for their patience and help in the preparation of these volumes.

October 2020

Lourdes Martínez-Villaseñor
Oscar Herrera-Alcántara
Hiram Ponce
Félix A. Castro-Espinoza

Organization

MICAI 2020 was organized by the Mexican Society of Artificial Intelligence (SMIA, Sociedad Mexicana de Inteligencia Artificial) in collaboration with the Universidad Panamericana, the Universidad Autónoma del Estado de Hidalgo, and the Universidad Autónoma Metropolitana.

The MICAI series website is www.MICAI.org. The website of SMIA, is www.SMIA.mx. Contact options and additional information can be found on these websites.

Conference Committee

General Chair

Félix A. Castro Espinoza   Universidad Autónoma del Estado de Hidalgo, Mexico

Program Chairs

Lourdes Martínez-Villaseñor	Universidad Panamericana, Mexico
Oscar Herrera-Alcántara	Universidad Autónoma Metropolitana, Mexico

Workshop Chair

Noé Alejandro Castro Sánchez   Centro Nacional de Investigación y Desarrollo Tecnológico, Mexico

Tutorials Chair

Roberto Antonio Vózquez Espinoza de los Monteros   Universidad La Salle, Mexico

Doctoral Consortium Chairs

Miguel González Mendoza	Tecnolóogico de Monterrey, Mexico
Juan Martínez Miranda	CICESE Research Center, Mexico

Keynote Talks Chair

Sabino Miranda Jiménez   INFOTEC, Mexico

Publication Chair

Hiram Ponce   Universidad Panamericana, Mexico

Financial Chair

Oscar Herrera-Alcántara   Universidad Autónoma Metropolitana, Mexico

Grant Chair

Félix A. Castro Espinoza   Universidad Autónoma del Estado de Hidalgo, Mexico

Local Organizing Committee

Local Chair

Hiram Ponce   Universidad Panamericana, Mexico

Local Logistics Chairs

Lourdes   Universidad Panamericana, Mexico
  Martínez-Villaseñor
Karina Pérez Daniel   Universidad Panamericana, Mexico
León Palafox   Universidad Panamericana, Mexico

Finance Chairs

Hiram Ponce   Universidad Panamericana, Mexico
Lourdes   Universidad Panamericana, Mexico
  Martínez-Villaseñor

Publicity Chairs

Hiram Ponce   Universidad Panamericana, Mexico
Monserrat Rosas Libreros   Universidad Panamericana, Mexico

Track Chairs

Natural Language Processing

Grigori Sidorov   CIC-IPN, Mexico
Obdulia Pichardo Lagunas   CIC-IPN, Mexico

Machine Learning

Alexander Gelbukh   CIC-IPN, Mexico
Navonil Majumder   CIC-IPN, Mexico

Deep Learning

Pierre Baldi   University of California, Irvine, USA
Francisco Viveros Jiménez   Eficiencia Informativa, Mexico

Evolutionary and Metaheuristic Algorithms

Laura Cruz Reyes   Instituto Tecnológico de Ciudad Madero, Mexico
Roberto Antonio Vázquez   Universidad La Salle, Mexico
  Espinoza de los
  Monteros

Soft Computing

Ildar Batyrshin	CIC-IPN, Mexico
Miguel González Mendoza	Tecnológico de Monterrey, Mexico
Gilberto Ochoa Ruiz	Tecnológico de Monterrey, Mexico

Image Processing and Pattern Recognition

Heydy Castillejos	Universidad Autónoma del Estado de Hidalgo, Mexico
Francisco Hiram Calvo Castro	CIC-IPN, Mexico

Robotics

Luis Martín Torres Treviño	Universidad Autónoma de Nuevo León, Mexico
Eloísa García Canseco	Universidad Autónoma de Baja California, Mexico

Intelligent Applications and Social Network Analysis

Helena Gomez Adorno	IIMAS-UNAM, Mexico
Iris Iddaly Méndez Gurrola	Universidad Autónoma de Ciudad Juárez, Mexico

Other Artificial Intelligence Approaches

Nestor Velasco Bermeo	University College Dublin, Ireland
Gustavo Arroyo Figueroa	Instituto Nacional de Electricidad y Energías Limpias, Mexico

Program Committee

Antonio Marín Hernández	Universidad Veracruzana, Mexico
Juan Martínez Miranda	CICESE, Mexico
Iskander Akhmetov	IICT, Kazakhstan
José David Alanís Urquieta	Universidad Tecnológica de Puebla, Mexico
Miguel Ángel Alonso Arévalo	CICESE, Mexico
Giner Alor Hernandez	Instituto Tecnológico de Orizaba, Mexico
Maaz Amjad	CIC-IPN, Mexico
Erikssen Aquino	ITSSMT, Mexico
Segun Aroyehun	CIC-IPN, Mexico
Gustavo Arroyo Figueroa	Instituto Nacional de Electricidad y Energías Limpias, Mexico
Ignacio Arroyo-Fernández	Universidad Tecnológica de la Mixteca, Mexico
Pierre Baldi	University of California, Irvine, USA
Alejandro Israel Barranco Gutiérrez	Cátedras CONACYT, Instituto Tecnológico de Celaya, Mexico
Ramon Barraza	Universidad Autónoma de Ciudad Juárez, Mexico
Ari Yair Barrera-Animas	Tecnológico de Monterrey, Mexico
Rafael Batres	Tecnológico de Monterrey, Mexico
Ildar Batyrshin	CIC-IPN, Mexico

Gemma Bel-Enguix	UNAM, Mexico
Igor Bolshakov	Russian State University for the Humanities, Russia
Vadim Borisov	National Research University, MPEI, Russia
Alexander Bozhenyuk	Southern Federal University, Russia
Ramon F. Brena	Tecnológico de Monterrey, Mexico
Davide Buscaldi	LIPN, Université Sorbonne Paris Nord, France
Alan Calderón Velderrain	CICESE, Mexico
Francisco Hiram Calvo Castro	CIC-IPN, Mexico
Ruben Cariño Escobar	INR, Mexico
J. Víctor Carrera-Trejo	Instituto Nacional de Astrofísica, Óptica y Electrónica, Mexico
Heydy Castillejos	Universidad Autónoma del Estado de Hildago, Mexico
Norberto Castillo García	Instituto Tecnológico de Ciudad Madero, Mexico
Félix A. Castro Espinoza	Universidad Autónoma del Estado de Hildago, Mexico
Noé Alejandro Castro Sánchez	Centro Nacional de Investigación y Desarrollo Tecnológico, Mexico
Ofelia Cervantes	Universidad de las Américas Puebla, Mexico
Haruna Chiroma	Federal College of Education (Technical) Gombe, Nigeria
Nareli Cruz Cortés	CIC-IPN, Mexico
Laura Cruz-Reyes	Instituto Tecnologico de Ciudad Madero, Mexico
Andre de Carvalho	University of São Paulo, Brazil
Andrés Espinal	Universidad de Guanajuato, Mexico
Edgardo Manuel Felipe Riverón	Centro de Investigación en Computación, Instituto Politécnico Nacional, Mexico
Denis Filatov	Sceptica Scientific Ltd., UK
Dora-Luz Flores	Universidad Autónoma de Baja California, Mexico
Juan José Flores	Universidad Michoacana, Mexico
Leticia Flores-Pulido	Universidad Autónoma de Tlaxcala, Mexico
Roilhi Frajo Ibarra Hernández	CICESE, Mexico
Anilu Franco-Arcega	Instituto Nacional de Astrofísica, Óptica y Electrónica, Mexico
Gibrán Fuentes-Pineda	UNAM, Mexico
Sofía N. Galicia-Haro	UNAM, Mexico
Eloisa García	UABC, Mexico
Vicente García	Universidad Autónoma de Ciudad Juárez, Mexico
Alexander Gelbukh	CIC-IPN, Mexico
Salvador Godoy-Calderón	CIC-IPN, Mexico
Claudia Gómez	Instituto Tecnológico de Ciudad Madero, Mexico
Helena Gómez	UNAM, Mexico
Eduardo Gómez-Ramírez	Universidad La Salle, Mexico
Gabriel González	TecNM, CENIDET, Mexico
Pedro Pablo González	Universidad Autónoma Metropolitana, Mexico
José Ángel González Fraga	UABC, Mexico

Luis-Carlos González-Gurrola — Universidad Autónoma de Chihuahua, Mexico

Miguel González Mendoza — Tecnológico de Monterrey, Mexico

Crina Grosan — Brunel University London, UK

Fernando Gudiño — FES CUAUTITLAN-UNAM, Mexico

Joaquín Gutiérrez Juaguey — Cibnor, Mexico

Rafael Guzmán Cabrera — Universidad de Guanajuato, Mexico

Yasunari Harada — Waseda University, Japan

Jorge Hermosillo — UAEM, Mexico

Yasmín Hernández — Instituto Nacional de Electricidad y Energías Limpias, Mexico

José Alberto Hernández — Universidad Auónoma del Estado de Morelos, Mexico

Paula Hernández Hernández — Instituto Tecnológico de Ciudad Madero, Mexico

Betania Hernández-Ocaña — Universidad Juárez Autónoma de Tabasco, Mexico

Oscar Herrera-Alcóntara — Universidad Autónoma Metropolitana, Mexico

Joel Ilao — De La Salle University, Philippines

Jorge Jaimes — Universidad Autónoma Metropolitana, Mexico

Olga Kolesnikova — Escuela Superior de Cómputo, IPN, Mexico

Nailya Kubysheva — Kazan Federal University, Russia

Diana Laura Vergara — CIC-IPN, Mexico

José Antonio León-Borges — UPQROO, Mexico

Mario Locez-Loces — Tecnológico Nacional de México, Instituto Tecnológico de Ciudad Madero, Mexico

Víctor Lomas-Barrie — IIMAS-UNAM, Mexico

Rodrigo López Faróas — CONACYT-CentroGeo, Mexico

Omar Jehovani López Orozco — Instituto Tecnológco Superior de Apatzingán, Mexico

Asdrubal López-Chau — CU UAEM Zumpango, Mexico

Gerardo Loreto — Instituto Tecnologico Superior de Uruapan, Mexico

Navonil Majumder — ORG, Mexico

Ilia Markov — Inria, France

Lourdes Martínez-Villaseñor — Universidad Panamericana, Mexico

Bella Citlali Martínez-Seis — IPN, Mexico

César Medina-Trejo — Tecnológico Nacional de México, Mexico

Iris Iddaly Méndez Gurrola — Universidad Autónoma de Ciudad Juárez, Mexico

Ivan Meza — IIMAS-UNAM, Mexico

Efrén Mezura-Montes — Universidad Veracruzana, Mexico

Sabino Miranda Jiménez — INFOTEC, Mexico

Daniela Moctezuma — CONACYT-CentroGEO, Mexico

Omar Montaño Rivas — Universidad Politécnica de San Luis Potosí, Mexico

Roman Anselmo Mora-Gutierrez — UNAM, Mexico

Saturnino Job Morales Escobar — Centro Universitario UAEM Valle de México, Mexico

Guillermo Morales-Luna — CINVESTAV-IPN, Mexico

Merlin Teodosia Suárez	Center for Empathic Human-Computer Interactions, Philippines
Israel Tabárez	Tecnológico de Monterrey, Mexico
Eric S. Téllez	CONACYT-INFOTEC, Mexico
David Tinoco	UNAM, Mexico
Aurora Torres	Universidad Autónoma de Aguascalientes, Mexico
Luis Torres Treviño	Universidad Autónoma de Nuevo León, Mexico
Diego Uribe	Instituto Tecnológico de la Laguna, Mexico
José Valdez	CIC-IPN, Mexico
Genoveva Vargas Solar	CNRS-LIG-LAFMIA, France
Roberto Antonio Vázquez Espinoza de los Monteros	Universidad La Salle, Mexico
Nestor Velasco Bermeo	University College Dublin, Ireland
Yenny Villuendas Rey	CIDETEC, Mexico
Francisco Viveros Jiménez	Eficiencia Informativa, Mexico
Fernando Von Borstel	Cibnor, Mexico
Saúl Zapotecas Martínez	Universidad Autónoma Metropolitana, Mexico
Alisa Zhila	NTENT, USA
Miguel Ángel Zúñiga García	Instituto Nacional de Electricidad y Energías Limpias, Mexico

Additional Reviewers

Alan Arturo Calderó-Velderrain	José Eduardo Valdez Rodríguez
Alberto Iturbe	Kazuhiro Takeuchi
Ángel Rodríguez	Marco Sotelo-Figueroa
Erick Ordaz	María Dolores Torres
Erik Ricardo Palacios Garza	Mario Aguilera
Fernando Von Borstel	Miguel Alonso
Gabriel Sepúlveda-Cervantes	Roilhi Frajo Ibarra Hernández
Hector-Gabriel Acosta-Mesa	Romeo Sanchez Nigenda
Joaquín Gutiérrez	Saúl Domínguez-Isidro
Jorge Alberto Soria-Alcaraz	Teodoro Macías-Escobar
José A. González-Fraga	Yasushi Tsubota

Contents – Part II

Best Paper Award, Third Place

Image Processing and Pattern Recognition

Intelligent Applications and Robotics

Contents – Part I

Evolutionary and Metaheuristic Algorithms

Natural Language Processing

An Algorithm to Detect Variations in Writing Styles of Columnists After Major Political Changes

Rodolfo Escobar[1], Luis Juarez[1], Erik Molino-Minero-Re[2],
and Antonio Neme[2(✉)]

[1] Postgraduate Program in Computer Science and Engineering, Universidad Nacional
Autonoma de Mexico, Yucatan Campus, Mérida, Yucatan, Mexico
[2] IIMAS, Universidad Nacional Autonoma de Mexico, Yucatan Campus, Mérida,
Yucatan, Mexico
{erik.molino,antonio.neme}@iimas.unam.mx

Abstract. Writers tend to follow a certain style that can be detected or at least sketched by an appropriate algorithm. Columnists in newspapers, being also writers, follow their specific style. The style tends to be stable once writers reach maturity, but it is subject to change when internal or external circumstances differ. Here, we apply a bag-of-words approach to approximate the style of several journalists working in Mexican newspapers, and we track their style for a long period of time with the aim of detecting changes when external circumstances, in particular political ones, change. This provided us with an environment for detecting variations in stylomics, which is the closest we can get to an experiment. In particular, we collected hundreds of writings of ten Mexican columnists from different newspapers, both previous to the Presidential Mexican elections of 2018 and posterior to it. We processed these documents on different supervised and not supervised learning algorithms, such as random forest, principal component analysis, and k-means. Likewise, we implemented different validation procedures. As a result, we detected that the style in all studied columnists suffered tangible changes in the frequency of use of some particular words, particularly at specific times, some of which may be related to the 2018 Mexican presidential elections.

Keywords: Stylomics · Bag of words · Political changes

1 Introduction

Writers tend to follow a certain style that can consists in a group of attributes and a function of those attributes [1]. Theoretically, the style of a writer is unique, and two writers, although may share some common aspects, will be properly tell apart by inspecting their styles. The study of patterns followed by an author is commonly known as stylistics or stylomics [2]. The identification of the actual style is an open-ended task, and its relevance covers a wide range, from

L. Martínez-Villaseñor et al. (Eds.): MICAI 2020, LNAI 12469, pp. 3–16, 2020.
https://doi.org/10.1007/978-3-030-60887-3_1

authorship attribution, digital forensics, and literary studies [3]. Of particular interest to us is detecting changes in the style of certain professional writers in response to external circumstances. In particular, our goal is to study the changes in what political and other columnists write once a regime has changed.

The Mexican Presidential Elections of 2018 are seen by several analysis as an abrupt variation in National politics [4,5], although some may have a different opinion [6]. In any case, it is relevant to detect changes in the perception of the overall political scenario in Mexico not only in certain indicators such as gross domestic product or some other economic and political -related attribute. A candidate for a proxy to detect national perception is to pay close attention to columnists in the media. In a democratic regime, columnists are though of as the voice of vast portions of the population. If this assertion is valid, then, conducting a close inspection of possible changes in what columnists write, a clearer picture can be obtained about the perception of politicians. The easiest way is to focus on those writing in newspapers, since it is straightforward to acquire the texts. In particular, those columnists that publish a periodic column, either daily or weekly, are the natural candidates to follow.

Our hypothesis is that a political change affects the perception of power, and it can be detected via changes in what columnists write. In other words, what journalists used to write before the elections may differ from what they publish after the elections. Our *ansatz* is that this is the case, and in the present contribution, we pursuit to corroborate it via machine learning tools. Our endeavour is to follow several columnists (or the equivalent in the Mexican journalism ecosystem) along several years before the elections. Then, we aim to obtain a summary of their writings up to that date, followed by the computation of a similar summary from their writings after that electoral process. Once we have a description of their writings before and after the inflection point, we can compare them. The comparison is via a classification algorithm.

Several approaches are available to corroborate our hypothesis. Sentiment analysis is a possible route to follow. It has shown its applicability and relevance in political contexts [12]. However, we are more interested in less abstract aspects than those offered by this tool. Writing style has also been defined as a particular cadence in the use of words [18], but it is possible that the style might not be affected after elections. We postulate that a simpler form is all what is needed to validate or reject our hypothesis. We focus here in a direct and easily measurable metric, in particular, the analysis of histograms or probability mass functions of the use of words. By inspecting an histogram, and comparing groups of them, we can highlight if columnists changed their writing patterns.

The Bag-of-Words (BoW) concept treats texts as structure-less instances, in which the only relevant attributes are the relative frequency of use or appearance of words (tokens), leaving grammar and any other aspect aside. The term was mentioned as early as more than fifty years ago, in a diminishing way to bold the relevance of structure in language [10]. Despite heavy criticism, BoW has shown interesting results in several contexts, such as spam detection [11], and also in authorship attribution [13].

Aligned with the BoW approach, the perspective offered by word2vec is a rather powerful one [14], based on coding of words as weights in neural networks. The tools under this umbrella allows a link from syntax to semantics. However, this technique requires very large datasets, and more importantly, might hide the changes in writing styles. Since we are interested in detecting changes in the general way an author writes after a given date, we postulate that relative frequency is enough. We will give evidence of this postulate while the paper unveils.

Different alternatives exist to BoW, such as that presented in [15]. There, a deep structure between words is obtained, and from it, inferences about the possible authorship are made. Neural networks with deep architectures offer also a possibility, at the expense of making inferences about the relevant attributes a task on itself [16].

Changes in writing styles over time is relevant not only from the literary point of view, but also, from medical aspects. The subtle shift of style in some novelists has been approached in [17]. By using descriptive statistics over the vocabulary and the use of some words, authors detected a change between texts written before and after a specific date. In [18], by using neural networks and information-theoretic related tools, authors detected discrepancies in the style of the last novel of the same author. In both contributions, the specific date as associated to a certain novelist being diagnosed with Alzheimer's disease. Although this event is deep since it affects the brain, it is a proof of principle that changes, in this case internal, affect the style. In the context of political opinion, which is closer to the research presented here, [7], authors infer policy positions derived from political texts. They studied texts within a period of five years with the aim of classifying whether the text states in favor or against certain policies.

In this contribution, we loosely refer to the style of a columnists as the probability mass distribution of words in his/her vocabulary. As we will describe in the rest of the contribution, when referring to some columnists, how they use words, and how frequent, changes not only as a function of time, but more importantly, as we argue latter on, as a function of a major political change. One of the advantages of using a BoW approach is its interpretability. By using the appropriate classification algorithm, the relevance of the words is directly computed.

In Sect. 2 we describe the proposed algorithm. We present the results of applying this algorithm to ten Mexican columnists in Sect. 3, and we discuss the potential utility of the algorithms and the results it offer in Sect. 4. In the same section, we offer some conclusions.

2 The Algorithm

First, all texts from the same columnist j are scanned to create his/her vocabulary, V_j. Then, in a second scan, each text t is mapped into a new space. This space is generated from the vocabulary, and it has $|V_j|$ dimensions. The coordinates of text t are linked to the relative frequency (probability) of use of each word in the vocabulary. Thus, the text t is a point in the space of relative frequencies. The N-gram approach, although simple, allows the identification of relevant patterns. We define γ_t as the location of text j in the vocabulary space V_j.

Each text t has associated a label, indicating whether it was written before or after the elections. This label is referred to as L_j. The values for L_j can be either 0, if the text was written before the election day, or 1, if it was written after. Now we have defined all the required elements to frame the problem we are attacking within classification theory. The attributes are the relative frequency of use of each word within the vocabulary, and the label is whether the text was written before or after the election date. What we are interested in solving is thus the classification task $L_j = \gamma_j$.

Figure 1 depicts the algorithm we followed to try to determine whether columnists are using a different style after the elections, when compared to the one they followed before the hypothetical inflection point. The error we considered is the average of false positives (FP) and false negatives (FN).

A classifier is an algorithm that, given a set of vectors or observations, described by several attributes, tries to link them to elements in a second set, namely the labels or classes. In other words, a classifier provides a function between elements in a set, with elements in a second one. This function can be either implicit, such as in the case of neural networks, or explicit, as in linear regression. The former tend to show very low errors (when provided by enough data both in quantity and quality), but tend to be only poorly interpretable. The latter are very interpretable, but tend to have very high errors, since almost no relevant phenomena is well described by a linear association. One classifier that maintains interpretability, as well as low errors is that of Random Forests (RF) [20]. It is an ensemble method that generates a group of decision trees [21], and for each one of them, randomly selects a subsample of the whole training set. The overall decision is computed considering the decision made by the majority [22] of the decision trees. In many implementations, the subsamples are randomly altered as to give more robustness to the ensemble.

We postulate that the harder it is for a classifier to tell apart the texts written before from those written after the elections, the less likely it is that the studied columnist changed her/his style or subjects of interest. That is, a measure of the difficulty to classify correctly is given by the number of attributes needed for that task. If the texts of a columnist are correctly classified using only a handful of attributes, then, his/her style suffered a change given by the modification of use on those attributes. On the other hand, if for a columnist it is required a large number of attributes to correctly classify the before and after dichotomy, then her/his style is more stable, since more attributes are needed.

Since there are several thousand possible attributes, a dimensionality reduction algorithm is needed to maintain the complexity as low as possible [8]. The simplest of those algorithms is based on the relative frequency of use of each word. We started with the k most frequent words and trained a classifier using exclusively those k attributes. We varied k from 5 to $|V_j|$. For each value of k, we

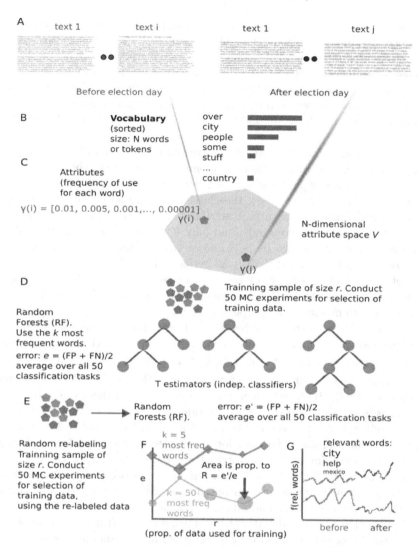

Fig. 1. The algorithm to detect changes in style before and after the Mexican elections of 2018. It starts by obtaining the vocabulary of all texts of a given columnist (A). The frequency of use of each word in the vocabulary is an attribute. The vocabulary is ranked in decreasing order by frequency of use (B). Each text is mapped to the attribute space, of dimension equal to the size of the vocabulary (C). From this space, a classification based on random forests (RF) is conducted. The classes are *before* or *after*. The k most common words are considered for the classification, and a fraction r of the total sample is used for training. 50 Monte Carlo (MC) experiments for each value of (r, k) were conducted (D). Since training data is lower than the number of attributes, 100 random permutation of labels were conducted and a new training as before was performed (E). The test error is reported for several values of k and r. The error of the classification of the permuted classed over the error obtained for the correct labeling is indicated by a geometric figures of size proportional to that ratio (F). RF identifies the most relevant attributes, and from these, we create a new variable displayed as the y-axis (G).

split the sample into training and test sets. In order to achieve a high degree of confidence, we varied the size of the training sample, r, from 40% to 90% of the total sample with increases of 10%, and the remaining percentage of the sample was used for validation. Unless otherwise stated, the results we report are based on test errors. For each combination of k and r, we conducted 50 Monte Carlo (MC) experiments, and we keep only the average test error. With this approach, we intend to decrease the likelihood that what we are observing is the result of pure chance.

Since the number of attributes V_j is greater than the number of samples, we might face spurious effects due to randomness, that are closely linked to the course of dimensionality [9]. Consider the following extreme case. If only one text belongs to each class ($L_0| = |L_1|$), the classification would be almost certainly straightforward, since the feature or attribute space (the relative frequency of words) is in the order of thousands, and almost any word would have a different frequency among the two texts. In order to fade down this possibility, we conducted a permutation test. This was performed as follows. For each combination of (r, k), that is, size of the training set given by r, and based on the k most frequent words, the label of each vector was randomly changed. The only constraint was to maintain the proportion of texts written before and after, for each columnist. With the new re-labeling, a RF was trained. This was conducted 100 times, and the average error was considered. Let e_j be the error obtained by the classifier for columnist j and using the correct labeling. Let e'_j be the average error over the 100 label permutations. Let $R_j = e'_j/e_j$. There are three relevant cases for R_j. If $R_j \approx 1$, then, the classification over the actual labeling is meaningless, since a classification over the same data with a different labeling is equally possible. If $R_j < 1$, the random labeling leads to an easier classification than the current one, since the error for the former is lower than for the latter. This indicates that the number of attributes is either too low, or the classifier is not adequate. Finally, if $R_j > 1$, it means that the original labeling is easier to classify than the randomly permuted case. We are interested in this last case, specially when $R_j \gg 1$. The rationale behind it is that the greater R, the most likely is that the considered attributes are relevant. R is a measure of the relevant signal (correct labeling) over background noise (random labeling).

Text clustering is a data mining technique that automatically groups a large amount of documents into meaningful categories, formally known as clusters [24]. One of our main objectives in to comprehend the changes before and after the elections in the writing style of the authors. A BoW approach was followed to represent the text samples through a set of variables, followed by a clustering algorithm was employed to label every document from the same author into two possible classes, each representing different writing styles.

The method selected for unsupervised learning was spectral clustering. This algorithm is able to create relevant clusters, and it has the capability to recognize non-convex distributions, as opposed to standard clustering algorithms [25] . Simultaneously, each document was manually assorted in one of the two classes, through the first class were represented all the texts that were written before a threshold date and with the second class those that were written after. From the automatic labeling generated by the clustering algorithm and manual labeling were obtained classification measures (Accuracy,F1-score, AUC-ROC) for different time thresholds and for both of the possible labels generated by the clustering method, choosing the date with the greatest classification measure.

3   Results

We studied ten columnists, all of them writing in Spanish, for a time span well before the National elections of 2018 (in some cases, up to ten years before that), and until the beginning of June, 2020. Table 1 shows a general description of the writings of the ten inspected columnists, indicating their id, number of texts, and range of publishing dates.

Table 1. Columnists documents details

Columnist	No. texts before	No. texts after	Oldest	Most recent	Vocabulary	Entropy
FMM	110	101	21.02.2016	29.05.2020	29409	12.6993
DD	41	37	02.03.2014	08.06.2020	12147	11.9905
PG	117	98	18.03.2016	29.05.2020	15023	11.6433
ML	53	49	27.06.2016	12.05.2020	11928	11.9569
JHL	122	98	13.01.2012	03.06.2020	19005	12.3522
EGO	135	100	13.01.2012	22.05.2020	18566	12.2066
CFV	170	119	13.01.2012	04.06.2020	17765	11.7841
MSA	61	69	16.03.2015	01.06.2020	12257	11.8604
EK	104	58	14.10.2015	20.04.2020	20567	12.455
JW	73	24	01.05.2007	21.05.2020	14818	12.1766

Figure 2-A shows the results of classifying texts in the dichotomy of before and after the elections. Once again, if such a date would be an irrelevant one for the purposes of the present work, the binary classification would be very hard for the classification algorithm. In each of the ten panels, it is shown the difficulty for the RF to correctly classify the texts from each columnist. In the x-axis, it is displayed r, that is, the proportion of the sample to be used as the training set. In the y-axis, it is shown the average false positive and false negative error. For each value of r, it is shown in y the average result of 50 Monte Carlo experiments, that is, the classification task was conducted 50 times, each time over a randomly selected set for the training and test datasets, satisfying the condition on the proportion r for the training set. For each columnist, it is shown the classification error for the k most frequent words.

Associated to each tuple (k, r), there is a geometric figure that varies in size. The area of this figure is proportional to R, the ratio of the error of the permuted-label case over the error of the correct labeling. As a reminder, the higher the value of R, the more likely is that the classification is relevant, since the error is much lower in the correct labeling than in the randomly labeling cases. The larger the associated geometric figure, the more likely is that the classification is meaningful.

We increased k from a value of 5 up to the first value for which the error is below 0.05. We refer to the lowest value of k with such a low error and a value of $R > 2$ as G. We further increased k to achieve errors below 0.01. Correspondingly, the lowest k for which this low error is obtained and a signal with respect to background is at least double the size ($R > 2$) is referred to as E. Figure 2-B shows both G and E for the ten columnists under analysis. Some of the columnists require a rather low number of words to be characterized as before or after. Since not all columnists are represented by the same number of texts, it is necessary to verify that the number of texts is not causing this effect. In Fig. 2-C, it is shown the relation between $log(G)$ and the number of texts for each columnist. As observed, there is no significant correlation between these two parameters ($R^2 = 0.1$). Several columnists have a similar G but achieved over a wide range of number of texts (JHL, MSA, JW). Similarly, columnists represented by a similar number of texts have a considerable difference in G, such as JW and ML.

It is observed that G and E are spread over a wide range. However, and quite surprisingly, for some of the columnists, G is rather low. This is an indication that exists a significant variation in the use of certain keywords within the two classes.

In Fig. 3, it is displayed a metric condensing the relevant attributes which allowed RF to classify texts as either written before or after the election day. For each columnist, we selected the RF trained from the combination of k words and training size r for which G was achieved. For each subject, it is also shown the most relevant words to classify the texts from him/her, ranked by RF. Instead of showing a map obtained by some dimensional reduction technique, we show, as a function of time, a function of the relevant attributes. Based on the most relevant attributes, a random variable was obtained via the method described in [19]. This random variable is presented as a function of publishing date (x-axis).

A first hypothesis about the nature of the words that are relevant for classification would be the name of the president. Once the new president is elected, his/her name will dominate the texts from the columnists, and the name of his/her predecessor would start to vanish from the texts. However, this was the case for only two columnists (see Fig. 3). Some of the relevant words are grouped in the context of politics, such as *power, president,* or *presidential.* Others denote change, such as *new* and *was.* Moreover, it is worth noting that some columnists seemed to change patterns before the election day, like EK, EGO, and JW, which may be related to the electoral campaign, that started a few months earlier, and it is a highly active period for the political news media.

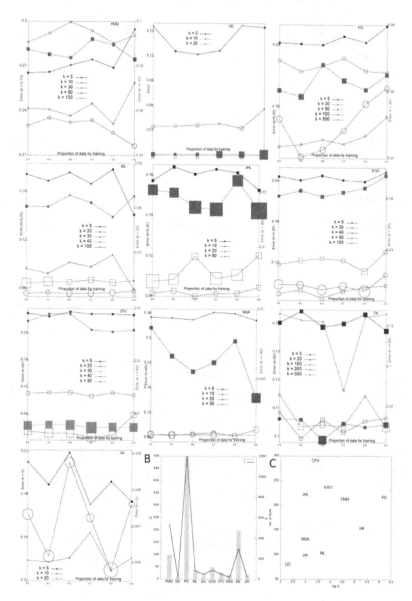

Fig. 2. The classification error for the ten studied columnists. It is shown the average test error (FP + FN) over 50 training experiments (Monte Carlo), with the proportion of the sample used for training as indicated in the x−axis. The classification was conducted by random forests (RF), using the k most frequent words for each author. Since the range of the error is large for varying k, two y-scales are in order, one at the left, and one at the right, coded by blue and red, respectively. To gain confidence in that the classification is meaningful, a randomly re-labeling was conducted over 100 Monte Carlo experiments, maintaining the proportion of texts written before and after the election day. The area of the geometric figure is proportional to the ratio of the average error for the re-labeled cases over the correct labeling. B. The number of words needed to achieve errors below 0.05 (G), and below 0.01 (E). (Color figure online)

12   R. Escobar et al.

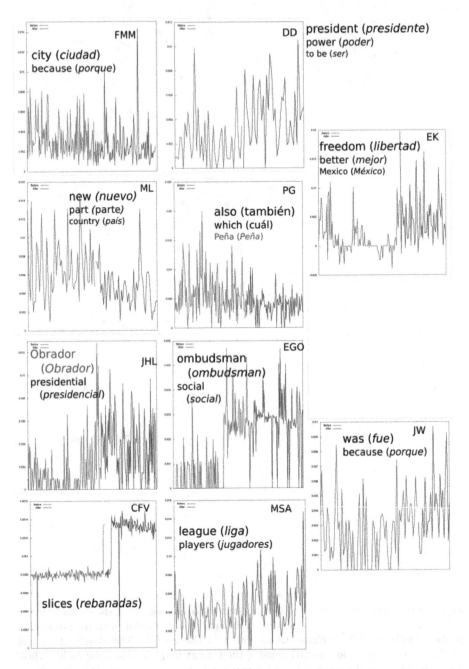

Fig. 3. The description of texts written before (red) and after (blue) the Mexican Presidential elections of July, 2018. For each of the ten columnists, it is shown the list of the most relevant words, obtained by RF, in a decreasing order. Green words indicate a personal name. The random variable (signal) was obtained from the relative frequency of the listed words, as described in [19] (Color figure online)

An alternative approach to detect classes changes before and after an event is via dimensionality reduction. In Fig. 4, it is shown, for three of the columnists, the results of applying principal component analysis (PCA) over the same list of attributes on Figs. 2 and 3. PCA'1 data into a new space, with the same number of dimensions, with the peculiarity that the new dimensions are created considering the maximum variance in the data. Under ideal conditions, the firsts two or three principal components explain most or the variance (dispersion) in data, which allows to plot projections in two or three dimensions [23]. Thus, here in Fig. 4 it is observed that in general, there is a clear separation between texts written before and after the 2018 election day.

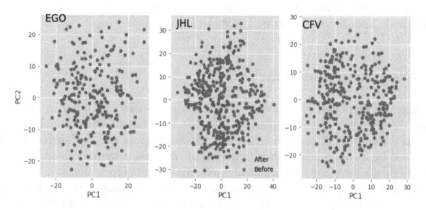

Fig. 4. Principal component analysis of three subsets of texts featured by the mutual information filtered bag of words method. Texts written before the elections are shown in blue and those written after that are shown in red. (Color figure online)

Analogously, we proposed a method based on spectral clustering to find the separation date in the writing style. What is shown in Table 2 is the date that separates each of the clusters, using different metrics. This algorithm gives us an idea of the authors that are suspected to be highly influenced by the 2018 presidential elections in Mexico, that is, through this approach, authors whose separation date is closer to the date of the presidential elections were more influenced by it. The performance of the classifiers is also shown in Table 2. Higher performances means that it is easy to distinguish between before and after in your texts. performances with low values, means that the author has more homogeneity in his writing style and it is more difficult for him to be influenced by external factors, or for him to have various writing styles over time.

Table 2. Spectral clustering results

Columnist	SD Acc	SD F1-Score	SD AUC-ROC	Accuracy	F1-Score	AUC-ROC
CFV	2016-12-24	2012-01-13	2012-01-13	0.61	0.71	0.78
DD	2019-12-27	2014-06-21	2020-02-04	0.72	0.81	0.73
EGO	2012-01-13	2012-01-13	2018-07-03	0.98	0.99	0.52
EK	2018-04-13	2018-04-13	2018-05-02	0.79	0.78	0.82
FMM	2016-11-06	2016-11-06	2016-02-21	0.63	0.74	0.79
JHL	2019-02-28	2012-01-13	2020-03-06	0.7	0.72	0.8
JW	2003-11-01	2003-11-01	2003-11-01	0.72	0.83	0.85
ML	2020-03-08	2016-06-27	2020-04-30	0.77	0.86	0.89
MSA	2018-07-16	2016-06-06	2020-02-03	0.64	0.71	0.76
PG	2016-03-18	2016-03-18	2016-03-18	0.72	0.83	0.86

4   Discussion and Conclusions

A change in the balance of power after elections can be measured in multiple forms. Some may be easily checked via hard facts, related to traditional metrics such as gross product or average income. However, more subtle changes can be perceived by a systematic observation of what columnists write. Political columnists in democratic regimes are though of reliable indicators of public opinion. In this contribution, we tested the hypothesis that columnists have undergone changes in their writings after the Mexican Presidential elections of July, 2018.

We studied hundreds of texts of ten columnists well before and up to 22 months after the cited elections. We approached the problem with a Bag of Words scheme to analyze texts. Each text was characterized by the relative frequency of words, and the vocabulary defined the feature space. Each dimension is the relative frequency of a word in the given text. In this space, each text was assigned its label, either written before, or after the elections. This allowed to frame the problem within a classification task. Using random forests and a robust scheme involving Monte Carlo and label permutation, we were able to classify with low errors texts written as before or after the election.

The texts of some columnists were easily classified, requiring only a small set of words. This means that with a low number of attributes, a change in what columnists write before and after the elections was detected. Other columnists were comparatively harder to classify, since the relative frequency of up to 200 words were needed. In any case, what is important in this context, is that there was a detectable change in the use of words before and after a relevant political event. In all cases, we controlled over both, text size and the number of texts for each columnist. Additionally, we validated these changes over random permutations of labels and compared the performance in both cases.

We conclude that it is possible to detect significant and stable changes in the use of words before and after the elections in several columnists. We have gathered strong evidence to support our hypothesis, that is, political changes may affect the style of what political columnists write.

As future work, we will apply sentiment analysis over the same context of political columnists. A change in the frequency of use of words was already demonstrated here, so the next natural step is to study the phenomena in a more abstract perspective, detecting changes in sentiments after political changes.

Acknowledgements. This work was partially supported by UNAM-PAPIIT IA103420. AN and EMMR thank SNI CONACyT.

References

1. Juola, P.: Authorship Attribution. NOW Press, Delft (2008)
2. Rocha, A.: Authorship attribution for social media forensics. IEEE Trans. Inf. Forensics Secur. **12**(1), 5–33 (2017). https://doi.org/10.1109/TIFS.2016.2603960
3. Varela, P., Justino, E., Britto, A., Bortolozzi, F.: A computational approach for authorship attribution of literary texts using sintatic features. In: 2016 International Joint Conference on Neural Networks (IJCNN), pp. 4835–4842 (2016). doi: 10.1109/IJCNN.2016.7727835
4. Mexico election: historic landslide victory for leftist AMLO. The Guardian. https://www.theguardian.com/world/2018/jul/02/mexico-election-leftist-amlo-set-for-historic-landslide-victory. Retrieved on 01 Jun 2020
5. Mexican general election. https://en.wikipedia.org/wiki/2018_Mexican_general_election. Retrieved on 01 Jun 2020
6. Flannery, N.: Political Risk Analysis: What To Expect After Mexico's 2018 Presidential Election. https://www.forbes.com/sites/nathanielparishflannery/2018/06/26/political-risk-analysis-what-to-expect-from-mexicos-2018-presidential-election/#73ebe8685a76. Retrieved on 01 Jun 2020
7. Laver, M., Benoit, K., Garry, J.: Extracting policy positions from political texts using words as data. Am. Political Sci. Rev. **97**(2), 311–331 (2003)
8. Kumar, S., Santosh, R.: Effective information retrieval and feature minimization technique for semantic web data. Comput. Electr. Eng. **81**, 106518 (2018). https://doi.org/10.1016/j.compeleceng.2019.106518
9. Hughes, G.F.: On the mean accuracy of statistical pattern recognizers. IEEE Trans. Inf. Theor. **14**(1), 55–63 (1968)
10. Harris, Z.: Distributional structure. Word **10**(2/3), 146–62 (1954). https://doi.org/10.1080/00437956.1954.11659520
11. Sahami, M., Dumais, S., Heckerman, D., Horvitz, E.: A Bayesian approach to filtering junk e-mail. In: AAAI'98 Workshop on Learning for Text Categorization (1988)
12. Ge, J., Alonso-Vazquez, M., Gretzel, U.: Sentiment analysis: a review. In: Sigala, M., Gretzel, U. (eds.) Advances in Social Media for Travel, Tourism, and Hospitality (2017)
13. Boughaci, D., Benmesbah, M., Zebiri, A.: An improved N-grams based model for authorship attribution. In: International Conference on Computer and Information Sciences (ICCIS), Sakaka, Saudi Arabia, pp. 1–6 (2019). doi: 10.1109/ICCISci.2019.8716391

14. Mikolov, T., Sutskever, I., Chen, K., Corrado, G., Dean, J.: Distributed representations of words and phrases and their compositionality. In: Advances in Neural Information Processing Systems. arXiv:1310.4546 (2013)
15. Gómez-Adorno, H., Sidorov, G., Pinto, D., Vilarino, D., Gelbukh, A.: Automatic authorship detection using textual patterns extracted from integrated syntactic graphs. Sensors **16**, 1374 (2016). https://doi.org/10.3390/s16091374
16. Shrestha, P., et al.: Convolutional neural networks for authorship attribution of short texts. In: Proceedings of the 15th Conference of the European Chapter of the Association for Computational Linguistics, vol. 2 (2017)
17. Garrard, P., Maloney, L.M., Hodges, J.R., Patterson, K.: The effects of very early Alzheimer's disease on the characteristics of writing by a renowned author. Brain **128**, 250–260 (2005)
18. Neme, A., Pulido, J.R.G., Muńos, A., Hernández, S., Dey, T.: Stylistics analysis and authorship attribution algorithms based on self-organizing maps. Neurocomputing **147**(5), 147–159 (2015)
19. Neme, A., Hernández, S., Nido, A., Islas, C.: Multilayer Perceptrons as Classifiers Guided by Mutual Information and Trained with Genetic Algorithms. In: Yin, H., Costa, J.A.F., Barreto, G. (eds.) IDEAL 2012. LNCS, vol. 7435, pp. 176–183. Springer, Heidelberg (2012). https://doi.org/10.1007/978-3-642-32639-4_22
20. Ho, T.K.: Random decision forests. In: Proceedings of the 3rd International Conference on Document Analysis and Recognition, Montreal, pp. 278–282 (1995)
21. Quinlan, J.R.: Induction of decision trees. Mach. Learn. **1**, 81–106 (1996). https://doi.org/10.1007/BF00116251
22. Breiman, L.: Random forests. Mach. Learn. **45**(1), 5–32 (2001). https://doi.org/10.1023/A:1010933404324
23. Principal, J.I., Analysis, C.: Springer Series in Statistics, vol. 195. Springer, New York (2002)
24. Mohammed, A.J., Yusof, Y., Husni, H.: Document clustering based on firefly algorithm. J. Comput. Sci. **11**(3), 453–465 (2015). https://doi.org/10.3844/jcssp.2015.453.465
25. Kamvar, K., et al.: Spectral Learning (2003)

Recognition of Named Entities in the Russian Subcorpus Google Books Ngram

Vladimir V. Bochkarev$^{(\boxtimes)}$ (iD), Stanislav V. Khristoforov (iD),
and Anna V. Shevlyakova (iD)

Kazan Federal University, Kazan 420008, Russia
vbochkarev@mail.ru, stnslv91@gmail.com, anna_ling@mail.ru

Abstract. This paper describes how to build a recognizer to identify named entities that occur in the Google Books Ngram corpus. In the previous studies, the text was usually input to the recognizer to solve the task of named entities recognition. In this paper, the decision is made based on the analysis of the word co-occurrence statistics. The recognizer is a neural network. A vector of frequencies of bigrams or syntactic bigrams including the studied word is fed at the input. The task is to recognize named entities denoted by one word. However, the proposed method can be further applied to recognize two- or multi-word named entities. The recognition error probability obtained on the test sample of 10 thousand words, which are free from homonymy, was 2.71% (F1-score is 0.963). Solving the problem of word classification in Google Books Ngram will allow one to create large dictionaries of named entities that will improve recognition quality of named entities in texts by existing algorithms.

Keywords: Named entities recognition · N-grams frequencies · Google Books Ngram · Syntactic bigrams · Neural networks

1 Introduction

Named entity recognition (NER) is one of the important tasks of automatic text processing used in a wide variety of application domains such as analysis of customer feedback [1]. A large number of works are devoted to research in this area, especially for the English language (see the literature review in [2]). Traditionally, all methods and approaches are divided into two main groups: the approach based on rules and dictionaries, and the approach based on training on a pre-annotated corpus. They are known as knowledge-driven and data-driven approaches, respectively. The first one requires considerable experts' efforts to create large dictionaries of names and rules for standard patterns of multi-word named entities. The second one is also time-consuming. However, it requires less qualified annotators that operate according to certain instructions. The main problems of this approach are the selection of automatically distinguished features of words (such as parts of speech) and the choice of training tools – classifiers. In general, it turned out to be a priority area of research.

L. Martínez-Villaseñor et al. (Eds.): MICAI 2020, LNAI 12469, pp. 17–28, 2020.
https://doi.org/10.1007/978-3-030-60887-3_2

To compare efficiency of different NER systems, standard test corpora were created, and several competitions were held as part of international conferences. The starting point can be considered a competition in the framework of the 1995 conference – Message understanding conference-6 [3]. Then, a new corpus [4] was created for the competition at the CoNLL 2003 conference, which has been subsequently most often used to evaluate new systems. According to the conditions of the competition, it was required to extract named entities of the following types: person, organization, location, and others. The mentioned first three types of named entities have been traditionally recognized by NER programs and used in competitions.

Different assessment methods are used to test the systems. F1 is a generally accepted measure. However, there are two types of this measure. The first type requires exact coincidence of the extracted name with the names marked by experts (strict F1-score), the second one allows extraction of part of the full name (relaxed F1-score).

The winner of the competition in the framework of CoNLL 2003 was the program [5], which showed a result of 88.76% according to the strict F1-score. It used a combination of a range of machine learning algorithms and numerous features. The best result (91.36%) obtained without the use of neural networks on the CoNLL 2003 corpus [6] was achieved using a large number of features and external resources.

However, the use of neural networks has become prevailing in this field. The neural network approach is attractive because it is independent on the subject area and does not require specific resources, such as dictionaries, ontologies, etc. Detailed reviews of the application of neural networks for NER, including lists of available corpora, systems and results can be found in [7, 8]. All approaches to the use of neural networks for solving the NER problem can be classified depending on the way the words are represented in the network: by words, word parts or letters. Neural networks of different architectures were tested for NER purposes: recurrent, convolutional, recursive and the recently proposed [9] network of hybrid architecture ("transformer").

It was concluded in the review by [7] that neural networks provide better results than other machine learning methods. However, the recognition accuracy of 96.6% was obtained in [10] without using neural networks. It was obtained on the MUC-6 corpus [11] created in 1995. Therefore, the estimates presented in [10] are not comparable to the ones obtained for the English language on the CoNLL 2003 corpus and described in the present paper. Currently, the best result of 93,5% accuracy on the CoNLL 2003 corpus was obtained in [12] by using pre-trains a bi-directional transformer model in a cloze-style manner.

As for the Russian language, there are not so many NER studies based on the Russian corpora [13]. The Russian corpus for NER is described in [14]. Ten newswires created from ten top cited "Business" feeds in Yandex "News" web directory were used as document sources. The corpus is marked by the experts according to the MUC 7 instructions [15]. Both knowledge-driven and data-driven approaches to NER are used in [14]. This paper describes that it is more difficult to perform NER for Russian than for English due to Russian complicated morphology.

The results of the NER competition FactRuEval 2016 (held in the framework of the Dialogue conference) were described in [16]. The best results obtained using different programs are the following: person - 93%, location - 91%, organisation - 79%, the average accuracy is 87%. The Skip-Chain CRF technique is used in [17] with different methods of calculating a word proximity (vector, etc.). Open-Corpora [18] was used for training and testing. The following results were obtained: location - 0.93, organization - 0.91, names 0.94, surnames - 0.92, patronymic names - 0.84, on average - 0.88. Apparently, these are currently the best results for the Russian language.

It is noted in [8] that using external resources may improve the results. It was proposed in [19] to extract POS markup available in Google Books Ngram as an additional piece of information. This paper also suggested creation and use of extra-large corpora for solving the NER problems.

The present study is based on the Google Books Ngram (GBN) corpus. It was presented in 2009 and includes data on frequency of words and phrases in 8 languages for the past five centuries [20]. The Russian subcorpus of GBN contains data on frequencies of words and n-grams based on the texts of 391 thousand books (published in 1607–2009) with a total number of more than 67 billion words. It exceeds manifold the size of any other Russian corpus. For example, the RNC corpus widely used in linguistic studies contains texts which total size is 130 less than that of GBN [21]. Data for the period 1920–2009 is used in this work to avoid difficulties associated with the spelling reform of 1918. For this period, the Russian subcorpus of GBN includes texts with a total number of 64.3 billion words.

The number of unique 1-grams in the Russian corpus is 4,863,328. They contain 4,090,861 1-grams consisting of letters of the Russian alphabet and, possibly, one apostrophe. It is obvious that most of these words are missed even in the most complete available dictionaries. Classification of words occurring in GBN would be very useful in various studies based on this corpus. Besides being used in purely linguistic works that study language evolution, it can increase efficiency of NER systems. Solving the problem of vocabulary classification in Google Books Ngram will allow one to create large dictionaries of named entities that will improve recognition quality of named entities in texts by existing algorithms.

Moreover, it is of interest to find out how reliably a word denoting a named entity can be identified by its distribution, especially by statistics of co-occurrence with other words.

The GBN corpus includes data on frequencies of 1-, 2-, 3-, 4- and 5-grams. Data on frequency of 5-grams are sometimes used to train neural network vector models. Such method allows one to solve various problems, for example, to study changes in meaning of words [22] or to estimate the concreteness rating [23], etc. A certain difficulty can be caused by the fact that n-grams whose total frequency of use is lower 40 throughout the studied period are not included in the corpus. As a result, 5-, 4- and even 3-grams including many rare words can be missed in the corpus. It is significant that the authors [23] had to exclude from consideration about half of the words they had because the small number of 5-grams including these words did not allow training the model. Since this study is aimed at working with a wide range of words, including rare ones, a vector representation of words based on 2-gram frequencies will be used in this paper.

The work objective is to build a recognizer which allows one to identify named entities in the GBN corpus. Statistics on the word co-occurrence is used as input data.

The task is to recognize named entities denoted by one word. However, the proposed method can be further applied to recognize two- or multi-word named entities.

2 Data and Method

Creating an annotated training sample is the most time-consuming part in the process of the NER system creation. The following fact can be considered while working with a large corpus. Words denoting named entities usually start with a capital letter in many languages including Russian; and words which do not denote named entities are lowercased in most cases.

The serious problem is homonymy, especially when a word has only one meaning denoting a named entity. For example, *lev, vera, avgust* (these and other examples are transliterated using the ALA-LC system).

When recognizing named entities in the text, it can be decided which meaning the homonymous word has based on the context. Working with the GBN corpus, the right decision is to identify the proportion of cases when the word denotes named entities. Such estimate can be obtained based only on the word co-occurrence statistics. However, to do this it is required to know how co-occurrence of words denoting named entities differs from co-occurrence of other words. It is desirable to have already trained recognizer of named entities. For this reason, this paper considers only words free from homonymy.

It should be noted that words denoting named entities and free from homonymy do not always start with the capital letters (for example, the word *maxim*). And in contrast, words that are not named entities can be capitalized (for example, the words *geroi, angel*). Therefore, to avoid ambiguity, words denoting named entities and starting only with a capital letter and words that always start with lowercase letters were included in the training sample.

There were 1185 thousand forms of words starting only with the lowercase letters and marked up as nouns in (at least in 90% cases) the corpus. There were also 563 thousand word forms marked up as nouns and started only with the capital letter. Fifty thousand most frequent words were selected from these two group of words for the period 1920–2009. The list included 18,085 word forms that always start with the capital letter and 32, 000 word forms starting only with the lowercase letters. The most frequent word in the sample was used $2.82 \cdot 10^7$ times in the corpus over the period 1920–2009 and the rarest word occurred only in 1982 cases.

The data used to train the neural network is a set of 2-gram frequency vectors taken from the GBN database. To avoid the difficulties associated with the spelling reform of 1918, all frequencies were calculated for the period 1920–2009. The number of unique 1-grams in the Russian corpus is 4,863,328. A certain word 'W' could potentially form 4,863,328 2-grams of the form Wx ('x' is a certain 1-gram), and the same number of 2-grams of the form xW. Therefore, using all data from the database to train a neural network is a computationally time-consuming task. In fact, only a very small percentage of all potential 2-grams is found in the language. Therefore, it is enough to use frequencies of combinations of the word 'W' with a limited set of "reference" words. The most frequent words in the corpus can be naturally regarded as the "reference" ones. At that, the word 'W' is represented by two 2-gram frequency vectors of the form Wx and xW, where 'x' is one of the "reference" word forms. A similar approach was used in [24] to determine POS, gender, and number of a word. There are 20 thousand of the most frequent words (1-grams consisting of letters of the Russian alphabet and, possibly, one apostrophe) used as reference words in the present paper. Therefore, a word is described by a pair of vectors of dimension 20,000. Both vectors are concatenated to obtain a single 40,000-dimensional vector representing the word.

Besides frequencies of ordinary 2-grams (pairs of words that go one after another in the text), the GBN corpus contains data on frequencies of syntactic bigrams [25]. Syntactic bigrams are units of syntactic structures denoting a binary relation between a pair of words in the sentence. In each syntactic bigram, one word is called the head, and the other is its dependent [26]. Recently, approaches based on the identification of syntactic bigrams and analysis of their frequency are used to solve various problems of natural language processing [26]. The representation of words by the frequency vectors of syntactic bigrams analogous to one described above was used in this paper.

To implement the named entity recognizer, the architecture of the classical direct distribution network was chosen. The network was a four-layer perceptron with 64, 128 and 128 neurons in the first three layers, respectively. The rectifier activation function (RELU) and ELU [27] were used as activation functions of all hidden layers. Using these activation functions allows one to achieve sparse activation [27] and thus to obtain a model of higher approximation power.

To make the recognition results independent on the corpus size, the input data (the frequency vector of the bigrams) should be normalized to one. Using the RELU activation function provides another interesting opportunity. If the activation functions of all RELU layers and the neuron bias are equal to zero, then such a network realizes a homogeneous function of the input data (the degree of homogeneity is equal to 1). That is, if all inputs are multiplied by the same number, all output values will also change by this factor, and the proportions between them will not change. This provides elimination of the explicit normalization of the input data. Two versions of the neural network were tested: 1) with the activation functions of the hidden RELU layers and zero values of neuron bias; 2) with the activation functions ELU and non-zero values of neuron bias.

The dimension of the last fourth layer is 2, in accordance with the number of classes presented in the training sample. The output layer was activated by the softmax function [28]. This ensures that the outputs of the neural network are non-negative, as well as the normalization of their sum to 1, which allows one to interpret the output data as the probability distribution on the target classes.

Moreover, since the dimension of the input vector is high (40,000 in our case), the number of weights between the input and the first hidden layer is also high, which can lead to overfitting of the model, especially with a small number of examples in the training set. To prevent overfitting, a dropout layer [29] with the parameter 0.5 was placed before the first hidden layer. Random 50% of the data from the input vector are cut out at each training iteration. This procedure gives a stochastic regularization of neural network learning. In the testing mode, information is no longer being cut out from the data vector. However, the output data of this layer are corrected in a certain way to avoid distortion in the process of using them by subsequent layers.

The model was trained using the error backpropagation method based on the Nadam algorithm, which is one of the varieties of the stochastic gradient descent. The review of gradient methods and features of their implementation are presented in [30]. The Nadam algorithm is based on a combination of the ideas described in [31, 32]. It adjusts the learning rate parameter according to the previous observations of the value of the gradient norm, as well as saves a certain analog of the inertia of the solution point motion, which allows training deep models in a fairly small number of iterations. In addition, the norm of the resulting gradient vector in the training process can be artificially bounded above so that when it hits the "steep" sections of the target function (this often occurs when moving in ravine functions), the solution point does not move too far from the studied area near a local minimum.

There are about 40 thousand examples in the training sample. Therefore, to ensure high efficiency of model training, the training sample is divided into a set of batches of a fixed size. The network weights are updated, and the error gradient vectors are aggregated after all the examples from the batch are provided. The batch size selected by us was 256. Thus, approximately 150 updates of network weights occur during one epoch. The neural network training was performed according to the criterion of the minimum cross-entropy. The neural network computing library PyTorch was used.

3   Results

In the available sample, 80% of the examples were selected for training the neural network (the training sample), and 20% – for testing of the obtained the result (the test sample). The data were divided into the training and test samples randomly. Both samples included approximately the same percentage of rare and frequent words. As already mentioned, the neural network had 2 outputs. The decision whether to consider a given word as a name or not is made depending on the ratio of the values obtained at the two outputs. It is decided that the word is a name if the ratio of the output values exceeds a predetermined threshold. The choice of the threshold may be different depending on the practical task to be solved. Different error values of the 1st and 2nd type will be obtained for each selected threshold value. Let us consider the case when the recognition is carried out based on the frequencies of syntactic bigrams by a neural network with the RELU activation functions and zero neuron bias. Figure 1, A shows the change in the error probabilities of the 1st (α) and 2nd type (β) depending on the choice of the threshold for this case. Figure 1, B shows the corresponding receiver operating characteristic curve.

To compare different recognizers, a threshold value will be chosen at which the error probabilities of the 1st and 2nd type are the same, i.e. $\alpha = \beta$ (this choice is shown in Fig. 1, B by a circle). This allows one to characterize the quality of the resulting recognizer with only one number. Let us consider the error probability choosing this threshold. As a rule, accuracy of the NER is estimated using the F1-score [3]. The value of the F1-score will be 0.9627 at the given threshold.

Let us consider what factors determine the recognition accuracy. Both the frequency vector of syntactic bigrams including the word under study, as well as the vector of ordinary bigrams (i.e., pairs of adjacent words) can be used as the input data. The test results showed that, ceteris paribus, slightly higher accuracy is obtained when using the

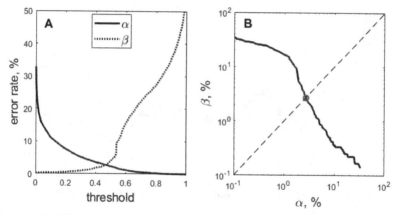

Fig. 1. A - probability of errors of the 1st (α) and 2nd type (β) for a different choice of a threshold for a neural network with the RELU activation functions and zero neuron bias. The estimation is performed based on the frequencies of syntactic bigrams; B - ROC for the same case

frequencies of syntactic bigrams than ordinary bigrams. For a network with the RELU activation functions and zero neuron bias, the error probability based on the frequencies of ordinary bigrams was 3.27% (versus 2.71% obtained based on the frequencies of syntactic bigrams). The result obtained based on syntactic bigrams may be better, since when they are used, the syntactic matching of words is considered but not just the immediate proximity of words in the text.

Besides a network with the RELU activation functions and zero neuron bias, the variant with the ELU activation functions and nonzero bias was also tested. In this case, the probability of error was 2.41% when using the frequencies of syntactic bigrams, and 3.14% when using ordinary bigrams. Thus, this variant shows slightly better results. However, in this case, one cannot be sure that the results will be reproduced for words whose frequencies lie outside the interval presented in the training sample. Therefore, the variant with the RELU activation functions and zero neuron bias is preferable.

Let us consider how recognition accuracy varies with the frequency of a word. Frequencies of bigrams that include rare words fluctuate more significantly which decreases recognition accuracy. To analyze changes in accuracy depending on frequency, the list of words in the test sample was ordered by frequency. The recognition error was estimated for each frequency range (according to the method described above) in a sliding window with a width of 1500 words. The results are shown in Fig. 2. It is seen that the error probability rises above 3% only for words with a frequency of 2000 uses over the entire period 1920–2009.

The adopted method of generating the training sample does not allow one to distinguish between different types of named entities. Nevertheless, available dictionaries can be used to analyse how an already trained recognizer identifies named entities of various types by using existing dictionaries. We used the OpenCorpora [18] morphological dictionary containing information on 391,268 lemmas of Russian words, including a total of 5,128,422 inflectional forms. Each word is identified as a named entity or not. Named entities are divided into 5 classes: name, surname, patronymic name, location

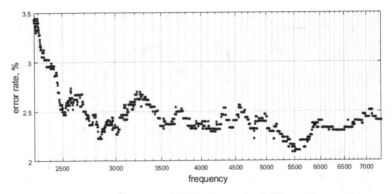

Fig. 2. Recognition error probability for words of different frequencies

and name of an organisation. There were 6559 word forms (out of 10 thousand) found in the OpenCorpora dictionary. This list contains 1430 word forms marked as belonging at least to one of the five classes. The recognizer made a mistake for 38 words from this list (in 2.657% of cases, which almost coincides with percentage for the entire test sample). Table 1 shows the number of cases in the sample, the number and percentage of errors for each class of words (the sums of the first and second lines in the table are not equal to the numbers indicated above, since some words may refer to several classes at once).

Table 1. The number of cases and the number of errors in the test sample for different classes

	Name	Surname	Patronym	Location	Organisation
The number of cases	498	360	102	483	29
Error number	16	13	0	11	1
Error percentage	3.21	3.61	0	2.28	3.45

It is of interest that all 102 patronymic names presented in the sample were recognized correctly (without any errors). Taking for the null hypothesis the assumption that (for this class) the error probability is equal to the average for the sample, it is easy to get a p-value equal to 0.0641. Thus, the available amount of statistics cannot allow one to conclude that the error probability for patronymics is less than for other classes of named entities. However, it is natural to assume that patronymic names can indeed be easier to recognize than other types of names. This may be due to the direct distribution of patronymics. As a rule, patronymics are used immediately after the first names although sometimes there are exceptions.

Analysis of recognition errors showed that all the words which were erroneously recognized by the model are used in objective cases, and half of the words (18 of 38) are words of foreign origin (*Vilnius, Mtskheta, Galatea*). In some cases, borrowed names are not grammatically assimilated in the Russian language and have endings not typical of the case forms of the Russian language (for example, the word *Filarete*). The name *Filaret* is recognized by the model since *Filaret* has endings characteristic of the Russian language. However, *Filarete* is not only the case form of the name *Filaret*, but also a surname in the nominative case (*Filarete byl khudozhnikom-arkhitektorom*). The surname *Filarete* has not fully assimilated and has not changed its form in objective cases (*barel'efy, vyleplennye Filarete*). Unusual endings of names are also often found in Church Slavonic texts (*Chto ty, Pilate, mechesh'sia; Prepodobie otche i Bogonose sviatiteliu Iakove*). Some erroneously recognized words have variant endings (*Shankhaiu, Kuntseve*). The list of 38 words includes 30 words that are ambiguous and can indicate a proper name and geographical location (*Tarasove, Babkine*), proper names and organization names (*Spartake*), geographical location and titles (*Brandenburg*), as well as denote plural and singular forms (*Tanechki, Vasiliev*). Also, some names are used figuratively (*preobrazovanie po abeliu*). There are relatively few errors (6 out of 38) that are not related to any of the described factors. Three of the 6 names are borrowings (*Vil'gel'me, Avelem, Eizenshteine*).

4 Conclusion

The conducted testing showed that the considered approach allows one to identify named entities in the large text corpus with high accuracy. The recognition error probability obtained on the test sample of 10 thousand words, which are free from homonymy and used at least 2000 times over 1920–2009, was 2.71% (F1-score is 0.963). These results were rechecked for a group of words (from the test sample) presented in the OpenCorpora morphological dictionary and marked in it as named entities. The NER accuracy for the described five classes of names presented in the OpenCorpora dictionary was no worse than 3.61%.

However, these values cannot be directly compared to the recognition accuracy results obtained in the well-known works on the recognition of named entities [1–10, 13–17], since a substantially different problem is solved in our work:

- As a rule, the initial data used in the works on NER is text. The initial data for the recognizer used in our work is statistics on the distribution (co-occurrence) of the word extracted from the large corpus.
- The work result of the traditional recognizer can be the decision that some words or phrases in the text are named entities and required disambiguation can be performed by analyzing the context of use of a word (phrase) in the text. The work result of our algorithm is classification of words found in a large corpus.

Another conclusion that can be drawn is that using frequencies of syntactic bigrams provides better recognition accuracy than the use of frequencies of ordinary bigrams.

There are also some problems which require further investigation. The proposed method for recognizing named entities has been tested on one-word named entities. However, this approach can also be used to recognize two- and multi-word named entities. To do this, frequency databases of 3-, 4- or 5-grams from the GBN corpus can be used. Consider, for example, the name of an object consisting of two words A and B. A vector can be constructed that includes 3-gram frequencies of the form Abx and xAB, where x (as described above) is one of the "reference" (most frequent) words. A recognizer for such 2-word named entities can be constructed using a vector representation completely similar to the one described above. However, this requires further experiments.

Another problem to be solved is classification of homonymous words. It is necessary to analyze features that allow one to distinguish the co-occurrence of a named entity and words that are not names. This will make it possible to estimate the percentage of use of polysemantic words denoting named entities.

The results presented in the work show that a combined approach based on statistics on the use of words starting with the capital and lowercase letters in the corpus, as well as on the analysis of the word form co-occurrence will be of great use for automatic classification of words in large corpora, including the GBN corpus.

Acknowledgements. This research was financially supported by the Russian Government Program of Competitive Growth of Kazan Federal University, and by RFBR, grant № 17-29-09163.

References

1. Solovyev, V., Ivanov, V.: Dictionary-Based Problem Phrase Extraction from User Reviews. In: Sojka, P., Horák, A., Kopeček, I., Pala, K. (eds.) TSD 2014. LNCS (LNAI), vol. 8655, pp. 225–232. Springer, Cham (2014). https://doi.org/10.1007/978-3-319-10816-2_28
2. Sharnagat, R.: Named entity recognition: a literature survey. Cent. Indian Lang. Technol. (2014)
3. Grishman, R., Sundheim, B.: Message understanding conference-6: a brief history. In: Proceedings of the 16th Conference on Computational Linguistics COLING'96, vol. 1, pp. 466–471 (1996)
4. Sang, E.F.T.K., De Meulder, F.: Introduction to the conll-2003 shared task: language-independent named entity recognition. In: Proceedings of the Seventh Conference on Natural Language Learning at HLT-NAACL 2003, vol. 4, pp. 142–147. Association for Computational Linguistics (2003)
5. Florian, R., Ittycheriah, A., Jing, H., Zhang, T.: Named entity recognition through classifier combination. In: Proceedings of the Seventh Conference on Natural language Learning at HLT-NAACL 2003, vol. 4, pp. 168–171. Association for Computational Linguistics (2003)
6. Agerri, R., Rigau, G.: Robust multilingual named entity recognition with shallow semi-supervised features. Artif. Intell. **238**, 63–82 (2016)
7. Yadav, V., Bethard, S.: Survey on recent advances in named entity recognition from deep learning models. In: Proceedings of the 27th International Conference on Computational Linguistics, pp. 2145–2158 (2018)
8. Li, J., Sun, A., Han, J., Li, C.: A survey on deep learning for named entity recognition. IEEE Trans. Knowl. Data Eng. (2020). https://doi.org/10.1109/TKDE.2020.2981314
9. Vaswani, A., et al.: Attention is all you need. In: 31st Conference on Neural Information Processing Systems (NIPS 2017), Long Beach, CA, USA, pp. 5998–6008 (2017)
10. Zhou, G., Su, J.: Named entity recognition using an hmm-based chunk tagger. In: proceedings of the 40th Annual Meeting on Association for Computational Linguistics, pp. 473–480. Association for Computational Linguistics (2002)
11. MUC-6, https://cs.nyu.edu/grishman/muc6.html. Accessed 29 Jun 2020

12. Baevski, A., Edunov, S., Liu, Y., Zettlemoyer, L., Auli, M.: Cloze-driven pretraining of self-attention networks. CoRR, arXiv:abs/1903.07785 (2019)
13. Solovyev, V., Ivanov, V.: Knowledge-driven event extraction in russian: corpus-based linguistic resources. Comput. Intell. Neurosci. **2016**, 4183760 (2016)
14. Gareev, R., Tkachenko, M., Solovyev, V., Simanovsky, A., Ivanov, V.: Introducing baselines for Russian named entity recognition. In: Gelbukh, A. (ed.) CICLing 2013. LNCS, vol. 7816, pp. 329–342. Springer, Heidelberg (2013). https://doi.org/10.1007/978-3-642-37247-6_27
15. Chinchor, N.A.: MUC-7 named entity task definition. In: Proceedings of the 7th Message Understanding Conference (MUC-7), Fairfax, VA, USA (1998)
16. Starostin, A., et al.: FactRuEval 2016: evaluation of named entity recognition and fact extraction systems for Russian. In: Computational Linguistics and Intellectual Technologies. Papers from the Annual International Conference "Dialogue", vol. 15, issue 22, pp. 702–720. RGGU, Moskow (2016)
17. Labutin, I.A., Firsov, A.N., Chuprina, S.I.: Raspoznavanie imenovannykh sushchnostei v tekstakh na estestvennom iazyke s ispol'zovaniem metoda probroso-tsepochnykh uslov-nykh sluchainykh polei. In: 23-ia Mezhdunarodnaia konferentsiia po komp'iuternoi lingvistike i intellektual'nym tekhnologiiam 31 May—3 June 2017, Studencheskaia sessiia, pp. 1–8 (2017). http://www.dialog-21.ru/media/3988/labutin.pdf. (In Russian)
18. Bocharov, V.V., et al.: Crowdsourcing morphological annotation. In: Computational Linguistics and Intellectual Technologies. Papers from the Annual International Conference "Dialogue", vol. 12, issue 1, pp. 109–115. RGGU, Moskow (2013)
19. Wen, Z., Huang, Z., Zhang, R.: Entity Extraction with Knowledge from Web Scale Corpora. arXiv preprint arXiv:1911.09373v1 (2019)
20. Michel, J.-B., et al.: Quantitative analysis of culture using millions of digitized books. Science **331**(6014), 176–182 (2011)
21. Russian National Corpus. http://www.ruscorpora.ru/. Accessed 14 Jun 2020
22. Kulkarni, V., Al-Rfou, R., Perozzi, B., Skiena, S.: Statistically significant detection of linguistic change. In: Proceedings of the 24th International Conference on World Wide Web, Florence, Italy, pp. 625–635 (2015)
23. Snefjella, B., Généreux, M., Kuperman, V.: Historical evolution of concrete and abstract language revisited. Behav. Res. Methods **51**(4), 1693–1705 (2018). https://doi.org/10.3758/s13428-018-1071-2
24. Khristoforov, S., Bochkarev, V., Shevlyakova, A.: Recognition of parts of speech using the vector of bigram frequencies. In: van der Aalst, W., et al. (eds.) Analysis of Images, Social Networks and Texts. AIST 2019. Communications in Computer and Information Science, vol. 1086, pp. 132–142. Springer, Cham (2020). https://doi.org/10.1007/978-3-030-39575-9_13
25. Lin, Y., et al.: Syntactic annotations for the google books Ngram corpus. In: Li, H., Lin, C.-Y., Osborne, M., Lee, G.G., Park, J.C. (eds.) Proceedings of the Conference on 50th Annual Meeting of the Association for Computational Linguistics 2012, vol. 2, pp. 238–242. Association for Computational Linguistics, Jeju Island, Korea (2012)
26. Sidorov, G., et al.: Syntactic dependency-based N-grams as classification features. In: Batyrshin, I., Mendoza, M.G. (eds.) Advances in Computational Intelligence. MICAI 2012. Lecture Notes in Computer Science, vol. 7630, pp. 1–11. Springer, Berlin, Heidelberg (2012). https://doi.org/10.1007/978-3-642-37798-3_1
27. Glorot, X., Bordes, A., Bengio, Y.: Deep sparse rectifier neural networks. In: Gordon, G., Dunson, D., Dudik, M. (eds.) AISTATS, JMLR.org, JMLR Proceedings, vol 15, pp 315–323 (2011)
28. Goodfellow, I., Bengio, Y., Courville, A.: Deep Learning. Adaptive Computation and Machine Learning. MIT Press, Cambridge (2016)

29. Srivastava, N., Hinton, G., Krizhevsky, A., Sutskever, I., Salakhutdinov, R.: Dropout: a simple way to prevent neural networks from overfitting. J. Mach. Learn. Res. **15**(1), 1929–1958 (2014)
30. Ruder, S.: An overview of gradient descent optimization algorithms. CoRR arXiv:abs/1609. 04747 (2016)
31. Kingma, D., Ba, J.: Adam: a method for stochastic optimization. arXiv preprint arXiv:141 26980 (2014)
32. Botev, A., Lever, G., Barber, D.: Nesterov's accelerated gradient and momentum as approximations to regularised update descent. CoRR arXiv:abs/1607.01981 (2016)

Candidate List Obtained from Metric Inverted Index for Similarity Searching

Karina Figueroa[1]([⊠]) [iD], Nora Reyes[2] [iD], and Antonio Camarena-Ibarrola[1] [iD]

[1] Universidad Michoacana de San Nicolás de Hidalgo, Morelia, Michoacán, Mexico
{karina.figueroa,camarena}@umich.mx
[2] Departamento de Informática, Universidad Nacional de San Luis,
San Luis, Argentina
nreyes@unsl.edu.ar

Abstract. Similarity searching consists of retrieving the most similar elements in a database. This is a central problem in many real applications, and it becomes intractable when a big database is used. A way to overcome this problem is by getting a few objects as a promissory candidate list of being part of the answer. In this paper, the most relevant and efficient algorithms for high dimensional spaces based on the permutations-technique are compared. Permutation-based algorithm is related to make a permutation of some special objects that allows us to *organize* the space of the elements in a database. One of the indexes related uses a complete permutation, and the second one utilizes a small part of the permutation and an inverted index.

Our research is focussed on two proposed ideas: the first consists in using a similar inverted index only with less information per object and computing the candidate list in a different way; and the second consists in changing a parameter during querying time in order to achieve a better prediction of the nearest neighbors. Our experiments show that our proposals do serve for implementing a better predictor and that the nearest neighbor can be found computing up to 45% fewer distances per query.

1 Introduction

Nowadays, most of the data generated are on multimedia databases where the concept of searching is basically to look for *similar objects*, also known as *similarity searching*. The *similar* or *close* objects are important for so many real applications; that is, on databases of images, fingerprints, audio signals, computational biology, etc. [3,11] For example, if you want to search for a similar image that you have. In this context, what does *similar* mean? Maybe you just want to retrieve one with similar colors, or you want one with some similar object/landscape that you have on your initial image.

Formally, similarity searching is the problem of looking for objects in a dataset similar to a given query. The similarity is defined with a distance function, or a metric, proposed by an expert on those fields, that considers specific features from data. When there are a database and a distance function, the problem can be model as a *metric space* [14], and it is called a metric database.

L. Martínez-Villaseñor et al. (Eds.): MICAI 2020, LNAI 12469, pp. 29–38, 2020.
https://doi.org/10.1007/978-3-030-60887-3_3

A metric space is defined as follows: there is a universe $\mathbb{X}$ of objects and a non-negative real valued distance function $d : \mathbb{X} \times \mathbb{X} \to \mathbb{R}^+$ defined among them. The distance functions must satisfies these properties: reflexivity, $d(x, x) = 0$; strict positiveness, $x \neq y \Rightarrow d(x, y) > 0$; symmetry, $d(x, y) = d(y, x)$; and the triangle inequality, $d(x, z) \leq d(x, y) + d(y, z)$. Unfortunatelly, there are many real databases where this distance function is expensive to compute.

The more *similar* objects, the smaller the distance between two objects is. The finite database used $\mathbb{U} \subseteq \mathbb{X}$, $|\mathbb{U}| = n$, is a subset of the universe. Our database will be *organized* with an index during preprocessing time (offline process) [6,12,14]. Later, when a query ($q \in \mathbb{X}$) is given, in an online process, the index will be used to retrieve a *candidate list* from the whole database with the most similar objects to that query. Notice that these two phases can be avoided if a sequential scan were used, but it can be expensive to compute, regarding time or resources.

Basically there are two main similarity queries: *Range queries* and *K-Nearest Neighbor queries*.

A *Range query* $R(q, r)$ retrieves those objects within a region centered on a given query object q with radius r; formally, $R(q, r) = \{u \in \mathbb{X}, d(u, q) \leq r\}$.

A *K-Nearest Neighbor* query $NN_K(q)$ retrieves the K elements of $\mathbb{U}$ that are closest to q, that is, $NN_K(q)$ is a set such that for all $x \in NN_K(q)$ and $y \in \mathbb{U} \setminus NN_K(q)$, $d(q, x) \leq d(q, y)$, and $|NN_K(q)| = K$.

There are several algorithms proposed for metric spaces, most of them work in spaces with low intrinsic dimension. As the intrinsic dimension grows the variance of data decreases, and the performance falls. This problem is known as *curse of dimensionality* [5,6,12,14]. In this paper, the permutation-based algorithm is studied because is one of the few algorithms which works well in high dimensional space . However, these algorithms can be further improved by appropriately choosing some of their characteristics such as: which function of distance between permutations is used, which part of the permutations is stored in the index, etc.

Therefore, the proposal is to improve the algorithm introduced in [2] and reduce the number of distances evaluations to get the nearest neighbors. In Sect. 2 are described the definition and principles of the permutation-based algorithm, and in Sect. 3 are introduced our two proposals: the first one is to change the information stored into the index, and the second one is to change one parameter used in [2] to compute the proximity by using the permutation-based algorithm technique. Section 4 shows how our proposals work on some synthetic databases . Finally, in Sect. 5, conclusions and future work are detailed.

2   Previous Work

2.1   Permutation-Based Algorithm

In [4] the authors introduced the *Permutation-Based Algorithm (PBA)*. The main idea was that every object in the database keeps how it sees the space according to some chosen points (called *permutants*).

During the preprocessing time, a subset of the database is choosen $\mathbb{P} \subseteq \mathbb{U}$, $\mathbb{P} = \{p_1, \ldots, p_k\}$ [1], called the set of permutants, where $|\mathbb{P}| = k$. Each object in the database $u \in \mathbb{U}$ computes the distance to all the permutants, $\forall p \in \mathbb{P}$, $d(u, p)$; that is, $D(u) = \{d(p_1, u), \ldots, d(p_k, u)\}$. Then, it sorts them in increasing order, where ties are broken in any consistent order. The *permutation* of u, Π_u is defined as the position of each $p \in \mathbb{P}$ after $D(u)$ was sorted, authors used $\Pi_u^{-1}(p_i)$ to identify the position of element p_i in the permutation Π_u. For example, let be $k = 5$, and $\mathbb{P} = \{p_1, \ldots, p_5\}$. Then, for an element u, let $D_u = \{4, 6, 2, 3, 1\}$ the distances between u and each $p \in \mathbb{P}$; that is, $d(p_1, u) = 4, d(p_2, u) = 6$ and so on. Π_u will be $(5, 3, 4, 1, 2)$, then $\Pi_u^{-1}(p_1) = 4, \Pi_u^{-1}(p_2) = 5$, and so on; that is, $\Pi_u^{-1} = (4, 5, 2, 3, 1)$. All the permutations are the index; so, the index can be kept with $n \times k$ integers.

The hypothesis is that two identical elements must have the same permutation, while two close or similar objects should have similar permutations. Therefore, when similar objects to a given query q are searched the key is to find similar permutations. There are several approaches to measure how similar two permutations are [13], but the authors recommend to use the Spearman Footrule metric [13].

$$F(u, q)_k = F(\Pi_u, \Pi_q) = \sum_{i=1}^{k=|\mathbb{P}|} |\Pi_u^{-1}(p_i) - \Pi_q^{-1}(p_i)| \tag{1}$$

For example, let be $\Pi_q = (1, 3, 5, 2, 4)$, according to Eq. 1 and using the Π_u previously computed considering $\mathbb{P} = \{p_1, \ldots, p_5\}$, $|\mathbb{P}| = 5$ then:

$$F(u, q)_{|\mathbb{P}|} = |1 - 4| + |2 - 2| + |3 - 1| + |4 - 5| + |5 - 3| = 8 \tag{2}$$

When a query q arrives, the most similar permutations to the permutation of q are searched in the database. Hence, every permutation in the index has to be compared with the query's permutation, by making $O(n)$ Footrule metrics [4]. Then, the elements are sorted in increasing order of the similarity given by this metric (Eq. 1). The authors showed that the most promissory elements to the query could be in the first positions of this order.

Metric Inverted File. In [2], the authors proposed using an inverted file to find the most similar permutations to the query's permutation. During the preprocessing time, they compute the permutation for each element in the database. Then, given a parameter $m_i \leq k$, the permutations are cut until the first m_i elements. In the inverted file, each permutant $p \in \mathbb{P}$ keeps a list of pairs (u, ψ). There will be a pair (u, ψ) for each element u that has this permutant between the m_i first elements of Π_u and where $\psi \leq m_i$ represents in which position is located the permutant p within Π_u. For example, let be $m_i = 4$ and consider our $\Pi_u = (5, 3, 4, 1, 2)$. The list of the permutant p_5 will have the pair $(u, 1)$, the list of the permutant p_3 will have the pair $(u, 2)$, and so on. The index in this proposal is this inverted file, which needs to keep $2 \times n \times m_i$ integers. Formally, each inverted index list is defined as follows:

$$\mathbb{I}_{p_j} = \{(u, \psi) \mid \Pi_u^{-1}(p_j) = \psi \le m_i\} \tag{3}$$

During the query time, a new parameter m_s is introduced, where $m_s \le m_i \le k$. The authors, in [2], propose to keep all the elements in the list (called inverted file) of the first m_s elements of the query's permutation (previously computed):

$$\mathbb{C} = \bigcup_{i=1}^{m_s} \mathbb{I}_p, \quad p \in \Pi_q(i), \tag{4}$$

In our running example, let be $m_s = 3$. Hence, the lists of permutants p_1, p_3, and p_5 will be processed. In order to compute the similarity between the permutations of u and q (see the example in Eq. 2), from the lists of p_1, p_3, p_5, it is possible to compute $|1 - 4|, |2 - 2|, |3 - 1|$. When some value is missing, the authors propose to use m_s. Basically, they compute $F'(u, q)$ for each element in the database, see Eq. 5.

$$F'(u, q) = \begin{cases} |\Pi_u^{-1}(p_i) - \Pi_q^{-1}(p_i)| & \text{if } \Pi_q^{-1}(p_i) \le m_s \text{ and } \Pi_u^{-1}(p_i) \le m_i \\ m_s & \text{otherwise.} \end{cases} \tag{5}$$

It can be noticed that for any permutant p_i, if $\Pi_q^{-1}(p_i) \le m_s$ then p_i is at the first m_s permutants in the permutation of q (Π_q); and if $\Pi_u^{-1}(p_i) \le m_i$ then there is a pair $(u, \psi = \Pi_u^{-1}(p_i))$ in the list of p_i in the inverted file.

Let see the complete example shown in Fig. 1. In this case, $\mathbb{P} = \{p_1, \ldots, p_6\}$, $k = 6$, $m_i = 4$, and $m_s = 3$. Notice that the inverted file proposed in [2] is depicted in Fig. 2. If a given query q with its permutation is $\Pi_q = \{2, 4, 1, 6, 5, 3\}$, just its first $m_s = 3$ permutants will be used. Hence, the shorted permutation of q will be $\{2, 4, 1\}$. Therefore, this short permutation will be used to compare with the information stored in the inverted file obtained from the short permutations, of length m_i, of all the database elements.

During the query time, as $m_s = 3$, he short permutation $\{2, 4, 1\}$ for q will be used. The permutant p_2 is at position number 1 in Π_q and from the inverted file it is possible to obtain, for example, that p_2 is at position number 4 for the element u_2. Hence, the difference between these positions can be calculated, i.e. $|4 - 1|$, in order to obtain one of the terms needed to compute the value of $F'(u_2, q)$. The complete calculus of F' for each $u \in \mathbb{U}$, according to Eq. 5, is shown in Table 1. It can be noticed that, with respect to F', the most similar elements to q are u_7 and u_4.

Other researchers have also used permutation-based methods, some use different data structures for storing the permutations [8, 10], but they are still using the concept described in [1, 4].

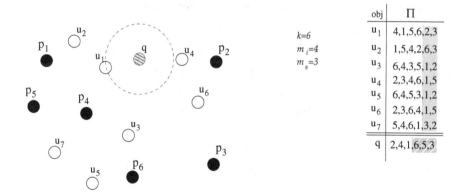

obj	Π
u_1	4,1,5,6,2,3
u_2	1,5,4,2,6,3
u_3	6,4,3,5,1,2
u_4	2,3,4,6,1,5
u_5	6,4,5,3,1,2
u_6	2,3,6,4,1,5
u_7	5,4,6,1,3,2
q	2,4,1,6,5,3

$k=6$
$m_i=4$
$m_s=3$

Fig. 1. Example of a database and the corresponding permutations.

perm	inverted file
p_1	(1,2),(2,1),(7,4)
p_2	(2,4),(4,1),(6,1)
p_3	(3,3),(4,2),(5,4),(6,2)
p_4	(1,1),(2,3),(3,2),(4,1),(5,2),(6,4),(7,2)
p_5	(1,3),(2,2),(3,4),(5,3),(7,1)
p_6	(1,4),(3,1),(4,4),(5,1),(6,3),(7,3)

Fig. 2. The inverted file [2] obtained for the database previously shown at Fig. 1.

3 Our Proposals

Although the inverted file is very helpful to determine the candidate list, the previous proposal [2] does not use all the possible information. Therefore, our proposal consists of using every available information and, at the same time, taking advantage of the inverted file. In [2], authors kept tuples (*object, position*) in the list of each permutant to store the information contained in the permutation of each database object. The first proposal is to keep an inverted file where the lists contain only those objects which have a permutant between its first m_i closer permutants; that is, the lists will store only the part (*object*) of these tuples.

Formally, our index will be as follows: let p_j a permutant, $p_j \in \mathbb{P}$, its list into the inverted index (called $\mathbb{I}'_{p_j}$) will have those objects $u \in \mathbb{U}$ which $\Pi_u^{-1}(p_j) \leq m_i$; that is:

$$\mathbb{I}'_{p_j} = \{(u) \mid \Pi_u^{-1}(p_j) \leq m_i\} \tag{6}$$

Figure 3 illustrates our new inverted file for the example of Fig. 1. Clearly, our new inverted file uses a half of needed space for the inverted file shown in Fig. 2,

Table 1. $F'(u, q)$ computed during the query time, with $m_s = 3$.

object	$F'(u, q)$
u_1	$F'(u_1, q) = m_s + \|1 - 2\| + \|2 - 3\| = 5$
u_2	$F'(u_2, q) = \|4 - 1\| + \|3 - 2\| + \|1 - 3\| = 6$
u_3	$F'(u_3, q) = m_s + \|2 - 2\| + m_s = 6$
u_4	$F'(u_4, q) = \|1 - 1\| + \|3 - 2\| + m_s = 4$
u_5	$F'(u_5, q) = m_s + \|2 - 2\| + m_s = 6$
u_6	$F'(u_6, q) = \|1 - 1\| + \|4 - 2\| + m_s = 5$
u_7	$F'(u_7, q) = m_s + \|2 - 2\| + \|4 - 3\| = 4$

perm	new inverted file
P_1	(1),(2),(7)
P_2	(2),(4),(6)
P_3	(3),(4),(5),(6)
P_4	(1),(2),(3),(4),(5),(6),(7)
P_5	(1),(2),(3),(5),(7)
P_6	(1),(3),(4),(5),(6),(7)

Fig. 3. The new inverted file obtained for the example of database.

however, our index is keeping the short permutations. Basically, both indexes are using exactly the same amount of memory.

During the query time, given a query q, its permutation Π_q is computed by ordering $Dq = \{d(q, p_1), \ldots, d(q, p_k)\}$, as it was explained previously. *We also use* the parameter m_s is still used at searches to short Π_q. In this case, the candidate list will be the union of the lists for each permutant p of $\Pi_q^{-1}(p) \leq m_s$ in the inverted file (as in Eq. 4). Formally:

$$\mathbb{C}' = \bigcup_{i=1}^{m_s} \mathbb{I}'_p, \quad p \in \Pi_q(i), \tag{7}$$

However, at this point, D_q is already computed and it is known the complete permutation Π_q of the query ($|\Pi_q| = k$). The list of candidates for nearest neighbors will be $u \in \mathbb{C}'$. The candidates will be sorted by similarity between permutations if the short permutations are kept of all the database elements u ($|\Pi_u| = m_i$). For this purpose, the Eq. 1 can be modified to evaluate the distance between permutations of different length, as follows:

$$F^*(u, q)_{m_i} = F^*(\Pi_u, \Pi_q) = \sum_{i=1}^{m_i} |i - \Pi_q^{-1}(\Pi_u(i))| \tag{8}$$

Considering the example depicted in Fig. 1 and applying our proposal, the results obtained are shown in Table 2. Noticeably, considering the order determined by F^*, as our proposal will do, the first place is for the actual nearest neighbor of q; this is the element u_1.

Table 2. $F^*(u, q)$ computed during the query time.

object	$F^*(u, q)$
u_1	$F^*(u_1, q) = \|2 - 1\| + \|3 - 2\| + \|5 - 3\| + \|4 - 4\| = 4$
u_2	$F^*(u_2, q) = \|3 - 1\| + \|5 - 2\| + \|2 - 3\| + \|1 - 4\| = 9$
u_3	$F^*(u_3, q) = \|4 - 1\| + \|2 - 2\| + \|6 - 3\| + \|5 - 4\| = 7$
u_4	$F^*(u_4, q) = \|1 - 1\| + \|6 - 2\| + \|2 - 3\| + \|5 - 4\| = 6$
u_5	$F^*(u_5, q) = \|4 - 1\| + \|2 - 2\| + \|5 - 3\| + \|6 - 4\| = 7$
u_6	$F^*(u_6, q) = \|1 - 1\| + \|6 - 2\| + \|4 - 3\| + \|2 - 4\| = 7$
u_7	$F^*(u_7, q) = \|5 - 1\| + \|2 - 2\| + \|4 - 3\| + \|3 - 4\| = 6$

3.1 Using m_i Instead of m_s

In order to evaluate how the parameters affect our proposal, we consider another option by using m_i instead of m_s during the query time. Hence, when a value is missing to compute $F'(u, q)$, the second proposal in this work is to replace m_i instead of m_s in the Eq. 5.

4 Experimental Results

Our proposal was empirically tested using a synthetic database composed by 100,000 vectors uniformly distributed in the unitary cube; that is, points in $\mathbb{R}^d$ with $d \in [16, 128]$. The vectors of these spaces were compared using Euclidean distance to measure how far or different they are to each other.

The results of this paper will be explain using the four variant of algorithms:

- **PP**. A short permutation of size m_i for each element is used as it was proposed in [4], that is without an index. In this case $F^*(u, q)_{m_i}$ was applied.
- **Ps**. A short permutation for each element $|\Pi_u| = m_i$, the inverted file, and $m_s \leq m_i$ are used during query time to complete the missing values for some elements, as it was proposed in [2].
- **Pm**. This is our proposal. A short permutation $|\Pi_u| = m_i$ and the inverted file, without saving the second element per pair; i.e. the position of a permutant within the permutation are used. In this case the Eq. 8 $F^*(u, q)_{m_i}$ is used.

- **Pi**. This is the other proposal in this paper, it use the short permutation of each element $|\Pi_u| = m_i$ and the inverted file used in [2], but, instead of m_s (as in **Ps**) for missing values, m_i is used.

In Fig. 4 the performance of our technique using vectors in $\mathbb{R}^{32}$ is depicted. It shows the average of 500 queries looking for the $K = 8$ nearest neighbors (on axis x). It can be noticed that the size of the complete permutation is ($k = 64$); but, in order to get a fair comparison, for each technique the same length of permutations in the index were used. However, the algorithms labeled by Pi, Ps, and Pm are using an inverted index. In Fig. 4 it shows that using all the permutations in the database (i.e. PP) achieves a better predictor than Pi, Ps, and Pm. On the other hand, the estimation proposed in [2] degrades the performance of Ps. As it can be seen, Pm is a better predictor of similarity between permutations than Ps. Besides, our second proposal Pi, which uses m_i instead of m_s, is a better predictor than Ps ([2]).

Permutation-based algorithm has excellent performance as the dimension is growing up. In Fig. 5 dimension 64 is shown. Notice that our experiments are using 64 permutants and our inverted files are using $m_i = 48$. In Fig. 5, on the

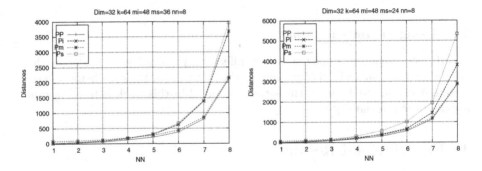

Fig. 4. Performance of our technique in dimension 32. 8 nearest neighbors per query were looking for.

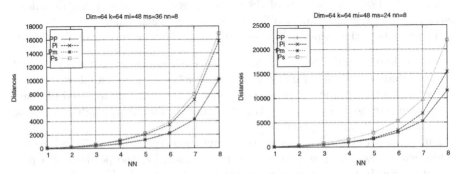

Fig. 5. Performance of our technique in dimension 64. 8 nearest neighbors per query were looking for.

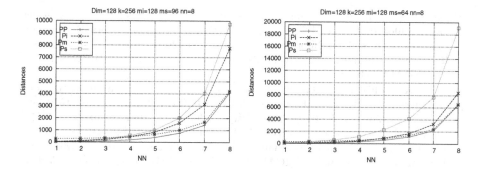

Fig. 6. Performance of our technique in dimension 128. 8 nearest neighbors per query were looking for.

left side it is depicted $m_s = 36$ and on the right side $m_s = 24$ is shown. Notice that using both proposals Pi and Pm it is possible to get better results than with Ps.

Finally, in the database with high dimension (R^{128}), showed in Fig. 6, our proposals get excellent results compared with Ps ([2]). In the case of PP ([4]), is using the whole information obtained from each permutation Π.

5   Conclusions and Future Work

Nowadays, similarity searching is one of the most important challenges in multimedia databases. Of course, the similarity or proximity is measured by a distance function defined for each database by an expert on those fields. This kind of problem can be modeled as a metric space.

In this paper it is proposed merging two powerful indexes for similarity searching based on the permutation technique and using differently the information stored in the index. Our results confirm that it is possible to improve the use of the inverted index. Experimentally it is shown that it is possible to avoid calculating up to 45% of the distances while using the same amount of memory of the previous proposal.

As future work, these new indexes can be combined with several similarity measures, as claimed in [9]. Also, instead of inverted indexes, other indexes might be used, such as the trie with prefix of permutations [7].

References

1. Amato, G., Esuli, A., Falchi, F.: A comparison of pivot selection techniques for permutation-based indexing. Inf. Syst. **52**, 176–188 (2015). https://doi.org/10.1016/j.is.2015.01.010

2. Amato, G., Savino, P.: Approximate similarity search in metric spaces using inverted files. In: Lempel, R., Perego, R., Silvestri, F. (eds.) 3rd International ICST Conference on Scalable Information Systems, INFOSCALE 2008, Vico Equense, Italy, June 4–6 2008, p. 28. ICST/ACM (2008). https://doi.org/10.4108/ICST. INFOSCALE2008.3486

3. Beyer, K., Goldstein, J., Ramakrishnan, R., Shaft, U.: When is "nearest neighbor" meaningful? In: Beeri, C., Buneman, P. (eds.) ICDT 1999. LNCS, vol. 1540, pp. 217–235. Springer, Heidelberg (1999). https://doi.org/10.1007/3-540-49257-7_15

4. Chávez, E., Figueroa, K., Navarro, G.: Proximity searching in high dimensional spaces with a proximity preserving order. In: Gelbukh, A., de Albornoz, Á., Terashima-Marín, H. (eds.) MICAI 2005. LNCS (LNAI), vol. 3789, pp. 405–414. Springer, Heidelberg (2005). https://doi.org/10.1007/11579427_41

5. Chávez, E., Navarro, G.: A probabilistic spell for the curse of dimensionality. In: Buchsbaum, A.L., Snoeyink, J. (eds.) ALENEX 2001. LNCS, vol. 2153, pp. 147–160. Springer, Heidelberg (2001). https://doi.org/10.1007/3-540-44808-X_12

6. Chávez, E., Navarro, G., Baeza-Yates, R., Marroquín, J.: Proximity searching in metric spaces. ACM Comput. Surv. **33**(3), 273–321 (2001)

7. Esuli, A.: MiPai: using the PP-index to build an efficient and scalable similarity search system. In: Proceedings of the 2nd International Workshop on Similarity Searching and Applications (SISAP 2009), pp. 146–148. IEEE Computer Society (2009)

8. Esuli, A.: Use of permutation prefixes for efficient and scalable approximate similarity search. Inf. Process. Manage. **48**(5), 889–902 (2012). https://doi.org/10.1016/j.ipm.2010.11.011

9. Figueroa, K., Paredes, R., Reyes, N.: New permutation dissimilarity measures for proximity searching. In: Marchand-Maillet, S., Silva, Y.N., Chávez, E. (eds.) SISAP 2018. LNCS, vol. 11223, pp. 122–133. Springer, Cham (2018). https://doi.org/10.1007/978-3-030-02224-2_10

10. Mohamed, H., Marchand-Maillet, S.: Quantized ranking for permutation-based indexing. Inf. Syst. **52**, 163–175 (2015). https://doi.org/10.1016/j.is.2015.01.009

11. Patella, M., Ciaccia, P.: Approximate similarity search: a multi-faceted problem. J. Discret. Algorithms **7**(1), 36–48 (2009)

12. Samet, H.: Foundations of Multidimensional and Metric Data Structures. Computer Graphics and Geometic Modeling, 1st edn. Morgan Kaufmann Publishers, Burlington (2006). University of Maryland at College Park

13. Skala, M.: Counting distance permutations. J. Discret. Algorithms **7**(1), 49–61 (2009). https://doi.org/10.1016/j.jda.2008.09.011

14. Zezula, P., Amato, G., Dohnal, V., Batko, M.: Similarity Search: The Metric Space Approach. Advances in Database Systems. Springer, Heidelberg (2006). https://doi.org/10.1007/0-387-29151-2

A Technique for Conflict Detection in Collaborative Learning Environment by Using Text Sentiment

Germán Lescano[1]([✉]), Carlos Lara[2], César A. Collazos[3], and Rosanna Costaguta[4]

[1] Institute of Research in Informatic and Information System/CONICET, National University of Santiago del Estero, Av. Belgrano (s) 1912, CP4200 Santiago del Estero, Argentina
gelescano@unse.edu.ar

[2] Unidad Tamaulipas, Cinvestav, Carretera Victoria- Soto la Marina Kilómetro 5.5, Ciudad Victoria - Soto la Marina, 87130 Cd Victoria, Tamps, México
c.alberto.lara@cinvestav.mx

[3] IDIS Group, Department of Computer Science, University of Cauca, Calle 5 No. 4-70, Popayán, Colombia
ccollazo@unicauca.edu.co

[4] Institute of Research in Informatic and Information System, National University of Santiago del Estero, Av. Belgrano (s) 1912, CP4200 Santiago del Estero, Argentina
rosanna@unse.edu.ar

Abstract. Computer-Supported Collaborative Learning (CSCL) can give many benefits to students such as promoting creativity and sense of community, sharing abilities, etc. However, when groups of people work together, conflict is inevitable. Generally, conflict in any CSCL situation is uncomfortable, time consuming and counterproductive. It is hard to characterize a conflict because it can involve many factors − e.g., environmental factors, member's differences, etc. This paper proposes a technique to recognize conflicts in a group and the members involved in them by focusing in the socio-emotional interactions. As disagreements between group members generally cause negative emotions, and members can induce negative emotions to other members; then, a conflict between two or more members can be recognized when there are bidirectional negative messages in the same conversation thread. The proposed technique represents chat interactions as a digraph in which the nodes represent users and the edges indicate the transference of negative sentiments during the interactions. Then, a matrix of scaled commute times is used to detect clusters (subgroups having conflict). The validation of the technique shows promising results. The proposed technique is able to detect conflicts automatically, reducing the human effort required to detect these conflicts by other means.

Keywords: Computer supported collaborative learning · Conflict recognition · Emotions

© Springer Nature Switzerland AG 2020
L. Martínez-Villaseñor et al. (Eds.): MICAI 2020, LNAI 12469, pp. 39–50, 2020.
https://doi.org/10.1007/978-3-030-60887-3_4

1   Introduction

Collaborative Learning (CL) is a situation where a group of persons, more or less at the same level, can perform the same actions, share common goals and work together with the aim to learn [11].

Interactions between group members influence their cognitive processes. In particular, CL gives many benefits (e.g., promotes creativity, sense of community, and sharing abilities) when the group work well [31]. A key factor in collaborative learning is the emotional stability of the group because it promotes a higher productive commitment in learning. However, groups are exposed to multiple social dynamics that could induce negative effects [16]. Conflicts are disagreements between two or more members in a group caused by individual dispositions and different aims, attitudes or previous experiences [2].

When a conflict come up, the cognitive system falls due to cognitive load so that information processing is blocked; hence, conflicts influence negatively the group performance [12]; these negative emotions can harm the group work [12,17,18,21]. Moreover, any conflict –regardless of its type– generates negative emotions [12,17,18,21].

Most collaborative learning groups can manage their interpersonal conflicts through negotiation which imply students should confront and persuade each other [24]. When they cannot resolve the conflict, the responsibility eventually fall on the teacher to determine how to resolve the conflict and the consequences that group members should face as a result of nonperformance [24]. Conflicts open the opportunity for students to learn to work in group; but, this is achieved with the help of teachers who should take an active role to guide groups to resolutions [6].

To have an active role, teachers should respond on those situations just in time providing recommendations to students related to the performance of roles played by them, the need to share the leadership, the balance of the task loads, the need to learn give feedback to partners; advices to make learners aware of behaviors, attitudes and actions to get good work relationships; and spaces to reflect about the experience of work in group to rescue the new learnings. However, tracking conflictive situations can be time consuming in Computer-Supported Collaborative Learning (CSCL), because teachers should revise interaction logs of each group. Hence, it is important to incorporate mechanisms that help teachers detecting conflicts.

The main contribution of this paper is a technique to recognize conflict in the interactions of a group and to identify those members that are involved in a conflict by focusing in the socio-emotional interactions. Any group member can influence the emotions of other members [33]. In the proposed technique a group working in a CSCL environment is represented by a digraph where nodes represent users and edges indicate socio-emotional relationships obtained from text sentiment analysis. In this way, group members that influence more in the group can be identified by using the digraph properties.

The rest of this paper is organized as follows: Sect. 2 reviews works related to the proposed approach. Section 3 explains the proposed approach. Section 4 describes the experimentation performed to analyze this proposal, and Sect. 5 presents and discusses the results. Finally, Sect. 6 depicts conclusions and future lines of work.

2 Related Work

Millar et al. [25] propose a technique to recognize conflicts in interactions by considering the power dimension of relationships. First, they assign one of three control directions to every utterance in a conversation: gaining ($\uparrow$), yielding ($\downarrow$), or neutralizing ($\rightarrow$) control; and then, they consider three consecutive attempts to gaining control ($\uparrow\uparrow\uparrow$) as a conflict pattern; i.e., it represents the first step toward an escalating, runaway spiral of interchange. This approach is based on the active opposition character of interpersonal conflict as manifested in the control implications of verbalization.

Bales [3] proposes the *Interaction Process Analysis* (IPA) that associates labels to messages for describing the participants' behavior. Once interactions are coded, six problems are inferred by considering the frequencies of each behavior. Authors suggest that behavior frequencies of a group without conflicts should be within certain lower and upper limits.

Bales and Steven [4] propose a methodology to analyze groups named SYM-LOG in which group members, and/or their behaviors, are mapped in three-dimensional space: dominant–submissive or up–down (UD), friendly–unfriendly or positive–negative (PN), and task-oriented–emotionally expressive or forward–backward (FB). In this method, group members answer a 26-item Likert-style questionnaire intended to know how often the members show a given behavior. Wall and Galanes [34] employ this method to study the amount of conflict experienced by a group. They showed that conflicts are not significantly related neither to FB nor UD dimensions; but, they found that the PN dimension is significantly, predictably, and negatively related to conflicts.

Social Network Analysis (SNA) techniques [8] can be used to analyze students' social interaction. By using graph theory, SNA techniques define metrics able to identify leaders, bridges, or isolated individuals [29]. Metrics can be categorized as global, individual, or related to clusters. Global metrics measure a single value for the whole network; individual metrics measure a value for each node; and cluster-related metrics allow identifying patterns or sub-groups into the graph [9].

Zachary [36] uses SNA to recognize fission – a phenomenon of subgroup formation due to differential sharing of sentiments. First, it defines a *capacitated network* (V, E, C) where nodes V are a set of individuals, E is the existence matrix, and C is the capacity matrix. Matrix E is symmetric and its entries are either 1 or 0 indicating whether two individuals have a relationship. Matrix C is symmetric and quantifies the relationships stated in E by considering the number of contexts (e.g., academic classes, bar, etc.) where two individuals interact. Then, the Ford-Fulkerson algorithm is used to predict subgroups and the locus of the fission.

Coviello et al. [10] propose a method for measuring the contagion of emotional expression in social networks. The approach is based on instrumental variables regression. This model assumes that emotional expression by a person at a given time is an additive linear function of other factors measured in the same time period including a time-specific factor (perhaps it is a holiday), an individual-specific factor (some people are always happier than others), the effect of an exogenous factor (like rainfall); the effect of an endogenous factor (the emotional expression of her friends), and an error term. They show that individual expression of emotions depends on what others in an individual's social network are expressing.

The proposed approach considers that opposite interactions are the key for detecting and analyzing conflicts just as [25]; but, it evaluates how positive or negative are the interactions instead of sequences of attempts to gaining control. Moreover, it can be used to analyze groups with two or more members while [25] only detects conflicts between two persons.

Aligned with the findings of [34], the proposed technique considers that positive/negative emotions expressed between group members are directly related to conflicts. It also considers the [10] study which confirms that emotions can be contagious also through computer-mediated communication. For automatically analyzing conflicts, the approach takes advantage of techniques that detect emotions in text messages; hence, it allows quantifying interactions within a CSCL context. Finally, inspired by those approaches that considers directionality of interactions −e.g., the power dimension [25] or antagonist behaviors [4]− the members' interactions are represented by a directed graph (digraph).

We consider that proposing a technique for conflict detection having into account how positive or negative are the messages interchanged, nowadays is possible thanks to the advances in sentiment analysis techniques. Sentiment analysis in text is a multidisciplinary research field which include fields such as natural language processing, computational linguistic, semantic, information retrieve, machine learning, and artificial intelligence [27]. The objective of sentiment analysis is to recognize the sentiments or emotions that people manifest when they write about products, organizations, persons or another topics [1,19]. The proposed technique for conflict detection considers exclusively the valence of messages transmitted.

3   Proposed Approach

As stated earlier, the approach uses a digraph to represent socio-emotional interactions of group members. This section first introduces the basic terms and concepts used in this work, then it explains the approach.

3.1   Preliminaries

Let $G = (V, E)$ be a weighted digraph defined on the node set $V = \{1, 2, \ldots, n\}$ and A be a $n \times n$ nonnegative, but generally asymmetric weight matrix such that

$a_{ij} > 0$ if and only if the directed edge $\langle i, j \rangle \in E$. The transition probability matrix of the Markov chain associated with random walks on G is defined as $P = D^{-1}A$ where $D = \mathrm{diag}(d_i)$ is a diagonal matrix of the node out-degrees, $d_i = \sum_{i=1}^{n} a_{ij}$. Each entry of P, $p_{ij} = a_{ij}/d_i$ is the probability to transit from node i to node j, if $\langle i, j \rangle \in E$.

When G is strongly connected; i.e., there is a (directed) path from any vertex i to any other vertex j. Then the Markov chain P is irreducible or ergodic, and it has a unique stationary probability distribution, π_i, where $\pi_i > 0$, $1 \leq i \leq n$. Namely, $\pi^\top P = \pi^\top$, where $\pi = [\pi_1, \ldots, \pi_n]^\top$ is the vector of stationary probabilities. In other words, over the long run, no matter what the starting state was, the proportion of time the chain spends in state j is approximately π_j for all j.

The normalized Laplacian $L^p = (I - P)$, has been used to analyze connectivity in terms of the mixing times or diffusion rate for the random walk as well as related expander constants, and in spectral graph partitioning [5].

The fundamental matrix Z is defined [14] as

$$Z = \left(I - P + \mathbf{1}\pi^\top\right)^{-1}, \tag{1}$$

where $\mathbf{1} = [1, 1, \ldots, 1]^T$.

If an ergodic Markov chain is started in state i, the expected number of steps to reach state j for the first time is called the *mean first passage time* or *hitting time*. The fundamental matrix can be used to compute h_{ij}, the mean first passage time from i to j as

$$h_{ij} = \frac{z_{jj} - z_{ij}}{\pi_j} \tag{2}$$

In matrix notation, H can be calculated as

$$H = \left(\mathbf{1} \cdot [\mathrm{diag}(Z)]^\top - Z\right) \cdot \Pi^{-1} \tag{3}$$

where Π is the diagonal matrix of stationary probabilities. The round-trip expected commute times matrix is calculated as

$$C = H + H^\top \tag{4}$$

Finally, the pseudo-inverse, M, can be calculated as

$$M = L^+ = \left(I - 1/n \cdot \mathbf{11}^\top\right) \cdot Z\Pi^{-1} \cdot \left(I - 1/n \cdot \mathbf{11}^\top\right). \tag{5}$$

The diagonal entries of M are a relative measure of centrality for the individual nodes [5].

3.2 Detecting Conflicts in CSCL Environments

The proposed approach allows automatic conflict recognition by representing chat interactions as a digraph G in which the nodes represent n users and the edges indicate the transference of negative sentiments during the interactions.

Henceforth, the nonnegative weight matrix $A = [t_{ij}]_{n \times n}$ is called the negative transfer matrix, or simply the transfer matrix.

The input for the proposed technique is a collaborative conversation composed of text messages where the sender and receiver(s) are known. Formally, a conversation is a sequence of m messages $[M_1, \ldots, M_m]$, the i–th message of the conversation is a tuple $M_i = \langle t_i, s_i, r_i \rangle$ where t_i is the text message, s_i is the sender, and r_i is the set of receivers.

Each text message is analyzed to estimate its valence $v_i \in [-1, 1]$; i.e., a very negative message has a valence of -1, and a very positive message has a valence of 1.

Let $V_{ij} = (v_{ij}^1, \ldots, v_{ij}^\ell)$ be the valence of the $\ell \leq m$ messages sent from member u_i to member u_j in the conversation.

The score t_{ij} for $i \neq j$ can be seen as a measure of how much student i can transfer negative sentiments to student j; the edge joining vertices i, j is weighted as

$$t_{ij} = \frac{1}{n-1} \exp\left(-\frac{(1 + \check{V}_{ij})^2}{2\sigma^2} \right) \qquad \text{for } i \neq j \qquad (6)$$

where

$$\check{V}_{ij} = \min V_{ij},$$

and σ is the decaying factor.

The weight of the self-loop, t_{ii}, is a measure of how i-th student prevents sending negative sentiments to the group

$$t_{ii} = 1 - \sum_{j \neq i} t_{ij}. \qquad (7)$$

The transfer matrix A, defined by (6) and (7), is used to calculate the mean first passage times H^A matrix from (3).

Let us introduce the diagonally scaled commute times matrix

$$\hat{C}^A = \hat{H}^A + (\hat{H}^A)^\top \qquad (8)$$
$$= H^A \Pi + (H^A \Pi)^\top,$$

that is used to detect subgroups having a conflict. It uses the mean first passage time matrix (3) scaled by the stationary probabilities to reveal the behavior of pairs of nodes in the graph. Here lower values indicate fastest flow of negative sentiments between two members.

As reference to detect conflicts and the members involved in it, we define the maximum diagonally scaled commute time for a given emotion interaction graph. By substituting $\check{V}_{ij} = 0$ (a neutral emotion) into (6), the reference matrix for any group of n members is $\hat{A} = [\hat{t}_{ij}]_{n \times n}$ with

$$\hat{t}_{ij} = \frac{1}{n-1} \exp \frac{-1}{2\sigma^2} \qquad (9)$$

Algorithm 1. Detecting members in conflict

Require: A conversation with m messages $(\langle T_1, R_1, M_1 \rangle, \ldots, \langle T_m, R_m, M_m \rangle)$ of a group of n members.
Ensure: Subgroups having a conflict and their centrality.

1: Estimate the valence V_i for each message M_i.
2: Calculate the transfer matrix A using (6) and (7).
3: $\hat{C}^A \leftarrow$ Calculate the diagonally scaled commute times matrix using (8)
4: $c_{\max} \leftarrow$ Calculate maximum scaled commute time (10)
5: **if** exists a conflict $-$condition (11)$-$ **then**
6:     $S \leftarrow$ Obtain k clusters from $\hat{C}^A$ with DBSCAN.
7:     $\bar{M}^A \leftarrow$ diag M, using (5) // {centrality}
8:     Associate the centrality $m : \mathbb{N} \rightarrow \mathbb{R}$ to each member in a cluster $s_i \in S$
9:     **return** $\{m(s_1), \ldots, m(s_k)\}$
10: **else**
11:     **return** $\{\}$
12: **end if**

---

for $i \neq j$ and $\hat{t}_{ii}$ is calculated with (7). The maximum scaled commute time is calculated as

$$c_{\max} = \min \hat{C}^{\hat{A}}. \tag{10}$$

where $\hat{C}^{\hat{A}}$ is the diagonally commute times for the reference matrix $\hat{A}$.

A conflict is detected by the following condition

$$\exists i, j \in V \mid c_{ij} < \tau c_{\max}. \tag{11}$$

where $0 < \tau \leq 1$ is a constant factor; finally, a clustering algorithm is used to detect subgroups involved in a conflict. For this purpose, the commute times is used as distance between points. The *Density-Based Spatial Clustering of Applications with Noise* (DBSCAN) [30] was selected for this purpose. Given a set of points in some space, it groups together points that are closely packed together (points with many nearby neighbors), marking as outliers points that lie alone in low-density regions (whose nearest neighbors are too far away). Here, subgroups are found by using the transfer score matrix. DBSCAN requires two parameters: ϵ and the minimum number of points required to form a dense region (minPts). The maximum commute time can be used for this purpose, we set $\epsilon = c_{max}/2$. As a conflict can be established from two or more members, then we set minPts = 2.

Once the subgroups having a conflict were detected, the pseudo-inverse M^A can be used to detect members that are central to the conflict $-$i.e., entries with lower values in M^A. Algorithm 1 describes the complete process to detect subgroups with conflicts.

4   Experimentation

For this study, student–student interactions with text during CSCL sessions were collected from July to November of 2018.

Students of programs in Informatics/Computing from two universities of Argentine (Universidad Nacional de Santiago del Estero UNSE, and the Universidad Católica de Santiago del Estero, UCSE) and two universities of Colombia (Corporación Universitaria Comfacauca, Unicomfacauca, and Institución Universitaria Centro de Estudios Superiores) participated in our experiments by using the COLLAB web application [22,32].

Students were grouped into small groups based on teachers' criteria to perform different weekly learning activities that require collaboration among students. In general, these activities consist in the production of essays on diverse information technology subjects. The students used COLLAB to communicate with their groups and to perform assignments. Teachers had access to chat sessions of groups and they can intervene when required; for example, when groups do not work. Figure 1(a) shows the proportion of groups according to the number members.

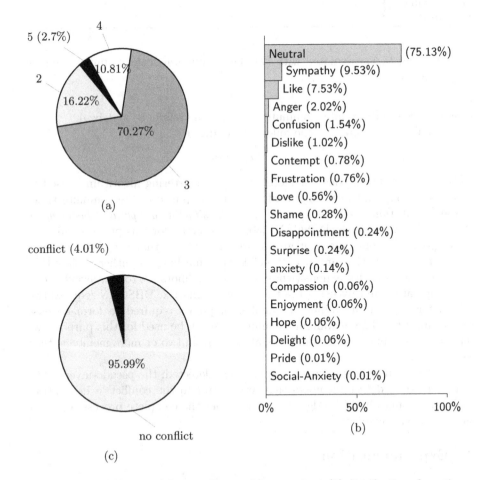

Fig. 1. General statistics of the experiment: (a) group sizes, (b) distribution of emotions recognized in the 7,145 sentences of the collaborative sessions, and (c) distribution of conflicts in the 2,717 windows.

Sentence Tags. Content analysis was applied to 7,145 text sentences following the Krippendorff methodology [20]. Three persons were trained to assign a tag to each sentence that indicates the emotion type according to the Pekrun's achievement emotions [28]. The alpha-reliability for the first tagging result was 0.58. Because of the low reliability, the tags were reviewed, the alpha-reliability after this process was 0.86. Figure 1(b) shows the distribution of emotions recognized in the sentences of the collaborative sessions.

Text Windows. Each collaborative session was segmented in 2,717 small windows. Each window is defined as the smallest sequence of text messages where a given user is the emitter of the message at the beginning and at the end of the window, but text messages in the middle have at least one sentence emitted by another participant. Each window was tagged as "conflict" or "no conflict" by two professors. A window was tagged as "conflict" if the professor recognizes conflict having into account the presence of disagreements, frictions, personality clashes, frustrations, and others indicators that the professors interpret as manifestation of conflict. The alpha-reliability for the tagged data was 0.91. As shown in Fig. 1(c), just a few windows have conflicts between participants (4.01%) and the majority of windows (95.99%) do not have any conflict.

Valence. The Senticnet 5 dictionary [7] was used to convert the Pekrun emotion tag of each sentence into a polarity.

Metrics. We ran the proposed algorithm with $\tau \in \{0.1, 0.2, \ldots, 0.9\}$ and $\sigma = \{0.3, 0.35, \ldots, 0.6\}$. The automatic labels obtained by a given pair of τ and σ values were contrasted with the manual tags to calculate the precision and recall.

5   Results and Discussion

Figures 2 and 3 show the ROC and precision-recall curves of the experiment. The ROC curve shows that the technique has a low false positive rate. The best result was obtained with $\tau = 0.6$ and $\sigma = 0.3$. In this case, the precision was 0.6 and the recall was 0.90.

We state that the proposed diagonally scaled commute times matrix can be used to evaluate the existence of conflicts in group interactions. When an entry of the commute times matrix tend to zero, it reflects that the members have send and received negative messages. This is key factor in the proposed technique. For instance, a member could send a negative message but the receiver could react positively to such stimuli; if this were the case, then there would be no conflict between group members. On the opposite, someone could send a positive message but the receiver could consider it improper, reacting in a negative way and confronting the sender. In any case, the proposed approach requires bidirectional negative messages. This behavior is in concordance with the theory which claim that emotional state of people condition their behavior and their way of interacting [13, 15, 23, 26, 35].

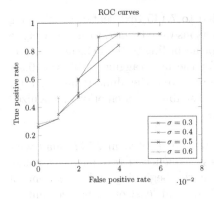

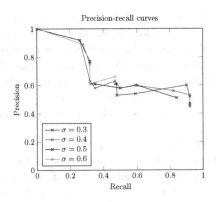

Fig. 2. ROC curve of the conflict detection approach using the commute times matrix

Fig. 3. Precision–Recall curve of the conflict detection approach using the commute times matrix

Also, this approach provides a centrality vector which allows recognizing members who are more negative in the interaction process. Knowing this information is helpful for professionals to correctly intervene and to offer support to the group for the mitigation of conflicts.

6   Conclusion and Future Work

Conflicts are an inherent characteristic of computer supported collaborative learning. Right handling of conflicts by learners and teachers are essential to keep an appropriated emotional environment that contributes to learning.

This study extends the application of Social Network Analysis (SNA) in Computer-Supported Collaborative Learning by allowing to recognize subgroups having a conflict. The algorithm proposed is based on the flow of sentiments in a social network that can be modeled as a random walk.

Results obtained show that the technique proposed can generate outputs that can be used as a good indicator of presence of conflict in a fragment of interactions between members of a group. As future work we include more interactions in the graph to improve the precision of the algorithm.

References

1. Appel, O., Chiclana, F., Carter, J., Fujita, H.: A hybrid approach to the sentiment analysis problem at the sentence level. Knowl.-Based Syst. **108**, 110–124 (2016)
2. Ayoko, O.B., Callan, V.J., Härtel, C.E.J.: Group Research. Small Group Res. **39**(2), 121–149 (2008)
3. Bales, R.: Interaction Process Analysis: A Method Forthe Study of Small Groups. Addison-Wesley Press, Cambridge (1950)

4. Bales, R., Steven, C.: SYMLOG: A System for the Multiple Level Observation of Groups. Free Press, New York (1979)
5. Boley, D., Ranjan, G., Zhang, Z.L.: Commute times for a directed graph using an asymmetric Laplacian. Linear Algebra Appl. **435**(2), 224–242 (2011)
6. Bolton, M.K.: The role of coaching in student teams: a "just-in-time" approach to learning. J. Manag. Educ. **23**(3), 233–250 (1999)
7. Cambria, E., Poria, S., Hazarika, D., Kwok, K.: SenticNet 5: discovering conceptual primitives for sentiment analysis by means of context embeddings. In: 32nd AAAI Conference on Artificial Intelligence, AAAI 2018, pp. 1795–1802 (2018)
8. Carringthon, P., Scott, J., Wasserman, S. (eds.): Models and Methods in Social Network Analysis. Cambridge University Press, Cambridge (2005)
9. Claros, I., Cobos, R., Collazos, C.: An approach based on social network analysis applied to a collaborative learning experience. IEEE Trans. Learn. Technol. **9**(2), 190–195 (2016)
10. Coviello, L., et al.: Detecting emotional contagion in massive social networks. PLoS ONE **9**(3), 1–6 (2014)
11. Dillenbourg, P., Baker, M., Blaye, A., Malley, C.O.: The evolution of research on collaborative learning. In: Spada, E., Reiman, P. (eds.) Learning in Humans and Machine: Towards an Interdisciplinary Learning Science, pp. 189–211. Elsevier, Oxford (1996)
12. Dreu, C.K.W.D., Weingart, L.R.: Task versus relationship conflict, team performance, and team member satisfaction: a meta-analysis. J. Appl. Psychol. **88**(4), 741–749 (2003)
13. Garcia-Prieto, P., Bellard, E., Schneider, S.C.: Experiencing diversity, conflict, and emotions in teams. Appl. Psychol. Int. Rev. **52**(3), 413–440 (2003)
14. Grinstead, C.M., Snell, J.L.: Introduction to Probability. American Mathematical Society, Providence (2012)
15. Heerdink, M.W., Kleef, G.A.V., Homan, A.C., Fischer, A.H.: On the social influence of emotions in groups : interpersonal effects of anger and happiness on conformity versus deviance. J. Pers. Soc. Psychol. **105**(2), 262–284 (2013)
16. Järvenoja, H., Järvelä, S.: Regulating emotions together for motivated collaboration. In: Affective Learning Together. Social and Emotional Dimensions of Collaborative Learning, Chap. 8, pp. 162–181. Routledge, New York(2013)
17. Jehn, K.A.: A qualitative analysis of conflict types and dimensions in organizational groups. Adm. Sci. Q. **42**(3), 530–557 (1997)
18. Jiang, J.Y., Zhang, X., Tjosvold, D.: Emotion regulation as a boundary condition of the relationship between team conflict and performance: a multi-level examination. J. Organ. Behav. **34**(5), 714–734 (2013)
19. Keshavarz, H., Abadeh, M.S.: ALGA: adaptive lexicon learning using genetic algorithm for sentiment analysis of microblogs. Knowl.-Based Syst. **122**, 1–16 (2017)
20. Krippendorff, K.: Content Analysis: An Introduction to its Methodology. Sage Publications, New York (2018)
21. Lee, D., Huh, Y., Reigeluth, C.M.: Collaboration, intragroup conflict, and social skills in project-based learning. Instr. Sci. **43**(5), 561–590 (2015). https://doi.org/10.1007/s11251-015-9348-7
22. Lescano, G., Costaguta, R.: COLLAB: conflicts and sentiments in chats. In: XIX International Conference on Human Computer Interaction (Interacción 2018) (2018)
23. Linnenbrink-Garcia, L., Rogat, T.K., Koskey, K.L.K.: Affect and engagement during small group instruction. Contemp. Educ. Psychol. **36**(1), 13–24 (2011)

24. Mello, J.: Improving individual member accountability in small work group settings. J. Manag. Educ. **17**(2), 253–259 (1993)
25. Millar, F.E., Rogers, L.E., Bavelas, J.B.: Identifying patterns of verbal conflict in interpersonal dynamics. West. J. Speech Commun. **48**(3), 231–246 (1984)
26. Molinari, G., Chanel, G., Bétrancourt, M., Pun, T., Bozelle, C.: Emotion feedback during computer-mediated collaboration: effects on self-reported emotions and perceived interaction. In: Rummel, N., Kapur, M., Nathan, M., Puntambekar, S. (eds.) 10th International Conference on Computer Supported Collaborative Learning, vol. 1, pp. 336–343. University of Wisconsin, Madison (2013)
27. Pang, B., Lee, L.: Opinion mining and sentiment analysis. Found. Trends Inf. Retrieval **2**, 1–135 (2008)
28. Pekrun, R.: Emotions and Learning. Educational Practices Series-24. UNESCO International Bureau of Education (2014)
29. Reffay, C., Chanier, T.: How social network analysis can help to measure cohesion in collaborative distance-learning. In: Wasson, B., Ludvigsen, S., Hoppe, U. (eds.) Designing for Change in Networked Learning Environments. Computer-Supported Collaborative Learning, vol. 2, pp. 343–352. Springer, Heidelberg (2003). https://doi.org/10.1007/978-94-017-0195-2_42
30. Schubert, E., Sander, J., Ester, M., Kriegel, H.P., Xu, X.: DBSCAN revisited, revisited: why and how you should (still) use DBSCAN. ACM Trans. Database Syst. **42**(3), 19:1–19:21 (2017)
31. Soller, A.: Supporting social interaction in an intelligent collaborative learning system. Int. J. Artif. Intell. Educ. **12**(1), 40–62 (2001)
32. de Santiago del Estero Facultad de Ciencias Exactas y Tecnologías, U.N.: Una herramienta para soportar la comunicación en entornos de aprendizaje colaborativo soportado por computadora (2020). http://chat.fce.unse.edu.ar/chat/web. Accessed 16 October 2020
33. Thompson, L., Fine, G.A.: Socially shared cognition, affect, and behavior: a review and integration. Pers. Soc. Psychol. Rev. **3**(4), 278–302 (1999)
34. Wall, V.D., Galanes, G.J.: The SYMLOG dimensions and small group conflict. Central States Speech J. **37**(2), 61–78 (1986)
35. Wosnitza, M., Volet, S.: Origin, direction and impact of emotions in social online learning. Learn. Instr. **15**(5), 449–464 (2005)
36. Zachary, W.W.: An information flow model for conflict and fission in small groups. J. Anthropol. Res. **33**(4), 452–473 (1977)

Mining Hidden Topics from Newspaper Quotations: The COVID-19 Pandemic

Thang Hoang Ta[1,2] , Abu Bakar Siddiqur Rahman[1(✉)], Grigori Sidorov[1(✉)], and Alexander Gelbukh[1(✉)]

[1] Instituto Politécnico Nacional, Mexico City, México
tahoangthang@gmail.com, bakar121107@gmail.com,
{sidorov,gelbukh}@cic.ipn.mx
[2] Dalat University, Da Lat, Viet Nam

Abstract. In this paper, we extract quotations from Al Jazeera's news articles containing keywords related to the COVID-19 pandemic. We apply Latent Dirichlet allocation (LDA), coherence measures, and clustering algorithms to unsupervisedly explore latent topics from the dataset of about 3400 quotations to see how coronavirus impacts human beings. By combining noun phrases as inputs before the training and C_v measure for coherence values, we obtain an average coherence value of 0.66 with a least average number of topics of 24.8. The result covers some of the top issues that our world has been facing against the COVID-19 pandemic.

Keywords: Topic model · Latent Dirichlet Allocation · Quotation mining · COVID-19

1 Introduction

Original from Wuhan, coronavirus (COVID-19) has quickly spread to 229 countries and territories, turn to the pandemic on a global scale just only in several months. No one could expect the massive impact of this virus on human society, initially thought as flu is able to lead to hundreds of thousands of deaths and a predictable economic recession in 2020 and even years later on. The policies to deal with this virus are not the same for all countries, from a lockdown (China, Italy, Spain, Germany) to a "herd immunity" model (Sweden, initially recommend in the UK). No matter what policies or methods are used and their effectiveness, the voice of famous people in a certain country probably has a weight that is big enough to orientate the public on how to counter the coronavirus.

A quotation (or quote) reflects someone's statement or thought regularly recorded when a famous individual declares something in interviews or public speeches. Direct quotations are obviously considered more subjective than interpretations from them (or indirect quotations) because they cover exacts words from the speakers or authors [1]. Hence, we use direct quotations to understand exactly what an individual thinks and his/her opinions about a certain problem, e.g., COVID-19. Note that we use quotations to refer to direct quotations for the remaining part of this paper.

© Springer Nature Switzerland AG 2020
L. Martínez-Villaseñor et al. (Eds.): MICAI 2020, LNAI 12469, pp. 51–64, 2020.
https://doi.org/10.1007/978-3-030-60887-3_5

From about 3400 quotations from Al Jazeera, we put them into a Latent Dirichlet allocation (LDA) model to mine hidden topics to know what these people, especially politicians care about COVID-19 pandemic. Several options are used to remove stopwords or high-probability words in quotations before/after training the LDA model. We select the best number of topics from combining coherence measures (C_V and C_{Umass}) with the sum of inverse topic frequency (ITF). Our purpose is to reduce the number of topics (topic number). Except for a topic number with the highest coherence value, we want to see other ones that are nearest to it by vector distances or any other metrics.

This paper is considered as "the first brick" of our research about quotation mining and opinion mining. We do hope our paper will be a valuable reference not only for relevant researches but also for those who are interested.

2   Literature Reviews

This paper's theme belongs to the topic model, which is used to exploit implicit topics in a set of documents or quotations as in our research. To solve problems of topic model, there are some prominent methods, such as LSA (LSI) [2], PLSA [3], LDA (HDP) [4, 5], word vectors or hybrid types [6–8], and graph neural networks [9, 10]. A lot of papers work with the comparison between LSA and LDA but it is still not clear which is the winner [23] and it basically depends on cases and corpora. WordNet is applied to LDA to improve the performance when it allows us to differentiate word senses and inherit hypernym and hyponyms hierarchy information to broaden the co-occurrence of terms in documents [24, 25]. Many recent methods based on word vectors (Word2Vec, GloVe) generally outperform traditional ones (LSA, PLSA) because they can capture the relatedness of words in topics beside word frequency distribution [11, 26]. Altszyler et al. found that LSA more effective than Word2Vec in a small corpus, less than 1 million words [12]. A latest research is a kind of multimodal when using GloVe, Wikidata, Wikipedia, and entity-topic co-occurrences all together to define topics [27]. In this research, we focus only on LDA to see how well it can work with quotations and no need to pre-define topics. We thus freely determine how many topics should be in the results and observe the features of output topics.

Stopwords removal is a popular step to pre-process input text before training the LDA model when stopwords are assumed to have a limited contribution [13]. However, this practice has no clear effect on the topic inference. Instead, dense high-probability terms should be removed due to their high chance appearing in every outcome topic [14]. We consider using different ways to remove stopwords and high-probability terms before/after the training process to see their impact on the output results. There are two ways to remove stopwords before training in quotations: (*) keep only noun phrases; (**) the same as (*) plus verbs, adjectives, and adverbs which are not listed as stopwords. Using (*) can produce a smaller number of topics but both ways are still difficult to label topic names where some topics are a mixture of sub-topics. Inspired by Alexandra et al. [14], we will consider using a third way, (***) remove dense high-probability terms after training, which is assumed to obtain more benefits than the previous two.

The number of topics is also important to LDA models. If this value is high, there will be more words that are repetitive in many topics. Otherwise, when the number of topics is low, it might be difficult to identify the topic name because a topic can cover many sub-topics inside. That is the reason why we must measure the topic coherence, representing in C_V and C_{Umass} values applied in our paper. C_{Umass} is the fastest way while C_V gives us the best performance [15]. At the output topics, we apply ITF [16, 17] to know how well we can discriminate words between topics or called the overlap rate of words between topics. Intuitively, if this value is low, it is easier for a person to identify and label the topic names that he is currently working on. Next, we combine the values of coherence measures and ITF to cluster them into groups by two algorithms, affinity propagation [18] and mean shift [19] to decide the best topic number (beside one with the best coherence value) for the LDA model.

3 Methodology

3.1 Quotation Extraction

We collect quotations by using a recursive algorithm to scan news articles from Al Jazeera by URLs. When working with a news article, we also extract URLs (use the regular expression) occurring inside that article and continue to scan these links later on. This process will repeat that until there will be no more link or we feel that we have enough quotations for our research.

With a typical news article, we extract text from HTML and organize it into paragraphs or sentences but we prioritize the former. Next, Spacy[1] is used to detect person names (named entity recognition – NER) from text. We call *PN* is a set of person's full name written by Western name order (First name, Middle name, and Last name) containing in a news article. Clearly, there is a culture clash for representing a person name in the world where different countries may have different styles to write a person name. In Al Jazeera, the Western name order is applied, so we will comply to this order to extract authors of quotations.

A quotation (direct quotation) is inside an open and a close quotation mark, called a quotation mark pair. There are several appearance types of quotation marks, such as '...', "...", '...', "...", and more which let us know which sentence or paragraph may contain quotations. When a sentence is detected to contain a quotation, we create a new sentence by replacing this quotation by a phrase, QUOTEx, whose x is the order occurrence (start from 0) of a quotation in the sentence. Normally, a sentence only contains a quotation but sometimes two quotations or more. Then, the new sentence is put to the dependency parsing process to detect a subject, a verb and an object for every single quotation. Let have a look at two examples below to understand how we extract quotations:

---

[1] https://spacy.io/

Example 1:
"My heart sank after hearing that," Zhang Hai said.
"QUOTE0," Zhang Hai said.

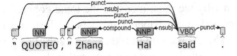

```
nsubj("said", QUOTE0)
nsubj("said", "Hai")
ncompound("Hai", "Zhang")
```

In Example 1, the quotation is "My heart sank after hearing that", the quotation author is "Zhang Hai" and the quotation verb is "said".

Example 2:
"Many pharmaceuticals so heavily rely upon active pharmaceutical ingredients from China and India," Huang said.
"QUOTE0", Huang said.

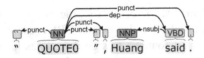

```
dep(QUOTE0, "said")
nsubj("said", "Huang")
```

In Example 2, the quotation is "Many pharmaceuticals so heavily...", the quotation author is "Huang" and the quotation verb is "said". The term "Huang" assumed to be last name of somebody like the Western name order. We prefer to get the full name of the quotation author so we back to set PN (1) to search for which one. If there are two authors have the same last name in PN, we discard this quotation.

Another type of quotations that the quotation author is not shown clearly, e.g., "He said: 'We will better day by day.'". To determine "he" is who, we can use neuralcoref[2], a coreference module to train text of news articles. However, we take it out of this research because it slows down heavily the speed of our quotation crawler.

3.2 Latent Dirichlet Allocation (LDA)

In this paper, we only use LDA to discover hidden topics from quotations. LDA is a generative probabilistic model that considers a document is a mixture of latent topics and each topic is represented by a distribution of probable words. A corpus D contains M documents where each document is a sequence of N words. LDA assumes the following generative process [4, 21]:

---

[2] https://spacy.io/universe/project/neuralcoref

1. Select $N \sim$ Poisson (ξ).
2. Select $\theta \sim$ Dir(α).
3. For each of the N words w_n:

    a. Select a topic $z_n \sim$ Multinomial (θ).
    b. Select a word w_n from $p(w_n|z_n,\beta)$, a multinomial probability conditioned on the topic z_n.

For more in details about equations used in the LDA model, please refer to [4] which we will not present in this paper.

Figure 1 shows the graphical display for the LDA model. Two parameters: α and β are the Dirichlet priors on the per-document topic distributions and the per-topic word distribution. θ is the topic distribution while z is the topic and w is a certain word.

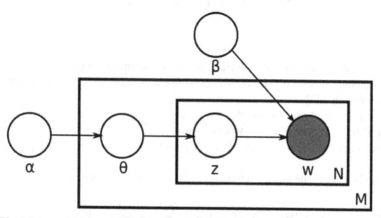

Fig. 1 The LDA model is represented by plate notation. M is a plate for documents while N is the repeated choice of topics and words in documents.

3.3 Topic Coherence Measures

Topic coherence is calculated by a score of the semantic analogy between top words in a given topic, helps to differentiate between topics. Although there are many coherence measures, we only apply C_{Umass} and C_v to this paper, one is famous for the runtime speed and the other for the best performance.

C_{Umass} is based on document co-occurrence and measured by a sum of component logarithm functions. We define the coherence of topic t with set W of top words as [15]:

$$C_{Umass} = C(t; W) = \sum_{i=2}^{L} \sum_{j=1}^{i-1} \log \frac{D(w_i, w_j) + \mu}{D(w_j)} \qquad (1)$$

with

- $D(w_i)$: document frequency of word type w_i
- $D(w_i, w_j)$: co-document frequency of word types w_i and w_j
- $W = \{w_1, w_2,...,w_L\}$: a list of the top L probable words in topic t
- μ: a parameter to avoid the case of zero logarithm. Originally, $\mu = 1$ [15] but the performance will be better if μ is relatively small [20].

Different from C_{Umass}, C_V is a combination between the normalized pointwise mutual information (NPMI) with the indirect cosine measure and the boolean sliding window. NPMI is an extension of pointwise mutual information (PMI), refers to single words in the corpus. PMI of word w_i and word w_j is defined as [15, 21]:

$$PMI(w_i, w_j)^\gamma = \left(\log \frac{P(w_i, w_j) + \mu}{P(w_i) \cdot P(w_j)} \right)^\gamma \tag{2}$$

with

- w_i, w_j: words belong to list W of top L probable words in the topic, $W = \{w_1, w_2,...,w_L\}$
- γ: a parameter for stressing more on context features. When it equals 2, it will produce the best results [22].
- $P(w_i, w_j)$: the joint probability of words x_i and x_j.
- $P(w_i)$, $P(w_j)$: the probability of words x_i or x_j.
- μ: a parameter to avoid the case of zero logarithm

From (2), here is the equation to define NMPI of words w_i and w_j.

$$NPMI(w_i, w_j)^\gamma = \left(\frac{PMI(w_i, w_j)}{-\log(P(w_i, w_j) + \mu} \right)^\gamma \tag{3}$$

3.4 Inverse Topic Frequency (ITF)

It is important that the word overlap rate between topics should be low so that we can easily identify topic names in the results. We borrow the idea of inverse topic frequency (ITF) [16, 17] to measure the frequency of words appearing in the output topics. Given a set T of output topics from the LDA model, we can define *ITF* of word w by T as:

$$ITF(w, T) = \frac{|\{t \in T : w \in t\}|}{N} \tag{4}$$

with

- N: total number of output topics, $N = |T|$.
- $T = \{t_1, t_2,...,t_N\}$: a set of output topics with the size is N.
- $|\{t \in T : w \in t\}|$: number of topics where the word w appears. If this word is not in the output topics, *ITF* will be zero.

From the Eq. (4), there is another variant of *ITF* which use the logarithm function as:

$$ITF(w, T) = \log \frac{N}{|\{t \in T : w \in t\}|} \tag{5}$$

Now, we can define the sum of ITF (ITFS) by the number of topics to compare which topic numbers gain the best overlap rate, the lowest value.

$$ITF(T) = \sum_{i=1}^{N \times M} ITF(w_i, T) \tag{6}$$

with

- N: total number of output topics, $N = |T|$.
- M: the number of words in each topic or the size of a topic. Note that all topics have the same size.
- $ITF(w_i, T)$: the inverse topic frequency of word w_i in topic number T.

3.5 Combine Coherence and ITFS Values

A topic number is represented as a vector $\vec{T} = (s, k)$ with two dimensions, ITFS (sum of ITF) and the coherence value. For the task of calculating the coherence value in Sect. 3.3, the number of topics always begins from 2. We thus define set T_N containing topic number vectors from 2 to m as:

$$\begin{aligned} T_N &= \left\{ \vec{T_2}, \vec{T_3}, \dots, \vec{T_m} \right\} \\ &= \{(s_2, k_2), (s_3, k_3), \dots, (s_m, k_m)\} \end{aligned} \tag{7}$$

with

- $\vec{T_i}$: vector of topic number i, $i \in [2, m] \wedge i \in \mathbb{N}$
- N: size of set T_N. Note that $N = m - 2 + 1$ with m is the last topic number in the set.
- $S = \{s_2, s_3,\dots, s_m\}$: set of topic ITFS values.
- $K = \{k_2, k_3,\dots, k_m\}$: set of coherence values.

Depending on coherence measures (C_{Umass} or C_V), we can define the topic number T_b with the best coherence value as:

$$T_b = \{itfs_b, k_b\} \tag{8}$$

with conditions

- $k_b = \max(K)$ if the coherence measure is C_V
- $k_b = \min(K)$ if the coherence measure is C_{Umass}. Note that we use `gensim`[3] for the LDA training so C_{Umass} values will be negative instead of fluctuating around 0.

---

[3] https://radimrehurek.com/gensim/

In (10), we will search for the best topic number T_b. Because the coherence values are negative when measuring by C_{Umass} so the best value will be the lowest one. In contrast, we will choose the maximum value of set K for the C_v measure.

Next, set T_N is visualized on a 2D plane with x-axis is for ITFS values and y-axis is for coherence values. There are two clustering algorithms, affinity propagation [18] and mean shift [19] used to cluster topic numbers into groups. We use these two because of their benefit of no need to pre-define the number of clusters. In the experiment, affinity propagation detects more clusters (with smaller number items in each cluster) comparing mean shift so we take it as the first priority.

4 Experiments

4.1 Dataset

The dataset contains 3400 direct quotations extracted from Al Jazeera news articles containing keywords ("coronavirus", "COVID-19", etc.) related to the COVID-19 pandemic. Each quotation has 11 fields: `quote_author`, `verb`, `object`, `quote`, `quote_sentence`, `news_url`, `news_title`, `date`, `update_date`, `publisher`, and `news_author`. We collect quotations in 2020 and only use field `quote`" for the experiments. Other fields and their combination with "`quote`", such as "`quote_sentence`", "`news_title`", or even articles' content will be used for mining hidden topics but not in this research.

4.2 Stopword Removals and Coherence Measures

We use three options of removing stopwords in quotations:

- *RV1*: keep only noun phrases, used before training the LDA model.
- *RV2*: *RV1* with non-stopwords (verbs, adjectives, adverbs), used before training the LDA model.
- *RV3*: remove high frequency words (no matter word types) after training the LDA model.

We observe that *RV3* has a little contribution to the results because our corpus is small. Therefore, we will not apply *RV3* in our experiments. Next, stopword removals are combined with two coherence measures (C_{Umass} and C_v) to calculate the average values of coherence, executed time and number of topics in 4 test cases. For each test case, we run 5 times with the topic number from 2 to 50 before calculating average values.

In Table 1, $C_{Umass} + RV1$ gains the best executed time in all test cases and look at all tests we can re-confirm that C_{Umass} always runs faster than C_v. However, the executed time is not so important when we focus more on the topic number with the lowest value as possible. In that case, C_v shows a better performance, with 24.8 is the average of number of topics. C_{Umass} or C_v goes with *RV1* produces better coherence values compared to when going with *RV2*. This could be explained that RV2 contains more terms in the corpus. Based on results in Table 1, we decide to choose $C_v + RV1$ for next experiments.

Table 1 The comparison between 4 combinations of coherence measures and stopword removals in 5 runtimes.

	Avg. coherence	Avg. executed time (s)	Avg. no. of topics
$C_{Umass} + RV1$	**−18.51**	**274**	32.6
$C_{Umass} + RV2$	−17.74	282	46.2
$C_v + RV1$	**0.66**	342	**24.8**
$C_v + RV2$	0.62	360	35.4

4.3 Topic Number Clustering

Although $C_v + RV1$ offers us the lowest topic number as possible, we still prefer to explore better values with the combination of coherence values and ITFS values. This section is a demo for Sect. 3.5.

Coherence-itfs values are organized into clusters by two algorithms: affinity propagation and mean shift. In Fig. 2 and Fig. 3, topic number 20 obtains the best coherence value, viewed as the center point for searching relevant topic numbers. Figure 2 (affinity propagation) shows a zoom that topic number 20 belongs to cluster 3. Besides cluster 3, topic numbers in cluster 2 can be chosen when they have a low ITFS (sum of ITF), less than 28. In this case, some topic numbers (9, 10, 11, 13, 14, etc.) may be candidates when gaining a low ITFS and the coherence value is higher than 0.63, a good enough value.

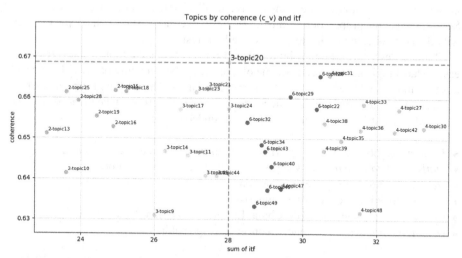

Fig. 2 A graphical display for clustering coherence-itfs values: affinity propagation

Similarly, we can see in Fig. 3 (mean shift), topic number 20 belongs to cluster 0 with more elements. Hence, we clearly see that affinity propagation is better than mean shift when it reduces the number of elements in a cluster. We manually choose relevant

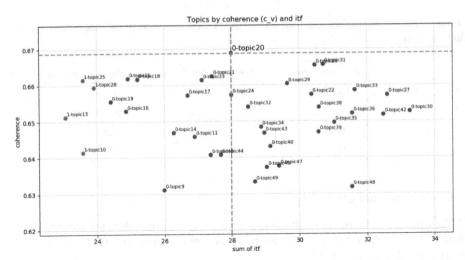

Fig. 3 A graphical display for clustering coherence-itfs values: mean shift.

topic numbers by using a greedy algorithm, the smaller topic numbers with high enough coherence values compared to the center point (the topic number with the best coherence value).

4.4 Explore Latent Topics

Depending on situations, if we want to have smaller topic numbers, we can do like in Sect. 4.3 to get the proper topic numbers. However, we should start with a topic number with the best coherence value to see how well our results are.

Table 2 describes 20 topics by terms and our guess for the topic names. It is not easy to label topic names because some of them could be a mixture of several topics or terms in topics or not enough clear to infer topic names. This may be from our corpus is small and we are working on a niche topic like COVID-19. We take the topic names from Table 2 and infer new ones, based on our understanding.

- *Economics*: jobs, employees, market, economic impacts, business, restaurants.
- *Statistics of coronavirus*: new cases, death cases, peaks, rate, outbreak.
- *Healthcare*: hospitals, patients, symptoms.
- *Policies*: lockdown, quarantine, time, risk.
- *Corona virus by countries*: US, China, Europe (Italy), Iran, Japan, Africa.
- *Origin of coronavirus*: evidences, concerns, responsibility.

We are a bit disappointed when some important terms related to the COVID-19 pandemic, such as "social_distancing", "masks", and "hand_washing" do not appear in the top probable words in output topics. This can be explained when many quotation authors are politicians, who may care more about economics and policies than medical and healthcare issues.

Table 2 Top 10 high probable terms by topics.

Topic	Terms	Topic name (our guess)
0	COVID-19, shops, government, cases, temperature, science, European_Union, World_Health_Organization, virus	–
1	Situation, government, jobs, thing, coronavirus, concern, evidence, world, spread, crisis	–
2	Information, opportunity, month, government, country, People, president, symptoms, confidence	–
3	People, number, day, death, way, cases, workers, today, millions, new_coronavirus	Coronavirus cases (death, new cases)
4	People, countries, Africa, coronavirus, community, South_Africa, EU, COVID-19, federal_government, deal	Coronavirus in Africa and EU
5	People, police, Japan, lot, Europe, People, COVID-19_pandemic, March, April, case	Coronavirus in Japan, EU
6	China, country, situation, calls, action, risk, care, groups, Chinese, Everybody	Coronavirus in China
7	Cases, China, virus, coronavirus_pandemic, things, water, Italy, middle, crisis, country	Coronavirus cases in China and Italy
8	Coronavirus, virus, health, number, lockdown, measures, residents, government, quarantine, people	Policies
9	Fact, hospital, disease, families, coronavirus, places, things, numbers, markets, stake	Healthcare
10	Virus, country, risk, lockdown, lives, people, Trump, new_cases, direction, time	Policies
11	Economy, people, data, hospitals, deaths, weekend, patients, virus, economic_activity, prisoners	Healthcare, economy
12	State, today, coronavirus, emergency, lot, public_health, health, authorities, people, possibility	Healthcare
13	Fed, people, virus, responsibility, war, demand, peak, effects, reality, restaurants	–

(continued)

Table 2 (*continued*)

Topic	Terms	Topic name (our guess)
14	Epidemic, market, spread, response, movement, order, Americans, vaccines, thousands, beginning	Market, healthcare
15	Year, week, coronavirus, economic_impact, life, matter, months, end, employees, March	Economics
16	People, COVID-19, impact, Iran, outbreak, behaviour, virus, coronavirus, country, threat	Coronavirus in Iran
17	China, virus, crisis, United_States, coronavirus, US, individuals, month, shock, number	Coronavirus in China and US
18	Government, pandemic, time, coronavirus, PPE, risk, people, decision, extent, home	Policies
19	Work, staff, food, women, people, cities, outbreak, reason, transmission, jails	–

From terms in output topics, we suggest to re-classify them into new topics, may be based on word embeddings and definitions from dictionaries. This method is expected to gain more accuracy when it captures the semantic meanings instead of the word distributions. It is also interesting if we remove keywords that we use for searching quotations in the output topics (coronavirus, COVID-19, etc.) because of their high occurrence. We thus may able to label the topic names easier from a set of remaining words. To solve the overlapping of topics, we suggest mapping all output terms to Wikidata or Wordnet (entity linking). Based on their hierarchy system (hypernyms and hyponyms), common terms will be decarded to gain a better topic recognition.

5 Conclusion

We present an unsupervised approach for mining hidden topics from quotations in Al Jazeera's news articles, related to the COVID-19 pandemic. The LDA model needs to remove stopwords or high-frequency words or both to gain better performance but it depends on the size of the corpus to make a proper decision. In our experiments, to keep only noun phrases works best for the model when it is able to obtain the highest coherence value and the lowest topic number at the same time. Clustering topics by coherence and ITFS values helps to reduce the topic number while still maintaining a good enough coherence value. We consider terms in output topics as a positive suggestion to manual label topic names although several topics may be a mixture of smaller topics. Our future work is to overcome this drawback by applying a multimodal method, combine LDA with word embeddings, word definitions (WordNet), Wikidata, Wikipedia, and DBSCAN to gain better performance.

Acknowledgements. The work was done with support of the Government of Mexico via CONA-CYT, SNI, CONACYT grant A1-S-47854, and grants SIP 20200797, SIP 20200859 of the Secretaría de Investigación y Posgrado of the Instituto Politécnico Nacional, Mexico.

References

1. Wade, E., Clark, H.H.: Reproduction and demonstration in quotations. J. Mem. Lang. **32**(6), 805–819 (1993)
2. Landauer, T.K., Foltz, P.W., Laham, D.: An introduction to latent semantic analysis. Discourse Process. **25**(2–3), 259–284 (1998)
3. Hofmann, T.: Probabilistic latent semantic analysis. arXiv preprint arXiv:1301.6705 (2013)
4. Blei, D.M., Ng, A.Y., Jordan, M.I.: Latent Dirichlet Allocation. J. Mach. Learn. Res. **3**, 993–1022 (2003)
5. Teh, Y.W., Jordan, M.I., Beal, M.J., Blei, D.M.: Sharing clusters among related groups: hierarchical Dirichlet processes. In: Advances in Neural Information Processing Systems, pp. 1385–1392 (2005)
6. Esposito, F., Corazza, A., Cutugno, F.: Topic modelling with word embeddings. In: CLiC-it/EVALITA (2016)
7. Wang, Z., Ma, L., Zhang, Y.: A hybrid document feature extraction method using Latent Dirichlet Allocation and word2vec. In: 2016 IEEE 1st International Conference on Data Science in Cyberspace (DSC), pp. 98–103. IEEE (June 2016)
8. Wang, Y., Xu, W.: Leveraging deep learning with LDA-based text analytics to detect automobile insurance fraud. Decis. Support Syst. **105**, 87–95 (2018)
9. Van, L.N., Tran, B., Than, K.: Graph Convolutional Topic Model for Data Streams. arXiv preprint arXiv:2003.06112 (2020)
10. Yang, L., et al.: Graph attention topic modeling network. In: Proceedings of the Web Conference 2020, pp. 144–154 (April 2020)
11. Naili, M., Chaibi, A.H., Ghezala, H.H.B.: Comparative study of word embedding methods in topic segmentation. Proc. Comput. Sci. **112**, 340–349 (2017)
12. Altszyler, E., Ribeiro, S., Sigman, M., Slezak, D.F.: The interpretation of dream meaning: resolving ambiguity using Latent Semantic Analysis in a small corpus of text. Conscious. Cogn. **56**, 178–187 (2017)
13. Wilson, A.T., Chew, P.A.: Term weighting schemes for Latent Dirichlet Allocation. In: Human Language Technologies: The 2010 Annual Conference of the North American Chapter of the Association for Computational Linguistics, pp. 465–473. Association for Computational Linguistics (June 2010)
14. Schofield, A., Magnusson, M., Mimno, D.: Pulling out the stops: rethinking stopword removal for topic models. In: Proceedings of the 15th Conference of the European Chapter of the Association for Computational Linguistics, Short Papers, vol. 2, pp. 432–436 (April 2017)
15. Röder, M., Both, A., Hinneburg, A.: Exploring the space of topic coherence measures. In: Proceedings of the 8th ACM International Conference on Web Search and Data Mining, pp. 399–408 (February 2015)
16. Usui, S., Palmes, P., Nagata, K., Taniguchi, T., Ueda, N.: Keyword extraction, ranking, and organization for the neuroinformatics platform. Biosystems **88**(3), 334–342 (2007)
17. Trnka, K.: Adaptive language modeling for word prediction. In: Proceedings of the ACL-08: HLT Student Research Workshop, pp. 61–66 (June 2008)
18. Wang, K., Zhang, J., Li, D., Zhang, X., Guo, T.: Adaptive affinity propagation clustering. arXiv preprint arXiv:0805.1096 (2008)

19. Cheng, Y.: Mean shift, mode seeking, and clustering. IEEE Trans. Pattern Anal. Mach. Intell. **17**(8), 790–799 (1995)
20. Vulić, I., De Smet, W., Moens, M.F.: Identifying word translations from comparable corpora using latent topic models. In: Proceedings of the 49th Annual Meeting of the Association for Computational Linguistics: Human Language Technologies, Short Papers, vol. 2, pp. 479–484. Association for Computational Linguistics (June 2011)
21. Mimno, D., Wallach, H.M., Talley, E., Leenders, M., McCallum, A.: Optimizing semantic coherence in topic models. In: Proceedings of the Conference on Empirical Methods in Natural Language Processing, pp. 262–272. Association for Computational Linguistics (July 2011)
22. Aletras, N., Stevenson, M.: Evaluating topic coherence using distributional semantics. In: Proceedings of the 10th International Conference on Computational Semantics, IWCS 2013, Long Papers, pp. 13–22 (March 2013)
23. Stevens, K., Kegelmeyer, P., Andrzejewski, D., Buttler, D.: Exploring topic coherence over many models and many topics. In: Proceedings of the 2012 Joint Conference on Empirical Methods in Natural Language Processing and Computational Natural Language Learning, pp. 952–961. Association for Computational Linguistics (July 2012)
24. Boyd-Graber, J., Blei, D., Zhu, X.: A topic model for word sense disambiguation. In: Proceedings of the 2007 Joint Conference on Empirical Methods in Natural Language Processing and Computational Natural Language Learning, EMNLP-CoNLL, pp. 1024–1033 (June 2007)
25. Li, C., Feng, S., Zeng, Q., Ni, W., Zhao, H., Duan, H.: Mining dynamics of research topics based on the combined LDA and WordNet. IEEE Access **7**, 6386–6399 (2018)
26. Moody, C.E.: Mixing dirichlet topic models and word embeddings to make lda2vec. arXiv preprint arXiv:1605.02019 (2016)
27. Bhargava, P., et al.: Learning to map Wikidata entities to predefined topics. In: Companion Proceedings of the 2019 World Wide Web Conference, pp. 1194–1202 (May 2019)

Analysis of COVID-19 Pandemic Using Artificial Intelligence

Maaz Amjad[1]([⊠]), Yuriria Rodriguez Chavez[2], Zaryyab Nayab[3], Alisa Zhila[4], Grigori Sidorov[1], and Alexander Gelbukh[1]

[1] Center for Computing Research (CIC), Instituto Politécnico Nacional (IPN), Mexico City, Mexico
maazamjad@phystech.edu, Sidorov@cic.ipn.mx, gelbukh@gelbukh.com
[2] Instituto Mexicano Del Seguro Social, IMSS, Mexico City, Mexico
yuri21rch@gmail.com
[3] University of Veterinary and Animal Sciences, Kasur, Pakistan
Nayyab08@gmail.com
[4] New York, USA

Abstract. The emerging issue of COVID-19, which is caused by the coronavirus SARS-CoV-2 is a significant problematic issue. This disease has impacted worldwide and covered the globe in its threat. Scientists and physicians have been investigating pathophysiological aspects of this pandemic to understand the virus structure for the treatment development. In this mini review, we briefly discuss the characteristics of SARS-CoV-2, its origin, diagnosis, and treatment. This study will provide an understanding of the role of artificial intelligence in this emerging COVID-19 pandemic along with the impact in our society, economy, health, and industrial level.

Keywords: Pandemics · COVID-19 · Artificial intelligence · SARS-CoV-2

1 Introduction

Coronavirus (COVID-19) is a syndrome that causes a respiratory illness with multiple signs and symptoms. The severity of respiratory illness may vary from mild respiratory distress to severe respiratory illness (SARS). COVID-19 belongs to the same family with different characteristics, which is now called severe acute respiratory syndrome coronavirus 2 (SARS-CoV-2) or 2019-nCoV. COVID-19 is a single-stranded, pleomorphic or spherical, positive-sense, and an enveloped RNA virus. This Virus has an envelope made up of lipid bilayer in which spike proteins, envelope proteins, and membrane proteins are embedded. The envelope exhibits club-shaped projections of glycoprotein. The RNA is embedded with nucleocapsid protein. Unlike coronaviruses group 2, SARS-CoV does not bear a hemagglutinin esterase glycoprotein [1] (Fig. 1).

A. Zhila—Independent Researcher.

© Springer Nature Switzerland AG 2020
L. Martínez-Villaseñor et al. (Eds.): MICAI 2020, LNAI 12469, pp. 65–73, 2020.
https://doi.org/10.1007/978-3-030-60887-3_6

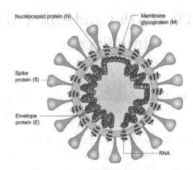

Fig. 1. Coronavirus 2019-nCoV. Source: Politis, Periklis. "Reporting Strategies in Crisis: The Case of Severe Acute Respiratory Syndrome."

1.1 Host Range Diversity of CoV

As far as the diversity of coronavirus is concerned, there exists a vast host range. There are four coronavirus classes based on host range; alphacoronaviruses, beta coronaviruses, gammacoronaviruses, and deltacoronaviruses. Alphacoronavirus and betacoronavirus infect humans while gamma and delta infect aves and fish, although few of these also cause disease in humans [2].

There were only six coronaviruses known as SARS-CoV, HKU1, HCoV-OC43, MERS-CoV, HCoV-NL63 and HCoV-229E, which cause a severe respiratory syndrome in humans before 2019 [3]. In 2019, there was an emerging new virus from beta coronaviruses, which causes respiratory infections and viral pneumonia in humans.

To get knowledge of the newly emerging 2019-nCoV, one should know its origin, where it comes from? The first Human coronavirus (HCoVs) was recorded in the late 1960s. That virus disease was a clinical syndrome that mainly affected the respiratory tract and enteric and central nervous system diseases. After that, there were new pandemics of human coronaviruses associated with a respiratory syndrome called SARS caused by severe acute respiratory syndrome coronavirus (SARS-CoV) in 2003. In addition to this, to rectify these pandemics' problems, the increased research had been conducted in 2004. As a result, two other human coronaviruses, HCoV-NL63 andCoV-HKU1 were discovered in the Netherlands and China, respectively [4].

Van der Hoek et al. took a nasopharyngeal fluid or aspirate from the child. That child was diagnosed with fever, bronchiolitis, and conjunctivitis. They isolated HCoV-NL63 from that child in March 2004. Furthermore, after one month, the same virus had been isolated in April 1988 from a child suffering from pneumonia [5]. Moreover, now in 2019, there is a massive outbreak of a novel coronavirus with various signs and symptoms in Wuhan, the city of China (Fig. 2).

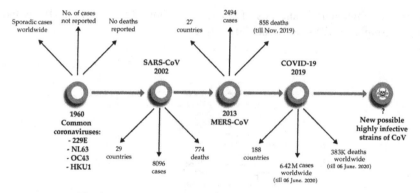

Fig. 2. The worldwide impact of coronavirus infections.

1.2 Comparison of SARS-CoV 2 (COVID-19), SARS-CoV

Each virus has to bind to the host cell to get its entry and infect the cell. Novel coronavirus binds a receptor of host cell called ACE2 (the angiotensin-converting enzyme 2) with the help of spikes to ensure its entrance in the human cell so that it can infect. A protease enzyme prepares the spikes to attach to cell receptors. Moreover, this enzyme may vary among different viruses, but SARS-CoV and COVID-19 use the same protease named as TMPRSS2 to get entry and proceed with the remaining replication process [6] (Fig. 3).

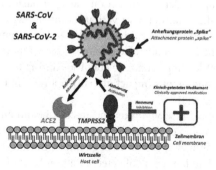

Fig. 3. The spikes of the both coronaviruses COVID-19 and SARS-CoV use (ACE2) receptor and the same protease enzyme TMPRSS2 [25].

After getting entry into the host cell virus uses the host machinery to prepare its proteins to increase its number, the virus assembles itself and buds off from the cell to attack the predisposing site [7].

Like COVID-19, MERS coronavirus and SARS coronavirus are also from beta coronaviruses group associated with a respiratory syndrome, which ranges from milder infection to severe infections. Moreover, current studies investigated that COVID-19 is more associated with milder respiratory infections [8]. Research proves that COVID-19 is more contagious and transmissible as it transmits in the community vert fast. COVID-19 is also associated with secondary infections and symptoms like pneumonia, enteric distress, etc. Phylogenetic analysis of coronavirus showed that (COVID-19) is more likely to resemble SARS-CoV about 70% [9] (Fig. 4).

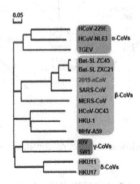

Fig. 4. Phylogenetic tree of coronaviruses, new coronavirus COVID-19 shown in red [9]. (Color figure online)

2   Worldwide Impacts of COVID-19

At the beginning of 2020, COVID-19 has caused the world to come into a particular state of standstill. It has an unprecedented impact on multiple businesses across industries, from the oil industry to the bankruptcy of various companies. A recent study [10] has investigated different factors that reveal how COVID-19 economically will impact in the coming few years. The same study also demonstrated a range of possible economic costs concerning different leading economic countries during the coronavirus outbreak. This health crisis has an outrageous impact globally not only in the business sector but also on the public mental health.

People around the world face vigorous and intense health emergencies during the current COVID-19 pandemic. People started to face social and mental health challenges induced by the outbreak of the COVID-19. For example, the latest study [11] conducted by the Imperial College Landon team collected a dataset based on the information like age-specific contact patterns and the outbreak of the coronavirus severances. The study highlighted the impact of the viral COVID-19 disease in 202 countries, such as China and high-income countries. The study revealed that in the absence of various organizations and government interventions, the coronavirus epidemic would have caused approximately 7 billion infections and 40 million worldwide mortalities in the present year.

Another study [12] investigated the Psychological impact based on Depression [15, 27], Anxiety and Stress during the outreach of the COVID-19. The study reported that people started to get depression due to the limitations of activities inside the home and the continuous fear of getting the virus themselves or family members.

Several countries had implemented national school closures to avoid high mortality rates in educational institutions. All the educational institutions, government, and private offices are either closed or function remotely. For instance, the following study [13] was conducted on school social distancing practices during the coronavirus epidemic. The study suggested that school closures prevent 2–4% of death casualties. However, governments are providing continuous efforts to make people's lives safer by taking many initiatives like lockdowns to combat this highly contagious disease. However, there is a need for vaccine development to ensure public health safety from a widespread outbreak of an infectious disease.

3 The Role of Artificial Intelligence During Covid-19 Pandemic

The emergence of technology and Artificial intelligence (AI) has made human life more comfortable, safe, and beneficial during the outbreak of the COVID-pandemic.

AI-based technologies have been used to combat and monitor the current epidemic that has brought our society at a standstill. In addition to this, AI-based models have been used to assist governments in ensuring public health planning by monitoring the epidemic situation. There have been numerous studies to investigate how AI is being used during the global pandemic. For example, the previous study [14] proposed an AI-based system to estimate the size, lengths, and ending time of Covid-19. The study used the AI system to analyze data collected from different Chinese cities, the first epicenter of this disease.

Artificial intelligence has been widely used in the health care sector. A recent report [13] demonstrated that AI-based technologies are extremely helpful in augmenting the genome sequencing analysis, quick diagnostics of multiple diseases especially identifying the COVID-19 symptoms, helping in scanner analysis, and for the research to evolve the treatment process, in particular, a vaccine development process.

To combat the COVID-19 epidemic, a team of researchers from China, the USA proposed an algorithm called "LinearDesign" to investigate protein folding [16]. The algorithm presents several benefits to study the structure of a virus' secondary ribonucleic acid (RNA) in terms of computational time complexity. The proposed algorithm is high-speed compared with other available methods. For example, a report published in the MIT technology review [17] published that the LinearDesign algorithm predicted the secondary structure of the RNA sequence of Covid-19 in 27 s instead of 55 min. It also provides new insights into the prevalence of coronavirus.

The leading tech companies have proposed various AI-based solutions to develop vaccines for the COVID-19. Multiple algorithms have been proposed by the tech companies to understand the protein's structure since understanding the protein structure helps to find out how it works. However, experiments that need to be performed for investigating protein structure require a considerable amount of time, like in months or longer.

Research has been done to design highly computational techniques to understand the protein structure from the amino acid sequence. For example, DeepMind also proposed an AI-based system known as the "AlphaFold AI system" to predict the coronavirus protein structures [18]. Similarly, other tech companies such as Google, Amazon, IBM, and Microsoft have also taken several initiatives to combat the pandemic. These companies have also made available computational sources (machines with acute computing power) to process and analyze a large amount of data related to molecular modeling, epidemiology, and bioinformatics for the development of the coronavirus vaccine development [19].

4 Diagnostic Methods for Covid-19

In such a public health crisis, fast and specific detection of COVID-19 is decisive to manage the pandemic. Molecular-based tests are the main techniques for laboratory diagnosis; these include reverse transcription loop-mediated isothermal amplification (RT-LAMP), reverse-transcription polymerase chain reaction (RT-PCR), and real-time RT-PCR (rRT-PCR). Furthermore, other methods, such as viral culture, are not profitable. The reason is that it requires a minimum of 3 days for SARS-CoV-2 to induce cytopath changes. All methods to isolate viruses require specific biosafety facilities for personal protection, which in most health care centers is not possible to maintain.

Serological antigen and antibody detection tests, on the other hand, have limitations because it may be cross-reactivity with SARS-CoV. Nonetheless, seroconversion takes some days (7–14 days), and titles are not related to viral load. Due to these restrictions, the most useful and standard laboratory diagnostic test to detect coronavirus (COVID-19) is still the reverse transcription-PCR (RT-PCR) [20, 21].

According to the World Health Organization (WHO) suggestions, patients suspected with SARS-CoV-2 should be sampled for molecular testing. These samples might be taken from the respiratory tract, mostly nasopharyngeal swab, but they can also be sputum or bronchoalveolar lavage fluid.

Specific probes and primers were developed according to the genomic sequence to create real-time RT-PCR diagnostic tests for SARSCoV-2. The open reading frames (ORF 1a and 1b), envelope (E), nucleocapsid (N), and RNA-dependent RNA polymerase gene (RdRp) are clue diagnostic targets for SARS-CoV-2 identification. Moreover, at least two molecular targets must be included in the assays, so potential cross-reaction with other coronaviruses and potential genetic drift are dodged [23].

RT-PCR results usually require approximately 2 to 8 days to turn into positive after several days, and some patients might have initial test results false-negative for virus infection [21, 22]. Therefore, patients presenting non-specific symptoms such as fever, coughing, myalgias, sore throat, arthralgias, asthenia, or dyspnea along with new COVID-19 contact, SARS-CoV2 infection should be diagnosed with typical chest computerized tomography (CT) changes even though RT-PCR results from negative [22].

For an efficient and accurate molecular diagnosis, appropriate sampling in time and location is fundamental [22]. In some countries, there are not enough sources to train appropriately and provide the right personal protective equipment to the people taking care of these patients. So this is a severe situation, how can authorities manage this crisis without putting in risk to their personal. There must be more research to find a comfortable, economical, and safe way to make a timely diagnosis.

5 Treatment

Specific antiviral drugs or vaccines to treat COVID-19 are not available nowadays. The medical management of patients focuses on supportive care. These include oxygenation, mechanical ventilation, thromboprophylaxis, antibiotics prophylaxis, and fluid therapy. However, the requirement of an effective antiviral treatment has forced physicians to use certain drugs just based on in vitro or extrapolated evidence [23]. Table 1 shows some drugs being investigated for COVID-19 treatment recently.

Table 1. The table shows the investigated medicines for the treatment of COVID-19 in vitro studies or in clinical trials [23].

Group	Drugs	Mechanism of action
Inhibitors of viral RNA polymerase /RNA synthesis	Remdesivir (GS-5734)	Adenosine nucleotide analogue, prodrug, RdRp inhibitor
	Favipiravir	Guanosinenucleotid analogue, prodrug, RdRp inhibitor
Inhibitors of viral protein synthesis	Lopinavir/ritonavir	Protease inhibitor
Viral entry inhibitors	Hydroxychloroquine	Increasing endosomal pH required for virus/cell fusion, as well as interfering with the glycosylation of cellular receptors of SARS-CoV (ACE-2)
	Chloroquine	
Imunomodulators	Nitazoxanide	Interference with host-regulated pathways involved in viral replication, amplifying cytoplasmic RNA sensing and type I IFN pathways
	Ivermectin	Inhibition nuclear import of host and viral proteins through inhibition of importin 1 heterodimer

Monoclonal antibodies are a trend in medical management mainly of autoimmune diseases. Having recognized that COVID-19 pathogenesis includes a cytokine storm. It had been suggested the use of Tocilizumab; a recombinant humanised monoclonal antibody against IL-6 receptor, as an effective therapeutic in covid patients. While shown improvement in symptoms, oxygen requirement and computer tomography severity score changes after the treatment with tocilizumab [24].

5.1 Convalescent Plasma Uses Evidence

Another therapeutic option that has strong evidence in the treatment of COVID-19 is the use of convalescent plasma. This therapy is found on the principles of passive immunization, getting immunoglobulins from recovered COVID-19 patients, and transfusing them to sick patients. In addition to this, plasma therapy is expected as prophylaxis for health care caregivers. Eventually, scientific evidence has demonstrated excellent results and fewer adverse events than other treatments [26].

6 Conclusion and Future Work

In this article, we aim to provide an overview of the Covid-19 pandemic and its global consequences. The paper shed light on the impact of COVID-19 and estimated the consequences that brought up to our society by the COVID-19 epidemic in terms of industrial, educational, economic, and health-related challenges. The study highlighted that to mitigate and overcome the consequences of the epidemic, we need to find the treatment as early as possible. Our studies found that spike proteins of SARS-CoV-2 are 10–20 times more likely to bind with the receptors of human cells, i.e., ACE2 than SARS-CoV of 2002, and that is why it spreads more rapidly human to human. Despite having all structural and genetic similarities with SARS-CoV, SARS-CoV-2 could not be captured by antibodies against SARS-CoV of 2002. So this study suggested that specific antibody-based treatment should be introduced. We hope these findings will substantially help make the design of potential vaccines and develop the treatments for SARS-CoV-2. We plan to increase the dataset related to Covid-19 by using data augmentation techniques to find people's attitudes towards such pandemic.

References

1. Petrosillo, N., Viceconte, G., Ergonul, O., Ippolito, G., Petersen, E.: COVID-19, SARS and MERS: are they closely related? Clin. Microbiol. Infect. **26**, 729–734 (2020)
2. Cui, J., Li, F., Shi, Z.-L.: Origin and evolution of pathogenic coronaviruses. Nat. Rev. Microbiol. **17**(3), 181–192 (2019)
3. Fehr, A.R., Perlman, S.: Coronaviruses: an overview of their replication and pathogenesis. Coronaviruses **1282**, 1–23 (2015). Humana Press, New York
4. van der Hoek, L., et al.: Identification of a new human coronavirus. Nat. Med. **10**(4), 368–373 (2004)
5. Vabret, A., et al.: Human coronavirus NL63, France. Emerg. Infect. Dis. **11**(8), 1225 (2005)
6. Hoffmann, M., et al.: SARS-CoV-2 cell entry depends on ACE2 and TMPRSS2 and is blocked by a clinically proven protease inhibitor. Cell **181**, 271–280 (2020)
7. McIntosh, K., Peiris, J.S.M.: Coronaviruses, 3rd edn. In: Clinical Virology, pp. 1155–1171. American Society of Microbiology (2009)
8. Munster, V.J., Koopmans, M., van Doremalen, N., van Riel, D., de Wit, E.: A novel coronavirus emerging in China—key questions for impact assessment. N. Engl. J. Med. **382**(8), 692–694 (2020)
9. Chen, Y., Liu, Q., Guo, D.: Emerging coronaviruses: genome structure, replication, and pathogenesis. J. Med. Virol. **92**(4), 418–423 (2020)

10. Walker, P., et al.: Report 12: The global impact of COVID-19 and strategies for mitigation and suppression (2020)
11. Wang, C., et al.: Immediate psychological responses and associated factors during the initial stage of the 2019 coronavirus disease (COVID-19) epidemic among the general population in China. Int. J. Environ. Res. Pub. Health **17**(5), 1729 (2020)
12. Viner, R.M., et al.: School closure and management practices during coronavirus outbreaks including COVID-19: a rapid systematic review. Lancet Child Adolesc. Health **4**(5), P397–P404 (2020)
13. Hu, Z., Ge, Q., Jin, L., Xiong, M.: Artificial intelligence forecasting of COVID-19 in China. arXiv preprint arXiv:2002.07112 (2020)
14. Chun, A.: In a time of coronavirus, China investment in AI is paying off in a big way. South China Morning Post, 18 March 2020
15. Iqra, A., Siddiqui, M.H.F., Sidorov, G., Gelbukh, A.: CIC at SemEval-2019 task 5: Simple yet very efficient approach to hate speech detection, aggressive behavior detection, and target classification in Twitter. In: Proceedings of the 13th International Workshop on Semantic Evaluation, pp. 382–386 (2019)
16. Zhang, H., et al.: LinearDesign: Efficient Algorithms for Optimized mRNA Sequence Design. arXiv preprint arXiv:2004.10177 (2020)
17. Baidu: How Baidu is Bringing AI to the Fight against Coronavirus. MIT Technology Review, 11 March 2020. https://www.technologyreview.com/2020/03/11/905366/how-baidu-is-bringing-ai-to-the-fight-against-coronavirus/
18. Computational Predictions of Protein Structures Associated with COVID-19. Deepmind. Accessed 6 June 2020. https://deepmind.com/research/open-source/computational-predictions-of-protein-structures-associated-with-COVID-19
19. Lardinois, F.: IBM, Amazon, Google and Microsoft Partner with White House to Provide Compute Resources for COVID-19 Research. TechCrunch, 22 March 2020. https://techcrunch-com.cdn.ampproject.org/c/s/techcrunch.com/2020/03/22/ibm-amazon-google-and-microsoft-partner-with-white-house-to-provide-compute-resources-for-covid-19-research/amp/
20. Ahn, D.-G., et al.: Current status of epidemiology, diagnosis, therapeutics, and vaccines for novel coronavirus disease 2019 (COVID-19). J. Microbiol. Biotechnol. **30**(3), 313–324 (2020)
21. Zhai, P., Ding, Y., Wu, X., Long, J., Zhong, Y., Li, Y.: The epidemiology, diagnosis and treatment of COVID-19. Int. J. Antimicrob. Agents **55**, 105955 (2020)
22. Tang, Y.-W., Schmitz, J.E., Persing, D.H., Stratton, C.W.: Laboratory diagnosis of COVID-19: current issues and challenges. J. Clin. Microbiol. **58**(6), e00512–e00520 (2020)
23. Şimşek Yavuz, S., Ünal, S.: Antiviral treatment of COVID-19. Turk. J. Med. Sci. **50**, 611–619 (2020)
24. Xu, X., et al.: Effective treatment of severe COVID-19 patients with tocilizumab. Proc. Natl. Acad. Sci. U.S.A. **117**(20), 10970–10975 (2020)
25. Hoffmann, M., et al.: SARS-CoV-2 cell entry depends on ACE2 and TMPRSS2 and is blocked by a clinically proven protease inhibitor. Cell **181**, 271–280 (2020)
26. Chen, L., Xiong, J., Bao, L., Shi, Y.: Convalescent plasma as a potential therapy for COVID-19. Lancet. Infect. Dis. **20**, 398–400 (2020)
27. Mustafa, R.U., Ashraf, N., Ahmed, F.S., Ferzund, J., Shahzad, B., Gelbukh, A.: A multiclass depression detection in social media based on sentiment analysis. In: Latifi, S. (ed.) 17th International Conference on Information Technology–New Generations (ITNG 2020). AISC, vol. 1134, pp. 659–662. Springer, Cham (2020). https://doi.org/10.1007/978-3-030-43020-7_89

Bot-Human Twitter Messages Classification

Carolina Martín-del-Campo-Rodríguez[✉], Grigori Sidorov,
and Ildar Batyrshin

Centro de Investigación en Computación (CIC), Instituto Politécnico Nacional (IPN),
Mexico City, Mexico
cm.del.cr@gmail.com, sidorov@cic.ipn.mx, batyr1@gmail.com

Abstract. Bots identification has gained relevance within social networks due to its ability to influence the opinion of users on political, consumer and ideological issues. This is why research related to bot identification has grown in recent years. Various models have been proposed for the identification of bots, but this is an issue that has not been resolved yet. In this article, a model is proposed that, through the use of specific preprocessing and a four-layer neural network, improves the bot-human classification accuracy of Twitter messages, reaching a precision of 0.9462, which represents an advance with respect to what is presented in the state of the art with the same corpus.

Keywords: Bots identification · Neural networks · Twitter · Text preprocessing

1 Introduction

Social networks nowadays play a significant role for political, commercial or religious purposes. Bots are used to impersonate humans and try to influence, either positively or negatively, on these issues on social media. For example, bots can be used to give a positive rating to a certain product or to give negative ratings to competitive products; they can also influence political campaigns to create a preference or dislike for a particular candidate. Also, these are frequently used to spread false news.

This is why the task of detecting bots, especially on Twitter, has gained relevance. This document explains the approach taken to improve the state of the art in the bot-human classification task in Twitter messages in Spanish, using the corpus described in [11].

In the Sect. 2 a review of the state of the art is presented, emphasizing those investigations where the same corpus was used; in the Sect. 3 the description of the corpus used is made; Sect. 4 explains how the text was pre-processed, vectorized and normalized, as well the specifications of the neural network used for the classification; the Sect. 5 describes the results obtained, as well as

© Springer Nature Switzerland AG 2020
L. Martínez-Villaseñor et al. (Eds.): MICAI 2020, LNAI 12469, pp. 74–80, 2020.
https://doi.org/10.1007/978-3-030-60887-3_7

the comparison of these with those obtained in the state of the art; finally, in the Sect. 6 the conclusions of the results obtained are written, as well as the improvements that the model and future work may have.

2   State of Art

Bots detection extends beyond its analysis on Twitter; in [5] Hall et al. used some basic characteristics of the articles of Wikipedia to identify bots, such as: article modifications, time on site, among others. Using *Random Forest* and *Gradient Boosting* he was able to detect bots with a precision of 0.88.

In [7] Kudugunta et al. used *LSTM* (Long Short-Term Memory) to detect bots at the account level on Twitter, with a hundred percent accuracy in the corpus created by them. As part of the preprocessing they used replacement of hashtags, url, user mentions with static labels, the text was transformed into lowercase, among others.

2.1   PAN 2019

PAN is an international forum that is responsible for promoting forensic investigation of digital text by organizing shared evaluation tasks. In *PAN 2019* [11] the task *"Bots and Gender Profiling"* was proposed, whose objective is the multi-language classification (between bots and humans) of Twitter messages. The task of identifying the gender (male-female) of the tweets classified as human was also proposed by PAN, but gender identification goes beyond this investigation. The precision was used to determine the best approach.

Several approaches were presented to solve this task. The best ones, in terms of bot vs human classification in Spanish, are:

In [10] Pizarro used as preprocessing the concatenation of the 100 tweets per author in a single string, using a tag to separate the tweets, later he converted the text to lower case and used *nltk* to tokenize. He replaced links, user mentions, and hashtags with tags, respectively. *SVM* was used for the classification, using character n-grams in a range from 3 to 5 and word n-grams in a range from 1 to 3. He obtained a precision of 0.9333.

Jimenez-Villar et al. [6] used the library *TweetTokenizer* to: concatenate the tweets of each author in a string separating them with the tag *END*, replace the line break with *NL*, convert to lower case, replace links with *URL*, and finally replace user mentions with *USER*. Then, they used the text distortion method *DV-MA*, where they consider W_k as the list of the k most frequent words of the language, any word not included in W_k is masked by replacing each character with a (*) and the digits are replaced with (#). To determine the most frequent words they used *DF* (*"Document Frequency"*), *FCE* (*"Frequently Co-occurring Entropy"*) and *IG* (*"Information Gain"*). As characteristics, they used character n-grams with n from 3 to 5, eliminated the terms with less than three occurrences and used *SVM* for classification. They obtained a precision of 0.9211.

3 Corpus Description

The corpus used for the detection of bots on Twitter is the one used in the task *Bots and Gender Profiling* of *PAN 2019* [11].

PAN as part of this task, made available a multi-language corpus of Twitter messages made by bots and by humans. For the purpose of this work, the corpus in Spanish of this task was used, which is divided as shown in Table 1. For each sample of the corpus there are 100 tweets, that is, for each author (bot-human), there are 100 example tweets. The corpus samples are in xml format.

Table 1. Spanish corpus used in PAN 2019 for the task *Bots and Gender Profiling*. The values presented in the column **Bots** and **Humans** refer to the number of authors (samples) for each category, **Total** is the sum of these. The column **Total Tweets** represents the total number of tweets we have (given 100 tweets per author -sample-)

	Bots	Humans	Total	Total Tweets
Train	2,060	2,060	4,120	412,000
Test	1,320	1,320	2,640	264,000
Total	3,380	3,380	6,760	676,000

4 Methods and Materials

As part of Tweets pre-processing: links, user mentions, and hashtags, respectively, were replaced with fixed tags.

Dates, times and numbers were replaced with fixed labels, respectively, just like the emojis. Special characters (unicode) were replaced by a fixed label. Line breaks were replaced by a fixed label and all punctuation and *stop words* were removed.

Tweets of the same author were concatenated into a single string, separated by a fixed tag and converted to lower case. Subsequently all the labels were replaced for later use by punctuation marks, respectively.

The vectorization was carried out using *TfidfVectorizer* of scikit-learn [8], specifying in the parameters character *n*-grams of size 3. In addition, those *n*-grams with an occurrence less than 2 were discarded.

4.1 Classification

The classification between bots and humans was done with a neural network, *tensorflow 2* [2] was used for its implementation.

For the construction of the model, four layers were used with 64, 32, 16 and 1 unit, respectively. After each layer a dropout of 0.8, 0.5 and 0.2 was used respectively. Figure 1 shows the model specifications.

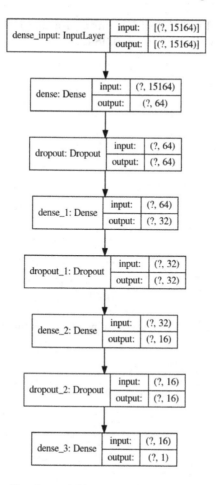

Fig. 1. Specifications of the proposed neural network model

The first three layers used *relu* as activation function (this to avoid the problem of gradient fading due to the number of layers [3]) and a kernel regularization *l2 (0.001)* to avoid *overfitting* [1].

The activation function for the last layer, *sigmoid* was used (in [9] it is shown that *sigmoid* can be as effective as *softmax* for binary classification).

The model was configured with *Adam* as optimizer, objective function *binary_crossentropy* and metric *accuracy*. The training was done with 150 epochs.

5   Evaluation

Precision was used as an evaluation metric, this in order to compare our approach with those presented in the state of the art with the same corpus. The program was run ten times. The average of the executions is shown in graphs Fig. 2 (error during learning) and Fig. 3 (precision during learning). The average value in precision at the end of the training is **0.9842**.

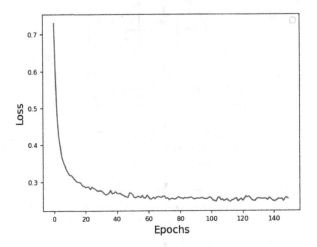

Fig. 2. Error during the training of the neural network

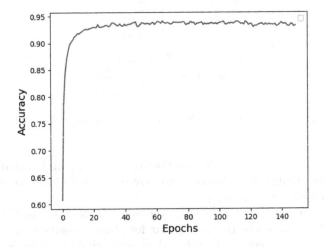

Fig. 3. Accuracy during the training of the neural network

The result obtained in the test corpus is shown in the Table 2. As can be seen, our proposal improves the result compared to the best state of the art approach.

Table 2. Accuracy in the test corpus for the task bot-human Twitter messages in Spanish classification

Our approach	**0.9462** ± 0.0045
Pizarro [10]	0.9333

6   Conclusion

The preprocessing used allow us to reduce to a label that information that is not relevant for the identification of bots-humans in tweets and helped us to improve the precision in the bot-human classification task. The substitution of dates, times and digits by labels helps to convert to a generic representation that information that does not contribute to the bots-human style identification. The use of neural networks, despite using a basic configuration (representation of a bag of words as input), helped to make a better generalization of the problem. The precision obtained by the methodology proposed in this article reaches 0.9462, improving by 0.0129 the state of the art of Pizarro [10].

As future work, the use of embedded words as input to neural networks is proposed, as well as the exploration of other neural network architectures. In addition to expanding the research to other languages and the identification of the genre of tweets made by humans.

Acknowledgement. The work was done with support of the Government of Mexico via CONACYT, SNI, CONACYT grant A1-S-47854, and grants SIP 20200797 and SIP 20200853 of the Secretaría de Investigación y Posgrado of the Instituto Politécnico Nacional, Mexico.

References

1. Overfit and underfit: Tensorflow core. https://www.tensorflow.org/tutorials/keras/overfit_and_underfit
2. Abadi, M., et al.: TensorFlow: large-scale machine learning on heterogeneous systems (2015). https://www.tensorflow.org/. Software available from tensorflow.org
3. Brownlee, J.: A gentle introduction to the rectified linear unit (ReLU) (2019). https://machinelearningmastery.com/rectified-linear-activation-function-for-deep-learning-neural-networks/
4. Daelemans, W., et al.: Overview of PAN 2019: author profiling, celebrity profiling, cross-domain authorship attribution and style change detection. In: Crestani, F., et al. (eds.) 10th International Conference of the CLEF Association (CLEF 2019). Springer (2019). http://ceur-ws.org/Vol-2380/
5. Hall, A., Terveen, L., Halfaker, A.: Bot detection in Wikidata using behavioral and other informal cues. In: Proceedings of the ACM on Human-Computer Interaction 2(CSCW) (2018). https://doi.org/10.1145/3274333

6. Jimenez-Villar, V., Sánchez-Junquera, J., Montes-y-Gómez, M., Villaseñor-Pineda, L., Ponzetto, S.: Bots and gender profiling using masking techniques. In: Cappellato, et al. [4]. http://ceur-ws.org/Vol-2380/

7. Kudugunta, S., Ferrara, E.: Deep neural networks for bot detection. Inf. Sci. **467**, 312–322 (2018). https://doi.org/10.1016/j.ins.2018.08.019, http://dx.doi.org/10.1016/j.ins.2018.08.019

8. Pedregosa, F., et al.: Scikit-learn: machine learning in Python. J. Mach. Learn. Res. **12**, 2825–2830 (2011)

9. Pedro, I.: Understanding the motivation of sigmoid output units (2020). https://towardsdatascience.com/understanding-the-motivation-of-sigmoid-output-units-e2c560d4b2c4

10. Pizarro, J.: Using N-grams to detect bots on Twitter. In: Cappellato, et al. [4]. http://ceur-ws.org/Vol-2380/

11. Rangel, F., Rosso, P.: Overview of the 7th author profiling task at PAN 2019: bots and gender profiling. In: Cappellato, et al. [4]. http://ceur-ws.org/Vol-2380/

Automatic Story Generation: State of the Art and Recent Trends

Brian Daniel Herrera-González, Alexander Gelbukh, and Hiram Calvo(✉)

Center for Computing Research, Instituto Politécnico Nacional, Av. Juan de Dios Bátiz s/n esq. Manuel Othón de Mendizábal, 07738 Mexico City, Mexico
bdherrerag@gmail.com, gelbukh@gelbukh.com, hcalvo@cic.ipn.mx

Abstract. Throughout history, human beings have used tales as a way of communicating ideas and transmitting knowledge. One of the great advances in this area was obtained thanks to writing, a fact that substantially facilitated the transmission of these stories. Being such an ancient activity in human history, it is not surprising that it has been the object of study of artificial intelligence, from a very early time in the development of the latter, leading, among other things, to the emergence of the automatic generation of stories. The automatic generation of stories has been a challenge that has been sought to be solved using different approaches. This review brings together some of these guidelines for carrying out this endeavour.

Keywords: Computational linguistics · Automatic story generation · Narration

1 Introduction

Narration has been a theme that has accompanied human beings for many centuries. Ancient evidence suggests the use of stories from a very primitive time of the human being, in fact the transmission of stories in an oral way had been, for many centuries, the predominant way to spread them. The written narrative was one of the great advances that this field obtained, since it allowed, among other things, the intercultural transmission of stories. In more recent times, the development of the printing press meant another progress for the propagation of these stories and in the last century the computer and the web have allowed that stories of any origin can reach almost any corner of the earth, however, all these stories have something in common, they have been the result of human's imagination.

Artificial intelligence has been used to solve different tasks pertaining to human beings, either reducing cost, time, effort, among other elements that have led to a new way of using computers, but among all this repertoire there is a question that has intrigued the scientific community: Can a computer be creative?

This work was done with support of the Government of Mexico via CONACYT, SNI, CONACYT, BEIFI grant A1-S-47854; grants SIP 2083, SIP 20200811 and SIP 20200859 of the Secretaría de Investigación y Posgrado of the Instituto Politécnico Nacional, Mexico, IPN-COFAA, and IPN-EDI.

© Springer Nature Switzerland AG 2020
L. Martínez-Villaseñor et al. (Eds.): MICAI 2020, LNAI 12469, pp. 81–91, 2020.
https://doi.org/10.1007/978-3-030-60887-3_8

This doubt seems to be far from having an answer that satisfies both the people that belong to this branch of science and those that do not, but one of the ways that can lead to clarify this enigma is the resolution of tasks that seem to appeal to man's creativity. One of these tasks is the automatic generation of stories, which has not yet a clear solution or a basic methodology to be solved, although some patterns have been identified [15].

In this review some of the approaches with which this task has been tried to be solved are contemplated; we provide an overview, addressing first the symbolic approach and then the neural approach.

2 Symbolic Approaches

In a very general way, there are two approaches for solving the task of automatic story generation, the symbolic approach and the connectionist one.

In the first works carried out to perform the activity of automatic generation of stories, the symbolic guideline was used. One of the main strengths of this approach is the ease with which it is possible to delimit the path followed by the actions of the story. However, one of its disadvantages is the lack of *creativity*, in other words, stories tend to be very repetitive and have a very reduced repertoire of events.

These works can be divided into two main areas, those case-based and those event-based.

2.1 Case-Based Methods

Case-based methods are so called because they are strongly restricted to solving a theme, either the stories domain, the author goals or both.

2.1.1 Methods for Specific Domain Stories

In 1987 Meehan [1] with TALE-SPIN, generates different characters and their respective goals. The system must look for an outcome for the resolution of these goals using an inference engine based on common sense reasoning theories; all the stories are based on King Arthur's stories. Then in 2010, Riedl et al. use the IPOCL methodology [6]. In that work, they write fables from POCL (Partial Order Causal Link) algorithms using operators, preconditions and effects of the operators.

In 2014, McCoy et al. [9] generate prom stories using knowledge bases about the social structure, its norms, the cultural aspect and the desires of the characters, as well as concepts of social interaction. Finally, in 2016, Calvo et al. [10, 11] use literary structures for writing a story and evaluate the stories generated, their results are presented in Table 1, in comparison with other works.

2.1.2 Methods for Solving Author's Goals

In 1987, Lebowitz [2] with UNIVERSE, [2] uses a hierarchical planner to turn the author's goals into a story. This is achieved with hierarchically structured rules to find related tasks that can achieve that goal. In 2009 Porteous et al. [5] use reference points obtained from the author for finding different events that lead to a goal established by the human agent.

Table 1. Comparison of models with a symbolic approach. First two columns from [4].

Metrics	MINSTREL [3]	MEXICA [4]	Literary structures [10, 11]
Sample	50	50	32
Cohesion	2.9	3.8	4.5
Coherence	3.2	3.7	3.6
Content	3.6	4.1	NA
Interestingness	3.3	3.8	4.0
Novelty	NA	NA	4.0
General	3.3	3.8	4.0

2.1.3 Methods for Specific Domain Stories with Author Goals

In 1993, Turner [3] with MINSTREL, generate new stories using existing story concepts as a basis. The system must adapt these concepts through a model of computational creativity that satisfies the objectives of the author and the story. Additionally in 2002, Pérez y Pérez et al. with MEXICA [4] use a cognitive method of engagement and reflection, in order to use already known narratives and generate a new story that maintains coherence and an increasing tension throughout the story. These methods are compared in Sect. 4.

2.2 Event-Based Methods

The event-based methods use a concept called events to construct the storyline. The events are the actions or the transitions form one state to another. In 2012, Onodera et al. [7] use different states generated in a virtual world from a storyline and values of characters, objects and places. Then the user selects the states they want in the story, the system transforms them into events and finally uses a circular generation process to transform these events into sentences.

In 2014, Gervás et al. [8] use states of the world and simulate the interactions between the world and the different characters, choosing the simulation that best corresponds to the established narrative characteristics.

In addition, Adolfo et al. in [12] propose a model with events that emphasizes the development of characters through the use of events. The objective of this model is to develop stories for children. Finally, in 2019 Farrell et al. [13] propose a system based on Indexter [14] to obtain the relevance of the events in different stories and to direct the story based on the relevance of those events.

3 Neural Approaches

Every big advance in the neural networks field has been used for automatic story generation. Methods with this approach solved the problem of the lack of novelty in stories, but it was difficult to maintain the coherence of the story [29].

Some of the proposals for generating text have been sequence-by-sequence models (seq2seq) with recurrent neural networks (RNN), in particular gated recurrent networks and the long short-term memories GRU and LSTM. Another proposal is the unsupervised learning by means of generative adversarial networks GAN or Bidirectional Associative Memories [39]. In the literature we have also found pre-trained language models. However, all the alternatives that use only this type of approach have failed in the task of writing a history that maintains coherence.

In this section, we will review methods divided according to the type of guideline used to generate the storyline, which is segmented into those who have an explicit scheme and those who do not.

3.1 Methods for Stories without an Explicit Outline

These methods do not have a delimited section for generating the storyline. They can be classified as those that write a story without a specific focus and those who have specific requirements.

3.1.1 Methods with a General Approach

Jain et al. [16] in 2017 start from small descriptions and use statistical machine translation and sequence-to-sequence networks to obtain a story. Their results contain very little cohesion; however, it has served as a basis for the use of hybrid systems. Furthermore Clark et al. [17] in 2018 use a generation model based on entities. This model takes into account the general content of these entities, the content of the previous sentence and the already written content of the same sentence and generates the next word. Then encodes these three elements and uses a recurrent neural network in order to generate a sentence. In this research a significant advance is achieved, however, human judges perceive a lack of fluency in the story.

Fan et al. [19] in 2018 also separate the problem into two parts: The generation of a premise and the generation of a story from that premise. The story base was obtained through instructions provided by the researchers to different writers. To obtain the premise, they use a seq2seq encoder-decoder convolutional model. In particular they use a non-context model with two attention elements, multiscale and closed. Finally, they use a fusion model to retrieve information in the hidden layers. The premise becomes a story using a top-k random sampling scheme on the models. With this approach, the stories have significant coherence but little cohesion.

Peng et al. [20] in 2018 focus on achieving controllability of automatic story generation. They are based on the closing valence and the story line. To obtain the closure validity they use a logistic regression classifier based on a LSTM and for the generation they use a conditional language model in the same way with LSTM. For the argument line they obtain the keywords and they obtain a graph of the document with weights for each keyword. The generation is also made with a LSTM with attention. An effective control of the end of an unfinished story is achieved. However, the generation of a story from some words is not yet a very coherent story.

In 2019, Yao et al. [24] proposed generating a story line from a title and then turning it into a story. For the creation of the static storyline and the generation of the text, they use a LSTM. The stories have a good performance in coherence terms. Also in 2019, Fan et al. [30] propose a semantic role labeling to improve their proposal in [19], obtaining much better results than in their previous work. The semantic role labeling achieves that two sentences containing two different structures and the same meaning can be interpreted as the same entity.

Among the last works carried out, is that of Guan et al. [31] who use a system that generates a pre-trained model from two common sense knowledge bases, Concept-Net and ATOMIC. First, they transform the ontological base into simple sentences and with this information they use the GPT-2 architecture [34] with fine tuning. With this work, they obtain decent results in terms of fluidity and improve the state of the art corresponding to coherence.

In 2020, Ippolito et al. [32] use a sentence level model. This model consists of the following phases: A sentence embedder obtained with BERT [35], a candidate sentence generator made with both an MLP and a residual MLP, and an auxiliary loss resulting in a distracting set. They conclude with some ideas to modify the model, since there is a strong dependence on the embedding space.

3.1.2 Methods Without a General Approach

The methods [21, 23, 25] and [27] try to solve specific problems, for example Roemmele et al. [21] in 2018 show a "creative" assistant to continue a story with a sentence that is the result of a model with a recurrent neural network and that uses a variable that adjusts the probability distribution of the model, showing that the most creative proposals were also the least useful for the human author.

Ding et al. [23] in 2018 use 3 components, the first is a seq2seq generator with a LSTM, which given a context, generates an ending, the second component is a binary discriminator that receives both the context and the ending and classifies it as a human-made or machine-generated ending, the third component is an adversarial training process between the two previous components. In this work "good" endings are obtained according to human evaluation.

Additionally, Luo et al. [25] in 2019 use two components for their system, the first one is a sentiment analyser that adopts three methods, an unsupervised rule-based method, a linear regression model and an adversarial learning model. For the second component they use a seq2seq model controlled by the intensity of the feeling. They manage to obtain a good result according to the intensity of the feeling sought, but like the previous papers, some sentences lack coherence. Likewise, we can observe the work in Guan et al. [27] in 2019 who use two components, the first one is an incremental encoder with a LSTM to obtain the clues that are used to reach a conclusion. The second element is a multi-source attention mechanism that is used to obtain the context of a common sense knowledge base. With this work, an increase in the fluidity of the story is obtained, but it does not improve the coherence of the story.

3.2   Methods for Stories Without an Explicit Outline

These methods show some kind of planning for the generation of the storyline. In 2018, Martin et al. [18] separate the task into two parts. The first consists in the generation of the series of events that make up the story, and the second one in the narrative that describes these events. Events are separated into tuples of subject, verb, object, a wild card, and literary genre; to find the next event they use a multi-layered recurrent code-decoder network. An LSTM network with Beam Search is used to transform events into sentences. They perform favourably in obtaining new events; however, the generation of sentences is deficient as it lacks a coherent structure.

Furthermore, Xu et al. [22] in 2018 use a scheme that they call *skeleton* and then convert it into sentences. The system consists of two modules, the first one obtains the skeletons from the database using a seq2seq model with encoder and decoder based on LSTM. The second module contains two sub-modules, which are *input to skeleton* and *skeleton to sentence*. To obtain the input to skeleton, a seq2seq structure is used, where the encoder is hierarchical and the decoder is attention based, the submodule in charge of transforming the skeleton into a sentence uses a seq2seq model with encoder and decoder based on a single layer LSTM with attention mechanism. These two modules are linked by a reinforcement learning algorithm. The results shown have a good coherence and a better fluency.

Tambwekar et al. [26] in 2019 propose to use verbs as history objectives. Reward is obtained by multiplying two parameters: the distance, that is, the number of verbs between the candidate verb and the target verb, and the frequency of the candidate verb before the target verb. To avoid a short story, they use a cluster of verbs. The output of the system are events in tuples of subject, verb, object and wildcard. That paper presents a rather unseen proposal, however, the absence of an event to sentence component limits the correct interpretation of the results.

Additionally, Ammanabrolu et al. [28] in 2019 use 5 different methods to transform events into sentences, using as a basis the events obtained with [25]. The first method is based on Hashimoto RetEdit [33] in 2018, the second method is a template filling, using a simplified grammar and a LSTM, the third method is a Monte Carlo Beam search applied to a seq2seq model, the fourth method is a restricted beam search with finite state machines as a guide. The result is an assembly of the five proposals based on the highest confidence obtained by the five methods. In that paper, a better performance for the translation of events into sentences is observed, although there is still a lack of development to obtain a coherent text.

4   Recent State of the Art Results

One of the biggest obstacles in this field is the evaluation of the work. There have been different proposals to carry out this work but the one that has been highlighted is the subjective evaluation. In most cases, stories generated by the systems proposed to human beings are presented, and the latter are in charge of judging them. However, some parameters of this evaluation vary from one to another. This represents a great difficulty to be able to compare them and establish new benchmarks in the state of the art.

Most of the methods with a strong symbolic focus aim to present a new model, often with the stories generated, but very few quantify their results in any way, although they use different metrics. In Table 1 we can see the results presented by the methods in this type of approach, the table is a complement of the one used in [4], which shows results of [3] and [4] and with the normalization of [10, 11].

As for the results of neural generation, most of them are compared with other models of neural generation, either controlled or uncontrolled, which makes the comparison very difficult. For this reason, it was decided to use the results from [31], which contain most of the advances in the area, except for those from [28].

In Table 2 we can see the perplexity. It can be inferred that the lesser the perplexity, the greater the cohesion of the story, although there are elements in which this does not apply. This is due to the fact that the perplexity calculated in these works is calculated with the tokens, while with GPT-2 and [31] it is calculated by means of byte-pair codings. We can also see BLEU [36] with BLEU-1 y BLEU-2, although its usefulness seems to be very low because it indicates how much the n-grams of the automatic history overlap with those of the human story. The coverage evaluates the percentage of common-sense knowledge triplets present in the story. The repetition [37] evaluates the percentage of generated stories that repeat at least once an n-gram, the author used n equal to 4. The distinction [38] shows the percentage of n-grams with respect to the other n-grams, he also used n equal to 4. In Table 3 we can see the human judgment they carried out in [31]. We can also see that in [28] a measure of 70,179 was obtained on the tokens and a measure of 0.0481 with BLEU-4. Figure 1 shows the human evaluation rated from 1 to 4.

Table 2. Automatic evaluation of models with a neural approach. Table from [31].

Models	Perplexity	BLEU 1	BLEU 2	Coverage	Repetition-4 (%)	Distinction-4 (%)
[19]	N/A	0.322	0.137	12.02	24.23	72.82
[25]	N/A	0.308	0.126	13.38	17.06	67.2
[22]	N/A	0.267	0.088	10.82	18.34	69.42
[30]	N/A	0.293	0.117	10.38	**15.36**	73.08
[34] (Scratch)	11.82	0.311	0.134	10.76	22.87	73.33
[34] (Pretrain)	33.5	0.257	0.085	8.04	39.22	64.99
[34] (Fine-tune)	7.96	0.322	0.141	12.4	29.41	73.85
[31]	**7.85**	**0.326**	**0.143**	**18.48**	21.93	**78.96**

Table 3. Human evaluation of models with neuronal focus. Table from [31].

Models	Grammaticality %			Logicality %		
	Win	Lose	Tie	Win	Lose	Tie
[31] vs. [19]	50	27	23	57	28	15
[31] vs. [30]	58	24	18	58	29	12
[31] vs. [34] (Scratch)	54	24.5	21.5	54	26	20
[31] vs. [34] (Pretrain)	52	31.5	16.5	56.5	32.5	11
[31] vs. [34] (Fine-tune)	42	28	30	51	27.5	21.5

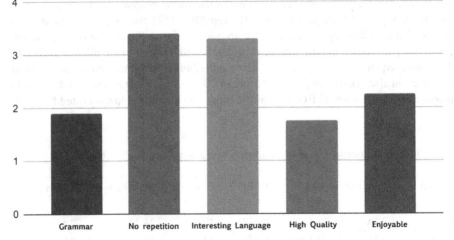

Fig. 1. Human evaluation [28]

5   Conclusions

The symbolic approach established a very high base in terms of the development of the story, with very good coherence and cohesion, but the lack of originality in these stories has caused it to be studied from a neural generation perspective. The best results in stories using the symbolic approach got better performance in some metrics than stories written by humans. They have a great level of coherence, and it was a major objective in the development of those stories. The same stories have a good performance in coherence, but only enough to understand the story without any problem.

Most of recent research uses the neural approach. The stories written using this approach have a novelty upgrade, but the coherence and cohesion is lower than those written with the symbolic approach. In the last papers we can see a great performance in cohesion but the difference in coherence between both approaches is still great.

Another point to highlight is the strong gap that still exists between the symbolic approach and the connectionist approach, but apparently using a hybrid approach is still an area under development.

Finally, better results have been obtained in terms of story coherence in recent works that generate events and translate them into sentences, so a good performance has been obtained in the controllability of the story without necessarily using a strong symbolic representation.

References

1. Meehan, J.R.: TALE-SPIN, an interactive program that writes stories. In: 5th International Joint Conference on Artificial Intelligence, pp. 91–98 (1977)
2. Lebowitz, M.: Planning stories. In: 9th Annual Conference of the Cognitive Science Society, pp. 234–242 (1987)
3. Turner, S.R.: MINSTREL: a computer model of creativity and storytelling. University of California, Computer Science Department (1993)
4. Pérez, R.P.Y., Sharples, M.: MEXICA: a computer model of a cognitive account of creative writing. J. Exp. Theor. Artif. Intell. **13**, 119–139 (2001)
5. Porteous, J., Cavazza, M.: Controlling narrative generation with planning trajectories: the role of constraints. In: Iurgel, I.A., Zagalo, N., Petta, P. (eds.) Interactive Storytelling. Lecture Notes in Computer Science, vol. 5915, pp. 234–245. Springer, Berlin (2009)
6. Riedl, M.O., Young, R.M.: Narrative planning: balancing plot and character. J. Artif. Intell. Res. **39**, 217–267 (2010)
7. Onodera, K., Akimoto, T., Ogata, T.: A state-event transformation mechanism for generating micro-structures of story in an integrated narrative generation system. In: Proceedings of the Annual Meeting of the Cognitive Science Society, vol. 34 (2012)
8. León, C., Gervás, P.: Creativity in story generation from the ground up: nondeterministic simulation driven by narrative. In: 5th International Conference on Computational Creativity (2014)
9. McCoy, J., Treanor, M., Samuel, B., Reed, A.A., Mateas, M., Wardrip-Fruin, N.: Social story worlds with comme il faut. IEEE Trans. Comput. Intell. AI Games **6**, 97–112 (2014)
10. Calvo, H., Daza-Arévalo, J.A., Figueroa-Nazuno, J.: Automatic story generation by learning from literary structures. Scholars' Press (2016)
11. Calvo, H., Daza-Arévalo, J.A., Figueroa-Nazuno, J.: Automatic text generation by learning from literary structures. In: Proceedings of the Fifth Workshop on Computational Linguistics for Literature, pp. 9–19 (2016)
12. Adolfo, B.T., Lao, J., Rivera, J.P., Talens, J.Z., Ong, E.C.J.: Generating children's stories from character and event models. In: Phon-Amnuaisuk, S., Ang, S.P., Lee, S.Y. (eds.) Multidisciplinary Trends in Artificial Intelligence. MIWAI 2017, vol. 10607. Springer, Cham (2017). https://doi.org/10.1007/978-3-319-69456-6_22
13. Farrell, R., Ware, S.G., Baker, L.J.: Manipulating narrative salience in interactive stories using Indexter's pairwise event salience hypothesis. IEEE Trans. Games **12**, 74–85 (2019)
14. Cardona-Rivera, R.E., Cassell, B.A., Ware, S.G., Young, R.M.: Indexter: a computational model of the event-indexing situation model for characterizing narratives. In: Proceedings of the 3rd Workshop on Computational Models of Narrative, pp. 34–43 (2012)
15. Propp, V.Y.: Morphology of the Folktale. University of Texas Press, Austin (1968)
16. Jain, P., Agrawal, P., Mishra, A., Sukhwani, M., Laha, A., Sankaranarayanan, K.: Story generation from sequence of independent short descriptions. In: SIGKDD Workshop on Machine Learning for Creativity (ML4Creativity) (2017)

17. Clark, E., Ji, Y., Smith, N.A.: Neural text generation in stories using entity representations as context. In: Proceedings of the 2018 Conference of the North American Chapter of the Association for Computational Linguistics: Human Language Technologies, vol. 1, pp. 2250–2260 (2018)

18. Martin, L.J., et al.: Event representations for automated story generation with deep neural nets. In: Thirty-Second AAAI Conference on Artificial Intelligence, pp. 868–875 (2018)

19. Fan, A., Lewis, M., Dauphin, Y.: Hierarchical neural story generation. In: Proceedings of the 56th Annual Meeting of the Association for Computational Linguistics, pp. 889–898 (2018)

20. Peng, N., Ghazvininejad, M., May, J., Knight, K.: Towards controllable story generation. In: Proceedings of the First Workshop on Storytelling, pp. 43–49 (2018)

21. Roemmele, M., Gordon, A.: Automated assistance for creative writing with an RNN language model. In: Proceedings of the 23rd International Conference on Intelligent User Interfaces Companion (2018)

22. Xu, J., Ren, X., Zhang, Y., Zeng, Q., Cai, X., Sun, X.: A skeleton-based model for promoting coherence among sentences in narrative story generation. EMNLP (2018)

23. Li, Z., Ding, X., Liu, T.: Generating reasonable and diversified story ending using sequence to sequence model with adversarial training. In: COLING, pp. 1033–1043 (2018)

24. Yao, L., Peng, N., Weischedel, R., Knight, K., Zhao, D., Yan, R.: Plan-and-write: towards better automatic storytelling. In: Proceedings of the Thirty-Third AAAI Conference on Artificial Intelligence (2019)

25. Luo, F., et al.: Learning to control the fine-grained sentiment for story ending generation. In: Proceedings of the 57th Annual Meeting of the Association for Computational Linguistics, pp. 6020–6026 (2019)

26. Tambwekar, P., Dhuliawala, M., Martin, L.J., Mehta, A., Harrison, B., Riedl, M.O.: Controllable neural story plot generation via reward shaping. In: Proceedings of the 28th International Joint Conference on Artificial Intelligence (2019)

27. Guan, J., Wang, Y., Huang, M.: Story ending generation with incremental encoding and commonsense knowledge. In: Proceedings of the AAAI Conference on Artificial Intelligence, pp. 6473–6480 (2019)

28. Ammanabrolu, P., et al.: Story realization: expanding plot events into sentences. arXiv preprint arXiv:1909.03480 (2019)

29. Li, J., Bing, L., Qiu, L., Chen, D., Zhao, D., Yan, R.: Learning to write stories with thematic consistency and wording novelty. In: Proceedings of the AAAI Conference on Artificial Intelligence, pp. 1715–1722 (2019)

30. Fan, A., Lewis, M., Dauphin, Y.: Strategies for structuring story generation. In: Proceedings of the 57th Annual Meeting of the Association for Computational Linguistics, pp. 2650–2660 (2019)

31. Guan, J., Huang, F., Zhao, Z., Zhu, X., Huang, M.: A knowledge-enhanced pretraining model for commonsense story generation. arXiv preprint arXiv:2001.05139 (2020)

32. Ippolito, D., Grangier, D., Eck, D., Callison-Burch, C.: Toward better storylines with sentence-level language models. arXiv preprint arXiv:2005.05255 (2020)

33. Hashimoto, T.B., Guu, K., Oren, Y., Liang, P.: A retrieve-and-edit framework for predicting structured outputs. In: 32nd Conference on Neural Information Processing Systems (2018)

34. Radford, A., Wu, J., Child, R., Luan, D., Amodei, D., Sutskever, I.: Language models are unsupervised multitask learners. OpenAI Blog 1(8), 9 (2019)

35. Devlin, J., Chang, M., Lee, K., Toutanova, K.: BERT: pre-training of deep bidirectional transformers for language understanding. In: Proceedings of the 2019 Conference of the North American Chapter of the Association for Computational Linguistics: Human Language Technologies, vol. 1, pp. 4171–4186 (2019)

36. Papineni, K., Roukos, S., Ward, T., Zhu, W.: BLEU: a method for automatic evaluation of machine translation. In: Proceedings of the 40th Annual Meeting on Association for Computational Linguistics, pp. 311–318 (2002)
37. Shao, Z., Huang, M., Wen, J., Xu, W., Zhu, X.: Long and diverse text generation with planning-based hierarchical variational model. EMNLP (2019)
38. Li, J., Galley, M., Brockett, C., Gao, J., Dolan B.: A diversity promoting objective function for neural conversation models. In: Proceedings of the 2016 Conference of the North American Chapter of the Association for Computational Linguistics: Human Language Technologies, pp. 110–119 (2016)
39. Acevedo-Mosqueda, M.E., Yáñez-Márquez, C., López-Yáñez, I.: Alpha-beta bidirectional associative memories: theory and applications. Neural Process. Lett. **26**(1), 1–40 (2007)

Comparative Methodologies for Evaluation of Ontology Design

Rafaela Blanca Silva-López[1] ⓘ, Iris Iddaly Méndez-Gurrola[2](✉) ⓘ,
and Hugo Pablo-Leyva[3] ⓘ

[1] Universidad Autónoma Metropolitana-Unidad Lerma, Estado de México, Mexico
r.silva@correo.ler.uam.mx
[2] Universidad Autónoma de Ciudad Juárez, Chihuahua, Mexico
iddalym@yahoo.com.mx
[3] Universidad Autónoma Metropolitana-Unidad Azcapotzalco, Ciudad de México, Mexico
hpl@correo.azc.uam.mx

Abstract. In general, it is advisable to evaluate ontology designed quality before developing an Information System based on Ontologies. Principles of ontology design, focus on ontologies design that can be reusable, easy-to-use, maintain and update over time. In this work a model for quality verification of the ontology design is proposed, it is based on ontology design principles of [4, 20, 22]. Methodology starts with an analysis of design principles, then principles are grouped into verification or evaluation collections and following verification techniques was established: 1) minimalist; 2) consistency; 3) flexibility; 4) standardization; 5) redundancy; and 6) efficiency. The main contribution of this work is a qualitative and quantitative model for the verification of an ontology applying design principles. As an application case, quality evaluation of ontological model for OntoPAA is performed, results show that ontology evaluated complies with design techniques that guarantee an adequate level of quality.

Keywords: Ontology design principles · Ontology verification techniques · Ontology evaluation · Ontology quality

1 Introduction

Ontology engineering is a branch of knowledge engineering that focuses on ontologies construction. It contemplates the study of ontology develop process, its life cycle, methods and methodologies to design ontologies, as well as the tools and languages for its construction. Before developing an Information System based on Ontologies, it is advisable to evaluate quality of an ontology design. The OntoPAA ontology [15] is used as an application case. The method is integrated by a set of techniques that group in six categories: minimalist, coherence, flexibility, standardization, redundancy and efficiency. The method is based on the design principles proposed by Gruber [20], Barry Smith [4], and Morbach, Wisner and Marquardt [22].

© Springer Nature Switzerland AG 2020
L. Martínez-Villaseñor et al. (Eds.): MICAI 2020, LNAI 12469, pp. 92–102, 2020.
https://doi.org/10.1007/978-3-030-60887-3_9

1.1 State of the Art

Since 1998, Guarino [11] defined term Information System based on Ontologies. Some authors such as Guarino [11], Colomb [17], Soares and Fonseca [3], Yildiz and Micksh [5] agree that the ontologies used in Information Systems contributes an improvement in the applications developed. For this reason, it is important to have a mechanism to evaluate the quality of an ontology design.

Literature on the ontology evaluation is fragmented, there are approaches that address specific evaluation issues, but not in a systematic way. Hartmann [10] introduces itself in the problems by providing a classification of network of ontological evaluation methods that allows to present methods in terms of structure, function, application, users types, and usability, among others. Relevant authors such as Porzel and Malaka [18], Gómez-Pérez [2], and Noy [13], proposed different methods to measure ontologies. Other authors made proposals associated with the evaluation of the quality of an ontology, for example, Tartir [19] presented a proposal based on metrics related to the ontology schema and the knowledge base; while Guarino and Welty [12] identify the problem areas that should be examined for rigidity, identity, unity and dependence in their work of OntoClean.

On the other hand, Gangemi [1] proposed three types of evaluations: 1) functional, it is focused on verifying that ontology meets its objective; 2) usability, it analyzes metadata and annotations; and 3) structural, that validates the structural properties of the ontology as a graph.

Other authors proposed methodologies that use and develop tools to support the ontology evaluation, for example, Corcho [14] proposed a tool called ODEval that automatically detects syntactic problems of ontology, such as cycles in inheritance tree of classes, inconsistency, incompleteness and redundancy. In 2005 Cross and Pal [21] integrated a plug-in in the editor Protegé that allows to evaluate the ontology quality based on the ontology definition and actual occurrences of ontological concepts.

Another approach is based on use of dimensions and metrics. Mostowfi and Fotouhi [7] proposed 8 metrics to evaluate an ontology, unlike other authors, they define a set of transformations to improve ontology quality. In addition, in OnQual, Gangemi [1] propose the ontology evaluation in three dimensions: the syntax and semantics of the ontology; the functional; and usability profiles, including 32 characteristics. Other authors approach the validation of ontology quality from point of view of Information Systems based on Ontologies Barchini [8], Fonseca and Martin [6], Colomb [17] and Colomb and Weber [16]. In general, they proposed dimensions and evaluation indicators to determine ontology quality level. As it is observed, there is a need to validate ontologies that are designed to identify points of improvement.

A proposal that contemplates some of principles of ontology design is the work of Barchini [9], author proposed 4 dimensions to evaluate operationally the ontology: a) descriptive, degree to which ontology provides information about its characteristics, meets a minimum ontological commitment, identifies the recipients, who is it?; b) structural, validated that the ontology expresses concepts explicitly, formal and consensual, associated with syntax and semantics, meets the specified requirements, what knowledge of domain contains? c) functional, valid if ontology does what end user intends; and d) operational, valid use capacity, can be used effectively.

Within structural dimension includes a sub-dimension called ontology in which it establishes indicators to evaluate the ontology design [8]. The main indicators are subdivided into classes and instances, relationships and axioms.

However, as we can noticed, there is a diversity in the criteria to determine characteristics that allow to evaluate ontology quality and way in which evaluation is done, most have an approach based on the components analysis (classes, relationships and properties). They not properly consider the design principles of ontologies. It is a priority that before realizing an Information System Based on Ontologies, to have the certainty that ontology is adequate and has quality.

In this context, this work proposes techniques for the evaluation or verification of the ontology design, based on ontology design principles of Gruber [20], Barry Smith [4] and Morbach, Wiesner and Marquardt [22].

2 Methodology

Methodology used starts with an analysis of design principles of three authors, then the principles are grouped into verification or evaluation collections to establish verification techniques: 1) minimalist verification; 2) consistency verification; 3) flexibility verification; 4) standardization verification; 5) redundancy verification; and 6) efficiency verification. Finally, techniques are applied to an application case, and the OntoPAA ontology [15] quality assessment is performed.

3 Principles of Ontology Design

To guarantee quality of ontology design, authors such as Gruber [20], Barry Smith [4] and Morbach, Wiesner and Marquardt [22] have defined quality criteria that guide the design and construction of ontology, this allows to evaluate quality of the design. The criteria also known as principles of ontology design, aim to design ontologies: reusable, easy-to-use, maintain and update over time. Reuse of an ontology refers to its ability to adapt to arbitrary application contexts, even in those not predicted at the time of its creation. Usability refers to effort required by a user to use ontology, its goal is to minimize the effort required and can be used by humans or machines under a specific application context.

Gruber [20] as part of his work proposes 5 criteria or design principles: clarity, coherence, extensibility, minimal coding tendency and minimum ontological adherence. On the other hand, for Barry Smith [4] is important that an ontology allows its adoption in the future, therefore, it emphasizes support for information exchange of the ontologies. This author proposes 14 principles for the design of an ontology: intelligibility, openness, simple tools, reuse of available resources, terminological moderation, intelligible definitions, terminological coherence, compound terms construction, instances types, non-circularity, singular nouns, consistency in use of operators for the terms construction, non-subjective definitions and non-redundant definitions. Finally Morbach, Wiesner and Marquardt [22] from construction of a enormous ontology of chemical domain called OntoCape, propose a set of recommendations to evaluate the quality of

OntoCape, they applied following design principles: consistency, concise terminology, intelligibility, reusability, adaptability, minimum ontological commitment and efficiency.

The principles for the ontologies design proposed by the authors, evidence similarities between them, per contra, there are important contributions that one author considers and another not, due to the nature of the knowledge domain in which they have developed their work and so principles that they conceptualized are related to their domain. Gruber has developed his work in the field of Computing, Smith in the Biomedical area and Morbach in the field of chemistry and chemical engineering.

4   Verification Techniques of the Ontology Design

Based on the ontology design principles of authors described in the previous section, a grouping of principles is performed to determine verification techniques of an ontology design.

The design principles of first group associated with minimalist verification technique, which focuses on validating the compliance with minimum indispensable principles that must be met by ontology design, including: clarity, intelligibility, homogeneity, non-subjective definition, intelligible definitions, non-redundant definitions, compound terms, consistency in the operators use and documentation.

The minimalist verification technique proposes:

1. To axiomatise to greatest extent possible the formal definitions.
2. Use defined classes and bounded constraints.
3. Use a homogeneous style that facilitates understanding of new concepts.
4. Document ontology considering: a) comments within the formal specification of ontology; b) elaborate a reference guide oriented to application developers based on ontology; c) develop a user manual; and d) develop a design document for developers who will maintain and make updates to the ontology.

The technique of consistency verification, focusing on validating compliance with coherence principle that is proposed by authors analyzed. Proposes to use ontology publisher's tools such as Protégé of Stanford University, to perform validation of syntax and logical consistency. There are reasoners such as Pellet, RacerPro, or FaCT++ with which more sophisticated consistency tests can be performed.

The flexibility verification technique, focused on validating compliance with principles of extensibility, personalization, openness and adaptability. Proposes to modularize an ontology in domains and sub-domains of application or conceptualization and that this facilitates adaptability, extensibility and personalization.

The standardization verification technique focuses on validating compliance with the principles of minimum coding trend, simple tools, and reuse of available resources. Proposes to use an ontology representation language that is standard, and is accepted by community. Where possible reuse and import ontologies that handle generic concepts such as time and measurements, among others.

The redundancy verification technique focused on validating compliance with the principles of concise terminology and terminological moderation. Proposes to use ontology editors such as Protégé that include mechanisms to perform the detection of redundant axioms, for example cardinality constraints that specify a minimum cardinality of zero. However, many of problems of redundancy in an ontology are caused by errors in the model design, to detect them a manual inspection must be done. Therefore, redundancies elimination in the design is achieved gradually, through continuous reviews and ontology re-engineering.

The efficiency verification technique focused on validating compliance with principles of minimum ontological commitment and efficiency. Proposes to comply with the concise terminology principle, which involves fewer axioms and is easier to process. The axioms number is one of many factors that influence efficiency. Also, type of axioms influences, some are more difficult to process than others.

A quantitative model is also proposed that admits evaluating each design principle considered in the verification techniques as shown in Table 1. Compliance with the principle implies the assignment of 1 point, if it is fulfilled to a lesser extent, a specific weight is determined according to with the covered.

Table 1. Quantitative model of minimalist verification techniques.

Verification technique	Design principle	Complies	Total
Minimalist	Clarity	+1	9 points
	Intelligibility	+1	
	Homogeneity	+1	
	Non-subjective definitions	+1	
	Intelligible definitions	+1	
	Definitions not redundant	+1	
	Compounds terms	+1	
	Consistency in operators use	+1	
	Documentation	+1	
Coherence	Coherence	+3	3 points
Flexibility	Extensibility	+1	4 points
	Customization	+1	
	Opening	+1	
	Adaptability	+1	
Standardization	Minimal encoding trend	+1	3 points
	Simple tools	+1	
	Reuse of available resources	+1	
Redundancy	Concise terminology	+1	2 points
	Terminological moderation	+1	
Efficiency	Minimum ontological commitment	+1	2 points
	Efficiency	+1	

5   Results: OntoPAA Application Case

In [15] an ontological model is designed for customization of learning activities called OntoPAA, this model is taken to evaluate its quality based on 6 minimum verification techniques. The objective of ontological model OntoPAA is to personalize learning activities of a course in such a way that student is assigned learning activities associated with his profile and act as motivators, and others that contribute to development of cognitive skills in accordance with the objectives of the course.

OntoPAA is composed of 4 ontologies: *Profiles, Students, Courses* and *Learning Activities* (see Fig. 1). Each ontology is independent of other in order to be able to reuse them.

The *Profiles* ontology are constituted by attributes with information from the Cognitive Theory of learning and thinking styles. The *Students* ontology is integrated by attributes that allow to know student learning style, seeks to characterize the student. While in the *Courses* ontology the information of available courses is concentrated, dividing content of course into a maximum of 10 sub-themes, each sub-theme having associated learning resources and supporting tools. Finally, *Learning Activities* ontology, concentrates learning activities of the course for each profile.

It is intended to develop an Information System based on OntoPAA, so it is important to do the evaluation of it to avoid problems with the system development that are generated by the ontologies. Table 2 shows the quality evaluation of the OntoPAA ontology.

Table 2. OntoPAA quality assessment.

Verification technique	Ontological model for the Personalization of Learning Activities (OntoPAA)
Minimalist	All classes are defined with bounded constraints A homogeneous notation is applied in names of ontologies, classes, DataProperty, etc. Comments are included in the specification It includes the reference guide and the ontology design document User manual required
Coherence	The consistency check with Pellet is applied from Protégé, guaranteeing the logical consistency
Flexibility	The ontological model is divided into 5 ontologies with the aim of modularizing and facilitating reuse
Standardization	No exist any ontology in the knowledge domain that could be reused given the discipline in which the ontology is focused
Redundancy	Two redundancy problems were identified between the Profiles ontology and those of Activities and Courses, design was adapted integrating relations between classes that generated redundancy
Efficiency	The axioms included are minimal and simple. However, this does not fully guarantee efficiency of the ontology

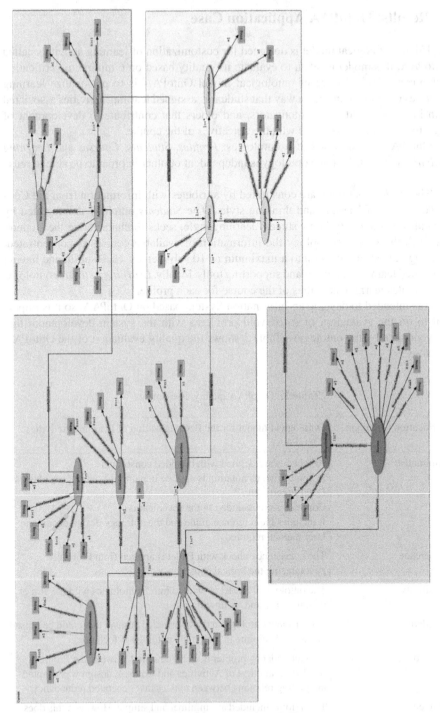

Fig. 1. OntoPAA ontology [15]

When evaluating the ontology quality based on quantitative model, missing values are established so that it is clear on which points to work to improve as shown in Table 3.

Table 3. OntoPAA quality assessment.

Verification technique	Design principle	Complies	Total
Minimalist	Clarity	+1	6.5 points
	Intelligibility	+1	
	Homogeneity	+1	
	Non-subjective definitions	+1	
	Intelligible definitions	+1	
	Definitions not redundant	+0.3	
	Compounds terms	0	
	Consistency in operators use	+0.5	
	Documentation	+0.7	
Coherence	Coherence	+3	3 points
Flexibility	Extensibility	+1	4 points
	Customization	+1	
	Opening	+1	
	Adaptability	+1	
Standardization	Minimal encoding trend	+1	2.5 points
	Simple tools	+0.5	
	Reuse of available resources	+1	
Redundancy	Concise terminology	+0.7	1.2 points
	Terminological moderation	+0.5	
Efficiency	Minimum ontological commitment	+1	1.2 points
	Efficiency	+0.2	

The quantitative model shows that the verification gives a level of 18.4/23. For this reason the areas where should work are shown in Table 4.

According to Table 4, the items with greatest problem have a value of −1, they are the ones that must be addressed first, in this case terms compound terms and efficiency. Furthermore, there is redundancy between *Profiles* ontology and *Learning Activities* ontology.

Table 4. Areas to work in OntoPAA.

Design principle	Value
Definitions not redundant	−0.7
Compounds terms	−1
Consistency in operators use	−0.5
Documentation	−0.3
Simple tools	−0.5
Concise terminology	−0.3
Terminological moderation	−0.5
Efficiency	−0.8

6 Conclusions

According to literature, there is a diversity in criteria to determine characteristics that allow to evaluating ontology quality and way in which evaluation is achieved, most of them have an approach based on the components analysis (classes, relations and properties), however, ontology design principles are not properly considered. The proposed model uses basic ontology design concepts, which in many cases are ignored and its impact is considerable in quality of ontology designed. In addition, it allows to validate ontology with the participation of knowledge engineer, knowledge domain expert and end user, by applying verification techniques: minimalist, consistency, flexibility, standardization, redundancy and efficiency.

The design principles of minimalist verification technique: clarity, intelligibility, homogeneity, non-subjective definition, intelligible definitions, non-redundant definitions, compound terms, consistency in use and documentation of operators, allow to validate compliance with the essential minimum principles in ontology design.

The minimalist verification technique considers in the first place to axiomatize formal definitions as much as possible, to use the defined classes and limited restrictions, as well as a homogeneous style that facilitates the understanding of new concepts, finally it proposes to document ontology considering: a) commenting on formal specification of ontology; b) develop a reference guide for application developers; c) develop a user manual and a design document for those responsible for maintaining and updating the ontology. Thus, it ensures that ontology design principles have been properly applied.

To establish the quality level of an ontology design, the assignment of points for each verification technique is contemplated, considering level 1 in case of complying with only 1 verification technique, while a level 6 will be obtained when complying with all of them.

In the case a checklist is shown that facilitates validation of ontology by applying verification techniques. The results show that OntoPAA has a level of quality level 6, as it complies with all verification techniques.

Although the evaluation of ontologies is not usually integrated in the methodologies for their construction, it allows to identify design errors before development of information system and therefore it should be one of the last steps to be carried out in process of designing an ontology.

There is a need for tools that facilitate the evaluation process, reasoners support consistency validation, however, there are many other variables to verify. As future work, the domains for which the verification technique is most suitable will be analyzed. As well as testing to expand its applicable principles.

References

1. Gangemi, A., Catenacci, C., Ciaramita, M., Lehmann, J.: Modelling ontology evaluation and validation. In: Sure, Y., Domingue, J. (eds.) ESWC 2006. LNCS, vol. 4011, pp. 140–154. Springer, Heidelberg (2006). https://doi.org/10.1007/11762256_13
2. Gómez-Pérez, A.: Ontology evaluation. In: Staab, S., Studer, R. (eds.) Handbook on Ontologies, pp. 251–274. Springer, Heidelberg (2003). https://doi.org/10.1007/978-3-540-24750-0_13
3. Soares, A., Fonseca, F.: Ontology-driven information systems: at development time. Int. J. Comput. Syst. Signals 8(2), 50–59 (2007)
4. Smith, B., Köhler, J., Munn, K., Rüegg, A., Skusa, A.: Quality control for terms and definitions in ontologies and taxonomies. BMC Bioinform. 7(212), 1–12 (2006)
5. Yildiz, B., Miksch, S.: Ontology-Driven Information Systems: Challenges and Requirements (2007)
6. Fonseca, F., Martin, J.: Learning the differences between ontologies and conceptual schemas through ontology-driven information systems. JAIS J. Assoc. Inf. Syst. 8(2), 129–142 (2007). Special Issue on Ontologies in the Context of IS
7. Mostowfi, F., Fotouhi, F.: Improving quality of ontology: an ontology transformation approach. In: Proceedings of the 22nd International Conference on Data Engineering Workshops (ICDEW 2006) (2006)
8. Barchini, G.E., Álvarez, M.M., Palliotto, D., Fortea G.: Evaluación de la calidad de los sistemas de información basados en ontologías. In: IX Congress ISKO-SPAIN, Valencia, España, pp. 268–288 (2009)
9. Barchini, G.E., Álvarez, M.M.: Dimensiones e indicadores de la calidad de una ontología. Revista Avances en Sistemas e Informática 7(1), 29–38 (2010)
10. Hartmann, J., et al.: Methods for ontology evaluation. Knowledge Web (2004)
11. Guarino, N.: Formal ontology and information systems. In: Proceedings of the 1st International Conference on Formal Ontologies in Information Systems, FOIS 1998, pp. 3–15. IOS Press (1998)
12. Guarino, N., Welty, C.: Evaluating ontological decisions with OntoClean. Comm. ACM 45(2), 61–65 (2002)
13. Staab, S., Gomez-Perez, A., Daelemana, W., Reinberger, M., Noy, N.: Why evaluate ontology technologies? Because it works!. IEEE Intell. Syst. 19(4), 74–81 (2004). https://doi.org/10.1109/MIS.2004.37
14. Corcho, Ó., Gómez-Pérez, A., González-Cabero, R., Suárez-Figueroa, M.C.: ODEval: a tool for evaluating RDF(S), DAML+OIL, and OWL concept taxonomies. In: Bramer, M., Devedzic, V. (eds.) AIAI 2004. IIFIP, vol. 154, pp. 369–382. Springer, Boston, MA (2004). https://doi.org/10.1007/1-4020-8151-0_32

15. Silva-López, R.B.: Modelo ontológico para la personalización de actividades de aprendizaje en ambientes virtuales, Ph.D. Dissertation, Sistema de Universidad Virtual, Universidad de Guadalajara (2016)
16. Colomb, R., Weber, R.: Completeness and quality of an ontology for an information system. In: Guarino, N., (ed.) Formal Ontology in Information Systems. International Conference on Formal Ontology in Information Systems (FOIS 1998), Trento, Italy, pp. 207–217. IOS-Press, Amsterdam (1998)
17. Colomb, R.M.: Quality of ontologies in interoperating information systems, Technical report 18/02 ISIB-CNR Pavoda, Italy (2002)
18. Porzel, R., Malaka, R.: A task-based approach for ontology evaluation. In: Proceedings of ECAI04 (2004)
19. Tartir, S., Arpinar, B., Sheth, A.: Ontological Evaluation and Validation (2007)
20. Gruber, T.R.: Toward principles for the design of ontologies used for knowledge sharing. Int. J. Hum Comput Stud. **43**(5-6), 907–928 (1995)
21. Cross, V., Pal, A.: Metrics for ontologies. In: Annual Meeting of the North American Fuzzy Information Processing Society (2005)
22. Marquardt, W., Morbach, J., Wiesner, A., Yang, A.: On-toCAPE A Re-Usable Ontology for Chemical Process Engineering. Springer, Heidelberg (2010). https://doi.org/10.1007/978-3-642-04655-1

Question Classification in a Question Answering System on Cooking

Riyanka Manna[1], Dipankar Das[1], and Alexander Gelbukh[2(✉)]

[1] Department of Computer Science and Engineering, Jadavpur University, Kolkata, India
riyankamanna16@gmail.com, dipankar.dipnil2005@gmail.com
[2] CIC, Instituto Politécnico Nacional, Mexico City, Mexico
gelbukh@gelbukh.com
http://www.gelbukh.com

Abstract. Question Answering is a very actively developing filed of Natural Language Processing. In this field, the most active research work is currently going on with relation to the Question Classification module of a Question Answering System. It plays a very important role in determining the expectations of the user. The aim of question classification is to identify the type of questions and based on question type the expected answer will be extracted from the data. In the past years, question classification was done by using only rule-based approach. This approach was too specific for the users and it was difficult to achieve the purpose. In this paper, we will classify the cooking questions into different classes; the overall system is able to identify the new question type that will able to give a perfect answer of the question by using machine learning approach.

1 Introduction

Question Answering (QA) system is one of the well-known applications of Natural Language Processing (NLP), which relies on response-oriented approach to access knowledge. QA can be found in a wide variety of domains, in which questions are submitted instead of simple keyword queries, and corresponding responses are returned by the system. The ultimate goal of a QA system is to generate the meaningful and useful response expressed in natural language to the high-end questioner. In QA task, the user is looking for the answers instead of documents. Automatic answer generation is certainly an important move forward in the state-of-the-art of information retrieval technology.

Complexity with respect to natural language processing and consequently cost associated with NLP based QA systems growth, resource consumption and low quality may explain the unwillingness of digital library and Web-based search platform designers to integrate QA technologies. Thesauri and lexicons are used in both fields to classify documents and categorize problems.

The ODQA [7] answers almost all queries and it can only concentrate on particular ontology. A particular ontology or common knowledge helpful in answering uncontrolled queries. The RDQA [8] addresses issues in a certain area, such as tourism, medicine etc. limited domain answering. One way to understand the QA task is language. In monolingual QA system, the query and their corresponding answer expressed in same language.

L. Martínez-Villaseñor et al. (Eds.): MICAI 2020, LNAI 12469, pp. 103–108, 2020.
https://doi.org/10.1007/978-3-030-60887-3_10

On the other hand, cross-language QA system, the query expressed in one language and answer expressed in some other language. In order to carry out the search, the question must be translated. Multicultural and multilingual systems deal with several target languages, which mean that a corpus contains data written in various languages.

Our proposed QA system is intended for cooking domain. Cooking is a dynamic, challenging, multi-modal domain of Question-Answering system. Exploration of cooking field in QA domain has started few years back. Some work has been done depending on recipes images; some work has been done depending on different recipes.

There is a huge chance to explore the question classification system in cooking domain. As this is a restricted field of question answering the profession like Chefs, bakers, food server managers, homemakers, nutritionists etc. are going to be very much benefited using this cooking QA system. The textual data contain a number of questions in the field of cooking, which show how the dishes are made, how long the preparatory work is, etc. Our proposed QA model is based on the Information Recovery system, as well as uses the food preparation Ontology to respond to the cooking related questions.

2 Related Work

An ontology-based cooking system [1] built to integrated models between different questions. The steps to the build the process consist mainly of specification, knowledge acquisition, conceptual model, implementation and evaluation. The ontology which consists of four primary modules that cover the cooking domain's are core concepts operation, food, recipes and utensils and three auxiliary modules namely units, measures, and categories of plate. The knowledge model was subsequently formalized by means of an ontology editor and an intelligent system-building framework. In a cafeteria, clinic/hospital or home provided a food based ontology-driven framework for food and menu planning.

In ontology, have product descriptions, substances, and nutrition details, daily diets suggested for various regions, plates and meals. A food recipes ontology design model offers a comprehensive and ontological model definition. The model has aimed at bridging diversity through symbolic choice through the creation of a general pattern for content ontology that allows information from different websites to be integrated [2].

A model to recognise supporting parts, such as trophies, guidelines, alerts and requirements, and for the guidance framework and metrics. Apart from contributing to the text analysis, the aim was to be availing the questions of procedure where the reply was a well-figured part of a text, not a small number of words, as to facto questions [3]. A question rating system has suggested the uses WordNet's formidable semantic meaning and the vast corpus of Wikipedia information specifically to define insightful words.

The question suggestion is discussed and the solution extraction method is in doubt. In order to determine the user interest distribution, the system uses a static language model and compiles the user list for the given questions. In the system presented in this paper, the candidate answers also are analysed, the question resemblance is determined by numerous lexical and semantic characteristics and the user is then given the likely response list, which will make it simpler for the user to select the best response need.

A program that analyses the user questions scans service reports from an inverted index and rates them by a personalized scoring feature is created in this article [5]. The

QA program has designed for definitive answers to public services questions in German. With recovery techniques, job trees and a rule-based method, the program effectively manages unclear issues.

A machine-learning approach [6], in which some modules of a system for answering questions has developed, i.e., a POS Tagger, a shallow parser, and a module for finding the goal of the question. In this approach, questions are analysed and the selected answers are exacted. Questions analyser utilizes these three aspects to extract the syntactical or semantic features and save them in question analytical documents. Answer selector compares the questions, evaluates the records and collected phrases, and delivers the full verdict.

Automatic interpretation of instructions such as cooking recipes into semantic representations that can make complex question replies easier. Automatic description into semantic depictions of how to use guidelines of cooking recipes to enhance advanced question reply. The remarkable performance with respect guidance on the sequence learning of the instructions, but these methods cannot manage the location and ambiguity of the Web. A used of pragmatic based approach on the rich representation of world nations to broaden those methods [9].

3 Dataset Description

There are no standard text corpora for specific cooking related questions are available for research. Therefore, we had no choice to use any standard data and we had to prepare experimental data for our own.

There is some website over Cooking, which contains information of Cooking. For example, several recipes, ingredients, the precaution one should take to make any recipes, the time taken to cook anything etc. There are some other websites where viewer can ask questions related to cooking. Stack-overflow, Facebook etc. are some websites where people can ask questions when they have a doubt, and people who know the answer can give answer. Like that, some website also available in cooking domain where people can ask their doubts and some other people can answer those questions. Among that website, top viewed websites are punjabi-recipes.com, www.tarladalal.com, www.all recipes.com etc. where people can ask question. Therefore, we choose this website to collect questions. We use Apache Nutch Crawler[1] to collect data.

Depending on the question type, we have divided the total question data set in 14 classes. Table 1 gives the detailed statistic of the total number of classes, as well as of the total data items in each class present in Cooking Dataset. In Table 2, we present an example for each one of the classes.

For the purpose of selection of the correct answer, we have used another website, which is Yahoo Answer.[2] More than 5000 questions in cooking domain of the Yahoo Answers site have been collected and 1668 recipes question data have been identified under human evaluators.

---

[1] http://nutch.apache.org/.

[2] http://answers.yahoo.com.

Table 1. Various question types

Sl No	Class	Statistics
1	QTY	60
2	ADV	120
3	ING	130
4	YESNO	210
5	PREP	40
6	DIR	198
7	WRN	50
8	SPLINFO	200
9	EQUIP	40
10	TIME	150
11	OBJ	250
12	JUST	50
13	DIFF	60
14	NAME	110

Table 2. Examples of various question types

Sl No	Class	Example
1	QTY	*How much water content is required for rice?*
2	ADV	*Give a diet rich in Vitamin D*
3	ING	*What are the ingredients for Chicken Biriyani?*
4	YESNO	*Is pasta good for health?*
5	PREP	*How to cook Mutton Biryani?*
6	DIR	*How to process chicken?*
7	WRN	*What are the precaution should be taken to preserve Lamb?*
8	SPLINFO	*How to check food quality?*
9	EQUIP	*What utensil is required for Omlet?*
10	TIME	*How much time is required to cook Pulao?*
11	OBJ	*What is Alu Parantha?*
12	JUST	*When could we use less sugar in kulfi?*
13	DIFF	*What is the difference between Pulao and Birayani?*
14	NAME	*Give recipes without grains*

4 Experiment and Evaluation: Question Classification

We have collected 1668 cooking recipe-related questions. We classified these questions in 14 classes. We have performed the classification task with various machine-learning algorithms such as Naïve Bayes classifier, Support Vector Machine (SVM) and Deep Neural Network for question classification task. We have used Bidirectional Encoder Representations from Transformers (BERT) [10] for pre-training the model on the corpus using the cloze task. It gives around 87.02% accuracy; see the details in Table 3.

Table 3. Experiment results on question types

Sl. No	Learning algorithm	No of questions	Accuracy (%)
1	SVM	1668	81.70
2	Naive Bays algorithm	1668	83.44
3	Bidirectional Encoder Representations from Transformers (BERT) [10]	1668	87.02

5 Conclusions

In this work, we have presented the module of question classification of our question-answering system for the cooking domain. Cooking question answering is a new and interesting topic in the area of Natural Language Processing. However, currently very few researchers are working in this domain, so there is a lack of standard data for cooking question-answering system. Our work is aimed at partially alleviating this data gap.

As we know, answer extraction from data is most important, and it requires a rich structured text data in order for the possibility of accurate answer extraction will be increased. In this set of experiments, we have worked on answer classification part and used less data, but still we have achieved good result in the question classification module of our system.

In our future work, we plan to increase the number of questions and question types in the dataset, in order to provide better training to the machine learning-based algorithms. We will experiment with various machine-learning classifiers in order to classify questions into different classes as a result we will get more accurate and reliable question types and based on that easy to extract proper answer from the dataset. In particular, we believe that using other deep-learning mechanisms, such as attention and memories, could increase the ability of the system to select the best answers for a given question. Another possible way of improvement is to research knowledge-representation mechanisms, in particular, improved domain ontologies, in our case, the ontologies related with the cooking domain.

References

1. Batista, F., Pardal, J.P., Mamede, P.V.N., Ribeiro, R.: Ontology construction: cooking domain. Artif. Intell.: Methodol. Syst. Appl. **41**, 1–30 (2006)
2. Snae, C., Bruckner, M.: FOODS: a food-oriented ontology-driven system. In: 2nd IEEE International Conference on Digital Ecosystems and Technologies, DEST 2008, pp. 168–176. IEEE (2008)
3. Delpech, E.: Investigating the structure of procedural texts for answering how-to questions. In: Language Resources and Evaluation Conference, LREC 2008, p. 544 (2008)
4. Ray, S.K., Singh, S., Joshi, B.P.: A semantic approach for question classification using WordNet and Wikipedia. Pattern Recogn. Lett. **31**(13), 1935–1943 (2010)
5. De Rijke, M.: Question answering: what's next? In: Sixth International Workshop on Computational Semantics, Tilburg (2005)
6. Yin, L.: Topic analysis and answering procedural questions. Information Technology Research Institute Technical Report Series, ITRI-04-14. University of Brighton, UK (2004)
7. Yang, H., Chua, T.S., Wang, S., Koh, C.K.: Structured use of external knowledge for event based open domain question answering. SIGIR, Association for Computational Linguistics (2003)
8. Diekema, A.R, Yilmazel, O., Liddy, E.D.: Evaluation of restricted domain question-answering systems. In: Proceedings of the ACL 2004 Workshop on Question Answering in Restricted Domain, Association for Computational Linguistics, pp. 2–7 (2004)
9. Malmaud, J., Wagner, E., Chang, N., Murphy, K.: Cooking with semantics. In: Proceedings of the ACL 2014 Workshop on Semantic Parsing. Association for Computational Linguistics, pp. 33–38 (2014)
10. Devlin, J., Chang, M.W., Lee, K., Toutanova, K.: Bert: pre-training of deep bidirectional transformers for language understanding. arXiv preprint arXiv:1810.04805 (2018)

Exploring the Use of Lexical and Psycho-Linguistic Resources for Sentiment Analysis

Rafael Guzmán Cabrera[1] and Delia Irazú Hernández Farías[2(✉)]

[1] División de Ingenierías, Campus Irapuato-Salamanca, Universidad de Guanajuato, Guanajuato, Mexico
guzmanc@ugto.mx
[2] División de Ciencias e Ingenierías, Campus León, Universidad de Guanajuato, Guanajuato, Mexico
di.hernandez@ugto.mx

Abstract. Twitter data have been used for monitoring customer opinions in different domains. In this paper, we investigated the role of a wide set of lexical and psycho-linguistic resources for discovering a set of features that can be potentially useful for this task. Different experiments were carried out, the obtained results allow us to demonstrate that by using a small set of features it is possible to obtain competitive results against more complex approaches. The experimental evaluation was carried out on a collection of tweets corresponding to the airline domain. The obtained results allow us to observe a significant reduction in the vector representation dimension, maintaining satisfactory results using lexical resources for capturing different kinds of information to perform sentiment analysis in tweets on the aforementioned domain.

Keywords: Sentiment analysis · Psycho-linguistic resources · Lexical resources

1 Introduction

The recent developments of smart technologies using mobile-based communication have entailed massive amount of data. The enormous amount of data generated requires having automatic systems that allow us to sort, classify, and select information. Nowadays, the powerful communication of social media has lead in to exploit them for achieving many objectives. The lure of social media is that it enables businesses to conduct real-time conversations directly with their customers very inexpensively.

User-generated data in Twitter represents a gold-mine for analyzing various and varied aspects such as, for instance, traces of individual behavior, or how a brand is perceived. Sentiment Analysis (SA) is a very popular natural language processing task which main aim is to determine the subjective component of a given piece of text. The important role of SA has been recognized beyond computer sciences. It has emerged as a trending topic in Industry due to the

© Springer Nature Switzerland AG 2020
L. Martínez-Villaseñor et al. (Eds.): MICAI 2020, LNAI 12469, pp. 109–121, 2020.
https://doi.org/10.1007/978-3-030-60887-3_11

wide range of application that can be exploited from its results, the outcomes of applying SA can be used for evaluating customer service, gathering consumer feedback, developing marketing campaigns, among others [1].

Identifying the opinions expressed by users of airline companies has been recognized as a powerful tool that can be used for these corporations in order to identify opportunities for improvement. Such a task has been investigated from different perspectives. In [2], the authors exploited a soft voting classifier approach that uses logistic regression and stochastic gradient descent with both traditional weighted schemes and pre-trained word-embeddings from text classification in order to categorize tweets in the airline companies domain. In [3], feature selection and class imbalanced techniques were used in order to classify comments of travelers' feedback regarding airlines. Particular aspects related to airlines such as punctuality, food and beverages quality, ticket prices, among others, were investigated in [4]. In [5], the authors analyzed the location of a set of tweets for determining how this aspect can help to airline companies.

In this paper, we are proposing to exploit not only the terms contained in the tweets but also a wide range of lexical resources for capturing different kinds of information that can be exploited in order to perform sentiment analysis in tweets reflecting opinions about airline companies. Aiming to propose a set of features for capturing the sentiment in such texts, we performed various experiments in order to identify potential aspects coming form different information sources. Our intuition is that, with a small number of features the sentiment analysis of these tweets can be performed with a comparative performance that when all the vocabulary is used. We experimented with a benchmark corpus in this domain, obtaining competitive results against the state-of-the-art. Furthermore, an analysis over this dataset was carried out.

The rest of the document is organized as follows. Section 2 introduces the methodology we propose for classifying tweets in the airlines domain by exploiting both the vocabulary in the data and also lexical-based information. In Section 3 we describe the experiments carried out as well as the obtained results. Finally, the conclusions and findings for future work are presented in Sect. 4.

2   Proposed Methodology

We are interested in to perform sentiment analysis in the context of Airlines comments. For doing so, we are proposing to exploit a wide range of lexical resources reflecting different aspects as well as traditional and word-embeddings representations. The SA task was performed as a text classification approach taking advantage of machine learning algorithms[1] such as: Naive Bayes (NB), Decision Tree (DT), K-Nearest Neighbors (KNN), Support Vector Machine (SVM), and Logistic Regression (LR). Besides, we also experimented with a majority voting ensemble of classifiers. For text representation, we exploited the methodologies described below.

---

[1] We used the sklearn implementation of these algorithms with default parameters, except for KNN, where the k value was established in 3.

2.1   Vocabulary-Based Experiments

We carried out a set of experiments considering only the content of the tweets for determining the polarity of each instance. Five different configurations with term frequency as weighted schema were used: *BOW_0*: text without any kind of pre-processing; *BOW_1*: text tokenized, lowercased, and stop-words removed, and discarding those terms with frequency lower than 3; *BOW_2*: text tokenized, lowercased, stop-words removed, and hashtags, mentions, and url replaced by corresponding labels, and leaving out those terms with frequency lower than 3; *BOW_3*: text tokenized, lowercased, stop-words removed, and considering only those terms with an Information Gain Rate (IGR) greater than 0.001; *BOW_4*: text tokenized, lowercased, stop-words removed, and hashtags, mentions, and url replaced by corresponding labels, and considering only those terms with an IGR greater than 0.001. We also used pre-trained word embeddings in order to generate a representation of the tweets by using an average vector with the words in each instance. We took advantage of three well-known pre-trained word embedding models: *word2vec*[2], *GloVe*[3], and *FastText*[4].

2.2   Lexicon-Based Experiments

The important role of lexical resources for performing SA has been widely recognized since they allow us to capture different nuances of affect ranging from sentiment polarity to finer-grained emotions [6]. The most basic approach in such resources involves the creation of lists of terms associated to two polarity strengths: positive and negative; there are other methods where a word is labeled with a score reflecting its value with regards to a particular aspect. We are interested in to evaluate the performance of such resources for determining the polarity of tweets in the airlines domain. For doing so, we selected a set of 14 lexical resources comprising different facets of affect. Two main groups can be distinguishing: those including information strongly related to sentiment and emotions, and those were psycho-linguistic information is also considered.

Sentiment and Emotions Resources. The first set of lexical resources can be further divided into two subgroups: (i) The **SA** group, we found: AFINN[7], Hu&Liu[8], SentiWordNet (SWN) [1], EffectWordNet [9], Semantic Orientation[10], Subjectivity lexicon[11]. We calculated three scores for each of the resources: positive (denoted as *pos_resource*), negative (denoted as *neg_resource*), and the sum of both (denoted as *tot_resource*); and, (ii) The **EMOT** group, that is composed by two sub-groups divided according to the main theories of emotions: *Categorical model* (**emotCat**): EmoLex [12], EmoSenticNet[13], and SentiSense[14]; and the *Dimensional model* (**emotDim**): SenticNet[15], ANEW (Affective Norms for English Words) [16], and the Dictionary of Affect in Language (DAL) [17] (it contains three dimensions namely,

---

[2] https://code.google.com/archive/p/word2vec/.
[3] https://nlp.stanford.edu/projects/glove/.
[4] https://fasttext.cc/docs/en/english-vectors.html.

Pleasantness, Attention, and *Imagery*). In the case of the **emotCat**, we calculated the frequency of words belonging to a given emotion in each tweet, while for **emotDim**, the sum of each dimension regarding the words in each instance was considered. It is important to mention that, EmoLex and SenticNet have also some aspects that were included into the **SA** group: a set of positive and negative words and an equation (denoted as *SN-eq*) for calculating the polarity of a given text in terms of affective dimensions defined in [18], respectively. Besides, we also considered the positive and negative categories included in the two psycho-linguistic resources that will be introduced in the following section. In the end, we have a vector composed of 60 features.

Psycho-Linguistics Resources. The second subset of lexical resources includes two dictionaries were a set of words are associated to different aspects reflecting the use of language from a psycho-linguistic perspective. Both of them have been successfully applied in different natural language processing tasks such as *Author Profiling* [19] and *Emotion Identification* [20]. The Linguistic Inquirer and Word Count (henceforth **LIWC**) [21] is a dictionary containing 64 categories such as *social* and *affective* processes, *personal concerns*, as well as *grammatical* (verbs, nouns, etc.). General Inquirer (henceforth **GI**) [22] is composed by 182 categories[5]. It was developed with the aim of analyze different aspects of language such as cognitive, emotions, interpersonal relations, among others. We used the **Category-based representation** as it was defined in [23]. Each resource was exploited individually, also we combined both of them into a single one (**GI+LIWC**). Then, we experimented with vectors of 64, 182, and 246 features for **LIWC**, **GI**, and **GI+LIWC**, respectively.

3   Results

3.1   Corpus Description

We experimented with the Twitter US Airline Sentiment (henceforth denoted as *TwAS*) corpus that is freely available[6]. It is a set of tweets posted in February 2015 regarding some well-known airline companies in the US. *TwAS* is composed by 14,485 tweets manually labeled according to three categories: *positive* (2332 instances), *negative* (9088 instances), and *neutral* (3065 instances). In *TwAS* there is a remarkable imbalanced class distribution towards the negative category. Besides the overall sentiment annotations, the tweets included in *TwAS* have there are other types of labels such as the target airline and the username of the author of each tweet.

3.2   Vocabulary-Based Experiments

Figure 1 shows the obtained results in terms of accuracy. As it can be observed, the best performance was achieved using the *BOW_0* with the ensemble of

---

[5] The description of all of them can be found in http://www.wjh.harvard.edu/~inquirer/homecat.htm.

[6] https://www.kaggle.com/crowdflower/twitter-airline-sentiment/data.

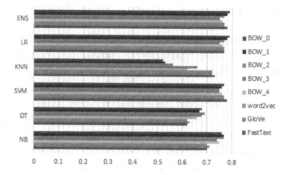

Fig. 1. Precision of the first set of experiments carried out on the airline corpus.

classifiers (denoted as (*ENS*) is composed by *NB*, *LR*, and *SVM*). Regarding the word-embeddings, *FastText* shows the highest rate.

Since we are interested in to determine an optimal set of features of the tweets by using different kinds of lexical resources, we plot a dimensionality reduction version of the BOW representation (see Fig. 2) of the instances by exploiting the TSNE technique[7]. Samples of the *negative* class are plotted with red color, of the *positive* class with green color, and of the *neutral* class with yellow color. The skewed amount of instances belonging to the *negative* class is clearly observed. Besides, there are not salient clusters of each class, instead, it is possible to observe a high rate of overlap among the classes. In Fig. 2, we can observe a high degree of overlap between the main components of each categories of the corpus under study.

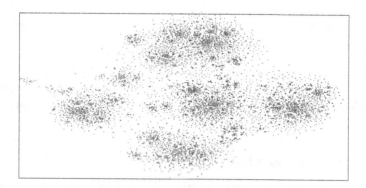

Fig. 2. TSNE dimension reduction using BOW representation (Color figure online)

---

[7] We exploited the sklearn implementation (https://scikit-learn.org/stable/modules/generated/sklearn.manifold.TSNE.html).

3.3   Lexicon-Based Experiments

Sentiment and Emotions Resources. We experimented with each group
of lexical resources on its own, and also by combining them into a single one
(**SA+EMOT**). Figure 3 shows the obtained results when the aforementioned
resources are exploited. Concerning the *Sentiment and Emotions*, the best performance is achieved when all these resources are used. Interestingly, using only
the sentiment related ones, there is a low decrease in the accuracy, confirming
the usefulness of them for characterizing the sentiment of a piece of text. With
respect to the Emotions, the subset regarding the categorical model shows a
slightly higher performance than the dimensional one, however, it is important
to emphasize that both models are composed of only a small number of features.
In terms of the *Psycho-linguistics* resources, overall, **LIWC** shows a better performance than **GI** and than **GI+LIWC**. The best result was obtained with
LIWC with the ensemble of classifiers.

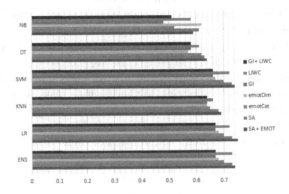

Fig. 3. Obtained results when the aforementioned resources are exploited

In Figure 4, we plot the TSNE representation based on lexical resources.
We can see more defined clusters, this is directly related to the best separation
between the classes, which, moreover, is reflected in the results shown in the
graph above.

3.4   Selecting the Most Relevant Features

With the aim of generating a representation with a lower dimensionality, we performed an Information Gain analysis[8] over the features obtained from both lexical and psycho-linguistics resources. Table 1 shows the best-ranked features for
each group. All the scores obtained from Afinn and Hu&Liu lexicons emerged as

---

[8] We exploited the implementation included in WEKA (https://www.cs.waikato.ac.
nz/ml/weka/).

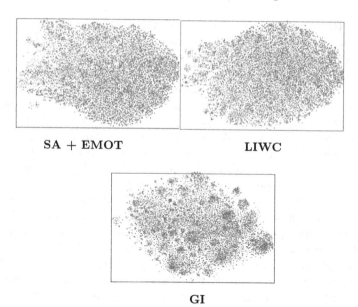

Fig. 4. Representation of the main components by class using lexical resources

very informative. Regarding the **emotCat**, it is observed that very opposite emotions serve to capture useful information. With respect to both psycho-linguistic resources, among the best-ranked dimensions we found those having a sort of negative connotation, it can be provoked due to the data skewed towards the *negative* class. Besides, we also identified some dimensions reflecting activities from the past, that is in line with the fact that users tend to post their experiences after traveling.

Table 1. Best ranked features for each group

	Features
SA	All the three scores from Hu & Liu and Afinn, the negative scores from SWN and EmoLex, the *objectivity* dimension from SWN, and *SN-eq*
emotCat	*positive* and *negative* from LIWC, *joy* and *sadness* from EmoSN, *like* and *disgust* from SentiSense, and *anger*, *fear*, and *disgust* from EmoLex
emotDim	All the dimensions from DAL, and *pleasantness* and *sensitivity* from SenticNet
GI	*Negativ*, *Negate*, *NotLw*, *Ngtv*, *Hostile*, *IAV*, *PosAff*, *Vice*, *Active*, and *Undrst*
LIWC	*posemo*, *funct*, *cogmech*, *affect*, *negate*, *negemo*, *preps*, *relativ*, *aux verb*, and *past*

Proposed Representations. We defined three different subsets of features coming from the lexical resources described before. The first one, denoted as **subset-1**, is composed of the fifteen best-ranked features according to the IGV obtained from the whole set coming from the *Sentiment and Emotions resources*. The features included on it are: All the three scores from Hu&Liu and Afinn, all the dimensions in DAL, the *negative* and *objectivity* dimensions from SWN, *pos_LIWC* and *tot_LIWC*, and *pleasantness* and *sensitivity* from SenticNet. The second one denoted as **subset-2**, comprised all the dimensions from GI and LIWC in Table 1. Finally, the last one denoted as **subset-3** includes all the features from SA, emotCat, and emotDim described in Table 1. In summary, the vector representation in each set is composed by 15, 20, and 24 features for the **subset-1**, **subset-2**, and **subset-3**, respectively. In addition to the experiments carried out with each set of features, we decided to combine each of them with the best performing representations based on vocabulary: *BOW_4* and *FastText*.

The obtained results are shown in Fig. 5. Regarding the proposed representations, the **subset-1** and **subset-3** show a higher performance than **subset-2**. When using the classifiers ensemble, it is possible to reach a 0.73 of accuracy. On the other hand, combining the proposed representations with *BOW_4*, the best performance is also achieved with the ensemble of classifiers with any of the subsets of features. With respect to the word-embeddings representation, the highest accuracy rate obtained is 0.79 with both the ensemble and SVM when it is merged with the **subset-1** and **subset-2**. It is important to highlight that, this result is the most similar to the baseline, i.e., the *BOW_0*, with the important difference that instead of using more than 10,000 features, only 315 and 320 were used. Fig. 6 shows the dimensionality reduction of each of the proposed representations. In this case, we can observe the effect of lexical resources on the definition of clusters by graphically representing the main components of each class.

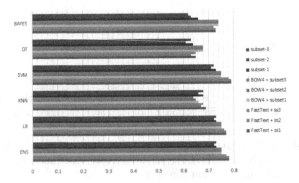

Fig. 5. Results obtained when combining each subset with the best performing representation based on vocabulary

The *TwAS* dataset has been used before for evaluating sentiment analysis methods, in [2] the highest accuracy rate reported was of 0.792 when a the

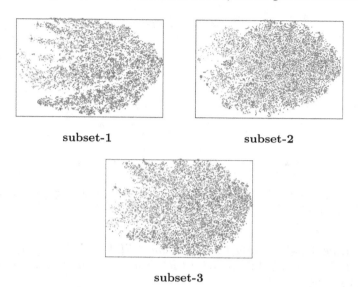

subset-1   subset-2

subset-3

Fig. 6. Clusters defined by each subset of features

proposed methodology was exploited with a TF-IDF schema, while a 0.783 using Word2vec pre-trained embeddings, and a 0.686 exploiting a LSTM classifier. As it can be observed, the obtained results with the different combinations we propose are very competitive even against more sophisticated techniques.

All the results presented until now are presented in terms of **Accuracy**. However, we decided to also include the outcomes obtained in terms of **F-score** for each class. We selected the seven best performing representations in the experiments carried out: **a**: BOW_0, **b**: BOW_4, **c**: FastText, **d**: SA+EMOT, **e**: LIWC, **f**: subset-1, and **g**: FastText+subset-1. Figure 7 shows the obtained results. As it can be observed, across the different configurations, the behavior is similar considering the performance among the classes. There is a significative drop in the performance for the *neutral* class while the F-score for the negative one remains almost the same. It is a slight improvement in terms of F-score for the *positive* and *negative* classes when the *FastText+subset-1* is used.

3.5 Data Analysis

Taking as starting point the overlapping of the instances we discovered along the TSNE-based representations proposed, a manual analysis of the instances in the *TwAS* was carried out. We identified some cases were instances composed by almost the same terms[9] were labeled with different, even contradictory classes.

- *@airline thank you* Labels: *neutral* and *positive*

---

[9] Most of the time they vary only in the last terms or in the URL.

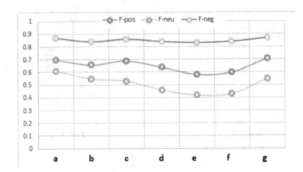

Fig. 7. Obtained results in terms of F-score on the three classes of tweets: positive, negative and neutral

- *@airline What a really GREAT & FLATTERING story about you! You should be very proud :) URL (via @mention) Labels: negative and positive*
- *@airline has getaway deals through May, from $59 one-way. Lots of cool cities URL #CheapFlights #FareCompare Labels: negative, positive, and neutral*

Then, attempting to remove those instances we applied two different sentiment analysis libraries namely *NLTK*[10] and *TextBlob*[11] in order to determine the sentiment of each tweet. Each tweet was "re-labeled" by considering the following criterion: When both the class assigned by each of the libraries and the original label of the tweet are equal, the instance is kept. We also considered both resources at the time, in this case, for selecting a given instance, the three labels must be the same. Table 2 shows the distribution of each subsample of data. Besides, in parenthesis we include the F-score obtained for each class when the classification task was performed over each subsample. The *FastText+subset-1* group of features was used.

Table 2. Distribution of each subsample of data

Class	Original	NLTK	TextBlob	BOTH
positive	2332	561 (0.91)	5415 (0.87)	365 (0.96)
negative	9088	87 (0.83)	3632 (0.9)	58 (0.79)
neutral	3065	13837 (0.98)	5438 (0.82)	1701 (0.99)

We also analyse the obtained results of re-annotating the tweets. Table 3 shows some samples. The first instance was labeled as *positive* during the manual annotation while both SA tools identified it as *negative*; correctly classifying

---

[10] https://www.nltk.org/api/nltk.sentiment.html.
[11] https://textblob.readthedocs.io/en/dev/.

such a complex expression is a challenge due to the fact that the sentiment expressed by the user is very subtle. The second and third samples have the *neutral* label, however both are clearly *positive*; these instances can be considered as the ones presented above, since in the *TwAS* we found tweets with almost the same content annotated with contradictory classes. The last two sentences reflect a negative connotation despite being annotated as *neutral*. Finally, we also identified some instances with irony and sarcasm, another important challenge for sentiment analysis [24] such as: *@airline never fails to disappoint.* and *@airline Another delay. Wow.*

Table 3. Obtained results of re-annotating tweets

Text	Original label	*NLTK*	*TextBlob*
@airline All flights Cancelled Flighted :(Trip refunded without difficulty, staff extremely helpful, no complaints! Way to handle bad weather!	Positive	Negative	Negative
@airline thanks!!	Neutral	Positive	Positive
@airline thanks for the help	Neutral	Positive	Positive
@airline in Bogota with no wallet is no fun. :-(	Neutral	Negative	Negative
@airline following!! My bad	Neutral	Negative	Negative

4 Conclusions

The use of lexical and linguistic resources can help in identifying subjective expressions in tweets. In this work, the evaluation of the proposed methodology was carried out using a corpus of tweets in the domain of commercial airlines. The advantages of managing to reduce the dimensionality in a significant (from more than 12000 to less than 200) way by incorporating lexical and psycho-linguistic resources are mainly in the computational cost as well as in the capability of obtaining similar results in terms of the performance rate obtained when using bag-of-words representation when carrying out sentiment analysis. As future work, we are interested in to further analyze this corpus considering the role of irony and sarcasm as well as to evaluate the performance of the proposed methodology over other domains.

References

1. Baccianella, S., Esuli, A., Sebastiani, F.: SentiWordNet 3.0: an enhanced lexical resource for sentiment analysis and opinion mining. In: Proceedings of the 7th International Conference on Language Resources and Evaluation, ELRA 2010 (2010)

2. Rustam, F., Ashraf, I., Mehmood, A., Ullah, S., Choi, G.S., Khan, Y.: Tweets classification on the base of sentiments for us airline companies. Entropy **21**(11), 1078 (2019)
3. Hakh, H., Aljarah, I., Al-Shboul, B.: Online social media-based sentiment analysis for US airline companies. In: New Trends in Information Technology, pp. 176–181 (2017)
4. Baker, D.M.A.: Service quality and customer satisfaction in the airline industry: a comparison between legacy airlines and low-cost airlines. Am. J. Tour. Res. **2**(1), 67–77 (2013)
5. Vadivukarassi, M., Puviarasan, N., Aruna, P.: An exploration of airline sentimental tweets with different classification model. Int. J. Res. Eng. Appl. Manage. (IJREAM) **4**(2), 72–77 (2018)
6. Nissim, M., Patti, V.: Semantic aspects in sentiment analysis, chap. 3. In: Pozzi, F.A., Fersini, E., Messina, E., Liu, B. (eds.) Sentiment Analysis in Social Networks, pp. 113–127. Morgan Kaufmann (2016)
7. Nielsen, F.Å.: A new ANEW: evaluation of a word list for sentiment analysis in microblogs. In: Proceedings of the ESWC2011 Workshop on Making Sense of Microposts: Big Things Come in Small Packages, vol. 718, pp. 93–98 (2011)
8. Hu, M., Liu, B.: Mining and summarizing customer reviews. In: Proceedings of the 10th ACM SIGKDD International Conference on Knowledge Discovery and Data Mining, KDD 2004, Seattle, WA, USA, pp. 168–177. ACM (2004)
9. Choi, Y., Wiebe, J.: +/-EffectWordNet: sense-level lexicon acquisition for opinion inference. In: Proceedings of the 2014 Conference on Empirical Methods in Natural Language Processing, pp. 1181–1191. ACL (2014)
10. Taboada, M., Grieve, J.: Analyzing appraisal automatically. In: Proceedings of the AAAI Spring Symposium on Exploring Attitude and Affect in Text: Theories and Applications, Stanford, US, pp. 158–161. AAAI (2004)
11. Wilson, T., Wiebe, J., Hoffmann, P.: Recognizing contextual polarity in phrase-level sentiment analysis. In: Proceedings of the Conference on Human Language Technology and EMNLP, pp. 347–354. ACL (2005)
12. Mohammad, S.M., Turney, P.D.: Crowdsourcing a word-emotion association lexicon. Comput. Intell. **29**(3), 436–465 (2013)
13. Poria, S., Gelbukh, A., Cambria, E., Hussain, A., Huang, G.-B.: EmoSenticSpace: a novel framework for affective common-sense reasoning. Knowl. Based Syst. **69**, 108–123 (2014)
14. de Albornoz, J.C., Plaza, L., Gervás, P.: SentiSense: an easily scalable concept-based affective lexicon for sentiment analysis. In: Proceedings of the LREC 2012, Istanbul, Turkey, pp. 3562–3567. ELRA (May 2012)
15. Cambria, E., Olsher, D., Rajagopal, D.: SenticNet 3: a common and common-sense knowledge base for cognition-driven sentiment analysis. In: Proceedings of AAAI Conference on Artificial Intelligence, pp. 1515–1521 (2014)
16. Bradley, M.M., Lang, P.J.: Affective Norms for English Words (ANEW): Instruction manual and affective ratings. Technical report, Center for Research in Psychophysiology, University of Florida, Gainesville, Florida (1999)
17. Whissell, C.: Using the revised dictionary of affect in language to quantify the emotional undertones of samples of natural languages. Psychol. Rep. **2**(105), 509–521 (2009)
18. Cambria, E., Livingstone, A., Hussain, A.: The hourglass of emotions. In: Esposito, A., Esposito, A.M., Vinciarelli, A., Hoffmann, R., Müller, V.C. (eds.) Cognitive Behavioural Systems. LNCS, vol. 7403, pp. 144–157. Springer, Heidelberg (2012). https://doi.org/10.1007/978-3-642-34584-5_11

19. Álvarez-Carmona, M.A., López-Monroy, A.P., Montes-y Gómez, M., Villaseñor-Pineda, L., Meza, I.: Evaluating topic-based representations for author profiling in social media. In: Advances in Artificial Intelligence, IBERAMIA 2016, pp. 151–162 (2016)
20. Yang, D., Lee, W.: Music emotion identification from lyrics. In: 2009 11th IEEE International Symposium on Multimedia, pp. 624–629 (2009)
21. Pennebaker, J.W., Francis, M.E., Booth, R.J.: Linguistic Inquiry and Word Count: LIWC 2001, vol. 71. Lawrence Erlbaum Associates, Mahway (2001)
22. Stone, P.J., Hunt, E.B.: A computer approach to content analysis: studies using the general inquirer system. In: Proceedings of the Spring Joint Computer Conference, AFIPS 1963, pp. 241–256 (1963)
23. Hernández Farías, D.I., Ortega-Mendoza, R.M., Montes-y-Gómez, M.: Exploring the use of psycholinguistic information in author profiling. In: Carrasco-Ochoa, J.A., Martínez-Trinidad, J.F., Olvera-López, J.A., Salas, J. (eds.) MCPR 2019. LNCS, vol. 11524, pp. 411–421. Springer, Cham (2019). https://doi.org/10.1007/978-3-030-21077-9_38
24. Farías, D.I.H., Rosso, P.; Irony, sarcasm, and sentiment analysis, chap. 7. In: Sentiment Analysis in Social Networks, pp. 113–127. Morgan Kaufmann (2016)

Ontological Model to Represent Information About Songs

Luis Yael Méndez-Sánchez⬤ and Mireya Tovar⁽✉⁾ ⬤

Faculty of Computer Science, Benemérita Universidad Autónoma de Puebla, 14 sur y Av. San Claudio C.U., Puebla, Puebla, México
luisyael_ms@outlook.com, mtovar@cs.buap.mx

Abstract. The purpose of this article is to carry out the design of an ontological model for the domain of songs, that is, its purpose is to cover the relevant information corresponding to the characteristic data currently sought to be known in the songs from different music genres, such as title, performer, album which it belongs, duration, release date, among many others. It is wanted for this model to answer queries made by users when they are facing the need to know a specific data. For this work realization, the Grüninger and Fox's methodology design phases, proper for the ontologies design, are followed, along with this, this document includes the proposal of a scenario, competition questions, definition of classes, properties, formalization and evaluation through the responses obtained through the SPARQL query language. Furthermore, the implementation of the presented ontological model was carried out using the software known as Protégé.

Keywords: Ontological model · Ontology design methodology · Songs · Query information

1 Introduction

Now days there are information sources on the web with search engines that respond to what the user requests, however, these engines work syntactically, that is, they reply results containing lexical elements related exactly to the query terms, so this is why the semantic web is used with the objective to overcome the current web limitations [3].

Gruber defines an ontology as an explicit formal specification of a shared conceptualization [12]. An ontology defines a hierarchy of concepts, relationships, restrictions, axioms and instances to describe a domain, which will serve for information exchange [3]. On the other hand, the use of ontologies has become a common interest area for some research groups, such as: the artificial intelligence line, knowledge engineering, natural language processing and knowledge representation, among others [1].

The ontology creation is a process made up from series of activities carried out in a certain order for a specific purpose. However, not all possible ontologies are already created, which is why it is necessary to develop and implement them in some way so that they can be used by the general community [2].

© Springer Nature Switzerland AG 2020
L. Martínez-Villaseñor et al. (Eds.): MICAI 2020, LNAI 12469, pp. 122–133, 2020.
https://doi.org/10.1007/978-3-030-60887-3_12

This is why, with the realization of this work, it is intended to carry out an ontology implementation that shows the information related to songs, which we can find today in our daily life, since, we all have a like, either for a specific genre or a song and many times we would like to know relevant data such as the artist, the album to which it belongs, the date of release, the duration, etc.

With the development of this ontology, the objective to be covered is to provide a songs library to the general public, that is, to make inference in the information corresponding to the songs according to their musical likes and thus make inquiries about the songs data, which becomes this tool, a very useful system for users who several times, in their daily lives, need to know this kind of information.

Following each of the steps specified in the used methodology, it is possible to develop an ontological model capable of handling the most important information regarding the domain of songs, the data being worked on is very extensive, so the use of a ontology is the most feasible solution to work with this type of context, this model allows queries to be made on the data, thus enabling the user to gather the information they require depending on their need, in addition to implementing the design of the ontology using the tool known as Protégé [6], therefore, a ready-to-use model is obtained.

The article is structured as follows: in Sect. 2 the related works are mentioned; Sect. 3 shows the design methodology used; Sect. 4 shows the ontology design; Sect. 5 shows the competence questions formalization and queries in SPARQL. Finally, the conclusions and references are presented.

2 Related Works

Some similar articles have been found in the literature and related to the use of methodologies for the ontology's construction, as well as proposed solutions focused on the domain of songs. Guzmán L. et al. [1] present a comparative analysis of methodologies and some existing methods mainly oriented to the design and implementation of ontologies, in addition, the specification of implementation tools typical of ontological models is included, for their application they use the domain oriented to the area of plastic arts; on the other hand, Bravo Contreras et al. [7] present a methodology for the design and construction of ontologies incorporating the most prominent design principles as well as a comprehensive evaluation process.

Flores et al. [3] present an ontological model design for the search of information within an institution of higher education and seeks to answer questions about the student's enrollment procedures and their respective release of professional practices, in addition they give a clear example of one Grüninger and Fox's methodology application for the ontologies design. In the same way, López R. et al. [2] present some methods and approaches published for the evaluation of ontologies, seeking to take up good practices from other existing methods and consider internal aspects of ontologies.

Uschold M. et al. [8] present a description of the emerging field related to the design and use of ontologies, aspects such as the benefits of using this kind of tools as ontological models are discussed, also in addition to mentioning how the methodologies are analyzed and then carrying out the ontology evaluation. Meanwhile, in another related work, Uschold M. together with King M. [9] present a work focused on modeling and emphasized in the process of capturing ontologies, describe ways to handle ambiguous terms so as not to fall into the problem that shared understanding is not adequate, since ontology must be understandable for its application.

In the same way, Banu A. et al. [10] propose an approach to reuse existing ontologies for the implementation of new ontological models that cover different domains, mainly based on retrieving ontologies based on important domain terms from online repositories using semantic web search engines.

Lachtar N. [11] presents a proposal for a musical repository based on the use of an ontology to index a collection of songs and the use of semantic links to allow the inference of all the relevant songs, in addition a refinement algorithm is proposed of queries to find songs, the development of a web application is carried out to index the songs by concepts, carry out conceptual searches for songs and exploit the semantic relationships that structure the ontology. Finally, Raimond Y. et al. [13] present and ontology development based on the domain of music production, including editorial, cultural and acoustic information regarding music, in the work presented by reusing "The Timeline Ontology" and "The Event Ontology" they are in charge of sound analysis which are involved in recording a piece of music, in addition to obtaining the different editorial information provided for the music.

Unlike the two works mentioned above, in Lachtar N. [11] they only make use of song titles to index or perform conceptual searches and Raimond Y. et al. [13] focus more on the field of music production, handling detailed information on music including acoustic aspects, in addition to making a more complex analysis to obtain the results they present; it is worth mentioning the contribution of this work is that, the presented ontology covers more information regarding a song, not mixing aspects of musical production or sound, that is the domain covered is broad in terms of information from a single song, since more data is worked in addition to the title, so the queries could be more complex to gather information, corresponding not only to a song but also to albums, interpreters, record labels, genres of music, composers and producers.

3 Ontology Design Methodology

It is proposed to use the Grüninger and Fox's methodology to carry out this work, which is inspired by the development of knowledge based on first-order logic [4]. Said methodology proposes an optimal design for the field of ontoloes, since it presents different phases that can be followed step by step to carry out its development.

Mainly, it was decided to use this methodology since the domain of songs is very extensive, it is necessary to have the description of the stage to know all the data that will be used regarding the songs, in addition to the fact that using questions that solve the needs of users when they want to know specific data is an optimal way to evaluate and keep in mind if the ontology fulfills its purpose, which is to provide more detailed information in the musical field, a reengineering is not used since other related works present similarity only in providing information about the songs, in this work the contribution is to provide a way to query information regarding albums, interpreters, record labels, composers and producers, that is, more detailed information and attributes related to songs are being handled.

As a first step, this methodology proposes to identify the main scenarios, that is, the possible applications in which the ontology will be used. As a second step it is proposed to ask a series of questions called "competence", which are used to determine the scope of the ontology. These questions and their answers are used to extract the main concepts and their properties, relationships and formal axioms of ontology [5]. Next, Sect. 3.1 shows the scenario approach; in Sect. 3.2 the proposed competence questions are shown and finally in Sect. 3.3 the classes description and their properties are presented.

3.1 Scenario Presentation

The ontological model will provide answers about the most relevant information you need to know about a song, in this case data such as the title, artist, album to which it belongs, song genre, language, release date, duration, recording year, related record label, composer and producer, this set of attributes of a song may be consulted depending on the data you want to know, the ontology will be able to provide the required information.

In the same way, specific information about the performers can be gathered, which can be a singer, a composer or a musical group, and information about the album related to the song, the recorded music genres, the musical instruments used by the performers mentioned above, related record labels, composers and producers, since all these attributes are related to each other because they are part of the data of a song, that is, why each of these terms are defined as classes, which will contain their own attributes (data) that depict them.

3.2 Competition Questions

Following the methodology used to carry out this ontological model, a series of questions were prepared, which are called "competition", this in order to have an idea of what a user could do as a search, that is, queries within the ontological model, in addition to being useful to identify classes, relationships, properties and axioms for the realization of this ontology of songs. Here are some elaborate competence questions:

1. What are the characteristics of a song?
2. What are the songs lasting more than four minutes?
3. What are the registered music genres?
4. What are the registered singers or songwriters' names with Mexican nationality?
5. Who are the members from the rock band called Queen?

3.3 Classes Description and Properties

For this phase, an analysis was carried out, in which, starting from the scenario and the competition questions, the different classes that the ontological model would contain were identified, in the same way the properties for each class were obtained together with their relationships. Table 1 describes some classes that were included in the ontological model, such as: song, musical group, singer-songwriter, composer, producer and music genre, but there are also classes such as album, singer, record label, person and musical instrument within the ontology.

Table 1. Ontological model classes.

Class	Description
Song	Main class containing the different data to be provided about a song
Band	Class corresponding to the song interpreters, in this case a group of singers
Singer_Songwriter	Subclass of Person, corresponds to the song interpreter, in this case it is the responsible person for composing and interpreting the song
Composer	Subclass of Person, corresponds to the song or songs composer. A composer can be a singer-songwriter too
Producer	Subclass of Person, refers to the song or the album producer. In the same way, it can be a singer or a singer-songwriter
Music_Genre	Class to list the different genres of music that exist and relate them to each song

Tables 2 and 3 describe the properties for each class, which are highly important, since they help to store the information to answer the competence questions above described.

Table 2. Data type properties.

Data type property	Domain	Range
has_title	Song	String
has_language	Song	String
has_publication_date	Song	String
has_duration	Song	Decimal
has_recording_year	Song	Int
has_name	Band	String
has_origin_place	Band	String
has_activity_status	Band	String
has_emergence_year	Band	Int
has_activity_period	Band	String
has_stage_name	Singer_Songwriter	String
has_occupation	Singer_Songwriter	String
has_activity_period	Singer_Songwriter	String
has_award	Singer_Songwriter	String
has_occupation	Composer	String

(continued)

Table 2. (*continued*)

Data type property	Domain	Range
has_activity_period	Composer	String
has_stage_name	Producer	String
has_award	Producer	String
has_name	Music_Genre	String

Table 3. Object type properties or relationships between classes.

Object property	Domain	Range
belongs_to	Song	Music_Genre
is_included_in	Song	Album
is_interpreted_by	Song	Singer Singer_Songwriter Band
is_composed_by	Song	Singer_Songwriter Composer
is_recorded_by	Song	Record_Label
is_produced_by	Song	Producer Singer Singer_Songwriter
has_member	Band	Singer Singer_Songwriter
interprets	Band	Music_Genre
works_with	Band	Record_Label
is_member_of	Singer_Songwriter	Band
uses	Singer_Songwriter	Musical_Instrument
interprets	Singer_Songwriter	Music_Genre
works_with	Singer_Songwriter	Record_Label
uses	Composer	Musical_Instrument
composes	Composer	Music_Genre
works_with	Producer	Record_Label
produces	Producer	Music_Genre

Table 2 shows the ontological model data type properties and Table 3, shows the object type properties, which are the relationships between classes.

4 Design

This section presents the design for the proposed ontology developed with the help of the Protégé tool, an open source software for building ontologies through its interface that helps the developer in the process [6]. In Fig. 1 the ontological model classes are depicted, the classes have relations between them, therefore, in Fig. 2 the object type properties diagram (relations between classes) is shown, which were described in Table 3.

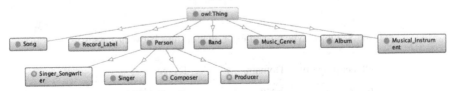

Fig. 1. Ontological model classes.

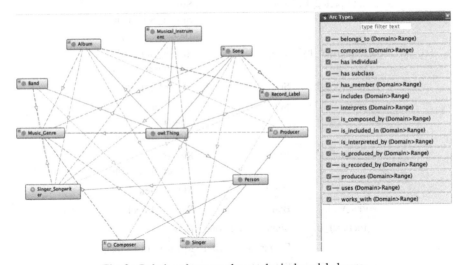

Fig. 2. Relations between the ontological model classes.

The same way, some rules were created for the classes as a singer-songwriter, who can be a singer and a composer at the same time, so it must be defined in the ontological model (see Fig. 3). The composer also, as being part of the singer-songwriter restriction, can be of both types, using Protégé the restriction is defined in the composer class as well. The producer can be a singer or a songwriter at the same time, since several times the same performers are in charge of being producers of their songs (see Fig. 4).

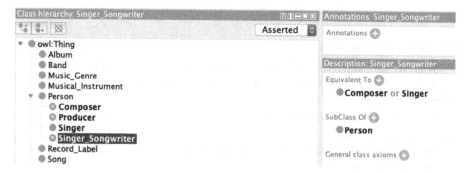

Fig. 3. Class restriction singer-songwriter.

Fig. 4. Producer class restriction.

5   Results

The following section shows the competence questions formalization (see Table 4), which were presented in Sect. 3.2. The ontology evaluation was carried out at the application level, that is, through the answers provided by the ontological system, in the same way these answers are provided when formalizing the competency questions, it is worth mentioning that the ontology meets the consistency criteria and the HermiT reasoner was used to test the ontological model from the Protégé tool.

Table 4. Competition questions formalization and answers in SPARQL.

N°	Question and formalization	Query SPARQL
1	**What are the characteristics of a song?** $\exists\$x\$m\$n\$p\$q\$r\$t\$u\$v\$w\$y$ (Song($\$x$) $\wedge$ has_title($\$x$, $\$m$) $\wedge$ belongs_to ($\$x$, $\$n$) $\wedge$ is_included_in ($\$x$, $\$p$) $\wedge$ is_interpreted_by ($\$x$, $\$q$) $\wedge$ has_language($\$x$, $\$r$) $\wedge$ has_publication_date($\$x$, $\$t$) $\wedge$ has duration($\$x$, $\$u$) $\wedge$ is_composed_by ($\$x$, $\$v$) $\wedge$ is_recorded_by ($\$x$, $\$w$) $\wedge$ is_produced_by ($\$x$, $\$y$))?	Figure 5

(continued)

Table 4. (*continued*)

Nº	Question and formalization	Query SPARQL
2	**What are the songs lasting more than four minutes?** $\exists\$x\$y\$z$ (Song($\$x$) $\wedge$ has_title ($\$x$, $\$y$) $\wedge$ (has_duration ($\$x$, $\$z$) > 4))?	Figure 6
3	**What are the registered music genres?** $\exists\$x\$y$ (Music_Genre ($\$x$) $\wedge$ has_name ($\$x$, $\$y$))?	Figure 7
4	**What are the registered singers or songwriters' names with Mexican nationality?** $\exists\$x\$p\$q\$w\$y\$z\$k\$m\$n\$o$ ((Singer_Songwriter($\$x$) $\wedge$ has_stage_name ($\$x$,$\$p$) $\wedge$ has_age ($\$x$,$\$q$) $\wedge$ has_occupation ($\$x$, $\$w$) $\wedge$ (has_nationality ($\$x$, $\$y$) = "Mexican")) $\vee$ ((Singer ($\$z$) $\wedge$ has_stage_name ($\$x$,$\$k$) $\wedge$ has_age ($\$x$,$\$m$) $\wedge$ has_occupation ($\$x$, $\$n$) $\wedge$ (has_nationality ($\$x$, $\$o$) = "Mexican"))?	Figure 8
5	**Who are the members from the rock band called Queen?** $\exists\$x\$p\$q\$w\$y\$z\$k\$m\$n\$o\$r\$t$ ((Singer_Songwriter ($\$x$) $\wedge$ has_stage_name ($\$x$,$\$p$) $\wedge$ has_activity_period ($\$x$,$\$q$) $\wedge$ has_nationality ($\$x$, $\$w$) $\wedge$ has_occupation ($\$x$,$\$y$) $\wedge$ (is_member_of ($\$x$, $\$z$) = "Queen")) $\vee$ ((Singer($\$k$) $\wedge$ has_stage_name ($\$k$,$\$m$) $\wedge$ has_activity_period ($\$k$,$\$n$) $\wedge$ has_nationality ($\$k$, $\$o$) $\wedge$ has_occupation ($\$k$, $\$r$) $\wedge$ (is_member_of ($\$k$, $\$t$) = "Queen"))?	Figure 9

For the realization of the queries, the proper language is used for the use of ontologies which is SPARQL and the answers to the competence questions are presented in Figs. 5–9 (see column 3 of Table 4).

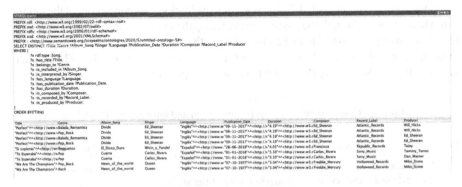

Fig. 5. Answer to question 1.

```
SPARQL query
PREFIX rdf: <http://www.w3.org/1999/02/22-rdf-syntax-ns#>
PREFIX owl: <http://www.w3.org/2002/07/owl#>
PREFIX rdfs: <http://www.w3.org/2000/01/rdf-schema#>
PREFIX xsd: <http://www.w3.org/2001/XMLSchema#>
PREFIX : <http://www.semanticweb.org/luisyaelms/ontologies/2020/5/untitled-ontology-5#>
SELECT ?Song_Title ?Duration
WHERE {
    {
        ?x rdf:type :Song.
        ?x :has_title ?Song_Title.
        ?x :has_duration ?Duration
    }
    FILTER (?Duration >= 4 )
}
ORDER BY ASC(?Song_Title)
```

Song_Title	Duration
"Estoy Enamorado"^^<http://www.w3.org/2001/XMLSchema#string>	"4.31"^^<http://www.w3.org/2001/XMLSchema#decimal>
"Perfect"^^<http://www.w3.org/2001/XMLSchema#string>	"4.19"^^<http://www.w3.org/2001/XMLSchema#decimal>
"Si supieras"^^<http://www.w3.org/2001/XMLSchema#string>	"4.01"^^<http://www.w3.org/2001/XMLSchema#decimal>

Fig. 6. Answer to question 2.

```
SPARQL query
PREFIX rdf: <http://www.w3.org/1999/02/22-rdf-syntax-ns#>
PREFIX owl: <http://www.w3.org/2002/07/owl#>
PREFIX rdfs: <http://www.w3.org/2000/01/rdf-schema#>
PREFIX xsd: <http://www.w3.org/2001/XMLSchema#>
PREFIX : <http://www.semanticweb.org/luisyaelms/ontologies/2020/5/untitled-ontology-5#>
SELECT ?Genre_Name
WHERE {
    ?x rdf:type :Music_Genre.
    ?x :has_name ?Genre_Name;
}
ORDER BY ASC(?Genre_Name)
```

Genre_Name
"Balada"^^<http://www.w3.org/2001/XMLSchema#string>
"Balada Romántica"^^<http://www.w3.org/2001/XMLSchema#string>
"Dance pop"^^<http://www.w3.org/2001/XMLSchema#string>
"Electrónica"^^<http://www.w3.org/2001/XMLSchema#string>
"Folk"^^<http://www.w1.org/2001/XMLSchema#string>
"Hard Rock"^^<http://www.w3.org/2001/XMLSchema#string>
"Hip Hop"^^<http://www.w3.org/2001/XMLSchema#string>
"Pop"^^<http://www.w3.org/2001/XMLSchema#string>
"Pop Latino"^^<http://www.w3.org/2001/XMLSchema#string>
"Pop rock"^^<http://www.w3.org/2001/XMLSchema#string>
"Reggae"^^<http://www.w3.org/2001/XMLSchema#string>
"Reggaetón"^^<http://www.w3.org/2001/XMLSchema#string>
"Rock"^^<http://www.w3.org/2001/XMLSchema#string>
"Trap"^^<http://www.w3.org/2001/XMLSchema#string>
"Ópera Rock"^^<http://www.w3.org/2001/XMLSchema#string>

Fig. 7. Answer to question 3.

```
SPARQL query
SELECT ?Stage_Name ?Age ?Nationality ?Occupation
WHERE {
    {
        ?x rdf:type :Singer_Songwriter.
        ?x :has_stage_name ?Stage_Name.
        ?x :has_age ?Age.
        ?x :has_nationality ?Nationality.
        ?x :has_occupation ?Occupation.
        FILTER( ?Nationality = "Mexicana")
    }
    UNION
    {
        ?x rdf:type :Singer.
        ?x :has_stage_name ?Stage_Name.
        ?x :has_age ?Age.
        ?x :has_nationality ?Nationality.
        ?x :has_occupation ?Occupation.
        FILTER( ?Nationality = "Mexicana")
    }
}
ORDER BY ASC(?Stage_Name)
```

Stage_Name	Age	Nationality	Occupation
"Alex Sar"^^<http://www.w3.org/2001/XMLSchema#string>	"39"^^<http://www.w3.org/2001/XMLSchema#integer>	"Mexicana"^^<http://www.w3.org/2001/XMLSchema#string>	"Músico, escritor"^^<http://www.w3.org/2001/XMLSchema#string>
"Carlos Rivera"^^<http://www.w3.org/2001/XMLSchema#string>	"34"^^<http://www.w3.org/2001/XMLSchema#int>	"Mexicana"^^<http://www.w3.org/2001/XMLSchema#string>	"Cantante, compositor, actor"^^<http://www.w3.org/2001/XMLS

Fig. 8. Answer to question 4.

```
SPARQL query
SELECT ?Stage_Name ?Activity_Period ?Nationality ?Occupation ?Band
WHERE {
    {
        ?x rdf:type :Singer_Songwriter.
        ?x :has_stage_name ?Stage_Name.
        ?x :has_activity_period ?Activity_Period.
        ?x :has_nationality ?Nationality.
        ?x :has_occupation ?Occupation.
        ?x :is_member_of ?Band.
        FILTER(?Band = :Queen)
    }
    UNION {
        ?x rdf:type :Singer.
        ?x :has_stage_name ?Stage_Name.
        ?x :has_activity_period ?Activity_Period.
        ?x :has_nationality ?Nationality.
        ?x :has_occupation ?Occupation.
        ?x :is_member_of ?Band.
        FILTER(?Band = :Queen)
    }
}
```

Stage_Name	Activity_Period	Nationality	Occupation	Band
"Brian May"^^<http://www.w3.org/2001/XMLSchema#string>	"1965-presente"^^<http://www.w3.org/2001/XMLSchema#string>	"Británica"^^<http://www.w3.org/2001/XMLSchema#string>	"Músico, compositor, cantante, productor, astrofísic	Queen
"Freddie Mercury"^^<http://www.w3.org/2001/XMLSchema#string>	"1969-1991"^^<http://www.w3.org/2001/XMLSchema#string>	"Británica"^^<http://www.w3.org/2001/XMLSchema#string>	"Cantante, compositor, pianista, diseñador gráfico y	Queen
"John Deacon"^^<http://www.w3.org/2001/XMLSchema#string>	"1965-1997"^^<http://www.w3.org/2001/XMLSchema#string>	"Británica"^^<http://www.w3.org/2001/XMLSchema#string>	"Músico"^^<http://www.w3.org/2001/XMLSchema#string>	Queen
"Roger Taylor"^^<http://www.w3.org/2001/XMLSchema#string>	"1968-presente"^^<http://www.w3.org/2001/XMLSchema#string>	"Británica"^^<http://www.w3.org/2001/XMLSchema#string>	"Cantante, músico y compositor"^^<http://www.w3	Queen

Fig. 9. Answer to question 5.

6 Conclusions

This document defines an ontology for the musical domain, created from the steps defined in the Grüninger and Fox's methodology. An ontological model has been obtained, which allows managing different relevant information data regarding songs, all this in order that different types of users can query these records when they have the need to know a song characteristic data, or his album, as well as its performers, related record labels, composers, producers, genres of music and musical instruments used by the performers.

The main contribution of this work is the application of a methodology for an ontology manual creation, applied to the songs domain for search information related to them, that is, songs specific data many times people want to know.

The methodology used has been of great help to solve the problems corresponding to structure, organization and information management through the use of semantic web tools such as ontologies. The Protégé tool in combination with the SPARQL query language has been used to answer the competence questions and to carry out the ontology design described in this work, since, by using this language the necessary queries are carried out within the ontological model designed in Protégé to gather the information you want to know; this makes the current ontology a practical method and as one option more for this area, with the aim of having an alternative to traditional databases.

Finally, as a proposal for future work, the creation of a web application that uses the ontology to make an automatic data population at the instance level of each class defined in the ontology is intended, query the registered songs information, delete desired records, as well as the presentation of the competition questions to an end user.

Acknowledgment. This work is supported by the Sectorial Research Fund for Education with the CONACyT CB/257357 project and by the VIEP-BUAP 100409344-VIEP2019 project.

References

1. Guzmán, L.J.A., López, B.M., Durley, T.I.: Metodologías y métodos para la construcción de ontologías. Scientia et Technica Año XVII, 50, 8 (2012)
2. López, R.Y.A., Hidalgo, D.Y., Silega, M.N.: Un método práctico para la evaluación de ontologías. Universidad de las Ciencias Informáticas, La Habana, Cuba, 10 (2018)
3. Flores, J.C., Tovar, M., Cervantes, A.-P.: Modelo ontológico para representar información sobre la práctica profesional en una institución educativa. Res. Comput. Sci. **145**, 14 (2017)
4. Bravo, M., Martínez-Reyes, F., Rodríguez, J.: Representation of an academic and institutional context using ontologies. Res. Comput. Sci. **89**, 9–17 (2014)
5. Gómez, P.A., Fernández, L.M., Corcho, O.: Ontological Engineering with Examples from the Areas of Knowledge Management e-Commerce and the Semantic Web. Springer, London (2004). https://doi.org/10.1007/b97353
6. Musen, M.A.: The Protégé project: a look back and a look forward. AI Matt. **1**(4), 4–12 (2015). http://doi.acm.org/10.1145/2757001.2757003
7. Bravo Contreras, M., Hoyos Reyes, L., Reyes Ortiz, J.: Methodology for ontology design and construction. Contaduría Y Administración **64**(4), e134 (2019). https://doi.org/10.22201/fca.24488410e.2020.2368

8. Uschold, M., Gruninger, M.: Ontologies: Principles Methods and Applications. Knowl. Eng. Rev. **11**(2), 93–136 (1996)
9. Uschold, M., King, M.: Towards a methodology for building ontologies. In: Workshop on Basic Ontological Issues in Knowledge Sharing, pp. 3–15 (1995)
10. Banu, A., Sameen, S., Rahman, K.: A re-usability approach to ontology construction. In: Proceedings of the Second International Conference on Computational Science, Engineering and Information Technology, pp. 189–193 (2012). https://doi.org/10.1145/2393216.2393248
11. Lachtar, N.: Using domain ontology to classify a song. In: Proceedings of the 7th International Conference on Software Engineering and New Technologies, pp. 1–5 (2018). Article no. 26. https://doi.org/10.1145/3330089.3330126
12. Gruber, T.R.: A translation approach to portable ontology specifications. Knowl. Acquis. **5**(2), 199–220 (1993). https://doi.org/10.1006/knac.1993.1008
13. Raimond, Y., Abdallah, S., Sandler, M., Giasson, F.: The Music Ontology. ISMIR, pp. 1–6 (2007)

Topic Modelling with NMF vs. Expert Topic Annotation: The Case Study of Russian Fiction

Tatiana Sherstinova[1,2]([✉]) [ID], Olga Mitrofanova[1] [ID], Tatiana Skrebtsova[1] [ID],
Ekaterina Zamiraylova[1] [ID], and Margarita Kirina[2] [ID]

[1] Saint Petersburg State University, 7/9 Universitetskaya Emb, St. Petersburg 199034, Russia
{t.sherstinova,o.mitrofanova,t.skrebtsova}@spbu.ru,
e.zamiraylova@gmail.com
[2] National Research University Higher School of Economics,
123 Griboyedova emb, St. Petersburg 190068, Russia
mkirina2412@gmail.com

Abstract. The paper presents an experiment aimed at comparison of results of topic modelling via non-negative matrix factorization (NMF) with that of manual topic annotation performed by an expert. The experiment was conducted on the annotated corpus of Russian short stories of the initial three decades of the 20th century, which contains 310 stories with a total of 1000000 tokens written by 300 Russian writers. The annotation scheme used in topic annotation includes 89 topics, further this list was reduced down to 30 generalized ones, the most frequent of which turned out to be the following: *death, relationships, love, social groups, social processes, family, money, human sins, nature, religion*, and *war*. Then, the corpus divided into three consecutive time periods was subjected to NMF topic modelling which provided a model including 24 topics. The results of both topic annotations were compared and described. The paper discusses the main findings of the study and the difficulties of fiction topic modelling which should be taken into account. For example, experimental results showed that topic modelling via NMF should be primarily recommended for the revealing of topics referring to general background of literary texts (e.g., *war, love, nature, family*) rather than for detecting topics related with some critical events or relations between characters (e.g., *death* or *relations*). The comparison of human and automatic topic annotation seems an important step for the improvement of artificial technologies techniques related with NLP.

Keywords: Corpus linguistics · NPL · Machine learning · Topic modelling · NMF · Fiction · Russian literature · Literary criticism · Digital humanities

1 Introduction

Topic modelling is one of the leading trends in contemporary computer science and data analysis, which inspires interdisciplinary research in computational linguistics [28], sociology [4], psychology [30], and other disciplines. There is a diverse range of topic

© Springer Nature Switzerland AG 2020
L. Martínez-Villaseñor et al. (Eds.): MICAI 2020, LNAI 12469, pp. 134–151, 2020.
https://doi.org/10.1007/978-3-030-60887-3_13

modelling algorithms (LSA, pLSA, LDA, NMF, etc.) [2, 5, 23] as well as their implementations (MALLET, Stanford Topic Modelling Toolbox, gensim, scikit-learn, BigARTM, etc.). A bulk of research is inspired by task-oriented topic models, namely, dynamic topic models [8] and author-topic models [32] which proved to be the most effective in social network analysis and media monitoring. A wide scope of their linguistic extensions was developed for topic modelling algorithms: such extensions allow to work with multilingual corpora [13], unigram and n-gram topic models [16] aggregating collocations alongside with single lemmas, to generalize topics by means of topic labels [14], etc.

Our study discovers novel facets in topic modelling as it is focused on the nonstandard empirical data, namely, fiction texts which often drop out of sight (with a few exceptions [3, 9, 25, 31]), and gives a pioneering treatment of the problem of topic generalization.

Topic modelling is an intricate fuzzy clustering technique, which is aimed at detection of latent interrelations covering all levels of text representation within a text corpus. A topic model represents any text of a corpus as a mixture of probabilistic topics (clusters of semantically related words), so that a topic embraces a cluster of texts with similar content. Strict computational formulation of the given bi-clustering model is discussed in detail in [38]. However, the problem often arises as regards linguistic interpretation of topics obtained by topic modelling and their generalization.

Possible solutions are provided by topic label assignment (or topic labelling) techniques [6]. Words in a topic are ranked in accordance with their relevance, so the first word may be assumed to be a label. However, top-first lemmas may be of concrete meaning, thus, failing to represent topic content. Therefore, researchers proposed several techniques of topic labelling which extract topic labels from a corpus by various ranking methods [15] or use external sources of data (web-collections, Wikipedia, WordNet, etc.) [12]. At the same time, automatically generated topic labels always require human evaluation as the output of almost any topic model is intended to be comprehensible by the users. Moreover, in case of specialized corpora processing, including that of fiction texts, manual topic label selection performed by experts is inevitable. Overlap of two alledgedly separate procedures, namely, automatic generation of labels and their human assessment, forms a crucial point in our research which gives rise to a more general task of topic annotation.

The paper presents an experiment aimed at comparison of results of topic modelling via non-negative matrix factorization with that of manual topic modelling performed by an expert. The experiment was conducted on the annotated corpus of Russian short stories of 1900–1930, which contains 310 stories with a total of 1000000 tokens written by 300 Russian writers [21]. Thereby, we are trying to pave the way for rapprochement of modern topic modelling techniques and traditional interpretive methods used by literary critics. Fiction topic modelling is a rather difficult and challenging task, as unlike nonfiction texts (scientific, mass media, business, etc.), in which several terms or keywords quite unambiguously determine the topics of the document, the topics of literary texts are often hidden. For example, a short story, which was tagged by a literary critic as a "love story", may not contain the word "love" or any of its derivatives. Because of that the comparison of human and automatic tagging of topics seems very important for the improvement of artificial technologies methods related with NLP.

The task of fiction automatic topic annotation belongs to the tasks of artificial intelligence and demands of algorithms much more complicated than the existing methods of statistical text processing. One can come to the solution of this problem through machine learning tools, trained on manual expert topic annotation.

2 The Corpus of Russian Short Stories and Its Expert Topic Annotation

The Corpus of Russian Short Stories of 1900–1930 is currently being developed in St. Petersburg State University in cooperation with National Research University Higher School of Economics, St. Petersburg [20, 21]. Corpus developers see their task in creation of a digital resource, which should provide the possibility to conduct a diverse stylometric analysis of fiction at different language levels (rhythmic and phonetic [10], lexical [17], syntactic [18], semantic, structural, etc.) [19]. Another important task that is to be solved is the adaptation of literary approaches to corpus practice and their involvement in the orbit of quantitative research [ibid.].

The corpus of Russian short stories of 1900–1930 contains the following subcorpora, which are important to be mentioned here:

Period I. Short stories of the beginning of the 20th century (1900–1913);

Period II. Short stories of the era of war and the acute social upheaval (1914–1922) – World War I, the February and October Revolutions and the subsequent Civil War, which may be further subdivided into two chronologically consecutive periods –

II-1. World War I before the Revolutions (1914–1916) and

II-2. Revolutions and the Civil War); and

Period III. Short stories of the post-revolutionary era (1923–1930).

The corpus size is more than 1 million words. It was lemmatized by MyStem [33] and annotated on morphological, syntactic and, selectively, rhythmic levels; the texts were segmented into fragments of narrator's speech, narrator's remarks and characters' speech. The literary annotation includes type of narrative, topics, and some structural features of texts.

In the first version of the corpus, which is currently being developed, the developers do not set the task of a full-fledged literary annotation, confining ourselves to its individual elements (the main topics, type of narration, general structure, and some others) [34], but it is assumed that the aspects of literary markup in the corpus of Russian short stories will be expanded in the future.

As there are no universally accepted algorithms of topic detection in works of fiction, a careful qualitative analysis is needed at the outset [35].

Topic annotation developed for the Corpus of Russian short stories presupposes the identification of all semantic components that contribute to the plot, determines the protagonist's motives and actions and directly bears on the conflict and its resolution [1, 40]. Topics are considered akin to keywords [37], so that each story is mapped onto a clustering of topics. Here, a parallel can be drawn with componential analysis that aims to present word meaning in a bundle of semantic features. The difference, though, is that while componential analysis tries its best to bring out the total semantic content of a word, a set of topics is not meant to fully define the short story plot [ibid.].

Topic annotation for the corpus of Russian Short Stories was done manually on the above-mentioned sample comprising 310 stories written by 300 different Russian writers and will be later used as training data for a learning model.

To detect typical topics of the whole timespan (1900–1930), an empirical approach was used. Topics identified manually in a bottom-up fashion (starting from individual stories) and then tested against the whole sample. In the end, a set of 89 topics ranging from political to personal, and from philosophical to mundane, united in 8 thematic groups was formed, the most frequent of them are given in Table 1. A rough set of topics was drawn from the stories of Period I (see above). It was subsequently tested against the stories of the two other periods (Period II and Period III), with inevitable corrections, deletions and additions.

Table 1. The list of main topics used in expert annotation.

Thematic group	Topics
Economic and social topics	Bright future; explorations and inventions; industrial advance; mass education; new lands development; new social order; the old vs. the new; women's emancipation; young people
God and religion	Christian God; religion as a social institution
Inner life	Art, creative activity; doppelgängering; dreams vs. reality; feeling of freedom; frustrated hopes, disillusionment; disturbed sleep; ideal vs. reality; insanity; loneliness; mysticism, hallucinations, nobility of character, magnanimity, self-sacrifice; presentiments; passion for life; readiness to forgive; remorse; shame; sleep vs. reality; spiritual rebirth; willingness to help, be of use, philanthropy
Interpersonal relations	Deceit; envy; fraternity, solidarity; friendship; greed; mentorship; pretense; revenge; rivalry; treachery
Love, sex, and family topics	Body life; children; fathers and sons; jealousy; marriage; mutual sexual love of a man and a woman; parental love; prostitution; rape; romantic love; unfaithfulness; unrequited love
Mundane aspects of life	Alcoholism; boredom; bribery; Christmas, Christmas tree; death from natural causes; monotonous life; poverty, hunger, hardships; sudden and accidental death; suicide; the rich vs. the poor; money; New Year's Eve
Political topics	Civil war; death in the war; execution; October revolution; pre-revolutionary unrest; punishments for political crimes (prison, hard labor, deportation); revolutionary movement; Russo-Japanese war; World War I
Social structure and lifestyle	Country life; country vs. city life; city vs. nature; emigration; land as property; non-peasant work; peasant life; pets and animals; violence; working class

Each story thus is mapped onto a set of topics deemed essential for its content. Annotation was done manually by one expert; however, we plan to perform additional assessment involving several independent experts. More detailed information on expert topic annotation may be found in [35].

The database of the corpus contains general information about Russian writers, short stories meta-information, and the results of text multilevel annotation.

Each story has a personal code and the following meta-information: the name of the story, the author, and the year of creation. Besides, each story has been additionally labeled with several tags, referring to information about the topics—the semantic cores that link the story. The tags have their unique identification codes as well, multiple tagging of the stories in case they are referring to more than one topic is allowed. Moreover, it is also possible to filter texts based on the tags they share for the further comparison.

As the initial annotation scheme proposed by an expert turned out to be rather detailed, it was decided to reduce the number of topics, i. e., to make normalization of this database field. For example, three distinct topics related with the wars of the correspondent historical period—the Russian Civil War, World War I, and Russo-Japanese War—got a common tag WAR.

When developing the normalization scheme, we tried to make it user friendly and understandable for an average reader [41]. The categorization we propose is cognitive by nature [36]. Too detailed or too abstract topics are undesirable since they can either lead to overgeneralization or and to the increase of the number of topics. That is why alongside with the normalized tag LOVE (*romantic love, unrequited love, mutual love, passion, sexual affection,* etc.) we distinguish the tag FAMILY (*fathers and sons, marriage, unfaithfulness,* etc.), including *parent's love* into the latter.

As a result of normalization the initial 89 topics suggested by an expert has been recategorized into 30 groups using the following tags: *future, mode of life, relations, war, city life, money, children, virtue, leisure time, art, beauty, love, dream, young people, violence, political struggle, human sins/vices, nature, progress, mental state, revolution, religion, freedom, family, death, sleep, social groups, social processes, labor,* and *fantasy.* Figure 1 illustrates the frequency of these tags counted for the stories in the annotated corpus.

Another peculiarity about the tagging of Russian short stories written in the beginning of the 20th century is the careful indication of the sociocultural context in the topics—the time when the text has been created. This specificity of our corpus led to the extraction of the tag POLITICAL STRUGGLE which includes the thematic elements of *revolutionary movement* and *punishments for political crimes* (*prison, hard labor, terror, deportation*). The second history-related tag is SOCIAL PROCESSES (*pre-revolutionary civil unrest, new mode of life, politically-induced changes in social roles*). The tag SOCIAL GROUPS unites stories about *Cossacks, peasantry, working people, relations and conflicts between different nations,* and *the Jewish question.* Since there was a lot of social and political changes in Russia in the considering time period from 1900 to 1930, two additional tags were used: WAR (*the Russian Civil War, World War I,* and *Russo-Japanese War*) and REVOLUTION.

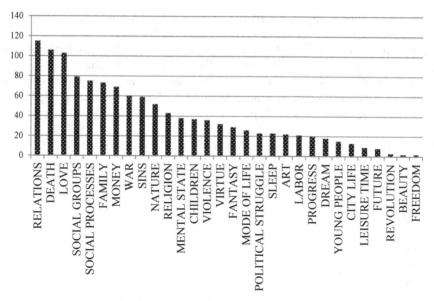

Fig. 1. Frequency distribution of topic tags

3 Topic Modelling with Non-negative Matrix Factorization

The experiment was based on topic modelling with non-negative matrix factorization (NMF), which is the first layer of dynamic topic modelling [https://github.com/derekgreene/dynamic-nmf; 7]. Preliminary experiments were performed with several topic modelling algorithms including LDA, author-topic model, dynamic topic model via NMF implementations in gensim, NMF proved to be the most reliable as regards linguistic interpretation of results.

In the first step, we lemmatized the original raw texts by Pymorphy2 Russian tagger [11], created "documents-lemmas" matrix, to which TF-IDF and document length normalization were applied before each matrix was made. This included building a document matrix for a window topic, where the topic model was created by applying NMF to each time window.

During the experiment the following parameters were used: the number of iterations—10; the number of topic words in the output—10 lemmas, and the number of topics—from 5 to 20.

The chosen model is considered as count-based but not as predictive by nature, it implies that distribution of lemmas and documents over topics depends on their observed frequencies, so that it is common words of a corpus that constitute topics. In this respect we can draw parallels between topic modelling and semantic compression (in particular, keyword extraction, summarization, etc.). NMF-based topics help to determine the most frequent plots of the corpus which are mostly represented by the nominative class.

The composition of topics was studied at the core level, namely the top part of the list, whose boundaries are set manually (in our case, the first 10 topic words were taken, since they are often considered the most informative [26, 27]). It is worth pointing out that topic labels are not generated automatically but are selected manually.

For tuning the parameters of topic modelling we used word2vec topic coherence measure described in [7, 29] as applicable to NMF, thus, the number of topics was determined in the following way [39]:

- top recommendations for number of topics for '1900–1913': 10 (see Fig. 2);

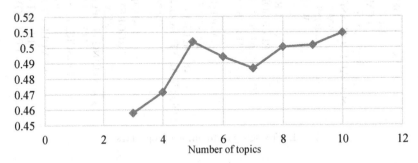

Fig. 2. Top recommendations for number of topics for '1900–1913': the number of topics is 10.

- top recommendations for number of topics for '1914–1922': 4 (Fig. 3);

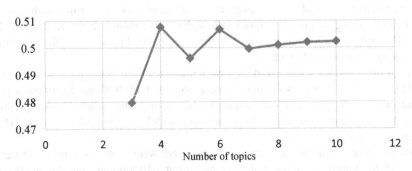

Fig. 3. Top recommendations for number of topics for '1914–1922': the number of topics is 4.

- top recommendations for number of topics for '1923–1930': 10 (Fig. 4).

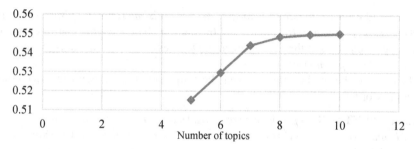

Fig. 4. Top recommendations for number of topics for '1923–1930': the number of topics is 10.

Now we are going to consider each period in accordance with the set of topics.

3.1 Early 20th Century: 1900–1913

As regards topic structure and content we can state that the nominative class represented by abstract and concrete nouns prevails in the NMF output. The nominative class includes: 1) description of people—*rebyonok* (*child*), *zhenshhina* (*woman*), *starik* (*old man*), etc.; 2) profession—*kupecz* (*merchant*), *izvozchik* (*horse-cab driver*), *soldat* (*soldier*), *krest'yanin* (*peasant*), etc.; 3) everyday realities—*komnata* (*room*), *izba* (*hut, house*), *dom* (*house, home*), *pis'mo* (*letter*), etc., 4) nature and animals—*les* (*forest*), *kust* (*bush*), *volk* (*wolf*), etc.; 5) abstract notions—*zhizn'* (*life*), *schast'je* (*happiness*), *mysl'* (*thought*), *smert'* (*death*), etc., and 6) collective nouns—*tolpa* (*crowd*).

The predicative class is represented by verbs and its amount is the biggest among other periods: *pisat'* (*to write*), *znat'* (*to know*), *lyubit'* (*to love*), *stoyat'* (*to stand*), *krichat'* (*to shout*), *bezhat'* (*to run*), *dumat'* (*to think*), *obedat'* (*to dine*), *pojti* (*to go*).

The attributive class includes qualitative *temny'j* (*dark*) and relative adjectives—*rabochij* (*working*), *russkij* (*Russian*).

The correlation of words in the topics reflects the diversity of paradigmatic and syntagmatic relations that organize the text [23, 24]. Linguistic links within the topics may be described with lexical functions in the model "Meaning ⇔ Text" [22] which allows to cover the predictable, idiomatic connections of a word and its lexical correlates.

Among paradigmatic relations in topics the following types prevail: synonyms (Syn), antonyms (Anti) and derivational (Der) relations, etc. For example: Syn: *mama* (*mom*)—*mat'* (*mother*)—*mama* (*mom*), etc.; Anti: *derevnya* (*village*)—*gorod* (*city, town*), *zhizn'* (*life*)—*smert'* (*death*), *stoyat'* (*to stand*)—*bezhat'* (*to run*), etc.; Der: *pis'mo* (*letter*)—*pisat'* (*to write*), *lyubov'* (*love*)—*lyubit'* (*to love*), *krik* (*shout*)—*krichat'* (*to shout*), etc. Partitive relations: *sem'ya* (*family*)—*rebyonok* (*child*), *muzh* (*husband*), *zhena* (*wife*), *otec* (*father*), etc., *priroda* (*nature*)—*prud* (*pond*), *reka* (*river*), *kust* (*bush*), *les* (*forest*), etc.; *derevnya* (*village*)—*izba* (*hut, house*), *krest'yanin* (*peasant*), *baba* (*country woman, peasant's wife*), *barin* (*lord*); *dom* (*house, home*)—*komnata* (*room*), *okno* (*window*), *kabinet* (*room, office*).

Syntagmatic relations are realized at the level of valency frames filled with words from the topic. Among lexical functions *Oper1,2* may be selected, which connect a verb, the name of the first or the second actant in the role of subject and the name of the situation as additions: *sup* (*soup*)—*obedat'* (*to dine*), *pis'mo* (*letter*)—*pisat'* (*to write*), *rebyonok* (*child*)—*krichat'* (*to shout*), etc. In addition, there are a number of examples for the implementation of the lexical function *Doc* (res) ("document that is the result"): *pisat'* (to write)—*pis'mo* (letter).

The topics of Period I (1900-1913) can be assigned to the following tags according to its topic words:

Topic 1 "LETTER"—*pis'mo* (*letter*), *ruka* (*hand, arm*), *pisat'* (*to write*), *znat'* (*to know*), *lyubov'* (*love*), *lyubit'* (*to love*), *komnata* (*room*), *zhenshhina* (*woman*), *scena* (*scene*), *rol'* (*role*),

Topic 2 "PRISON"—*tyur'ma* (*prison*), *zhizn'* (*life*), *my'sl'* (*thought*), *ruka* (*hand, arm*), *zvuk* (*sound*), *okno* (*window*), *kaplya* (*drop*), *smert'* (*death*), *kazalos'* (*to seem*), *slovo* (*word*)),

Topic 3 "MAINTENANCE OF ORDER IN A VILLAGE"—*soldat* (*soldier*), *barin* (*lord*), *starik* (*old man*), *oficer* (*officer*), *derevnya* (*village*), *blagorodie* (*honour*), *krest'yanin* (*peasant*), *post* (*post*), *ogonek* (*little fire*), *izba* (*hut, house*),

Topic 4 "HOUSE IN THE FOREST"—*lesa* (*forest*), *les* (*forest*), *loshad'* (*horse*), *izba* (*hut, house*), *ded* (*grandfather, old man*), *kust* (*bush*), *solnce* (*sun*), *volk* (*wolf*), *noga* (*leg*), *temny'j* (*dark*),

Topic 5 "FAMILY HAPPINESS"—*rebyonok* (*child*), *muzh* (*husband*), *zhizn'* (*life*), *drug* (*friend*), *god* (*year*), *zhena* (*wife*), *zhenshhina* (*woman*), *schast'je* (*happiness*), *kabinet* (*room, office*), *den'ga* (*money*),

Topic 6 "CROWD"—*tolpa* (*crowd*), *ulicza* (*street*), *rabochij* (*working*), *krik* (*shout*), *stoyat'* (*to stand*), *krichat'* (*to shout*), *rebyonok* (*child*), *golos* (*voice*), *drug* (*friend*), *bezhat'* (*to run*),

Topic 7 "ROMANTIC RENDEZVOUS"—*student* (*student*), *devushka* (*young lady*), *prud* (*pond*), *reka* (*river*), *lodka* (*boat*), *noch'* (*night*), *dyadya* (*uncle*), *doroga* (*road*), *mgnovenie* (*instant*), *dom* (*house, home*),

Topic 8 "PRIEST WITH PARISHIONERS IN A VILLAGE"—*batyushka* (*father, priest*), *ruka* (*hand, arm*), *zhena* (*wife*), *delo* (*affair*), *khozyain* (*landlord*), *baba* (*country woman, peasant's wife*), *pojti* (*to go*), *den'ga* (*money*), *matushka* (*mother, priest's wife*), *tetka* (*aunt*),

Topic 9 "LIFE IN A TOWN/CITY"—*russkij* (*Russian*), *parokhod* (*steamer*), *gorod* (*city, town*), *kupecz* (*merchant*), *prikazchik* (*manager*), *noch'* (*night*), *znat'* (*to know*), *starik* (old man), *izvozchik* (*horse-cab driver*), *chasy* (*clock, watch*),

Topic 10 "DINING ARRANGEMENTS"—*otecz* (*father*), *sestra* (*sister*), *mama* (*mom*), *komnata* (*room*), *doktor* (*doctor*), *dom* (*house, home*), *mat'* (*mother*), *sup* (*soup*), *obedat'* (*to dine*), *dumat'* (*to think*).

3.2 World War I and Revolutions: 1914–1922

The number of unique terms is not numerous in the second time window (1914–1922), which is due to the fact that this is a revolutionary time and the description of life is minimal (see Fig. 3).

Among paradigmatic relations there is Der: *nemecz* (*German*)—*nemeczkij* (*German*), etc.; partitive relations: *armiya* (*military*)—*oficer* (*officer*), *soldat* (*soldier*), *rota (troop)*, *komanda* *(command, team)*, etc.; *priroda* *(nature)*—*reka* *(river)*, *veter* *(wind)*, *more* *(sea)*, *nebo* *(sky)*, *bereg* *(bank)*, *derevnya* *(village)*—*izba* *(hut, house)*, *muzhik* *(country man, peasant man)*, *baba* *(country woman, peasant's wife)*, etc.; religion—*Bog* *(God)*, *svyatoj* *(saint)*, *batyushka* *(father, priest)*; *chasti tela* *(parts of body)*—*ruka* *(hand, arm)*, *noga* *(leg)*, *telo* *(body)*.

The nominative class includes: 1) description of people—*zhenshhina* *(woman)*, *starik* *(old man)*, *muzhik* *(country man, peasant man)*, *baba* *(country woman, peasant's wife)*, etc.; 2) profession—*student* *(student)*, *soldat* *(soldier)*, *oficer* *(officer)*, *batyushka* *(father, priest)*, etc.; 3) everyday realities—*komnata* *(room)*, *dver'* *(door)*, etc.; 4) nature—*veter* *(wind)*, *more* *(sea)*, *nebo* *(sky)*, *solnce* *(sun)*, *noch'* *(night)*, *bereg* *(bank)*, etc., 5) abstract notion—*zhizn'* *(life)*, 6) collective nouns—*komanda* *(command, team)*, *rota* *(troop)*.

The predicative class is represented by verbs: *znat'* *(to know)*, *dumat'* *(to think)*, *strelyat'* *(to shoot)*. The attributive class: *nemeczkij* *(German)*, *svyatoj* *(saint)*.

Topics may be labelled in the following way:

Topic 11 "LETTER-2"—*pis'mo* *(letter)*, *znat'* *(to know)*, *komnata* *(room)*, *zhizn'* *(life)*, *ruka* *(hand, arm)*, *slovo* *(word)*, *student* *(student)*, *dumat'* *(to think)*, *zhenshhina* *(woman)*, *dver'* *(door)*,

Topic 12 "WAR"—*soldat* *(soldier)*, *nemecz* *(German)*, *oficer* *(officer)*, *tovarishh* *(comrade)*, *vintovka* *(gun)*, *nemeczkij* *(German)*, *strelyat'* *(to shoot)*, *ruka* *(hand, arm)*, *komanda* *(command, team)*, *rota* *(troop)*,

Topic 13 "FAITH"—*otecz* *(father)*, *muzhik* *(country man, peasant man)*, *baba* *(country woman, peasant's wife)*, *starik* *(old man)*, *batyushka* *(father, priest)*, *Bog* *(God)*, *devka* *(maid)*, *svyatoj* *(saint)*, *izba* *(hut, house)*, *uchitel'* *(teacher)*,

Topic 14 "NATURE AND LOVE"—*ruka* *(hand/arm)*, *veter* *(wind)*, *more* *(sea)*, *nebo* *(sky)*, *solnce* *(sun)*, *noch'* *(night)*, *bereg* *(bank)*, *noga* *(leg)*, *reka* *(river)*, *telo* *(body)*.

3.3 New Time, or the Early Soviet Period: 1923–1930

In the post-revolutionary period the vocabulary increases again, there are unique words that reflect the "new life"—*tovarishh* *(comrade)*, *fabrika* *(plant)*, *rabochij* *(working)*, *Lenin* *(Lenin)*.

Among paradigmatic relations in topics prevail the following: synonyms (Syn), antonyms (Anti) and derivational (Der) relations, etc. For example: Syn: *mama* *(mom)*—*mat'* *(mother)*, *chernyj* *(black)*—*temnyj* *(dark)*, etc.; Der: *rabota* *(work)*—*rabochij* *(working)*—*rabotat'* *(to work)*, *komanda* *(command, team)*—*komandir* *(commanding officer)*, *ded* *(grandfather, old man)*—*dedushka* *(granddad, old man)*, etc.; partitive relations: *sem'ya* *(family)*—*mama* *(mom)*, *syn* *(son)*, *mat'* *(mother)*, *otec* *(father)*, *dedushka* *(granddad, old man)*; *priroda* *(nature)*—*sosna* *(pine)*, *ptica* *(bird)*, *lesa* *(forests)*, *zayac* *(hare)*, *zver'* *(animal)*, *boloto* *(swamp)*; *okhota* *(hunting)*—*ruzhje* *(rifle)*, *zver'* *(animal)*, *lesa* *(forests)*; *derevnya* *(village)*—*muzhik* *(country man, peasant man)*, *baba* *(country woman, peasants wife)*, *izba* *(hut, house)*, *telega* *(telega, horse wagon)*; *dom* *(house, home)*—*komnata* *(room)*, *dver'* *(door)*, *okno* *(window)*, *lampa* *(lamp)*, *kukhnya* (*kitchen*); *peredvizhenie na poezde* (*go by train*)—*vagon* (*coach*), *passazhir* (*passenger*), *poezd* (*train*), *stanciya* (*station*); *fabrika* (*plant*)—*rabochij* (*working*), *rabotat'* (*to work*), *stanok* (*lathe*), etc.

There are a number of examples for the implementation of the lexical function Cap: *komanda* (*command, team*)—*komandir* (*commanding officer*); *otryad* (*squad*)—*komandir* (*commanding officer*), *cerkov'* (*church*)—*pop* (*priest*), etc.; the lexical function *Equip* ("personnel, staff"): people (*folk*)—man (*country man, peasant man*), etc.

The following topics may be tagged with the following generalized labels:

Topic 15 "MEETING IN A ROOM"—*komnata* (*room*), *ruka* (*hand, arm*), *devushka* (*young lady*), *dver'* (*door*), *doktor* (*doctor*), *okno* (*window*), *khotet'* (*to want*), *lampa* (*lamp*), *tovarishh* (*comrade*), *lyubit'* (*to love*),

Topic 16 "VILLAGE"—*muzhik* (*country man, peasant man*), *baba* (*country woman, peasant's wife*), *ruka* (*hand, arm*), *loshad'* (*horse*), *izba* (*hut, house*), *telega* (*telega, horse wagon*), *noga* (*leg*), *xleb* (*bread*), *telo* (*body*), *derevnya* (*village*),

Topic 17 "MAN IN A FOREST"—*ded* (*grandfather, old man*), *bol'shoj* (*big*), *sosna* (*pine*), *ptica* (*bird*), *lesa* (*forests*), *gorbatyj* (*hump-backed*), *boroda* (*beard*), *chernyj* (*black*), *dedushka* (*granddad, old man*), *shapka* (*cap*),

Topic 18 "WAR-2"—*komissar* (*commissioner*), *komandir* (*commanding officer*), *otryad* (*squad*), *parokhod* (*steamer*), *vintovka* (*gun*), *soldat* (*soldier*), *komanda* (*command, team*), *shtab* (*headquarter*), *tovarishh* (*comrade*), *kazak* (*cossack*),

Topic 19 "WORK AT THE FACTORY"—*rabochij* (*working*), *rabota* (*to work*), *delo* (*affair*), *fabrika* (*plant*), *rabotat'* (*to work*), *tovarishh* (*comrade*), *rebyonok* (*child*), *zhit'* (*to live*), *dumat'* (*to think*), *bol'shoj* (*big*),

Topic 20 "HUNTING"—*parokhod* (*steamer*), *sneg* (*snow*), *dyadya* (*uncle*), *prostranstvo* (*space*), *temnyj* (*dark*), *zvezda* (*star*), *zayac* (*hare*), *ruzhje* (*rifle*), *bol'shoj* (*big*), *lesa* (*forests*),

Topic 21 "PEOPLE ON SWAMPS"—*starik* (*old man*), *starukha* (*old woman*), *zver'* (*animal*), *boloto* (*swamp*), *sneg* (*snow*), *passazhir* (*passenger*), *Lenin* (*Lenin*), *ruzhje* (*rifle*), *syn* (*son*), *mat'* (*mother*),

Topic 22 "FEAST IN A VILLAGE"—*pop* (*priest*), *cerkov'* (*church*), *otec* (*father*), *muzhik* (*country man, peasant man*), *telega* (*telega, horse wagon*), *narod* (*folk*), *starukha* (*old woman*), *baba* (*country woman, peasant's wife*), *svad'ba* (*wedding*), *prazdnik* (*feast*),

Topic 23 "TRAVELLING BY TRAIN"—*vagon* (*coach*), *prazdnik* (*passenger*), *poezd* (*train*), *meshok* (*bag*), *stanciya* (*station*), *masterskij* (*skillful*), *stanok* (*lathe*), *uchyonyj* (*scientist*), *ruka* (*hand, arm*), *vskochit'* (*to jump up*),

Topic 24 "MOTHER AND HER CHILD"—*mal'chik* (*boy*), *mama* (*mom*), *veshhi* (*belongings*), *rebyonok* (*child*), *starshij* (*senior*), *soldat* (*soldier*), *dom* (*house, home*), *kukhnya* (*kitchen*), *ruka* (*hand, arm*), *kurs* (*course*).

4   Comparison of the Results of Two Topic Tagging Approaches

The comparison of results of topic modelling via NMF with that of manual topic annotation performed by an expert was made by combining two types of annotation in the corpus database. As mentioned above, the experiment was conducted on the annotated Corpus of Russian short stories written in the period of 1900–1930 by 300 Russian writers, which contains 310 texts of different sizes with a total of 1000000 tokens.

The output results of topic modelling via NMF presented in the form of relational tables were imported into corpus database. The tables consist of two ranged datasets—topic terms (the words describing a topic) and topic documents (short stories relevant to this topic) for each of 24 automatically determined topics.

For this pilot experiment we decided to limit ourselves to 10 core elements on both dimensions. That is, it was agreed to consider just the first 10 top words for each topic model (as it is presented in previous section) and to leave in the correspondent cluster of documents just 10 topic-forming texts. Then, for each of 24 topics automatically determined topics (A-topics) frequency lists of topics appointed to the corresponded text documents by an expert (E-topics) were compiled. The examples of this frequency distributions are given in Table 2. The frequency indicates the number of stories out of 10, which were tagged with a correspondent tag.

Table 2. Top frequency lists of E-topics for selected A-topics.

A-Topic	E-Topic	Freq.	%
Period I: Early 20th century: 1900–1913			
Topic 1 "LETTER"	LOVE	7	21.88
	SINS	4	12.50
	RELATIONS	3	9.38
	MENTAL STATE	3	9.38
Topic 2 "PRISON"	DEATH	6	14.29
	RELATIONS	5	11.90
	POLITICAL STRUGGLE	5	11.90
	MENTAL STATE	4	9.52
Topic 4 "HOUSE IN THE FOREST"	NATURE	5	16.67
	PROGRESS	5	16.67
	MONEY	4	13.33
	SOCIAL GROUPS	3	10.00
Topic 5 "FAMILY HAPPINESS"	FAMILY	7	15.22
	LOVE	5	10.87
	RELATIONS	4	8.70
	MONEY	4	8.70
	CHILDREN	4	8.70
Topic 6 "CROWD"	LOVE	4	11.43
	DEATH	4	11.43

(*continued*)

Table 2. (*continued*)

A-Topic	E-Topic	Freq.	%
	SINS	3	8.57
	MONEY	2	5.71
Topic 10 "DINING ARRANGEMENTS"	FAMILY	7	16.67
	CHILDREN	4	9.52
	LOVE	4	9.52
	RELATIONS	3	7.14
Period II: World War I and Revolutions (1913–1922)			
Topic 11 "LETTER-2"	LOVE	7	15.91
	RELATIONS	6	13.64
	MENTAL STATE	5	11.36
	DEATH	5	11.36
Topic 12 "WAR"	WAR	8	22.86
	RELATIONS	5	14.29
	DEATH	4	11.43
	SOCIAL GROUPS	4	11.43
Topic 13 "FAITH"	RELIGION	7	24.14
	NATURE	3	10.34
	SOCIAL GROUPS	3	10.34
	WAR	2	6.90
Topic 14 "NATURE AND LOVE"	LOVE	5	13.16
	RELATIONS	3	7.89
	NATURE	3	7.89
	DEATH	3	7.89
	WAR	2	5.26
Period III: New time, or the Early Soviet Period: 1923–1930			
Topic 16 "VILLAGE"	SOCIAL GROUPS	8	18.60
	DEATH	7	16.28
	MONEY	4	9.30
	RELATIONS	3	6.98
Topic 17 "MAN IN A FOREST"	SOCIAL GROUPS	6	15.00
	NATURE	5	12.50
	DEATH	5	12.50
	RELATIONS	4	10.00

(*continued*)

Table 2. (*continued*)

A-Topic	E-Topic	Freq.	%
Topic 18 "WAR-2"	WAR	10	27.78
	RELATIONS	5	13.89
	DEATH	5	13.89
	SOCIAL GROUPS	4	11.11
Topic 19 "WORK AT THE FACTORY"	SOCIAL PROCESSES	9	19.15
	SOCIAL GROUPS	6	12.77
	RELATIONS	5	10.64
	MONEY	4	8.51
	LABOUR	4	8.51
Topic 23 "TRAVELLING BY TRAIN"	LOVE	4	11.11
	SINS	4	11.11
	DEATH	4	11.11
	WAR	3	8.33
Topic 24 "MOTHER AND HER CHILD"	FAMILY	6	12.50
	CHILDREN	5	10.42
	WAR	4	8.33
	RELATIONS	3	6.25

The obtained results are quite heterogeneous. Thus, for certain topics, there is a good correlation between key terms of A-topic (see 3.1-3.3) and E-topics, between some other topics there is almost no correlation, most of the topics are characterized by partial coincidence.

The best match observed refers to the A-topics WAR (#12 and #18), which are common for Periods II and III. The topics of NATURE, FAMILY, RELIGION could be detected rather well, too. On the other hand, there are obvious difficulties with automatic revealing of the topics DEATH, RELATIONS, and SOCIAL GROUPS.

In the first and second periods, A-topic LETTER is automatically detected with the very similar sets of E-topics. Perhaps this is a "traditional" topic of Russian prerevolutionary literature, like Pushkin's Tatiana Larina from "Eugene Onegin". Curiously, in the Soviet period, this topic is absent.

An interesting observation is that in the Period II of WWII and revolutions, when only 4 topics were detected by topic modelling, a good correlation between all A-topics and E-topics can be seen. Therefore, we may assume that experiments with the number of topics can help to achieve a better correlation between A-topics and E-topics. Moreover, it seems that the results of better matching between automatic topic annotation and the expert one should be a criterion for choosing the number of topics.

The analysis of reverse frequency lists (from E-topics to A-topics) led approximately to the same result: the topics WAR, NATURE, FAMILY, RELIGION, as well as the topic CHILDREN are well distinguished. In addition, it was noted that the topic WAR in Russian short story does stand apart—it actually does not occur with such "popular" topics like LOVE, MONEY or SINS.

A preliminary conclusion that can be made is that the highest correlation between expert and automatic topic annotation is observed for those topics that are background for the whole story (WAR, NATURE, FAMILY). Topics or events narrowly localized in the narrative space of the story (e.g., DEATH) are not captured automatically.

Poor correspondence between E-topic and A-topic is also observed when the expert did not mark as a topic the background against which the action is taking place, whereas the topic modelling algorithm proposes this background to be the central topic of the text. An example of such a mismatch is the A-topic CROWD (#6), which unites the stories tagged by an expert with E-topics LOVE, DEATH, SINS, and MONEY. Here, the *crowd* represents just the background for the story where the action takes place (e.g., at political rally or in the opera house), but it could not be considered by an expert to be a valid topic for the story in general.

5   Conclusion

For the corpus of Russian short stories, a novel approach to text annotation, namely, topic annotation was introduced. In the given research we are the first to propose and implement a novel type of text annotation—a hybrid (expert and automated) topic annotation. This type of annotation is specific for genres of fiction texts as it allows to describe relevant aspects of the plot, features characterizing main characters of the text, etc. Topic annotation proves to be a significant extension to tradition multilevel annotation schemes accepted in contemporary corpus linguistics.

The experiment aimed at comparison of results of topic modelling via non-negative matrix factorization (A-topics) with that of manual topic modelling performed by an expert (E-topics) has shown that A-topics better reveal general background of stories (e.g., WAR, LOVE, NATURE, FAMILY) than the important plot-forming topics (e.g., DEATH or RELATIONS). Therefore, topic modelling via NMF should be primarily recommended for the revealing of topics referring to general background of the story rather than for detecting topics related with some critical events or relations between characters.

Another hypothesis that can be proposed is that the experiments with the number of A-topics for each period can help to achieve a better correlation between A-topics and E-topics. The results of better matching between automatic topic annotation and the expert one should be considered to be a criterion for choosing the number of A-topics. As for revealing the wide set of other E-topics, they need further consideration and additional tuning of computational algorithms. The experiments on comparison of manual and automatic tagging of topics should be further continued as the obtained results seem to be important for the improvement of artificial technologies methods, especially for processing of texts referring to "real life", such as fiction or transcripts of everyday spoken speech.

Acknowledgements. The research is supported by the Russian Foundation for Basic Research, project #17-29-09173 "The Russian language on the edge of radical historical changes: the study of language and style in prerevolutionary, revolutionary and post-revolutionary artistic prose by the methods of mathematical and computer linguistics (a corpus-based research on Russian short stories)".

References

1. Bakhtin, M.M.: Estetika slovesnogo tvorchestva. Iskusstvo, Moscow (1979)
2. Blei, D.M., Ng, A.Y., Jordan, M.I.: Latent Dirichlet allocation. J. Mach. Learn. Res. 3(4–5), 993–1022 (2003)
3. Blummer, B., Kenton, J.M.: Academic libraries' outreach efforts: identifying themes in the literature. Public Serv. Quart. **15**(3), 179–204 (2019)
4. Bodrunova, S., Blekanov, I., Kukarkin, M.: Topic modelling for twitter discussions: model selection and quality assessment. In: Proceedings of the 6th SGEM International Multidisciplinary Scientific Conference on Social Sciences and Arts SGEM2018, Science and Humanities, pp. 207–214. STEF92 Technology Ltd, Sofia, Bulgaria (2018)
5. Daud, A., Li, J., Zhou, L., Muhammad, F.: Knowledge discovery through directed probabilistic topic models: a survey. Front. Comput. Sci. China (2010)
6. Erofeeva, A., Mitrofanova, O.: Automatic assignment of topic labels in topic models for Russian text corpora. In: Structural and Applied Linguistics, vol. 12, pp. 122–147. St. Petersburg University (2019)
7. Greene, D., Cross, J.P.: Unveiling the political agenda of the European parliament plenary: a topical analysis. In: Proceedings of the ACM Web Science Conference (WebSci'15), Oxford, UK (2015)
8. Greene, D., Cross, J.P.: Exploring the political agenda of the european parliament using a dynamic topic modelling approach. Polit. Anal. **25**(1), 77–94 (2017)
9. Iyyer, M., Guha, A., Chaturvedi, S., Boyd-Graber, J., Daumé III, H.: Feuding families and former friends: unsupervised learning for dynamic fictional relationships. In: Proceedings of the 2016 Conference of the North American Chapter of the Association of the Computational Linguistics, Association for Computational Linguistics, San Diego, California, pp. 1534–1544 (2016)
10. Kazartsev, E., Davydova, A., Sherstinova, T.: Rhythmic structures of Russian prose and occasional iambs. In: A Diachronic Case Study. SpeCom 2020. LNCS (LNAI), vol. 12335 (2020, in print). https://doi.org/10.1007/978-3-030-60276-5_20
11. Korobov, M.: Morphological analyzer and generator for Russian and Ukrainian languages. Commun. Comput. Inf. Sci. **542**, 320–332 (2015)
12. Kriukova, A., Erofeeva, A., Mitrofanova, O., Sukharev, K.: Explicit semantic analysis as a means for topic labelling. In: Ustalovet, D., et al. (eds.) Artificial Intelligence and Natural Language, vol. 930, pp. 167–177. Springer, Cham (2018). https://doi.org/10.1007/978-3-030-01204-5_11
13. Krstovski, K., Kurtz, M.J., Smith, D.A., Accomazzi, A.: Multilingual Topic Models. https://arxiv.org/pdf/1712.06704.pdf. Accessed 21 May 2020
14. Lau, J.H., Grieser, K., Newman, D., Baldwin, T.: Automatic labelling of topic models. In: Proceedings of the 49th Annual Meeting of the Association for Computational Linguistics: Human Language Technologies, vol. 1, pp. 1536–1545. Association for Computational Linguistics, Stroudsburg, PA (2011)

15. Lau, J.H., Newman, D., Karimi, S., Baldwin, T.: Best topic word selection for topic labelling, In: Proceedings of the 23rd International Conference on Computational Linguistics, pp. 605–613. Association for Computational Linguistics, Stroudsburg, PA (2010)
16. Loukachevitch, N., Nokel, M., Ivanov, K.: Combining thesaurus knowledge and probabilistic topic models. In: van der Aalst, W., et al. (eds.) Analysis of Images, Social Networks and Texts. AIST 2017. Lecture Notes in Computer Science, vol. 10716, pp. 59–71. Springer, Cham (2018). https://doi.org/10.1007/978-3-319-73013-4_6
17. Martynenko, G., Sherstinova, T.: Corpus of Russian short stories of the first third of the 20th century: theoretical issues and linguistic parameters. Strukturnaya i prikladnaya linguistika 14. St. Petersburg State University, St. Petersburg (in print)
18. Martynenko, G., Sherstinova, T.: Emotional waves of a plot in literary texts: new approaches for investigation of the dynamics in digital culture. In: Alexandrov, D., Boukhanovsky, A., Chugunov, A., Kabanov, Y., Koltsova, O. (eds.) Digital Transformation and Global Society. DTGS 2018. Communications in Computer and Information Science, vol. 859, pp. 299–309. Springer, Cham (2018). https://doi.org/10.1007/978-3-030-02846-6_24
19. Martynenko, G., Sherstinova, T.: Linguistic and stylistic parameters for the study of literary language in the corpus of Russian short stories of the first third of the 20th century. In: R. Piotrowski's Readings in Language Engineering and Applied Linguistics, Proceedings of the III International Conference on Language Engineering and Applied Linguistics (PRLEAL-2019), CEUR Workshop Proceedings, vol. 2552, pp. 105–120 (2020)
20. Martynenko, G.Y., Sherstinova, T.Y., Melnik, A.G., Popova, T.I.: Methodological issues related with the compilation of digital anthology of Russian short stories (the first third of the 20th century). In: Proceedings of the XXI International United Conference 'The Internet and Modern Society', IMS-2018, Computational linguistics and computational ontologies, Issue 2, pp. 99–104. ITMO University, St. Petersburg (2018)
21. Martynenko, G.Y., Sherstinova, T.Y., Popova, T.I., Melnik, A.G., Zamirajlova, E.V.: O printsipakh sozdaniya korpusa russkogo rasskaza pervoy treti XX veka. In: Proc. of the XV Int. Conf. on Computer and Cognitive Linguistics 'TEL 2018', pp. 180–197. Kazan Federal University, Kazan (2018)
22. Melchuk, I.A.: Experience of the Theory of the Linguistic Models "Meaning ⇔ Text". Moscow (1974/1999)
23. Mitrofanova, O.A.: Topic modelling of special texts based on LDA algorithm. In: Proceedings of XLII International Philological Conference. Selected works, pp. 220–233. St. Petersburg State University, St. Petersburg (2014)
24. Mitrofanova, O.A., Shimorina, A.S., Koltsov, S.N., Koltsova, O.Yu.: Modelling semantic links in social media texts using the LDA algorithm (based on the Russian-language segment of the LiveJournal). Strukturnaya i prikladnaya lingvistka **10**, 151–168 (2014)
25. Mitrofanova, O.A., Sedova, A.G.: Topic modelling in parallel and comparable fiction texts (the case study of english and Russian prose). In: Information Technology and Computational Linguistics (ITCL 2017), ICPS Proceedings, IMS2017: Proceedings of the International Conference IMS-2017, pp. 175–180 (2017)
26. Mitrofanova, O.A.: Topic modelling of the Corpus of 'Russian folk tales by A. N. Afanasiev'. Strukturnaya i prikladnaya linguistika **11**, 146–154 (2015)
27. Mitrofanova, O.A.: Verojatnostnoje Modelirovanije Tematiki Russkojazychnyh Korpusov Tekstov s Ispol'zovanijem Kompjuternogo Instrumenta GenSim. In: Proceedings of the International Conference 'Corpus Linguistics – 2015'. St. Petersburg State University, St. Petersburg (2015)
28. Nikolenko, S., Koltcov, S., Koltsova, O.: Topic modelling for qualitative studies. J. Inf. Sci. **43**(1), 88–102 (2017)
29. O'Callaghan, D., Greene, D., Carthy, J., Cunningham, P.: An analysis of the coherence of descriptors in topic modelling. Expert Syst. Appl. (ESWA) **42**(13), 5645–5657 (2015)

30. Panicheva, P., Litvinova, O., Litvinova, T.: Author clustering with and without topical features. In: Salah, A., Karpov, A., Potapova, R. (eds.) Speech and Computer. SPECOM 2019. Lecture Notes in Computer Science (LNAI), vol. 11658, pp. 348–358. Springer, Cham (2019). https://doi.org/10.1007/978-3-030-26061-3_36

31. Rhody, L.M.: Topic modelling and figurative language. J. Digit. Hum. **2**(1) (2012)

32. Rosen-Zvi, M., Griffiths, T., Steyvers, M., Smyth, P.: The author-topic model for authors and documents. In: Uncertainty in Artificial Intelligence, pp. 487–494 (2004)

33. Segalovich, I., Titov, V.: MyStem, https://yandex.ru/dev/MyStem/ (2011). Accessed 12 May 2020

34. Sherstinova, T., Skrebtsova, T.: Russian literature around the October revolution: a quantitative exploratory study of literary themes and narrative structure in Russian short stories of 1900–1930. In: CompLing (2020, in print)

35. Skrebtsova, T.G.: Thematic tagging of literary fiction: the case of early 20th century Russian short stories. In: CompLing (2020, in print)

36. Stockwell, P.: Cognitive Poetics: An Introduction. Routledge, London (2002)

37. Todd, R.W.: Discourse Topics. John Benjamins, Amsterdam & Philadelphia (2016)

38. Vorontsov, K., Potapenko, A.: Tutorial on Probabilistic Topic Modeling: Additive Regularization for Stochastic Matrix Factorization. In: Ignatov, D., Khachay, M., Panchenko, A., Konstantinova, N., Yavorsky, R. (eds.) Analysis of Images, Social Networks and Texts, vol. 436, pp. 29–46. Springer, Cham (2014). https://doi.org/10.1007/978-3-319-12580-0_3

39. Zamiraylova, E., Mitrofanova, O.: Dynamic topic modelling of Russian fiction prose of the first third of the XXth century by means of non-negative matrix factorization. In: R. Piotrowski's Readings in Language Engineering and Applied Linguistics, Proceedings of the III International Conference on Language Engineering and Applied Linguistics (PRLEAL-2019), CEUR Workshop Proceedings, vol. 2552, pp. 321–339 (2020)

40. Zhirmunskii V.M.: Teoriya literatury. Poetika. Stilistika. Leningrad, Nauka (1977)

41. Zholkovsky A., Shcheglov, Y.: K Ponyatiyam 'Tema' i 'Poeticheskiy Mir'. Trudy po znakovym systemam 7, 143–167. Tartu University, Tartu (1975)

Thesaurus-Based Methods for Assessment of Text Complexity in Russian

Valery Solovyev[1]([✉]), Vladimir Ivanov[2], and Marina Solnyshkina[1]

[1] Kazan Federal University, 2, Tatarstan Street,
Room 467, Kazan, The Republic of Tatarstan 420021, Russian Federation
`maki.solovyev@mail.ru, mesoln@yandex.ru`
[2] Innopolis University, st. Universitetskaya, 1, Innopolis, Republic of Tatarstan
420500, Russian Federation
`nomemm@gmail.com`

Abstract. The study explores the problem of assessing complexity of Russian educational texts. In this paper, we focus on measuring conceptual complexity which is rarely selected as a research question and propose to use a thesaurus (or a linguistic ontology) to this end. We also compiled an original corpus of school textbooks on Social Studies, History used in high school, and textbooks for elementary school specifically for this set of text complexity experiments. On the first stage of the research, RuThes-Lite thesaurus, a linguistic knowledge base with the total size of 100,000 concepts, was used to elicit concepts in the texts of schoolbooks and represent them as graphs. To the best of our knowledge, we a new method for text complexity assessment using RuThes-Lite graphs and identify graphs-based semantic characteristics of texts that impact complexity. The most significant findings of the research include identification of statistically significant correlations of the selected features, such as node degree, with complexity of educational texts.

Keywords: Text complexity · Thesaurus · Russian language

1 Introduction

Of the three generally accepted levels of text complexity, i.e. lexical, syntactic and semantic/informational/conceptual, the third one is evidently more intricate to scrutinize and is universally recognized as the least explored [21]. The semantic level of a text, defined as the amount of background knowledge required to comprehend a text, to a great extend facilitates text comprehension. Automatic measurement of lexical and syntactic complexity of Russian texts has been proposed in a number of studies [11,20,22] and it is not unexpected as these two levels are easier to formalize. The predominant approaches in previous work on Russian text complexity combine lexical and syntactic features [2]. As for the influence of the semantic level on text complexity, the studies conducted are still few and mostly in English.

L. Martínez-Villaseñor et al. (Eds.): MICAI 2020, LNAI 12469, pp. 152–166, 2020.
https://doi.org/10.1007/978-3-030-60887-3_14

In this article, we are developing a new approach to defining the conceptual complexity of texts through knowledge databases such as WordNet. The conceptual complexity of the text is viewed as the amount of knowledge (in particular, from the thesaurus) necessary for understanding a text.

Evaluation of the proposed approach requires a set of texts of different conceptual complexity. One possible way to solve the problem is to abridge original texts and use the abridged versions as part of the Corpus. This idea has been implemented in [23]. However, it seems important to assess the method on real, not artificial texts. A natural example of such a set of texts of different conceptual (not just formal, lexical and grammatical) complexity are school textbooks of different grades which we use as the material in the current study.

An adequate conceptual level of a text complexity is especially important for readers with an insufficient level of knowledge [6,18], in particular, for schoolchildren. In the situations when schoolchildren experience a lack of necessary knowledge, it may lead to difficulties in understanding textbooks. Thus, while developing educational materials for a specific audience, a designer is expected to be aware of the approximated amount of theoretical and practical knowledge the audience can employ. For this purpose we make use of the corpus of school textbooks as a kind of a reference corpus.

2   Related Works

A deeper level of semantic analysis, also referred as conceptual analysis [23], implies taking into account semantic and pragmatic links between concepts in a text. Although it is the conceptual level that presents a real complexity of a text, so far the possibility and methods of measuring the conceptual level of text complexity have remained unexplored.

The notion of the conceptual complexity of a text is viewed as related to the number of abstract concepts verbalized in a text, i.e. abstract words incidence, in a text [21]. The correlation between text abstractness and "linguistic complexity" was convincingly proved in [26] where the author used Russian texts as the material for the study. In a similar study, A. Laposhina [13] found out that only one of the four groups of text lexis which she identified with the help of ABBYY COMPRENO, i.e. words denoting abstract concepts, could be used as an indicator of text complexity. The groups A. Laposhina classified included the following: (1) 'lex_physical', i.e. nouns denoting specific material objects, including people (e.g..'cutlet', 'table', 'mom'); (2) 'lex_virtual', i.e. virtual, intangible objects, e.g.. 'base', 'internet'; (3) 'lex_abstract', words denoting abstract concepts including terms (e.g.. 'avantgarde', 'whim', 'affacement'), and (4) 'lex_substance', names of substances, e.g. .'silver', 'vinegar'. Elements of the semantic approach, similar to that in the afore-cited paper, are implemented in [17] in which the authors apply latent semantic analysis to determine semantic proximity of text fragments. However, in most studies published, researchers analyze not a corpus or a text as a whole, but adjacent sentences or paragraphs only.

An important component of text complexity is the text cohesion that has been studied in a number of studies [5,17]. In [17], the notion of cohesion is defined as a concept uniting referential cohesion and deep cohesion. The first indicates how concepts in a sentence or adjacent sentences overlap and is manifested in repeated words, stems, pronouns, etc. The second establishes cohesion due to the sequence of tenses, use of subordinate and connecting conjunctions and other means. In [17,25], cohesion is viewed as a notion verbalized with lexical chains. The latter refer to a number of adjacent words with similar meanings. Thornbury (2005) emphasizes the importance of lexical chains for maintaining text cohesion arguing that *"The lexical connectors include repetition and the lexical chaining of words that share similar meaning"* [25]. The author also provides an example of a set of isolated sentences that have matching grammatical categories, but do not form a coherent text: *"The university has got a park. It has got a modern tram system. He has got a swimming pool."*

As for semantic similarity of words, there are two different approaches to quantifying the extent to which words have similar meanings. The first approach (Latent Semantic Analysis) uses statistical characteristics of words in a text: frequency, co-occurrence. It was successfully used in a number of papers on text complexity, however, as noted above, mainly for local analysis.

Another possible approach is to explore semantic relations between words in a text. To this end, a researcher can use information presented in semantic networks. The theory of semantic networks began developing half a century ago [4] with the purpose of explaining the structure and functioning of human memory and by present there have been developed numerous models of networks and there is an extensive bibliography on the topic.

The research performed in [24] indicates that many of the semantic networks studied, including those built on the basis of psycho-semantic associative experiments, have similar statistical characteristics.

In modern studies aimed at word processing, the two most popular and more frequently used resources are thesauri (lexical ontologies), such as WordNet [7], and DBpedia (https://wiki.dbpedia.org/), i.e. a knowledge base with multiple semantic relationships between concepts. WordNet was originally created for research of human memory. This gives reason to expect that the use of structures like WordNet will make it possible to advance in understanding the problem of text comprehension complexity.

The semantic proximity of words is determined by their closeness in the structure of the knowledge base or thesaurus. Thesauri, however, are also one of the types of knowledge bases, since, unlike traditional dictionaries, they contain not only linguistic, but also extralinguistic information, i.e. world knowledge. The latter is registered in thesauri with hypo-hyperonymic connections: for example, a tomato is connected to a class of vegetables and not garments. Thus, this is a knowledge-based approach, in which the knowledge of many experts participating in the development of such resources is being exploited.

RuWordNet, the Russian language thesaurus, as other thesauri, presents the concept of "vegetables" as a member of a synonymic set, a set of hyponyms and a set of hyperonyms (https://ruwordnet.ru/ru/). In RuWordNet the concepts TOMATO and CUCUMBER are registered as co-names related to the concept of VEGETABLES. The easiest way to determine the distance between concepts in a thesaurus is to compute the number of steps from one concept to another moving along the edges of the tree of hypo-hyperonymic connections. The distance between TOMATO and CUCUMBER in RuWordNet is 2.

The Wordnet thesaurus is widely used to measure semantic proximity between words. For example, by September 25, 2019, Google Scholar registered 58900 articles mentioning Wordnet as a tool and word similarity as a research objective. Over 3150 works on the topic appeared in Google Scholar only in September 2019.

In [3], the authors offer the first systematic comparison of various measures of word proximity based on WordNet. In [19] the authors offer a broad survey of proximity metrics such as path length based measures, information content based measures, feature based measures, and hybrid measures. The innovative information theoretic approach to measure the semantic similarity between concepts of Word Net has been proposed in [9]. In [1], it was proposed to assign weights to the edges of WordNet and determine the proximity of words based on "weighted WordNet".

A comprehensive review of Russian thesauri is presented in [12]. It includes, in particular, information about RuThes and RuWordNet thesauri, i.e Russian WordNet. In [14,27], these thesauri are used to solve the problem of establishing semantic proximity of words. To the best of our knowledge, for establishing the degree of conceptual coherence of a text, and hence its complexity, this approach has been used so far in very few studies. In article [23], DBpedia was used as a knowledge base, and Newsela article corpus (https://newsela.com/data) as a corpus of texts, containing original and artificially simplified texts. Entities expressed in the text by nominal groups were mapped onto knowledge base concepts. All DBpedia concepts, together with the semantic relationships linking them to each other, form a graph. As a result, the text was displayed on a subgraph of the complete DBpedia graph. The authors of the article reviewed 13 parameters of the graph and calculated their values for original and simplified texts (total number of texts = 200). The research shows that all the parameters studied have a statistically significant relationship between metrics of the parameters of the graph and text complexity (at least in the case of significant differences in text complexity). Thus, this approach is viewed as reliable to assess texts complexity. In [10] the same authors propose a mechanism of distributing activation in a network of concepts which may be implemented to model the effect of priming. As priming is viewed as a mechanism accelerating text comprehension [8], it offers researchers another instrument to evaluate conceptual complexity of texts.

In this paper, we propose a novel approach to assess conceptual complexity of texts. It differs from the approach proposed in [23] in many significant aspects: (1) using a WordNet-like thesaurus as a knowledge base, rather than a DBpedia-type knowledge base; (2) using a set of structural features of the graph; (3) using natural texts of different conceptual complexity for testing the approach rather than artificially generated texts.

3   Materials and Methods

In this section we describe the datasets and methods used in the experiments held for the study.

3.1   Datasets

Russian Readability Corpus compiled for the current research comprises three sets of books, i.e. Social Science textbooks, History textbooks, Elementary school texts. Initially [22], Russian Readability Corpus (RRC) was compiled of two sets of textbooks on Social studies for secondary and high school for Russian students. It contained 45380 sentences from 14 textbooks: edited by Bogolyubov (BOG) and by Nikitin (NIK). Later, a dataset of 17 elementary school texts (1st – 4th grades) along with a dataset of 6 textbooks on History (10th – 11th grades) were added. The descriptive statistics of the whole corpus (of 37 books in total) is presented in Tables 1, 2 and 3.

Table 1. Properties of the pre-processed corpus in social studies. The asterisk (*) marks textbooks for advanced students.

	Tokens		Sentences		Words per sent	
Grade	BOG	NIK	BOG	NIK	BOG	NIK
5-th	–	17,221	–	1,499	–	11.49
6-th	16,467	16,475	1,273	1,197	12.94	13.76
7-th	23,069	22,924	1,671	1,675	13.81	13.69
8-th	49,796	40,053	3,181	2,889	15.65	13.86
9-th	42,305	43,404	2,584	2,792	16.37	15.55
10-th	75,182	39,183	4,468	2,468	16.83	15.88
10-th*	98,034	–	5,798	–	16.91	–
11-th	–	38,869	–	2,270	–	17.12
11-th*	100,800	–	6,004	–	16.79	–

Pre-processing of the RRC. All texts in the corpus were pre-processed in the very same manner. Pre-processing included tokenization, splitting text into sentences and Part-of-Speech tagging (using the TreeTagger for Russian[1]. During the pre-processing step we excluded all extremely long sentences (longer than 120 words) as well as too short sentences (shorter than 5 words) which we consider outliers. Thus, the pre-processing step decreases the total number of sentences.

---

[1] http://www.cis.uni-muenchen.de/~schmid/tools/TreeTagger/.

Table 2. Properties of the preprocessed corpus in history

Author	Grade	Tokens	Sentences	Words per sentence
Soboleva	10-th	81544	7116	11.46
Klimov	10-th	40949	3676	11.14
Guryanov	11-th	100331	9393	10.68
Petrov	11-th	85409	8536	10.01
Plenko	11-th	63804	5292	12.06
Ponomarev	11-th	44833	4003	11.2

Table 3. Properties of the preprocessed Corpus from elementary school

Author	Grade	Tokens	Sentences	Words per sentence
Kurevina	1-st	3300	368	8.97
Lutseva	1-st	19154	2026	9.45
Ragozina	1-st	8793	834	10.54
Rogovtseva	1-st	12763	1548	8.24
Rudskiy	1-st	1622	198	8.19
Vakhrushev	1-st	17855	2188	8.16
Rudskiy	2-nd	2285	340	6.72
Vakhrushev	2-nd	24975	2232	11.19
Rudskiy	3-rd	8089	896	9.03
Vakhrushev	3-rd	19779	1813	10.91
Benenson	4-th	5389	739	7.29
Goryachev	4-th	11099	1305	8.50
Vakhrushev	4-th	41877	3637	11.51

Clearly, extremely long or short sentences can exist in other domains, but in high school textbooks sentences shorter than 5 words most likely are outliers. Extremely short sentences mostly appear as names of chapters and sections of the textbooks or as a result of incorrect sentence splitting. Obviously, in Russian sentences with five – seven words can still be viewed as short sentences; the average sentence length (in our corpus) is higher than ten. Extremely long sentences (more than 120 words) are either errors in sentence splitting or represent uncommon situations (such as very long excerpts from juristic documents). Imprecise Part-of-Speech tagging might affect the quality of further steps of text processing.

Russian Thesaurus RuThes-Lite. RuThes Thesaurus of the Russian language [15] is typically referred to as a linguistic knowledge base for natural language processing. The thesaurus provides a hierarchical network of concepts. Each concept has a name and is related to other concepts and to a set of language signs (words and phrases), the meanings of which correspond to the concept.

The conceptual relations in RuThes include the following:

- the class-subclass relation;
- the part-whole relation;
- the external ontological dependence, and others

RuThes includes 54 thousand concepts, 158 thousand unique text entries (75 thousand single words), 178 thousand concept-text entry relations, more than 215 thousand conceptual relations. The first publicly available version of RuThes (RuThes-lite) is available from http://www.labinform.ru/ruthes/index.htm. The process of generation of the RuThes-Lite from the RuThes is described in [16].

As it was mentioned above, for this research we use RuThes-Lite thesaurus as a linguistic knowledge base to estimate text complexity. The structure of a thesaurus can be represented as a graph (G_0): with nodes derived from the concepts and edges derived from relations. While processing a text (or a fragment of a text) we match words from the text with RuThes-Lite concepts. Preforming a correct matching of thesaurus concepts with raw text entries is an important step that usually involves disambiguation processes. However, the problem of automatic disambiguation of a word sense in Russian has not been solved yet. Thus, when matching RuThes-Lite concepts with a text we use a simple string thus matching normalized words from the text and the text entries that correspond the thesaurus concepts. We keep all the matching concepts in a temporary list. This process may produce a lot of false positives in the temporary list, i.e. concepts that were not used in the text fragment. On the next step, i.e. while building a subgraph, we filter out all isolated nodes. It is during this procedure the vast majority of false positives are excluded. The procedure of building a subgraph is also straightforward: we use the matched concepts to produce a new graph G_S which is a subgraph of G_0. If two nodes are connected in G_0 then they are connected in G_S too. The complexity metrics are assessed based on the subgraph provided in Sect. 4.1. In case of a false positive match, the false positive will remain in subgraph G_S if and only if it has a connection with another falsely matched concept in the same text. So, having two interconnected false positives is still possible, but having more than three (interconnected false positives) is a much more rare event. Limitations of the described procedure are obvious as some of the isolated nodes may still contain valuable information for further analysis.

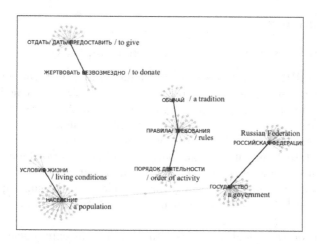

Fig. 1. A subgraph G_S derived from the RuThes-Lite and a text fragment. The subgraph contains only nodes and (hypernymy-hyponymy) edges between them (such edges are dark). Nodes without labels are kept in the figure as an illustration of the related RuThes concepts that did not appear in the text fragment, but they are linked to nodes from the subgraph G_S

3.2   Sampling from RRC

RRC contains 37 documents and thus hardly be viewed as a representative sample of the population of all school textbooks. However, for the purposes of text complexity studies we could split each document in multiple non-overlapping parts. If each part (or a sample) is 'long enough' it can serve as a good representative of the whole document and at the same time will keep certain variability. As we have no assumptions on the idea of "big enough" in terms of the sample, we will denote the size of each sample as parameter S. This parameter is measured in tokens. In experiments, both dependent variables (such as readability value) and independent variables measured for a given sample. The sample size (S) was set to different values: 200, 500 1000 and 2000 tokens. During sampling, keeping the order of tokens and sentences is important, otherwise the sampled texts will be less natural, even though they could carry the main features of the documents from the corpus. Thus, we sample S token sequences from each document[2]. We calculate all features for readability analysis using the described sampling technique. Using the technique we can estimate the mean and range of a feature metrics with several samples taken from RRC.

---

[2] The last sentence is not truncated, hence the size of a sample in experiments is at least S tokens and at most (S+k) tokens, where k tokens are used to keep the last sentence in the sample.

4   Experiments

4.1   Thesaurus-Based Features for Text Complexity

Sampling from RRC produces a set of text fragments and the terms a text sample and a text fragment are used as synonyms. Subgraph G_S is designed from each sample of the kind. An important stage of the research is pre-processing the set of nodes in graph G_S. Typically, a text contains many concepts from the thesaurus; a text fragment of 1000 tokens usually contains around 500 concepts from RuThes-Lite. Some of the text concepts could be spurious matches due to lexical ambiguity; some of the concepts found in a text could be just accidental and noisy matches. Thus, in the subgraph G we keep in the subgraph G_S only those nodes which have a connection in RuThes-Lite and remove all isolated nodes from the subgraph G_S.

Below, we describe features of G_S that we use in experiments with text complexity. In our experiments, for each G_S we calculate the following features:

- number of RuThes concepts (Co);
- number of components (NC);
- average component size (CS);
- maximum node degree (MND);
- total number of connected nodes (TCN);
- average shortest path (ASP).

Number of RuThes concepts (Co). This feature is calculated as a total number of RuThes-Lite concepts found in a text fragment. The value grows with the size of a fragment, but the proportion of concepts found in a text fragment with N tokens is approximately the same ($N/2$ concepts). All the following features are calculated on the subgraph G_S after removing all isolated nodes.

Number of components (NC). This feature is calculated as the total number of connected components in the subgraph G_S. The subgraph in Fig. 1 has NC = 4.

Average Component Size (CS). This feature is calculated as the total number of nodes in all connected components in the subgraph G_S divided by the number of components.

Maximum Node Degree (MND). This feature is calculated as a maximum number of edges incident to a node in the subgraph G_S. The subgraph in Fig. 1 has MND = 2.

Total Number of Connected Nodes (TCN). This feature is calculated as a total number of all nodes in the subgraph G_S (in Fig. 1 TCN = 9).

Average Shortest Path (ASP). This feature is calculated in the following way. We retrieve all possible shortest paths that connect nodes from the subgraph G_S in RuThes-Lite (i.e. in the graph G_0). Then we calculate average length of the path.

4.2 RuThes-Based Metrics

We computed the correlation between features based on RuThes-Lite and the grade level (Table 4). In this table we highlight the values greater than 0.7 of the correlation coefficient. For each row of Table 4 we measured 50 random samples per a textbook from the RRC.

Table 4. Pearson correlation between features based on the RuThes-Lite and the grade level

S	Co	NC	CS	MND	TCN	TCN/Co	ASP
200	0.45	**0.86**	0.32	0.53	**0.84**	**0.76**	0.27
500	0.62	**0.86**	0.68	0.45	**0.88**	**0.79**	−0.25
1000	0.56	0.38	**0.73**	0.40	**0.85**	0.69	−0.36
2000	**0.88**	0.40	**0.81**	0.68	**0.86**	**0.70**	−0.54

Box-plots of the TCN feature ("Total number of connected nodes") on Fig. 2 show that correlation is very small in the elementary school (1st - 4th grades) and becomes more obvious in the higher grades (6th – 11th grades). This trend stays the same for larger sizes of a text fragment (S = 1000) as it is shown in Fig. 3. It was expected that values of the TCN feature will depend on the size of a fragment (S). However, in contrast to the number of RuThes concepts (Co), the TCN feature demonstrated a stronger correlation with the target variable ("Grade level").

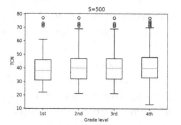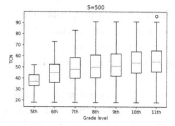

Fig. 2. Values of the "Total number of connected nodes" across grade levels (for S = 500 tokens).

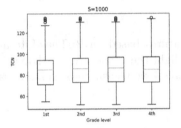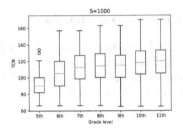

Fig. 3. Values of the "Total number of connected nodes" across grade levels (for S = 1000 tokens).

5   Discussion

As it was shown in the previous section, it is evident that, except for the average shortest path, most of the parameters under study correlate well with the grade level, i.e. text complexity (Table 4). Correlations of TCN and TCN/Co do not depend on the text size. It turned out that the level of correlation varies with the size of the excerpt processed. For the parameters "Number of concepts" and "Average component size", the level grows with the increase of the excerpt size, for the parameter "number of connected components", correlation decreases, and for the "Maximum degree of a node" it undergoes significant fluctuations with no visible tendency. For the parameter "Number of concepts", the correlation with the complexity seems to be quite natural: in the textbooks for the senior classes the information density increases. The parameter "number of connected components" has the highest correlation coefficient among all considered variations - 0.86 (if text size is 200). This also seems completely expected. Unrelated components correspond to different topics of the text. The more different themes there are in the text, the greater the cognitive work of the reader is expected as he has to memorize the themes and establish distant connections between them (in the knowledge base graph).

It is natural to compare our results with the results of [23] for the parameters studied in both papers. Our data confirm the hypothesis expressed in [23] that the number of connected components of a graph grows with text complexity. On the other hand, we did not reveal a correlation between text complexity and the parameter of "an average length of the shortest path". As for the parameter "the degree of node", ours were the opposite results to those of [23]: according to our data, "the node degree" has a direct correlation with text complexity, while according to the [23], it is reversed.

Given the numerous differences in the research methods applied, the discrepancy in the results is not unexpected. Below are the main differences in methodologies used.

1. The authors in [23] use sentences and paragraphs as text units of the analysis. In our work we explore significantly longer text excerpts: from 200 to 2000 words.

2. The authors in [23] use data corpora of different languages and, original texts were compared with artificially generated simplified texts. In our research we studied only natural texts of various levels of complexity. If we view example # 9 from Table 1 of the above work. The original sentence: "... said Mark Southerland, a private consulting ecologist who has worked with the Department of Natural Resources." was simplified to, "... said Mark Southerland, an ecologist." One way to simplify a text is to divide sentences, following the principle "one main idea per sentence". Thus, apparently, in many cases, simplification of the text was accompanied with a decrease of its length. Given that in [23] the unit of analysis was either a set of sentences or paragraphs, the values of the graph parameters were influenced not only by the conceptual complexity of the text itself, but also by the length of sentences. To eliminate this factor in our work, we selected segments of analysis of a fixed length.
3. As a knowledge base, we use the thesaurus RuThes – the Russian-language analogue of WordNet, and in [23] – the knowledge base DBpedia.
4. In our work, words of all parts of speech in the text are displayed in the thesaurus graph, while in [23] they display nouns only.
5. In [23], most of the parameters chosen for analysis are directly related to some type of text simplification. Therefore, it seems natural to expect a correlation of the selected parameters with the complexity of the texts under study. In our work there are no abridged versions of texts and, accordingly, parameters of the graph are not tied to anything. These are the parameters usually presented and studied in mathematical graph theory and network analysis literature. There are other less significant technical differences, too.

Direct extrapolation of the results received in [23] onto real texts are unlikely to result in a success due to the automated enforced 'simplification' the authors of the research omplemented. In our opinion, high values of the "node degree" demonstrated in simple texts in [23] may be features of smaller sizes of simplified and shortened texts. In short texts (a la Twitter or TV news ticker), a researcher has to limit himself to the most important concepts. In longer texts (e.g.. The Forsyte Saga), it is impossible to write only about the most important things: nominations of significant concepts function in texts inseparably, i.e. connected with a variety of secondary ones.

6   Conclusion

Text complexity is of utmost importance both for textbooks authors and students looking for educational materials. Modern methods and approaches of artificial intelligence, including knowledge bases, allow to assess conceptual complexity of texts thus providing educators and students with instruments they need. The method of displaying texts on the hierarchical structure of a thesaurus (lexical ontology) is also used for the purpose of defining various characteristics of the concept graph thus obtained. We present a number of characteristics of the graph correlating with text complexity.

In this study we explore 6 typical parameters of graphs, 5 of which are proved to correlate with the conceptual complexity of the texts (greater than 0.7) and, thus, can be used to assess conceptual complexity of texts. One of the parameters has high correlation with the conceptual complexity (greater than 0.8) across different sizes of texts. For three of these parameters, we also revealed a dependence on the sizes of the analyzed text fragments. Thus, the specific methodology for assessing the conceptual complexity of texts should be further refined in the light of the revealed patterns. Due to the limited number of textbooks in one subject area, statistical analysis was not implemented. However, the general patterns can be traced very clearly.

The combined application of methods of computational linguistics and artificial intelligence can be successfully used when determining text complexity and thus contribute to significant progress in understanding the complexity at a deeper conceptual level. Though our study was conducted on the material of the Russian language only, it is expected that similar results can be achieved for other languages with the help of the same methods as on the conceptual level, languages reflect the real world using the same cognitive mechanisms. With the help of WordNet, the same approach can be applied to English and other languages. Thus, while switching to another language the researcher changes nothing but the linguistic ontology and low level pre-processing programs.

Many research questions still present a niche. The perspective of the study lies in applying this approach to the knowledge base DBPedia and comparing the effectiveness of lexical ontologies and knowledge bases in performing the same task. An increase in the number of graph parameters may allow to detect new relevant complexity features. Topic modeling methods can be combined with information from thesauri in hierarchical Topic models. The application of this approach is also a matter of future research. Finally, we plan to expand the size of the corpus and apply this approach to the texts of other genres, styles and target audience. Our work in a number of aspects creates a benchmark, which can be used in further studies of conceptual complexity of texts.

Acknowledgements. The study was supported by the Russian Science Foundation, grant 18-18-00436.

References

1. Ahsaee, M.G., Naghibzadeh, M., Naeini, S.E.Y.: Semantic similarity assessment of words using weighted wordnet. Int. J. Mach. Learn. Cybern. **5**(3), 479–490 (2014)
2. Biryukov, B., Tyukhtin, B.: O ponyatii slozhnosti [about the concept of complexity]. V kn.: Logika i metodologiya nauki. Materialy IV Vsesoyuznogo simpoziuma, pp. 219–231 (1967)
3. Budanitsky, A., Hirst, G.: Evaluating wordnet-based measures of lexical semantic relatedness. Comput. Linguist. **32**(1), 13–47 (2006)
4. Collins, A., Quillian, M.: Retrieval time from semantic memory. J. Verbal Learn. Verbal Behav. **8**, 240–248 (1969)

5. Crossley, S., McNamara, D.: Text coherence and judgments of essay quality: models of quality and coherence. In: Proceedings of the Annual Meeting of the Cognitive Science Society, vol. 33 (2011)

6. Denton, C.A., et al.: Text-processing differences in adolescent adequate and poor comprehenders reading accessible and challenging narrative and informational text. Read. Res. Quart. **50**(4), 393–416 (2015)

7. Fellbaum, C., (ed.) WordNet: An Electronic Lexical Database. MIT Press, Cambridge (1998)

8. Gulan, T., Valerjev, P.: Semantic and related types of priming as a context in word recognition. Rev. Psychol. **17**(1), 53–58 (2010)

9. Hong-Minh, T., Smith, D.: Word similarity in wordnet. In: Bock, H.G., Kostina, E., Phu, H.X., Rannacher, R. (eds.) Modeling, Simulation and Optimization of Complex Processes, pp. 293–302. Springer, Heidelberg (2008). https://doi.org/10.1007/978-3-540-79409-7_19

10. Hulpus, I., Štajner, S., Stuckenschmidt, H.: A spreading activation framework for tracking conceptual complexity of texts. In: Proceedings of the 57th Conference of the Association for Computational Linguistics, pp. 3878–3887 (2019)

11. Ivanov, V., Solnyshkina, M., Solovyev, V.: Efficiency of text readability features in Russian academic texts. Komp'juternaja Lingvistika i Intellektual'nye Tehnologii **17**, 277–287 (2018)

12. Lagutina, N.S., Lagutina, K.V., Adrianov, A.S., Paramonov, I.V.: Russian language thesauri: automated construction and application for natural language processing tasks. Modelirovanie i Analiz Informatsionnykh Sistem **25**(4), 435–458 (2018)

13. Laposhina, A.: Relevant features selection for the automatic text complexity measurement for Russian as a foreign language. In: Computational Linguistics and Intellectual Technologies: Papers from the Annual International Conference "Dialogue", pp. 1–7 (2017)

14. Loukachevitch, N., Alekseev, A.: Summarizing news clusters on the basis of thematic chains. In: Ninth International Conference on Language Resources and Evaluation (LREC-2014), pp. 1600–1607 (2014)

15. Loukachevitch, N., Dobrov, B.: RuThes linguistic ontology vs. Russian wordnets. In: Proceedings of the Seventh Global Wordnet Conference, pp. 154–162 (2014)

16. Loukachevitch, N., Dobrov, B., Chetviorkin, I.: Ruthes-lite, a publicly available version of thesaurus of Russian language ruthes. In: Computational Linguistics and Intellectual Technologies: Papers from the Annual International Conference "Dialogue", vol. 2014 (2014)

17. McNamara, D., Graesser, A., McCarthy, P., Cai, Z.: Automated evaluation of text and discourse with Coh-Metrix. Cambridge University Press, New York (2014)

18. McNamara, D.S., Graesser, A., Louwerse, M.M.: Sources of text difficulty: across genres and grades. In: Advances in How We Assess Reading Ability, Measuring Up, pp. 89–116 (2012)

19. Meng, L., Huang, R., Gu, J.: A review of semantic similarity measures in wordnet. Int. J. Hybrid Inf. Technol. **6**(1), 1–12 (2013)

20. Reynolds, R.: Insights from Russian second language readability classification: complexity-dependent training requirements, and feature evaluation of multiple categories. In: Proceedings of the 11th Workshop on Innovative Use of NLP for Building Educational Applications, pp. 289–300 (2016)

21. Solnyshkina, M., Kiselnikov, A.: Slozhnost' teksta: etapy izucheniya v otechestvennom prikladnom yazykoznanii [text complexity: study phases in Russian linguistics]. Vestnik Tomskogo gosudarstvennogo universiteta. Filologiya [Tomsk State Univ. J. Philol.] **6**, 38 (2015)

22. Solovyev, V., Ivanov, V., Solnyshkina, M.: Assessment of reading difficulty levels in Russian academic texts: approaches and metrics. J. Intell. Fuzzy Syst. **34**(5), 3049–3058 (2018)

23. Štajner, S., Hulpus, I.: Automatic assessment of conceptual text complexity using knowledge graphs. In: Proceedings of the 27th International Conference on Computational Linguistics. Association for Computational Linguistics, pp. 318–330 (2018)

24. Steyvers, M., Tenenbaum, J.B.: The large-scale structure of semantic networks: statistical analyses and a model for semantic growth. Arxiv preprint cond-mat/0110012 (2001)

25. Thornbury, S.: Beyond the sentence: introducing discourse analysis. ELT J. **60**(4), 392–394 (2006)

26. Tomina, Y.A.: Ob'ektivnaya otsenka yazykovoy trudnosti tekstov (opisanie, povestvovanie, rassuzhdenie, dokazatel'stvo) [an objective assessment of language difficulties of texts (description, narration, reasoning, proof)]. Abstract of Pedagogy Cand. Diss, Moscow (1985)

27. Ustalov, D.: Concept discovery from synonymy graphs. Vychislitel'nye tekhnologii [Comput. Technol.] **22**, 99–112 (2017)

Evaluating Pre-trained Word Embeddings and Neural Network Architectures for Sentiment Analysis in Spanish Financial Tweets

José Antonio García-Díaz[1]([✉])[iD], Oscar Apolinario-Arzube[2][iD], and Rafael Valencia-García[1][iD]

[1] Facultad de Informática, Universidad de Murcia, Campus de Espinardo, 30100 Murcia, Spain
{joseantonio.garcia8,valencia}@um.es
[2] Facultad de Ciencias Matemáticas y Físicas, Universidad de Guayaquil, Cdla. Universitaria Salvador Allende, 090514 Guayaquil, Ecuador
oscar.apolinarioa@ug.edu.e

Abstract. Sentiment Analysis supports decision making in the financial domain by gaining rapid insights regarding brand image as well as the perceived quality of their services and products in the market. The state of the art regarding Sentiment Analysis consists of using pre-trained word embeddings from large unannotated corpora in order to capture rich and meaningful properties of the words together with their semantic relationships. These rich semantic representations are used to feed a neural network in order to learn to distinguish among positive, negative or neutral texts. However, although pre-trained word embeddings have been applied to different domains and languages, as far as our knowledge goes, there are no studies regarding their reliability applied to the financial domain in Spanish. Consequently, we compiled and labelled a corpus composed of 7,435 tweets from economists and financial news sites and we evaluated the performance of different pre-trained word embeddings, some well-known neural network architectures and linguistic features. Our results indicate that the fastText model, trained with the Spanish Unannoted Corpora and in conjunction with linguistic features, achieved the best accuracy of 58.036% using a Gated Recurrent Unit. As an extra contribution, the compiled corpus was released to the scientific community.

Keywords: Sentiment analysis · Deep learning · Word embeddings

1 Introduction

Assessing the mood and attitude of the general public towards the financial market can help identify new opportunities that can benefit active traders and long-term investors. In this sense, social networks are a popular meeting place in which users share their personal interests and tastes. This fact has lead to companies

© Springer Nature Switzerland AG 2020
L. Martínez-Villaseñor et al. (Eds.): MICAI 2020, LNAI 12469, pp. 167–178, 2020.
https://doi.org/10.1007/978-3-030-60887-3_15

to continuously monitor social networks in order to gain better understanding regarding the interests of potential customers as well as a mean to strengthen their presence on the Internet for a more direct communication strategy [1]. However, extracting data from social networks is not an easy task as most data sources, such as user reviews, blog posts, or opinion pieces, are generally stored in an unstructured manner. Fortunately, recent advances on Natural Language Processing (NLP) and deep learning have mitigated these barriers.

For a computer to process text written in natural language, text must be encoded in a mathematical language. The current state of the art regarding text representation are word embeddings, a technique in which sentences and words are encoded as dense vectors which contains semantic and contextual information [2]. Word embeddings provide a mean in which knowledge can be transferred from a domain to another, so word vectors can be learned from unannotated large corpora and use this knowledge to solve other NLP tasks. It is possible to find pre-trained word embeddings from high-resource languages such as English [3]. In case of Spanish, although the number of available resources it is not so large, it is possible to find some resources which have been trained from social networks, free encyclopedias, or news sites.

In this paper we compare some available Spanish pre-trained word embeddings in a Sentiment Analysis task regarding the financial domain. Regarding NLP, the financial domain is challenging because some words and expressions depends heavily on the context; so it is possible to find texts with similar words but with opposite sentiments. For example, the Spanish sentences: "sube el desempleo" (*unemployment rises*), "sube la prima de riesgo" (*the risk premium rises*), and "suben las acciones en la empresa" (*shares in the company rise*) share the work "suben" (*to rise*), but they commonly represent subjective opposite facts. Moreover, what annotators know about the financial domain greatly influences the examples used to train a supervised machine learning classifier.

The rest of the paper is organised as follows. Section 2 contains background information regarding Sentiment Analysis. Section 3 describes the corpus compilation process and summarises some relevant statistics. Then, the results of the pre-trained word embeddings combined with different neural network architectures and linguistic features are described in Sect. 4 and discussed in Sect. 5. Finally, the reader can find the conclusions of the paper as well as some further work lines in Sect. 6.

2   Background Information

Sentiment Analysis (SA) is the NLP task in which the subjective sentiment of a piece of writing is obtained [4]. According to the degree of specificity needed, SA can be classified into three main categories. The most basic is known as the document-level, in which each document is labelled with a single sentiment. Document-level is accurate on small documents and provides general insights concerning the user's attitudes. However, there are documents in which more detailed insights can be extracted. For example, in product reviews found on specialised magazines or online stores, it is possible to find different opinions, and even contradictory ones, about the same product or service. A straightforward method to obtain more detailed insights is known as the sentence-level, in which the document is divided into sentences, and a sentiment is calculated for each one. However, sentence level approaches require manual revision in order to determine what are the subtopics the users are talking about. This inconvenient can be solved applying Aspect-based Sentiment Analysis (ABSA), the most sophisticated approach for conducting SA, in which texts are divided into subtopics and a sentiment is assigned for each one [5].

In order to extract the polarity of a text, there are two typical approaches: (1) lexicon based methods, and (2) machine learning [6]. On the one hand, lexicon based methods rely on calculating the polarity from specific sentiment words in the text, in which a polarity was calculated beforehand. These keywords, and their respective polarities, are typically obtained from available resources such as SentiWordNet [7]. These approaches are straightforward to apply but they are weak against certain linguistic phenomena, such as polysemy or ambiguity. To solve these issues, some authors have proposed the usage of domain specific lexicons, such as the work described in [8]. Machine learning based approaches, on the other hand, rely on building a machine learning classifier that learns how to distinguish among positive, negative and neutral documents. These approaches require examples annotated beforehand, which is a very time and effort consuming task.

To be able to apply SA applying machine learning techniques, it is necessary to extract meaningful features from the documents. This process is known as feature engineering and these features can generally be categorised as: (1) statistical, (2) linguistic, and (3) contextual. First, statistical features consist of representing word distributions. Bag of Words, for example, is a popular model in which a text is represented as the frequency of the words within a vocabulary. The state of the art regarding statistical features are word embeddings, that are a novel representation technique in which words are represented as dense vectors. Words with similar semantics are clustered together and the distance between clusters capture interesting semantic relationships. Second, linguistic features are related to linguistic phenomena such as stylistic features. The major drawback of these features are that they are language dependant and the linguistic are not easily shared between languages and cultures. In order to extract linguistic features, a common approach is to use the Linguistic Inquiry and Word Count (LIWC) [9] tool which categorised words into content and style words. LIWC has been tested on complex text classification tasks, such as satire detection [10]. Finally, contextual features include external information such as author gender or the time in which the text was written. However, as these features are not always available most of the time, it is more difficult to find studies that use them.

In the financial domain SA can be applied in several areas, such as market research, product reputation, customer experiences or market research, just to name a few. It is possible to find in the bibliography some works regarding this domain, such as the described in [11], in which the authors conducted a SA experiment regarding market analysis and they found a strong correlation between market and public sentiments. Another example can be found in [12], in which the authors analysed tweets between 2010 and 2011 regarding volatility, trading volume and stock prices. These tweets were compiled from big cap technological stocks. The authors found a high correlation between the extracted sentiments and stock prices. In the same line, in [13] the authors proposed a predictive model based on sentiments from news articles as well as other indicators and metrics from a fundamental analysis perspective.

3   Corpus

The corpus used in this experiment was compiled from Twitter, a micro-blogging social network in which users can send and receive short posts called tweets of less than 280 characters. The way in which tweets are shared between users are based in a subscription mechanism in which a user can follow others. The popularity of this social network has led many organisations to create Twitter accounts to communicate with their audience. Twitter is widely used as data source for conducting SA. Specifically in the financial domain, it is possible to find research regarding marketing and stock prediction [14–16].

To compile the corpus we selected ten accounts from popular Spanish economists and ten accounts from news sites focused on economic news. The tweets were compiled between November 2017 and December 2107 and were manually labelled by four different annotators with the following labels: *very-positive, positive, neutral, negative, very-negative, out-of-domain*, and *do-not-know-do-not-answer*. A tweet can be receive labels from different annotators; however, not all tweets were rated by all the annotators. Specifically, the annotators performed a total of 11,793 annotations achieving a Krippendorff's alpha of 0.73% [17]. An example of a tweet from the corpus is shown in Fig. 1[1]. It is worth noting that a first version of this corpus was released in [18] that consisted in 1000 positive and 1000 negative tweets. In that experiment, the authors performed a SA binary classification applying Support Vector Machines achieving a F1 measure of 73.2% with a combination of linguistic and statistical features.

To avoid a strong label imbalance, we discarded those tweets labelled as *out-of-domain* and we merged the tweets labelled as *positive* with the tweets labelled *very-positive* as well as those tweets labelled as *negative* with the tweets labelled as *very-negative*. It is worth noting that if a tweet received two scores, one positive and one negative, the tweet was considered *neutral*. At the end of this process we obtained 7435 tweets: 2741 positive tweets, 2045 neutral tweets, and

---

[1] In English: Opening — The Ibex 35 starts on Tuesday aimlessly and with little change.

elEconomista.es ✓
@elEconomistaes

#Apertura | El Ibex 35 arranca el martes sin rumbo fijo y sin apenas cambios bit.ly/2gBqoyC

9:01 a. m. · 5 sept. 2017 · Twitter Web Client

Fig. 1. An example of the corpus

2649 negative tweets. The corpus is available at http://pln.inf.um.es/corpora/economics/economics-2020.rar and it contains the Twitter's IDs of the tweets divided into three files: positive, neutral and positive. We only included the Twitter IDs accordingly to Twitter guidelines (https://developer.twitter.com/en/developer-terms/more-on-restricted-use-cases) so the authors of the posts can conserve the rights to delete their content on the internet.

4   Results

In this experiment we evaluate the reliability of different pre-trained word embeddings and neural network architectures. The Spanish pre-trained word embeddings are word2vec, GloVe, and fastText. Word2Vec was the first proposal regarding word embeddings. It is based on two different approaches: skip-grams, which objective is the prediction of context words based on the target word and a sliding window; and CBoW, which objective is the prediction of the current words give context words [19]. However, word2vec does not handle polysemy so the resulting vector of a word is the same regardless of its context. In addition, word2vec does not handle unknown words. Some approaches have arisen in order to solve these drawbacks. GloVe, for example, extracts the meaning of the words from word-word co-occurrences [20]. This learning strategy captures sub-linear relationships in the vector space which provides a better representation of word analogies. Other approach is fastText, which is more similar to Word2Vec but with the difference that fastText handles each word as a set of char-n-grams [21]. In this sense, a vector of a word is made up as a combination of its different char-n-grams. Taking into account char-n-grams for learning word embeddings cause that fastText can generalised better for made-up words, misspellings or rare words. For this experiment, we identified two versions of the (1) the Spanish Unannoted Corpora (SUC) [22], trained with fastText and word2vec, and the (2) Spanish Billion Word Corpus and Embeddings (SBWC) [23], trained with fastText and GloVe. In addition to word embeddings, we also evaluate the reliability of applying linguistic features trained with a Multilayer Perceptron (MLP). Linguistic features were extracted with UMUTextStats [24], a tool inspired in LIWC that handles 317 linguistic features including stylistic and morphological features, a fine-grained detail of part of speech categories and subcategories and different sentiment lexicons.

The neural network architectures evaluated are: (1) Convolutional Neural Network (CNN), (2) Bidirectional Long-Short Term Memory (BiLSTM), and (3) Gated Recurrent Unit (GRU). CNNs are widely applied in computer vision tasks but they have also shown good results in text classification. The key advantages of CNNs in text classification rely in the fact that CNNs handle word features across the spatial dimension; that is, meaningful joint words regardless their position in the sentence. Both GRUs and BiLSTM, on the other hand, are Recurrent Neural Networks (RNNs). RNNs handle word embeddings as a sequence, so they take into account the temporal dimension. They solve the short-term memory in different ways by incorporating a cell state and various gates in case of LSTM and with a reset gate and a update gate in case of GRU [25]. In this work we applied an special case of LSTM called BiLSTM in which the network is connected both to previous and next words.

We use TensorFlow [26] and Keras [27] in order to create the deep learning classifiers. We took the 75% of the corpus for training, leaving the rest to validating it. Prior to evaluate the pre-trained word embeddings and neural network architectures, we performed a scan to tuning some of the hyper-parameters of the neural networks, including (1) the size batch_size, selecting between 16, 32 and 64; (2) the learning rate, with values between a 0.1 and 1; and (3) the relevance of varying the weights of the pre-trained word embeddings during training. These hyper-parameters were evaluated with the Talos tool [28]. We set a maximum number of epochs of 100 along with an early stopping mechanism to avoid time wasting in unpromising combinations of features. After the analysis of the results of the scan, we selected a batch size of 64, a fixed learning rate of 0.7, and Adam as optimisation function [29].

The results of the experiments are grouped according to the pre-trained word embeddings used, namely, (1) SUC with FastText, (2) SBWC with Word2Vec, (3) SBWC with GloVe, and the (4) Spanish model from fastText trained from the Spanish Wikipedia. Each table contains the neural network architecture, the feature set including the pre-trained word embeddings (PTWE) and Linguistic Features (LF), the accuracy metric (see Eq. 1) of each model and the loss function based on the categorical_cross entropy. In cases when linguistic features were analysed in isolation, we used a Multilayer Perceptron (MLP).

$$Accuracy = TP + TN/(TP + TN + FP + FN) \qquad (1)$$

From the results obtained with SUC and FastText (see Table 1), it can be observed that the best accuracy is 58.036%, applying GRU and combining PTWE with LF. It draw our attention that the results in which LF and PTWE are combined are less uniform that the ones in which only word embeddings are applied. In this sense, GRU, BiLSTM, and CNN applying only PTWE, obtained similar accuracy (55.912%, 56.766%, and 56.290% respectively). However, when looking the results of the same classifiers but combining PTWE with the LF, the accuracy drops significantly. For example, CNN decreases the

Table 1. Results obtained with SUC trained with FastText

Architecture	Features	Loss	Accuracy
GRU + MLP	PTWE + LF	1.11228	**0.58036**
BiLSTM + MLP	PTWE + LF	1.05159	0.51426
CNN + MLP	PTWE + LF	1.09002	0.47648
GRU	PTWE	1.01902	0.55912
BiLSTM	PTWE	1.00396	0.56766
CNN	PTWE	1.03280	0.56290
MLP	LF	1.13923	0.39749

accuracy from 56.290% to 47.648%. This fact can be explained due to the poor reliability obtained with LF in isolation, which achieved an accuracy of 39.749%.

Table 2. Results obtained with SBWC with Word2Vec

Architecture	Features	Loss	Accuracy
GRU + MLP	PTWE + LF	1.06431	0.50147
BiLSTM + MLP	PTWE + LF	1.08714	0.48561
CNN + MLP	PTWE + LF	1.14133	0.42439
GRU	PTWE	1.02888	0.55074
BiLSTM	PTWE	1.07427	0.43809
CNN	PTWE	1.04751	**0.56572**
MLP	LF	1.11411	0.39222

Then, we calculate the accuracy with SBWC trained with Word2Vec (see Table 2). We can observe that the best accuracy is obtained with CNN with PTWE, achieving an accuracy of 56.572%. The second best experiment is GRU (also without LF) with an accuracy of 55.074%. Regarding the combination of PTWE and LF, GRU is the deep learning architecture with greater accuracy.

Table 3. Results obtained with SBWC with GloVe

Architecture	Features	Loss	Accuracy
GRU + MLP	PTWE + LF	1.05984	0.49095
BiLSTM + MLP	PTWE + LF	1.04625	0.49547
CNN + MLP	PTWE + LF	1.06644	0.50024
GRU	PTWE	1.01865	**0.57431**
BiLSTM	PTWE	1.00552	0.55337
CNN	PTWE	1.00380	0.56389
MLP	LF	1.13639	0.40306

The next pre-trained word embeddings analysed is SBWC with GloVe (see Table 3). We obtained the best accuracy of 57.431% by using GRU with PTWE. CNN and BiLSTM obtained a similar accuracy of 56.389% and 55.337% respectively without LF. In this case, the combination of LF reduced the accuracy for all the architectures to values near the 50% of accuracy.

Table 4. Results obtained with the Spanish model from fastText trained from the Spanish Wikipedia

Architecture	Features	Loss	Accuracy
GRU + MLP	PTWE + LF	1.14064	0.51617
BiLSTM + MLP	PTWE + LF	1.10796	0.43322
CNN + MLP	PTWE + LF	1.14406	0.47580
GRU	PTWE	1.03338	**0.57560**
BiLSTM	PTWE	1.03494	0.53716
CNN	PTWE	1.03728	0.56402
MLP	LF	1.13676	0.39721

Finally, in Table 4 we can observe the results of the Spanish pre-trained model of fastText. In this case, GRU and CNN, both without LF, get the best two results with an accuracy of 57.560% and 56.402% respectively. When looking the results of combining PTWE and LF, we can observe that GRU is the deep learning architecture which benefits most although the accuracy is lower than without LF.

5   Analysis

After the analysis of the results described in Sect. 4, we achieve the following insights:

– **Linguistic Features are Not Relevant for SA in the Financial Domain.** Although the best accuracy was achieved merging PTWE from fastText trained with SUC combined with LF (see Table 1), this behaviour was not observed in the rest of the architectures and feature sets. Moreover, we observed that the combination of LF and PTWE downplays in the majority of the cases the accuracy of the same architecture but only with PTWE. In order to understand the poor reliability of the LF we calculated the Information Gain (IG) [30] from the whole data set. This metric provides the weight in which each LF contributes in the class prediction. We can observe in Fig. 2 that those features related to negative sentiments are the ones which provided more information for sentiment classification. We also noticed that there is an important difference between the best LF and the second best LF. When looking the rest of the LFs, we observed that the presence of the symbol %

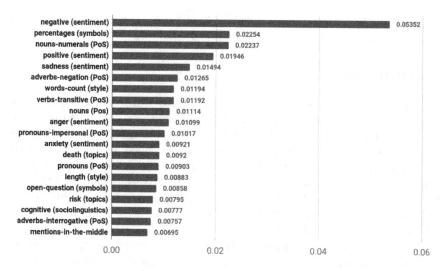

Fig. 2. Information gain of the twenty best linguistic features.

and numerals are important to determine the class; but as the IG values are obtained with the whole corpus (instead of the training set) we argue that these values provide useless or contradictory information according to the training and test split. For example, the model can learn incorrectly to classify as negative tweets with facts and numbers; whereas the test dataset contains tweets with facts and numbers classified as positive. Anyway, one of the benefits of applying LF is that they provide interpretability and we can find, therefore, the presence of interesting features: topics such as risk or death, other sentiments such as anxiety, sadness, or positive sentiments, and the usage of the open interrogation symbol, which suggests formal communication.

– **Deep Neural Networks Based on RNNs Provided Better Accuracy than CNNs.** As far as our knowledge goes, it is not clear in which cases CNNs can perform better than RNNs in text classification. In some comparisons and reviews, such as [31] or [32], the authors observed than some architectures based on RNNs perform slightly better than others based on CNN but it is not possible to generalise this behaviour. In our experiment, the results obtained with Word2Vec with SBWC obtained the best accuracy but with the rest of the pre-trained word embeddings GRU achieved better accuracy.

– **SUC Trained with fastText Provided the Best Results.** We calculated the average accuracy of each model PTWE and observed that fast-Text with SUC obtained an average accuracy of 52.261%. The next PTWE ordered by average accuracy are: SBWC, with GloVe with an average accuracy of 51.161% accuracy; the Spanish pre-trained model from fastText, which an average accuracy of 49.988% and the worst average accuracy is 47.974, obtained by SBWC with Word2Vec. Spanish is as agglutinative language and makes an intensive use of prefixes, grammatical gender, gender and number agreement. This fact explains why fastText provides better accuracy than word2vec, because it takes into account char-n-grams.

6   Conclusions

In this paper, a evaluation of different pre-trained word embeddings, several deep learning architectures, and different feature sets, was performed in a case-study involving Sentiment Analysis in the financial domain. Moreover, a corpus composed of tweets related to economics is described and shared with the community. Our results show that deep learning architectures based on GRU and pre-trained word embeddings learned with the Spanish Unannoted Corpora achieved an accuracy of 58.036% with a Gated Recurrent Unit and Linguistic Features. However, these results should be viewed with caution due to the complexity of the financial domain.

As further work we will evaluate the reliability of combining all the pre-trained word embeddings by averaging their weights as suggested in [33]. Other line of improvement is to put the focus in the linguistic features, which they do not improve the accuracy of the pre-trained word embeddings. However, it is possible to find in the bibliography that the combination of linguistic features and word embeddings generally increases the classification accuracy. In this sense, we will focus in performing a more detailed linguistic analysis of the corpus in order to determine why the linguistic features do not provide information in this domain and how this problem can be resolved. In addition, we will try to explore different combinations of layers and activation functions for the MLP. Finally, regarding the corpus quality, not all the tweets were labelled for all volunteers. We consider that it would be interesting to observe if the results remain stable as well as what are the best classifier and word embeddings model if we can provide an online tool that allows to other volunteers to classify and then analyse the corpus based on the inter-agreement score.

Acknowledgements. This work has been supported by the Spanish National Research Agency (AEI) and the European Regional Development Fund (FEDER/ ERDF) through projects KBS4FIA (TIN2016-76323-R) and LaTe4PSP (PID2019-107652RB-I00). In addition, José Antonio García-Díaz has been supported by Banco Santander and University of Murcia through the Doctorado industrial programme.

References

1. Zaglia, M.E.: Brand communities embedded in social networks. J. Bus. Res. **66**(2), 216–223 (2013). http://www.sciencedirect.com/science/article/pii/S014829631200210X
2. Conneau, A., Kruszewski, G., Lample, G., Barrault, L., Baroni, M.: What you can cram into a single vector: probing sentence embeddings for linguistic properties. CoRR abs/1805.01070 (2018). http://arxiv.org/abs/1805.01070
3. Wang, A., Singh, A., Michael, J., Hill, F., Levy, O., Bowman, S.R.: GLUE: a multi-task benchmark and analysis platform for natural language understanding. CoRR abs/1804.07461 (2018). http://arxiv.org/abs/1804.07461

4. Bakshi, R.K., Kaur, N., Kaur, R., Kaur, G.: Opinion mining and sentiment analysis. In: 2016 3rd International Conference on Computing for Sustainable Global Development (INDIACom), pp. 452–455. IEEE (2016)
5. Schouten, K., Frasincar, F.: Survey on aspect-level sentiment analysis. IEEE Trans. Knowl. Data Eng. **28**(3), 813–830 (2015)
6. Kolchyna, O., Souza, T.T.P., Treleaven, P.C., Aste, T.: Twitter sentiment analysis: Lexicon method, machine learning method and their combination. CoRR abs/1507.00955 (2015). http://arxiv.org/abs/1507.00955
7. Baccianella, S., Esuli, A., Sebastiani, F.: Sentiwordnet 3.0: an enhanced lexical resource for sentiment analysis and opinion mining. In: Lrec, vol. 10, pp. 2200–2204 (2010)
8. Ruiz-Martínez, J.M., Valencia-García, R., García-Sánchez, F., et al.: Semantic-based sentiment analysis in financial news. In: Proceedings of the 1st International Workshop on Finance and Economics on the Semantic Web, pp. 38–51 (2012)
9. Tausczik, Y.R., Pennebaker, J.W.: The psychological meaning of words: liwc and computerized text analysis methods. J. Lang. Soc. Psychol. **29**(1), 24–54 (2010)
10. del Pilar Salas-Zárate, M., Paredes-Valverde, M.A., Rodriguez-García, M.Á., Valencia-García, R., Alor-Hernández, G.: Automatic detection of satire in twitter: a psycholinguistic-based approach. Knowl. Based Syst. **128**, 20–33 (2017). https://doi.org/10.1016/j.knosys.2017.04.009
11. Mittal, A., Goel, A.: Stock prediction using twitter sentiment analysis. Standford University, CS229 15 (2012)
12. Rao, T., Srivastava, S.: Analyzing stock market movements using twitter sentiment analysis. In: ASONAM 2012, pp. 119–123 (2012). https://doi.org/10.1109/ASONAM.2012.30
13. Picasso, A., Merello, S., Ma, Y., Oneto, L., Cambria, E.: Technical analysis and sentiment embeddings for market trend prediction. Expert Syst. Appl. **135**, 60–70 (2019)
14. Pak, A., Paroubek, P.: Twitter as a corpus for sentiment analysis and opinion mining. LREc **10**, 1320–1326 (2010)
15. Ghiassi, M., Skinner, J., Zimbra, D.: Twitter brand sentiment analysis: a hybrid system using n-gram analysis and dynamic artificial neural network. Expert Syst. Appl. **40**(16), 6266–6282 (2013)
16. Nisar, T.M., Yeung, M.: Twitter as a tool for forecasting stock market movements: a short-window event study. J. Financ. Data Sci. **4**(2), 101–119 (2018)
17. Krippendorff, K.: Reliability in content analysis: some common misconceptions and recommendations. Hum. Commun. Res. **30**(3), 411–433 (2004)
18. García-Díaz, J.A., Salas-Zárate, M.P., Hernández-Alcaraz, M.L., Valencia-García, R., Gómez-Berbís, J.M.: Machine learning based sentiment analysis on Spanish financial Tweets. In: Rocha, Á., Adeli, H., Reis, L.P., Costanzo, S. (eds.) World-CIST 2018. AISC, vol. 745, pp. 305–311. Springer, Cham (2018). https://doi.org/10.1007/978-3-319-77703-0_31
19. Mikolov, T., Chen, K., Corrado, G., Dean, J.: Efficient estimation of word representations in vector space. arXiv preprint arXiv:1301.3781 (2013)
20. Pennington, J., Socher, R., Manning, C.D.: Glove: global vectors for word representation. In: Proceedings of the 2014 Conference on Empirical Methods in Natural Language Processing (EMNLP), pp. 1532–1543 (2014)
21. Grave, E., Bojanowski, P., Gupta, P., Joulin, A., Mikolov, T.: Learning word vectors for 157 languages. arXiv preprint arXiv:1802.06893 (2018)
22. Cañete, J.: Compilation of large Spanish unannotated corpora. https://doi.org/10.5281/zenodo.3247731. Accessed 24 Aug 2020

23. Cardellino, C.: Spanish Billion Words Corpus and Embeddings. https://crscardellino.github.io/SBWCE/. Accessed 24 Aug 2020

24. García-Díaz, J.A., Cánovas-García, M., Valencia-García, R.: Ontology-driven aspect-based sentiment analysis classification: an infodemiological case study regarding infectious diseases in latin America. Future Gener. Comput. Syst. **112**, 614–657 (2020). https://doi.org/10.1016/j.future.2020.06.019

25. Sherstinsky, A.: Fundamentals of recurrent neural network (RNN) and long short-term memory (LSTM) network. Phys. D Nonlinear Phenom. **404**, 132306 (2020). http://www.sciencedirect.com/science/article/pii/S0167278919305974

26. Abadi, M., et al.: TensorFlow: large-scale machine learning on heterogeneous systems (2015). https://www.tensorflow.org/. Software available from tensorflow.org

27. Chollet, F., et al.: Keras (2015). https://keras.io

28. Autonomio talos [computer software] (2019). http://github.com/autonomio/talos

29. Kingma, D.P., Ba, J.: Adam: a method for stochastic optimization. arXiv preprint arXiv:1412.6980 (2014)

30. Patel, N., Upadhyay, S.: Study of various decision tree pruning methods with their empirical comparison in weka. Int. J. Comput. Appl. **60**(12), 20–25 (2012)

31. Yin, W., Kann, K., Yu, M., Schütze, H.: Comparative study of CNN and RNN for natural language processing. CoRR abs/1702.01923 (2017). http://arxiv.org/abs/1702.01923

32. Minaee, S., Kalchbrenner, N., Cambria, E., Nikzad, N., Chenaghlu, M., Gao, J.: Deep learning based text classification: a comprehensive review. arXiv preprint arXiv:2004.03705 (2020)

33. Onan, A.: Sentiment analysis on product reviews based on weighted word embeddings and deep neural networks. Concurr. Comput. Pract. Exp. e5909 (2020)

Speech to Mexican Sign Language for Learning with an Avatar

Fernando Barrera Melchor[✉], Julio César Alcibar Palacios,
Obdulia Pichardo-Lagunas, and Bella Martinez-Seis

UPIITA-IPN, Insituto Politécnico Nacional, Mexico City, Mexico
barreramelchorf@gmail.com, jcesarap1@gmail.com,
{opichardola,bcmartinez}@ipn.mx

Abstract. This work proposes the combination of natural language processing techniques and the use of graphic development engines to create an avatar to display information translated into Mexican Sign Language (MSL). It describes the application of the guarantee analysis carried out on Mexican Sign Language in the implementation of a rules-based automatic translation system. Taking into account the graphic character of sign language, the generated translation is represented through the use of an avatar developed in a virtual environment Unity. The translation is based on a restricted vocabulary associated with the learning of history of Mexico for the fourth level of primary education. The grammatical structures most used in the history book were also analyzed and included in the rules system. The application was developed that, through a cloud server that will use Natural Language Processing (NLP) and Automatic Translation (AT), transforms the voice into MSL. The signs will be made by a digitally animated avatar on a mobile device .

Keywords: Mexican Sign Language · Application · Natural Language Processing · Database · Inclusive education · Avatar · Hearing impairment

1 Introduction

In Mexico, the number of deaf persons amounts a total of 2.4 million, of which 84,957 are known to be under the age of 14. According to the national survey of demographic dynamics 67% attend school and data of the Sectoral Coordination of Primary Education show that 38,418 attend the 514 primary special education centers.

The sign language is the tool deaf people use to communicate with the world. Mexican Sign Language (MSL) is a natural language with its own vocabulary, semantic and grammatical structures that distinguish it from the rest of the languages, whether spoken or not.

The development of technological tools focused on MSL that facilitate the integration of members of the deaf community it is a task to which we have led our efforts.

© Springer Nature Switzerland AG 2020
L. Martínez-Villaseñor et al. (Eds.): MICAI 2020, LNAI 12469, pp. 179–192, 2020.
https://doi.org/10.1007/978-3-030-60887-3_16

Automatic translation from any spoken language to a sign language has several challenges and requires expertise in fields like natural language processing, computer vision and the design of computer graphics.

Natural language processing techniques allow to carry out a linguistic analysis of sign language, identifying its structure and characteristics for creating models. In our case the developed models LET the creation of rules for automatic translation or generation of Mexican Sign Language.

Virtual environment design engines today are capable of creating animations of the human figure articulated enough to represent sign languages. An ideal character should be able to make the necessary movements of the hands, arms, face, head and even the eyes. However, the design of the animations is not enough yet, you must know a set of rules and patterns with which the system can generate a text representation in sign language.

In the case of MSL, it is important to translate using the proper grammar of the language and the appropriate vocabulary, although this not fully regulated due to the lack of a written representation of signs. Other challenge is the display of the sign which includes a lot of movements and body parts.

This work proposes the development of an application that, through a cloud server that uses Natural Language Processing (NLP) and Automatic Translation (AT), transforms the voice into Mexican Sign Language (MSL). The signs will be made by a digitally animated avatar on a mobile device; so that the student can understand more clearly the class taught by the teacher.

2   Automatic Translation into Sign Languages

This work combines sign language translation with avatar technology to improve inclusive education. In this sense, this section explores some works of automatic translation from spoken or written languages to signs. There are different works that approach and propose a different solution to help people with hearing disabilities using machine translation and natural language processing, developed both in national and international universities, as well as associations that they are dedicated to this and have products in distribution in the market.

Automatic translation methods based on corpus is useful for spoken languages. In sing languages most of the works focus on rule-based methods. It has been developed and applied for many languages such as English [15], Arabic [10], French [7], and Spanish [12]. There are attempts to generate rules automatically for French to French Sign Language (FSL) [7], and some other works that explore the statistic translation [2]

It is well known that the Spanish has varieties according to the country and region. San-Segundo, et al. [14] translate from speaking Spanish to Spain Sing Language (LSE: Lengua de Signos Esañola) represented by an 3D Avatar for people applying for Identity Card. In [12] is presented a translation from written Spanish to Mexican Sing Language (LSM: Lengua de Señas Mexicana) represented by sequences of video for restricted grammar structures. For LSM, in [3] is presented a classification of signs with artificial data.

Once the automatic translation process is done, the output is usually a word sequence that should represent by signs. This representation can be done by images, videos, animations or avatars. Because of the advance technology in 3D animation, recent works use avatars to display the signs. Avatars are used in [5] to display translation from written Greek to GSL, from German to Swiss German Sign Language (DSGS) [6], from Spanish to LSE [14], from English to ASL [15] using an avatar with Inverse Kinematic, among others. Some other works, not only use avatars to display the signs: KAZOO [1] allows automatic sign production with a 3D avatar, [11] discusses avatar optimizations that can lower the rendering overhead in real-time displays focused on ASL, and some evaluations in facial expressions of avatars has be done [16].

Applications and algorithms for automatic translations to signs are mainly focus on a topic because of the vocabulary and grammar structures. Using rule-based methods, [14] develop an application for official explanations for identity card for Spain Sing Language (SSL); for bus information, [9] translates to SSL with a 3D avatar animation module. [15] focus on railway station announcements translating to American Sign Language. Most of the topics focus on mobility or public services, missing the educational field.

The Salamanca Declaration (UNESCO, 1994) [4], a political document that defends the principles of inclusive education, recommends that all students have the right to develop according to their potential and develop skills to participate in society. Several international documents, the Salamanca Declaration (UNESCO, 1994) mentioned above, and the Standard Rules for the Equalization of Opportunities for Persons with Disabilities, reflect the need to use sign language as a vehicular language in the education of deaf students.

Sign language is a tool that leads us to interact, communicate, think and learn as well as being part of everyone from a very young age. Therefore, it is necessary to include this in educational programs in order to awaken in society the importance of inclusion for the benefit of all.

3 Translation of Voice to Signs in History Class

A system was developed to help with the communication of people with hearing disabilities in an inclusive environment. In this case for children between 9 and 11 years old who are studying the 4th grade History course at the elementary level. This work was developed for Android smart mobile devices. Figure 1 shows the main processes carried out by the system in order to get successful translations.

3.1 Automatic Translation to MSL

The subject of History has grammatical structures that it implies are simple and in common use, that is: Subject + Verb + Complement, for example:

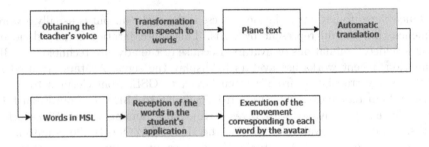

Fig. 1. System block diagram.

"Miguel Hidalgo started independence of Mexico on September 16, 1810"
Subject: *Miguel Hidalgo*
Verb: *started*
Completion Begin: *the independence of Mexico on September 16, 1810.*

History uses common vocabulary, with particular names but not as technical as in other subjects, such as mathematics. It is useful because sign language is not very developed in this type of vocabulary, the SEP's dictionary for the deaf mute reaches only 535 signs [8], this means that there are many words to document that are missing, taking into account that the Spanish Royal Academy has approximately 93,000 [13] words.

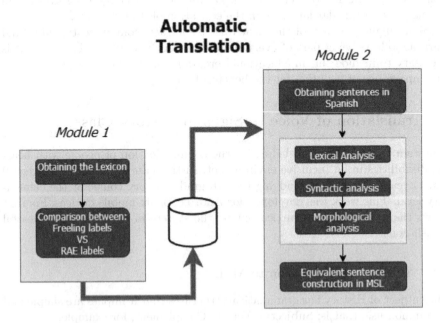

Fig. 2. Interpretation process to restricted vocabulary [12].

Figure 2 shows the representation of the automatic translation. There are two main modules. Module 1 was used to construct the corpus. There were analyzed several common phrases in History classes to get the lexicon. In order to save the vocabulary, there was also done a comparison between grammar labels of the words. Module 2 focus on the translation, 13 sentence structures were studied that resulted in translation rules for 6 syntactic trees, these trees allow the translation of up to 52 grammatical structures. First we decode the speech to sentences in Spanish, then we do a natural language processing by a lexical, syntactic and morphological analysis. We use a rule based method to translate the sentences to a equivalent sentence construction in MSL.

Here is an example of the translation.
The person says: *Children play with the ball.*
It would be interpreted as: *Child many ball play.*
Avatar. *Represent each word through movements.*

3.2   Signal Representation with Avatar

Once the final translation of the sentence entered into the system has been obtained, it will be sent to the student's mobile device, which represents the movements of the avatar thanks to the animation engine Unity Engine. The system will be in constant communication with a server using a Restful API programmed in Flask, which is a Python Framework dedicated to creating Web applications, which is connected to a database that stores all the corpus information, from restricted vocabulary, machine translation rules, as well as synonyms and collocations for this vocabulary in MSL, this data is found in relational tables stored on an SQL server. It should be mentioned that being a prototype and having a restricted vocabulary, the teacher must speak loudly, clearly and at a low speed, omitting words that are outside the restricted vocabulary.

3.3   System Interaction for Automatic Translation

Figure 3 shows the parts that make up the system. The teacher has a mobile device to speak as in a regular History class. He uses the microphone that is integrated in smartphones and tablets, or by means of an external microphone the system will take the teacher's voice to transform it into plain text using Google's speech recognition API.

Once the plain text has been obtained, it is sent by means of HTTP requests to the server, which is interpreted into a restricted vocabulary, performing a lexical, syntactic and morphological analysis (see Illustration 3), a process currently used by the "Direct Translation System of Spanish to MSL with marked rules" [12]; once the restricted vocabulary is obtained, it is automatically translated into MSL, once the words in MSL are obtained, the database is searched to verify the existence of collocations or synonyms, which are replaced by their respective equivalents in MSL.

The avatar representation is show in an mobile device. The interface is focus on students. They can access the class with a key given by the teacher.

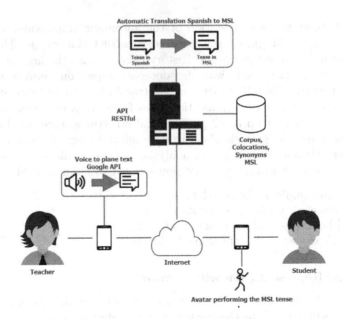

Fig. 3. General diagram of the system.

4   Tests and Results

Figure 4 shows the 3D avatar that the student displays once they enter their email associated with their account and a valid class code. It is here where the student appreciates all the corresponding movements with the AT from Spanish to MSL of each phrase said by the teacher, it should be remembered that the application has 2 parts, teacher and student, as shown in the General diagram of the system.

Fig. 4. 3D avatar.

Table 1. BLEU analysis.

Sentence	Server Response	Time ms	Expected response	BLEU
Â¡Hola, buenos días!	["hola", "buenos-días"]	351	hola buenos-días	1
Â¿Cómo están, niños y niñas?	["como-estar","niño", "mucho","niño", "mujer", "mucho"]	438	como-estar niño mucho niño mujer mucho	1
Vamos a iniciar nuestra clase de Historia	["nuestro", "clase", "historia", "empezar"]	272	nuestro clase historia empezar	0,4347
El tema es El inicio de la guerra de independencia y la participación de Hidalgo y allende	["tema", "empezar", "guerra", "independencia", "participación", "hidalgo", "Allende"]	617	tema empezar guerra independencia participacion hidalgo Allente	1
Observaremos una imagen de su libro de texto	["dibujo", "libro", "ver"]	319	dibujo libro ver	1
Página 166	["página", "num-cien", "num-sesenta", "num-seis"]	169	pagina num-cien num-sesenta num-seis	1
Página 166	["página", "num-cien", "num-sesenta", "num-seis"]	169	pagina num-cien num-sesenta num-seis	1
Después contesten las siguientes preguntas	["después", "siguiente", "preguntar", "mucho", "contestar"]	297	después siguiente pregunta mucho contestar	1
Â¿Qué tíulo pondrís a esta pintura?	["qué", "título", "dibujo", "poner"]	396	qué título dibujo poner	1

(*continued*)

Table 1. (*continued*)

Sentence	Server Response	Time ms	Expected Response	BLEU
Â¿Qué están haciendo estos hombres?	["qué", "hombre", "mucho", "hacer"]	412	qué cara explicarr	1
Â¿Qué expresan sus rostros?	["qué", "cara", "explicar"]	321	qué cara explicar	1
Â¿Cómo están vestidos?	["como-estar", "vestir"]	286	como vestir	0,5
Explica porque lo elegiste.	["porque", "explicar", "elegir"]	228	explicar porque elegir	1
Ya vimos en clases anteriores	["ahora", "en", "clase", "mucho", "pasado", "ver"]	309	ahora clase mucho pasado ver	0,53
Recordaremos las causas de la guerra de independencia	["causa", "mucho", "guerra", "independencia", "recordar"]	371	causa mucho guerra independencia recordar	0,53
Anoten lo que recuerden en el pizarrón	["qué", "en", "pizarrón", "anotar", "recordar"]	346	en pizarrón anotar	0,66
Anoten lo que recuerden en el pizarrón	["qué", "en", "pizarrón", "anotar", "recordar"]	346	en pizarrón anotar	0,66
Leeremos la página 168, 169 y 170 del libro de historia	["página", "num-cien", "num-sesenta", "num-ocho", "num-cien", "num-sesenta", "num-nueve", "num-cien", "num-setenta", "libro", "historia", "leer"]	691	página num-cien num-sesenta num-ocho num-cien num-sesenta num-nueve num-cien num-setenta libro historia ver	1
			Average BLEU : 0.87	

While Table 1 shows the comparison of the expected translation against the translation thrown by the system, a python tool called "sentence_bleu" from the Nltk library was used to obtain the BLEU score, Bilingual Evaluation Unserstudy, which is a method of evaluating the quality of translations performed by machine translation systems. A translation has higher quality the more similar it is with respect to another reference, which is supposed to be correct. BLEU can be calculated using more than one reference translation. This allows for greater robustness to size compared to free human translations.

As the system only works with restricted vocabulary to successfully translate a history class, the BLEU assessment was carried out in a total of 20 sentences with a total of 53 words and 27 vowels, it must be taken into account that the system spells any word outside the restricted vocabulary, so 27 vowels were included to be able to interpret words unknown to the system.

4.1   Response Time and Words in the Avatar Vocabulary

Tables 2, 3, 4, 5 and 6 below show the execution times of the main processes carried out by the system, such as AT carried out on the server and animation of the words carried out on the mobile device, as well as the difference in percentage of the "dictation" that would be the equivalent of speaking normally, against the process carried out by the translation system.

Table 2. Response time 1.

Word	Time in seconds					
	AT	Animation	Spell	Total	Total traduction	% difference
Hola	0.59	1.5	1.5	3.59	2.09	39.33%
Cómo están	0.59	1.8	1.5	3.89	2.09	39.33%
Clase	0.59	1.3	1.5	3.39	2.09	39.33%
Historia	0.59	2.88	1.5	4.97	2.09	39.33%
Allende	0.59	2.44	1.5	4.53	2.09	39.33%
Hoy	0.59	3.4	1.5	5.49	2.09	39.33%
Ver	0.59	1.34	1.5	3.43	2.09	39.33%
Tema	0.59	1.34	1.5	3.43	2.09	39.33%
Hidalgo	0.59	2.39	1.5	4.48	2.09	98.67%
Average	0.59	2.043333	1.5	4.1333333	2.63333	75.56%

Table 2 shows a comparison of time between sentences and words that are represented in MSL with a single sign. As you can see in the last column, the signs that are made up of more than 1 movement (History, Allende), are the ones that take the longest time when animating, but in the Automatic Translation process they all take the same time. 0.59 s.

Table 3 shows a comparison of the signs that represent the alphabet, mainly useful for spelling words that are outside the restricted vocabulary. These translations take the same time to process simple phrases and words as the ones exemplified in Table 2, but the animation time is significantly reduced to 0.4 s, but as you can see the time increase is 98%, due to that the translation time added to the animation time far exceed the normal dictation that a personal one would do when speaking.

Table 3. Response time 2.

Word	Time in seconds					
	AT	Animation	Spell	Total	Total traduction	% difference
A	0.59	0.4	0.5	1.49	0.99	98.00%
B	0.59	0.4	0.5	1.49	0.99	98.00%
C	0.59	0.4	0.5	1.49	0.99	98.00%
D	0.59	0.4	0.5	1.49	0.99	98.00%
E	0.59	0.4	0.5	1.49	0.99	98.00%
F	0.59	0.4	0.5	1.49	0.99	98.00%
G	0.59	0.4	0.5	1.49	0.99	98.00%
H	0.59	0.4	0.5	1.49	0.99	98.00%
I	0.59	0.4	0.5	1.49	0.99	98.00%
Average	0.59	0.4	0.5	1.49	0.99	98.00%

As a complement to Table 3, Table 4 shows a set of vowels and consonants, which represent the translation time for words that the system will have to spell when outside the restricted vocabulary. The difference time is less than the spoken time because the system takes very little time to perform the Automatic Translation and the signs of the letters are mostly fast movements and with little body movement. As a result the increase in the time of translation and animation compared to speaking, is only 52.35%.

Table 4. Response time 3.

Word	Time in seconds					
	AT	Animation	Spell	Total	Total traduction	% difference
ABC	0.69	1	1.5	3.19	1.69	12.67%
ABCDE	0.75	2.39	2	5.14	3.14	57.00%
ABCDEFG	0.77	3.19	2.5	6.46	3.96	58.40%
ABCDEFGHI	0.84	4.6	3	8.44	5.44	81.33%
Average	0.7625	2.795	2.25	5.8075	3.5575	52.35%

Table 6 used longer sentences as a reference, in which translations with longer sign lengths were expected, as it can be seen, the time of the Automatic Translation increased considerably on the server as they were more complex grammatical structures as well as the time of execution of the signs on the mobile device so the average percentage difference against a common class was 296.24%

Table 5. Response time 4.

Sentences	Time in seconds					
	AT	Animation	Spell	Total	Total traduction	% difference
Clase de historia de Hidalgo y Allende	6.5	8	6	20.5	14.5	141.67%
Hoy hay clase de historia de Hidalgo y Allende	8	11	6	25	19	216.67%
El tema de hoy es el inicio de la guerra de inde-pendencia y la participación de Hidalgo y allende	18	20	7	45	38	442.86%
Hola ¿Cómo estás hoy?	3.4	9.2	3	15.6	12.6	320.00%
El tema de hoy es el inicio de la guerra de independencia	10	13	5	28	23	360.00%
Average	9.18	11.64	5.4	26.82	21.42	296.24%

And as a last test to compare the translation times of the system, long sentences with words outside the system were included in the system restricted vocabulary, as we can see this mainly increased animation time by having to run the words in spelling mode, which increased the total time against a common class by 430.08%

Table 6. Response time 5.

Sentences	Time in seconds						
	Total words	AT	Animation	Spell	Total	Total traduction	% difference
abcde Hay clase de historia abcedefg	2	6.5	8	6	20.5	14.5	141.67%
abcd tema de hoy es la independecia	1	4	8.4	3	16.4	12.4	413.33%
abcdefghijk	3	8	18.9	6	35.9	26.9	448.33%
Average	2	6.166	11.766	5	24.266	17.933	430.08%

4.2 Field Tests

Figure 5 shows the field tests carried out in "La casa del Sordo", there you can see an instructor of the "Casa del Sordo" evaluating the precision of signs and the translation that was done by the system. In the image he is doing the sing of Ignacio Allende, a leader in Mexican Independence.

Fig. 5. Field tests 1.

5   Conclusions and Future Work

Sign language presents various peculiarities, since it is adaptable to different situations, moments and places; therefore, the study is complicated, as is the creation of a mobile application to translate the Mexican sign language. Because to carry out a translation to MSL of the central zone it is not possible to generate word by word, since the sentences can be represented in multiple ways depending on the region, zone, country, it was concluded that in this case they will be represented with that of Mexico City. An algorithm capable of simulating a grammatical tree can be created, which is in charge of generating word arrangements to translate, however, it must be executed faster to achieve a more realistic translation. The processing speed of the text coming from the voice, together with the reproduction of each animation, exceeds for a considerable time the human voice and translation. Natural language processing techniques make it easy to categorize the words that make up a sentence. The open source Freeling language natural processing library facilitates the processing of all the sentences put in this work, in this way, it was possible to establish a limited vocabulary to teach a history class of 4th grade of primary school. The results obtained after the development and implementation of this work, the scientific advance to help the inclusion of deaf people in education, the incorporation of increasingly sophisticated and accurate software, the growing expectation that users place on technology to supporting the solution of problems in your life, allows us to suggest, as a future work, the incorporation of the following functionalities into the System like the incorporation of more sign animations, that is, to strengthen not only history classes, but all kinds of conversations and situations, improve the processing time of machine translation, and design signs with greater precision and better animation technologies.

References

1. Braffort, A., et al.: KAZOO: a sign language generation platform based on production rules. Univ. Access Inf. Soc. **15**(4), 541–550 (2016)
2. Bungeroth, J., Ney, H.: Statistical sign language translation. In: Workshop on Representation and Processing of Sign Languages, LREC, vol. 4 (2004)
3. Cervantes, J., et al.: Clasificación del lenguaje de señas mexicano con SVM generando datos artificiales. Revista vínculos **10**(1), 328–341 (2013)
4. Domínguez, A.B.: Servicio de información sobre discapacidad. https://sid.usal.es/idocs/F8/ART11921/educacion_para_la_inclusion_de_alum_sordos.pdf. Accessed 2019
5. Efthimiou, E., et al.: From grammar-based MT to post-processed SL representations. Univ. Access Inf. Soc. **15**(4), 499–511 (2016)
6. Ebling, S., Glauert, J.: Building a Swiss German sign language avatar with JASigning and evaluating it among the deaf community. Univ. Access Inf. Soc. **15**(4), 577–587 (2016)
7. Filhol, M., Hadjadj, M.N., Testu, B.: A rule triggering system for automatic text-to-sign translation. Univ. Access Inf. Soc. **15**(4), 487–498 (2016)

8. Hernández, M.T.C.: Diccionario Español-Lengua de señas Mexicana (DIELSEME), SEP, México
9. López-Ludeña, V., et al.: Translating bus information into sign language for deaf people. Eng. Appl. Artif. Intell. **32**, 258–269 (2014)
10. Luqman, H., Mahmoud, S.A.: Automatic translation of Arabic text-to-Arabic sign language. Univ. Access Inf. Soc. **18**(4), 939–951 (2019)
11. McDonald, J., et al.: An automated technique for real-time production of lifelike animations of American Sign Language. Univ. Access Inf. Soc. **15**(4), 551–566 (2016)
12. Pichardo Lagunas, O., Partida Terrón, L., Martínez Seis, B.C., Alvelar Gallegos, A., Serrano Olea, R.: Sistema de traduccion directa del español a LSM con reglas marcadas, Ciudad de México: IPN (2016)
13. RAE, Real Academia Española, Real Academia Española. http://www.rae.es/. Accessed 2019
14. San-Segundo, R., et al.: Speech to sign language translation system for Spanish. Speech Communi. **50**(11-12), 1009–1020 (2008)
15. Shaikh, F., et al.: Sign language translation system for railway station announcements. In: 2019 IEEE Bombay Section Signature Conference (IBSSC). IEEE (2019)
16. Smith, R.G., Nolan, B.: Emotional facial expressions in synthesised sign language avatars: a manual evaluation. Univ. Access Inf. Soc. **15**(4), 567–576 (2016)

Poincaré Embeddings in the Task
of Named Entity Recognition

David Muñoz[1,2(✉)], Fernando Pérez[1], and David Pinto[2]

[1] Department of Computing, Technological University Dublin, Blessington Rd,
Tallaght, Dublin D24 FKT9, Ireland
devasc26@gmail.com, fernandopt@gmail.com
[2] Faculty of Computer Science, Benemerita Universidad Autonoma de Puebla,
Av. San Claudio, Blvrd 14 Sur, CU, 72592 Puebla, Mexico
davideduardopinto@gmail.com

Abstract. Hyperbolic embeddings have become important in many natural language processing tasks due to their great ability to capture latent hierarchical data and to encode valuable syntactic and semantic information. We study and consider the ability of Poincaré embeddings to get the most similar nodes to a given node when trying to recognize named entities in a set of text documents. In this paper, we propose a classifier model for the NER (Named Entity Recognition) task by implementing Poincaré embeddings and by using the most frequent n-grams and their Part-of-Speech (POS) structures from the training dataset. We found that POS structures and n-grams help to map possible named entities, while using Poincaré embeddings manage to affirm and refine this recognition, improving the recognition of named entities.

Keywords: Poincaré embeddings · NER · Named entity recognition · Part-of-Speech · N-grams

1 Introduction

Nowadays, the amount of textual information available electronically is incredibly abundant, causing an information overload when searching for something specific. It makes difficult to obtain the most relevant information on a specific topic and it is time consuming. To deal with this problem there are information retrieval (IR) systems, which search for information in a collection of documents and retrieve the most relevant resources based on a specific information need [1]. To achieve this some techniques are required such as the recognition of named entities (relevant information) through which a text document can be represented, reducing the information overload [2]. In this sense NER is a subtask of Information Extraction (IE) where the information we are looking for is known beforehand.

In the present work, a NER classifier model is proposed by including Poincaré embeddings in order to get the closest words to a given word. By using the closest words it is possible to know if a tag (class) is near to that given word, and

L. Martínez-Villaseñor et al. (Eds.): MICAI 2020, LNAI 12469, pp. 193–204, 2020.
https://doi.org/10.1007/978-3-030-60887-3_17

then assign the tag to that word. But also the use of representative features of each class is considered to improve the correct recognition of the named entities. The named entities that make up each class are converted to n-grams just like their POS structures. Then the most representative n-grams of each class as well as the POS structures are used to refine the correct recognition of the named entities, previously achieved by the Poincare models. It is through this proposal that it is intended to improve the recognition of named entities which can represent a document with the most important information and thus facilitate some tasks such as the classification of texts and retrieval information. To create the Poincaré models we created a Gold Standard by manually tagging a set of documents (Job Descriptions) in the IT field, and then converted to an structured format (tab-separated columns) where the first column contains every token in the document and the second column contains its belonging tag. Then Poincare models can be used for getting vector representations of words and try to recognise the tags of every word.

In summary, this paper presents a classifier model for the task of named entity recognition by employing Poincaré embeddings, n-grams and Part-of-Speech. The rest of the paper is structured as follows. Section 2 describes the background related to our proposal, such as NER Task, Poincaré embeddings, N-grams, and Part-of-Speech. Section 3 explains the design of the proposed classifier model. Section 4 shows the experiments and results carried out, and how Poincaré models, POS structures and n-grams help in the task of NER. Section 5 presents a conclusion of the results obtained.

2   Background

2.1   Named Entities

The Named Entity (NE) task is a subtask of information extraction (IE) and was developed by the committee in the Sixth Message Understanding Conference in 1995. It became an important task in IE field since it aims to locate and classify named entities in a text into categories previously defined (e.g. People, Organizations, Geographic Locations). In this way, texts can be represented by their named entities (most relevant terms) [3]. Another definition for named entity is, roughly speaking, anything that can be referred to with a proper name: a person, a location, an organization. However, the term is commonly extended to include things that are not entities such as, including dates, times, and other kinds of temporal expressions, and even numerical expressions like prices [10].

2.2   Poincaré Embeddings for Word Representation

Hyperbolic geometry is a non-Euclidean geometry which studies spaces of constant negative curvature. It is, for instance, associated with Minkowski space time in special relativity. In network science, hyperbolic spaces have started to receive attention as they are well-suited to model hierarchical data. For instance, consider the task of embedding a tree into a metric space such that its structure is reflected in the embedding. Due to the underlying hyperbolic geometry, this allows us to learn parsimonious representations of symbolic data by simultaneously capturing hierarchy and similarity [4].

Word embeddings are effective for many natural language processing tasks because they are flexible and encode valuable syntactic and semantic information. Word embeddings are motivated by the concept that semantic similarities between words are based on their distributional properties in the large amount of text. The idea of distributional properties is called distributional hypothesis [5], meaning that linguistic items with similar distributions have similar meanings. Popular word embeddings such as GloVe, Word2Vec, and FastText are widely used in various tasks and have shown great success. Although these embedding methods have proven successful, very few methods exist that are able to encode tree-like or graph-like hierarchical relationships of the data. Poincaré embeddings are better at capturing latent hierarchical information than traditional Euclidean embeddings.

2.3 Part-of-Speech (POS) Tagging

POS tagging is the process of marking a word in a text with a tag carrying the corresponding information to a particular part of speech, based both on its definition and its context (its relation to adjacent and related words in a phrase, sentence or paragraph) [7]. This tagging is done using computational algorithms that associate discrete terms, as well as hidden parts of the speech, through a set of descriptive labels. These POS tagging algorithms are divided into two types: rule-based and stochastic.

2.4 N-gram Language Model

An n-gram is a subsequence of n elements of a sequence of words given in a text and is widely used in the study of natural language. An n-gram model is a type of probabilistic model that allows a statistical prediction to be made of the next element of a certain sequence of elements that has occurred so far. An n-gram model can be defined by a Markov chain of order $n-1$, due to its relative simplicity to increase the study context by increasing the size of n [6].

3 Proposed Classifier Model

This section explains the characteristics considered in the construction of the classifier model as well as the way in which they were calculated.

3.1 Definition of Used Named Entities

As mentioned in Sect. 2.1 the term Named Entity is commonly extended to include more things that result important for some tasks. In this case we focus on extracting relevant concepts for our case study where our training set refers to Job Descriptions in the IT field. The classes we found most important are defined as *Role, Knowledge, Skill, Talent, Character* and *Responsibility*[1].

- **Role.** The position or purpose that someone or something has in a situation, organization, society or relationship. Some examples are: *"Enterprise Technical Support Engineer"* and *"Senior IT Project Manager"*.
- **Knowledge.** Awareness, understanding, or information that has been obtained by experience or study, and that is either in a person's mind or possessed by people generally. Some examples are: *"Understanding of Microsoft Azure"* and *"Programming languages"*.
- **Skill.** A particular ability that you develop through training and experience and that is useful in a job. Some examples are: *"Good analytical skills"*, *"Good troubleshooting skills"* and *"Excellent communication skills"*.
- **Talent.** Someone who has a natural ability to be good at something, especially without being taught. Some examples are: *"Enable business capabilities through innovation"* and *"Self-managing"*.
- **Character.** A person, especially when you are describing a particular quality that they have. Some examples are: *"Independent"* and *"Team player"*.
- **Responsibility.** The things you are in charge of in your job. Some examples are: *"Responsible for providing technical support to customers"* and *"Responsible for providing after-hours support as part of an on-call rotation"*.

3.2 N-grams Used in Gold Standard

The corpus is composed of 160 Job Descriptions in the field of IT, collected from the website www.jobs.ie and labelled with the 6 classes previously described. However, we want to know what n-grams are used most in each class. To achieve this we have measured the length of the named entities (n-grams) that make up each class, immediately we grouped them by n size, and then we counted their frequencies in the corpus. Figure 1 shows the distribution of the existing n-grams in each class.

With the graphic representations of the n-grams can be seen that most of the classes are mainly made up of bigrams, with the *Role, Skill* and *Knowledge* classes being the most common. However, the Knowledge and Character classes are mainly represented by unigrams, with a large percentage difference from the rest of the classes. That means that these two classes are characterized by being unigrams and bigramas. On the other hand, it is possible to observe that the trigrams occupy the third position in representing each of the classes, being the only one more balanced group based on the number of occurrences. Four-grams represent from 6% to 12% of the composition of each class, ranking fourth in the main named entity sizes. As the rest of the n-grams have such a low presence in the classes, with the exception of the Responsibility class, being the only class

---

[1] Definitions of used classes are explained in the original source and can be found in https://dictionary.cambridge.org/dictionary/english/.

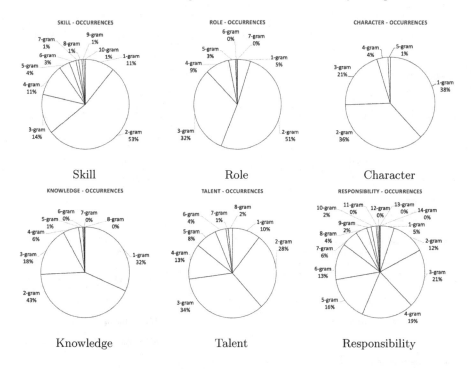

Fig. 1. Distribution of n-grams in each class

that represents a less disproportionate distribution of named entities of short and long length. Furthermore, the Responsibility class has the longest n-grams existing in the corpus, being something characteristic of this class.

3.3 POS Structures of N-grams

POS tagging was applied by using the tagger provided by the *Natural Language Toolkit* library, then all the sequences found in the named entities were grouped (and the occurrences of each one were counted). The default function POS tagset was used because it offers the possibility of tagging with 35 different classes [8], making the structure of a named entity more precise and thus recognizing its similarity with another. In Fig. 2 it can be seen which are the POS-tags of the most representative named entities for each class.

The class *Skill* is mainly made up of bigrams, which is something that characterizes the class and it is worth looking at its component structures: *NN NN, JJ NN, NN NNS, NN VBG, JJ JJ, NN RB, NNP NNP, VBG NNS, JJ NNS* and *NNP NN*, where nouns and adjectives are the most used POS tags and resulting characteristic of their named entities.

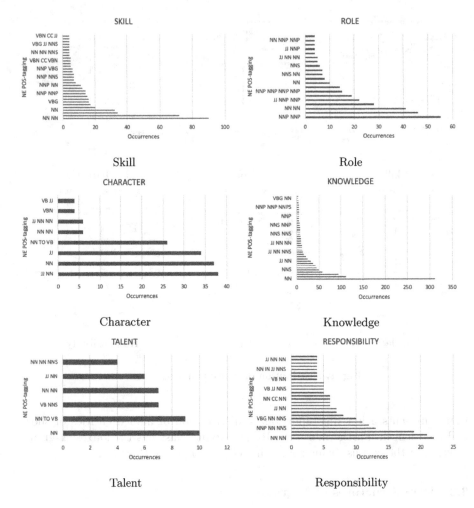

Fig. 2. Most representative POS-tag for each class

On the other hand, the class **Role** is made up mostly of bigrams and trigrams, so it would be very rare for an named entity in this class to be just one word. The main structures of bigrams in this class are: *NNP NNP, NN NN, JJ NN* and *NN NNP*, while the main structures of trigrams are: *NNP NNP NNP* and *JJ NNP NNP*, being something characteristic of this class that its named entities are mainly made up of proper nouns in the singular form.

In the class **Character** mainly unigrams are used (*NN* and *JJ*), while the most present bigrams have the structure *JJ NN*, and most of the trigrams have the structure *NN TO VB*.

While the class **Knowledge** is strongly represented by unigrams of the *NN* form, followed by the *NNS* form, and bigrams with the structures *NN NNS*,

NNP NNP, NN NN, NNP NNS and *NNP NN.* So it would be rare to find a trigram in this class, or a higher-order n-gram.

The class **Talent** is made up of unigrams, bigrams and trigrams, with the bigrams being the most representative and with the following structures: *VB NNS, NN NN* and *JJ NN.* The main trigrams have the structures: *NN TO VB* and *NN NN NNS,* and unigrams are represented with the form *NN.* This class strongly uses nouns, which are part of any of its named entities.

Finally, the class **Responsibility** is mainly made up of trigrams that have the following structures: *NNP NN NNS, NN NN NNS, VBG NN NNS, VBG JJ NN, JJ NN NNS, VBG JJ NNS, NN NN NN, VB JJ NNS, VBG NN NN* and *NNP JJ NNS.* And those trigrams are formed with singular and plural nouns, adjectives and verbs in gerund (present participle). The main bigrams that conform the class have the structures: *NN NN, NN NNS, JJ NN, VBG NNS* and *NNP NN,* where it is possible to realize that these bigramas are formed in the same way as the trigrams (with nouns, adjectives and verbs in gerund). And as for the unigrams they are characterized for being *NN* and *VBG,* concluding that the use of verbs in present participle is something characteristic of the class *responsibility.*

3.4   Poincaré Models

The corpus consist of 121,293 tokens, and a vocabulary of 5,297 unique words. The file format in which the information is stored is double column, saving the relation of a token in the first column to its assigned class in the second column. By using the gensim library was possible to train the Poincaré model with the corpus file. The number of dimensions of the trained model is set to 50 by default and it also needs a number of iterations which is set to 50 by default. As suggested in [4] all embeddings were randomly initialized from the uniform distribution U $(-0.001, 0.001)$, causing embeddings to be initialized close to the origin. The main focus for Poincare embeddings is its capability to embed data that exhibits latent hierarchical structures. Once the model is trained it is possible to find the closest words to a user-specified word through the most_similar method implementation where the top-N most similar nodes to the given node (word) is returned in increasing order of distance.

This characteristic is used to get the most similar words for each one of the words conforming to a named entity. Between the most similar words are also

Table 1. Assigning classes to NEs using Poincaré embeddings

Tested NE	Original class	Predicted classes	Most frequent class
Integration developer	Role	Role, Knowledge, Skill, Role, Role	**Role**
Providing support	Responsibility	Responsibility, Responsibility, Knowledge, Skill, Role	**Responsibility**
Excellent communication	Skill	Skill, Responsibility, Skill	**Skill**

included classes belonging to each word. Then the most frequent class in every named entity is selected, discarding others that occur just a few times. In Table 1 are showed some examples for named entities and the closest words returned by the Poincaré model implementation. So, for the named entity *Excellent communication* the classes *Skill* and *Responsibility* are predicted, however the class *Skill* occurs more times. That is the reason of selecting the class *Skill* as the assigned class to the named entity which is correct. And the same happens for every one of the named entities, showing excellent results as the predicted class is the original assigned class in the corpus.

3.5   Process of Construction of the Classifier Model

By making use of Poincaré embeddings our classifier intends to improve the recognition of named entities which represent Job Description documents with the most relevant information and thus facilitate some tasks such as the classification of texts and retrieval information. The classifier works considering the features described before when n-grams and part-of-speech were studied over the corpus.

In Sect. 3.2 we found that most of the classes are conformed by unigrams, bigrams and trigrams, being cuatrigrams also present in good percentage. N-grams with five or more elements are unusual, that is why when finding named entities we consider a window size of 4 words in order to find a relation between the current word and the next 4 words.

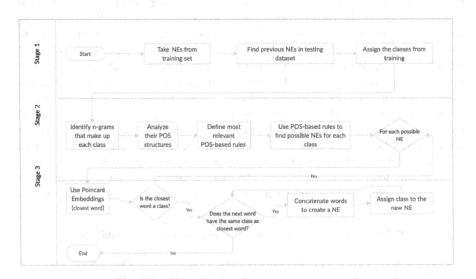

Fig. 3. Workflow diagram implemented on the classifier model

When studying the representative part of speech structures for each class, some patterns were found. However the most interesting structure was found for the class *Role* where named entities are normally proper nouns in singular form (NNS), and that is something very characteristic of that class. Figure 3 shows a diagram of the workflow operation implemented on the classifier model developed. The process consists of 3 stages for the recognition of named entities.

Classifier Design

- **Match Gold Standard named entities in test dataset (Stage 1).** From the original set of named entities in training dataset, if they exist in the test dataset then they are tagged. In other words, it consists in searching the entire list of training named entities in the test set and assign them the corresponding tags.
- **Search POS structures (Stage 2).** In stage 2, first an analysis of the named entities of each class is performed, thus determining the number of most used n-grams in each class, as well as their POS structures. Thus, only those structures with the most occurrences for each class will be considered. Once these most representative features of each class have been selected, they will be searched in the test dataset, and then there will be possible named entities to be tagged.
- **Use Poincaré embeddings and most frequent n-gram sizes (Stage 3).** By employing Poincaré embedding models, with the most frequent n-gram sizes it is verified if the words in that windows size have the same tag as one of the 5 closest neighbors. If all of the words on the window size have the same tag, then they are tagged and considered as a named entity. A more detailed explanation of how this stage works is shown in Fig. 4.

In stage 2, named entity candidates are found considering features of each class, but is in stage 3 that these named entities are tagged or discarded. In stage 3, for each of the words that make up a possible named entity, its closest neighbor is obtained, and if the closest neighbor for any of these words is one of the classes used in the corpus, then this word is tagged with that class and becomes a seed word or root word. Then a window of 4 words on the left and 4 on the right is used, and the 5 closest neighbors are obtained for each of those words and if the same class as the root word is found within the closest neighbors, then this word is added to the root word, and so on until the words inside the window are finished. At the end, a named entity is obtained with the words that share the same tag within the range of the window.

4   Experiments and Results

To know how the proposed classifier model developed performs, an experiment is carried out demonstrating the help of the Poincare embeddings when predicting named entities. For this experiment we have used a Gold Standard that we

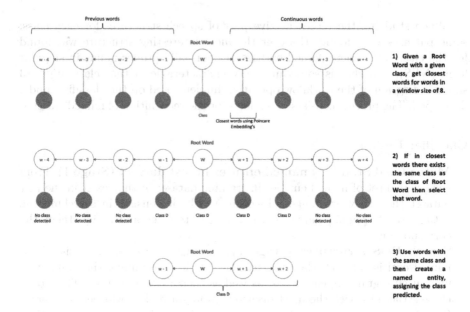

Fig. 4. Representation of how Poincaré models are used to predict named entities.

have manually tagged with a text annotation tool called *Web-based Tagger tool*[2] developed in [9]. This annotation tool allow users to highlight the most relevant concepts of a set of text documents, and then the tool transforms this data to an structured format ready to be used in the training of some predictive models. The Gold Standard is composed of a set of 160 Job Description documents in the field of IT and was collected from the website www.jobs.ie and tagged with the 6 classes described previously.

Table 2. Performance of classifier when using different stages of the workflow process

Performance		Classifier model		
Class	Measure	Stage 1	Stage 1 & 2	Stage 1 & 2 & 3
Character	F-1	0.6381	0.6710	0.8057
Knowledge	F-1	0.5037	0.5800	0.7681
Responsibility	F-1	0.1760	0.3250	0.4578
Role	F-1	0.4433	0.5480	0.6553
Skill	F-1	0.5068	0.5553	0.7046
Talent	F-1	0.4224	0.4695	0.5719

---

[2] The Web-based Tagger Tool can be found in https://lkesymposium.tudublin.ie/Tagger/.

The use of features improves the performance of the proposed classifier thanks to the fact that these characteristics allow named entities to be identified more accurately. To demonstrate the usefulness of these features, we present a series of experiments, where the first experiment only considers the first stage of the workflow diagram, the second experiment considers the first and second stages of the workflow diagram, and the third experiment considers all the stages of the workflow diagram showed in Fig. 3.

V-fold cross validation method was used to measure the classifier's performance considering different features presented on the workflow diagram, and the averaged obtained results are presented in Table 2. It can be seen that Stage 3 plays an important role in recognizing named entities. The main reason for seeing a notable improvement here is the fact of using Poincaré embeddings, since in stages one and two the most common and highly secure named entities have been tagged, so in stage 3 the amount of untagged information has decreased considerably and thus it is easier to analyze the remaining data and determine through the closest words to a root word potentially identified as belonging to a class.

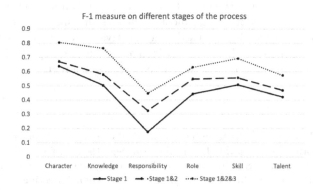

Fig. 5. F1 measure of classifier performance when using different stages

Figure 5 shows a comparison of the model performance when implementing different stages of the classifier. In stage one all those named entities that exist in the training corpus are tagged, that is why the recognition level is not so high, in addition to that the Precision in this stage is generally high as observed in Table 2. However, the second stage allows POS structures to be recognized based on the POS structures of each class in the training corpus, thus increasing the Recall level and decreasing Precision. Finally, the third stage allows to create a better balance between Precision and Recall, when using the Poincare embeddings since they establish a relationship between a root word with an assigned class and allows to know which words around it belong to the same class and therefore to the same named entity.

5   Conclusions

As it is possible to observe, the use of Poincaré embeddings improves the performance of the proposed classifier thanks to the fact that this feature allows named entities to be identified more accurately by using the closest words to a given word. The word representations obtained by the Poincaré models effectively captured the existing relationships between words, helping us to always identify if for a given word there is a class close to it.

The previous analysis of the POS structures of the n-grams allowed us to know which structures are most representative of each class and, therefore, to take them into consideration to determine possible named entities. In this way, the recognition of named entities could be expanded compared to stage 1.

On the other hand, having carried out a previous analysis of the n-grams that make up each one, could lead us to consider mainly those n-grams that occurred mostly in each class. In this way, it was possible to define a window size of words contiguous to a word to be analyzed, and then be able to know if those words together with the root word can form a named entity. In this way, for each contiguous word, the Poincaré model was called to determine if the same class of the root word was found within the closest words, achieving an improvement in prediction. But the results showed that most of the classifier improvement occurs in stage 3 where Poincaré models are used. The effectiveness of Poincaré embeddings to capture the relationships between words in a context in a hyperbolic space are helpful when we want to recognize the named entities in a text.

References

1. Jansen, B.J., Rieh, S.: The seventeen theoretical constructs of information searching and information retrieval. J. Am. Soc. Inf. Sci. Technol. **61**, 1517–1534 (2010)
2. Hasegawa, T., Sekine, S., Grishman, R.: Discovering relations among named entities from large corpora. In: Proceedings of the 42nd Annual Meeting on Association for Computational Linguistics, p. 415. Association for Computational Linguistics (2004)
3. Grishman, R., Sundheim, B.: Design of the MUC-6 evaluation. In: Proceedings Sixth Message Understanding Conference (MUC-6), pp. 1–11 (1996)
4. Nickel, M., Kiela, D.: Poincaré embeddings for learning hierarchical representations. In: Advances in Neural Information Processing Systems, pp. 6338–6347 (2017)
5. Harris, Z.S.: Distributional structure. Word **10**(2–3), 146–162 (1954)
6. Lay, D.C.: Linear Algebra and Its Applications, 5th edn, pp. 255–262. Addison Wesley Publishing Company, Boston (2018)
7. Sketch Engine: POS tagger (2020). https://www.sketchengine.eu/my_keywords/pos-tagger/
8. Bird, S., Klein, E., Loper, E.: Natural Language Processing with Python. O'Reilly Media, Sebastopol (2009)
9. Muñoz, D., Pérez-Téllez, F., Pinto, D.: Collaborative web-based tagger for named entities in the task of information extraction. Pistas Educativas **40**, 877–893 (2018)
10. Jurafsky, D., Martin, J.H.: Information extraction. In: Speech and Language Processing: An Introduction to Natural Language Processing, Computational Linguistics, and Speech Recognition, pp. 725–743 (2009)

Tweets Monitoring for Real-Time Emergency Events Detection in Smart Campus

Jorge Ramírez-García[1]🆔, Rodolfo E. Ibarra-Orozco[1]🆔,
and Amadeo J. Argüelles Cruz[2](✉)🆔

[1] Instituto Tecnológico del Valle de Oaxaca, Nazareno, Xoxocotlán, Oaxaca, Mexico
jorgeramirezgarciaa@gmail.com, rodolfo.io@voaxaca.tecnm.mx
[2] Instituto Politécnico Nacional, Centro de Investigación en Computación,
Nueva Industrial Vallejo, G.A. Madero, 07738 Mexico City, Mexico
jamadeo@cic.ipn.mx

Abstract. An intelligent campus has the purpose of improving the quality of life of students, making intensive, global, sustainable and efficient use of information technologies to interconnect all the actors and services for the benefit of the entire community, to establish an intelligent environment of teaching, learning and living [2]. In such smart environments, the role of users is becoming increasingly relevant, going from passive beneficiaries of services to participants assets through their social media activities. In this project, a system for detecting emergency events was developed for the IPN-Zacatenco Intelligent Campus for detecting emergency events, through analyzing messages (tweets) from Twitter users near of the area of interest. Tweets were classified under 4 categories: Mobility, Fire, Health and None (to discard unrelated tweets). In this article, we compare the machine learning models got with the Bayes Multinomial, Vector Support Machines and k-Nearest Neighbors algorithms.

Keywords: Support Vector Machines · Naive Bayes · k-Nearest Neighbors

1 Introduction

An emergency is a duly proclaimed existence of conditions of disaster to the safety of persons or property caused by air pollution, fire, floodwater, storm, epidemic, earthquake, intruder or other causes. This may be beyond the control of the services, personnel, equipment and facilities of the academy and require the combined efforts of the State or other political subdivisions. Academy facilities must be prepared to respond to an emergency or traumatic event in an organized and timely manner so that students and staff can continue to function effectively without additional trauma or the development of additional emergencies [1].

© Springer Nature Switzerland AG 2020
L. Martínez-Villaseñor et al. (Eds.): MICAI 2020, LNAI 12469, pp. 205–213, 2020.
https://doi.org/10.1007/978-3-030-60887-3_18

The Smart Campus project is carried out at the Centro de Investigación en Computación (Computing Research Centre or CIC), belonging to the Instituto Politécnico Nacional (National Polytechnic Institute or IPN), which faces the same challenges and needs as modern universities and cities.

Machine learning systems have the potential to predict and detect diseases early, so they can be treated more and prevent the occurrence of larger disasters.

We propose an architecture for an intelligent Emergency Event Detection System. The proposed architecture is based on the following four main building blocks: (i) geographically delimited Tweet acquisition, (ii) intelligent processing for knowledge discovery and emergency detection, (iii) intelligent response to send alerts in a timely, coordinated and effective manner, (iv) privacy and security to enhance a reliable and secure system, which guarantees data integrity, privacy and anonymity of users. In this document, we present our advances in the emergency detection block.

2   Machine Learning Algorithms

2.1   Support Vector Machines

Support Vector Machines (SVMs) are a popular machine learning method for classification, regression, and other learning tasks. The problem of learning the Linear Discriminant Function corresponding to SVM is posed as a convex optimization problem. This is based on the intuition that the hyperplane separating the two classes is learned so it corresponds to maximizing the margin or some kind of separation between the two classes. They are also known as maximum-margin classifiers. Some important properties of SVM [5] are:

- The problem of learning the Linear Discriminant Function corresponding to SVM is posed as a convex optimization problem. This is based on the intuition that the hyperplane separating the two classes is learnt so that it corresponds to maximizing the margin or some kind of separation between the two classes. So, they are also called as maximum-margin classifiers.
- Another important notion associated with SVMs is the kernel trick which permits to perform all the computations in the low-dimensional input space rather than in a higher dimensional feature space.

Training a Support Vector Machine requires the solution of a quadratic programming optimization problem [6]. The problem is subject to box constraints and to a single linear equality constraint; it is dense and, for many practical applications, it becomes a large-scale problem. The main idea behind the pattern classification algorithm SVM is to separate two point classes of a training set, with a surface that maximizes the margin between them. This separating surface is obtained by solving a convex quadratic program, see Eq. (1):

$$Maximize \ \ F(\Lambda) = \Lambda \cdot 1 - \frac{1}{2}\Lambda \cdot H\Lambda^T \qquad (1)$$

subject to $\Lambda \cdot y = 0$ and $0 \leq \Lambda \leq C$ with the output vector $y = (y_1, \ldots, y_n)$ and the Lagrangian multipliers vector $\Lambda = (\lambda_1, \ldots, \lambda_n)$. H is the Hessian Matrix, and it's defined as $H_{iji} = y_i y_j k(x_i \cdot x_j)$, where $k(x_i \cdot x_j)$ denotes a kernel function. C is the penalization constant. The nonzero components in the solution of Eq. 1 correspond to the training examples that determine the separating surface; these special examples are called support vectors (SVs) and their number is usually much smaller than number of instances [10].

2.2 Naive Bayes

Naive Bayes method [11], is a supervised learning algorithms based on applying Bayes' theorem with the "naive" assumption of conditional independence between every pair of features given the value of the class variable. Bayes' theorem states the following relationship, given class variable y and dependent feature vector $x_1, \ldots, x_n$:

$$P(y|x_1, \ldots, x_n) = \frac{P(y)P(x_1, \ldots, x_n|y)}{P(x_1, \ldots, x_n)} \qquad (2)$$

Given Eq. 2 and using the naive conditional independence assumption (Eq. (3))

$$P(x_i|y, x_1, \ldots, x_{i-1}, x_{i+1}, \ldots, x_n) = P(x_i|y) \qquad (3)$$

for all i, this relationship is simplified as shown in Eq. (4).

$$P(y|x_1, \ldots, x_n) = \frac{P(y) \prod_{i=1}^{n} P(x_i|y)}{P(x_1, \ldots, x_n)} \qquad (4)$$

Since $P(x_1, \ldots, x_n)$ is constant given the input, we can use the classification rule displayed in Eq. (5) and Eq. (6).

$$P(y|x_1, \ldots, x_n) \approx P(y) \prod_{i=1}^{n} P(x_i|y) \qquad (5)$$

$$y = argmax P(y) \prod_{i=1}^{n} P(x_i|y) \qquad (6)$$

Naive Bayes Multinomial. Naive Bayes Multinomial (MultinomialNB) [7], implements the naive Bayes algorithm for multinomially distributed data, and is one of the two classic Naive Bayes variants used in text classification (where the data are typically represented as word vector counts, although tf-idf vectors are also known to work well in practice).

2.3   k-Nearest Neighbors

The k-Nearest Neighbors rule is among the simplest statistical learning tools in density estimation, classification, and regression [3]. Trivial to train and easy to code, the nonparametric algorithm is surprisingly competitive and fairly robust to errors given good cross-validation procedures.

The k-nearest neighbour classifier works based on the distance calculated between the query point and each data points in the training dataset. Then we choose the K closest points and take a vote of their class labels to decide the label of the query point [12].

Nearest neighbor algorithms are attractive because they are easy to implement, nonparametric, and learning-based.

3   Term Frequency – Inverse Document Frequency

TF-IDF stands for Term Frequency-Inverse Document Frequency [8]. This technique was originally developed as a metric for ranking functions for showing search engine results based on user queries and has come to be a part of information retrieval and text feature extraction now.

Mathematically, TF-IDF is the product of two metrics and can be represented as $tfidf = tf \times idf$, where term frequency tf and inverse-document frequency (idf) represent the two metrics.

Inverse document frequency denoted by idf is the inverse of the document frequency for each term. It is computed by dividing the total number of documents in our corpus by the document frequency for each term and then applying logarithmic scaling on the result. In our implementation we will be adding 1 to the document frequency for each term just to indicate that we also have one more document in our corpus that essentially has every term in the vocabulary. This is to prevent potential division-by-zero errors and smoothen the inverse document frequencies. We also add 1 to the result of idf computation to avoid ignoring terms completely that might have zero. Mathematically the implementation for idf is represented in Eq. (7).

$$idf = 1 + log\frac{C}{1 + df(t)} \tag{7}$$

where $idf(t)$ represents the idf for the term t, C represents the count of the total number of documents in our corpus, and $df(t)$ represents the frequency of the number of documents in which the term t is present.

Thus the term frequency-inverse document frequency can be computed by multiplying the above two measures together.

The normalized version of the $tfidf$ matrix is obtained dividing it with the L2 norm of the matrix, also known as the Euclidean norm, which is the square root of the sum of the square of each term's $tfidf$ weight. Mathematically, the final $tfidf$ feature vector is represented in Eq. (8).

$$tfidf = \frac{tfidf}{||tfidf||} \tag{8}$$

4   Twitter

Twitter [4], is an online social networking and blogging service that allows users to read the tweets of other users by "following" them. Social networks have an enormous number of users. As of the first quarter of 2019, Twitter averaged 330 million monthly active users [9]. With user numbers in the millions, inevitably, massive amounts of data are generated through short messages called "tweets" which are sent in real time using any digital device such a cellphone. Tweets were originally restricted to only 140 characters, but this limit was doubled in November 2017. One of the main features of Twitter is the use of the hashtag symbol # followed by a keyword related to the subject topic. The ability to search Twitter by topics that are mentioned within the tweets allows us to access and interpret large amounts of data.

Twitter allows to search for tweets published in the past 7 days using the Twitter Search API.

5   Alert System for Emergency Events

The procedure used for the data analysis is shown in Fig. 1, and in the following we describe how each of the stages was developed.

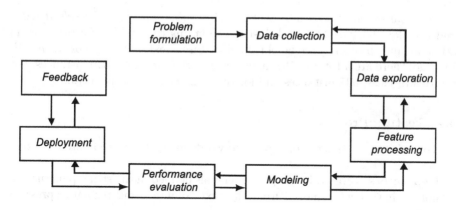

Fig. 1. Proposed methodology for the Machine Learning process.

5.1   Problem Formulation

Increasing urbanization brings new challenges to reducing the threat of emergencies. Yet emergencies are often ignored until they strike, when the damage has been done and relief is the only response. The situation is compounded by the separation of a campus programming from emergencies management. Alerting students from emergency events happening near the campus could help reduce the risk of damage to the student's integrity.

For the IPN-Zacatenco smart campus, the following categories of emergency events were identified to be classified automatically:

- Mobility.
- Fire.
- Health.
- None.

5.2 Data Collection

To use the resources provided by the Twitter API tool, a developer account was requested. In this way, access to the collection of tweet information was achieved. A Python script using the tweepy library was used to create a streaming connection with the Twitter Streaming API. To delimit the geographic area to be monitored, the Bounding Box tool was used. Finally, the date, time, text, location and verification of the account collected by the API were stored in a database.

5.3 Data Exploration

The database is made up of a total of 2,013 instances, with start date 03/10/2020 and end date 06/25/2020. Each instance contains the text of the tweet as an attribute and the event type label (Mobility, Fire, Health, None). For the mobility class, there are a total of 534 instances. For the Fire class, 510 instances. For the Health class, 507 instances and for the None class, 462 instances.

5.4 Feature Processing

- **Tokenization.** Tokens correspond to words in the tweet, separated by commas.
- **Characters Removal.** All special symbols defined in the string.punctuation library have been removed from the text column using regular expression matches.
- **Stopwords Removal.** It consists of the elimination of words that provide little or no information for the analysis of texts. Empty words are usually articles, conjunctions, prepositions, pronouns, etc.
- **Instances Labeling.** Each Tweet was manually reviewed and a tag with the class Mobility, Fire, Health or None was added. Table 1 shows examples of the collected tweets after tagging.
- **Feature Extraction.** For each term in our data set, a $tf\text{-}idf$ measure was calculated, which is represented in a numerical matrix.

Table 1. Examples of tagged tweets

Tweet	Label
buenas tardes requerimos el apoyo para un choque automovilistico entre un camion de reparto y un sedan negro en Patriotismos y Alfonso Reyes col Hipodromo Condesa	Mobility
seguimos en riesgo de contagio hay que mantener las medidas preventivas para poder trabajar	Health
hubo un incendio en un depto del edificio de la entrada y solo crisis nerviosas e intoxicación leve	Fire
nadie entra a nuestras vidas por accidente	None

5.5 Modeling

Modeling with the SVM, k-NN, and MultinomialNB algorithms was implemented using the Python's Scikit Learn library.

For the training stage, the data set was divided into a training set (with 70% of the data) and a test set (with 30% of the data). With the SVM algorithm, tests were performed with the polynomial, gaussian and linear kerner, having the best results with the linear kernel. For training with k-NN, the best results were obtained with k = 10. The MultinomialNB algorithm was used with $\alpha = 0$.

5.6 Performance Evaluation

Figure 2 shows the confusion matrices of tests performed with cross validation.

Figure 3 compares, using box diagrams, the accuracy of each algorithm at the testing stage. The accuracy for the model built with the SVM algorithm is 0.888791, for the MultinomialNB algorithm it is 0.865955 and for k-NN it is 0.80481.

Taking into account the results obtained, the model generated with the SVM algorithm was selected to carry out the implementation.

5.7 Deployment

In order to carry out the implementation, a display system was developed where the most recent tweets of an emergency event near the campus are displayed. The information displayed to the user for each tweet is the date and time, the text of the tweet, location (registered by the user in their account), and account verification.

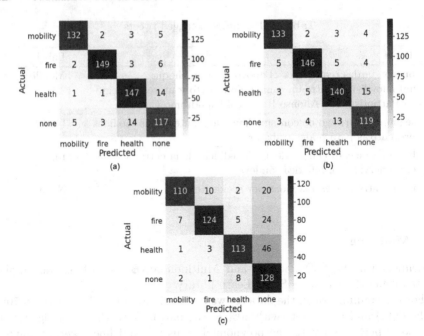

Fig. 2. Confusion matrices of the three learning models: (a) MSV, (b) MultinomialNB, (c) k-NN.

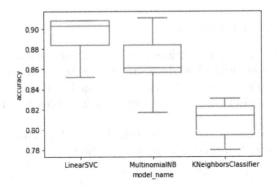

Fig. 3. Box plot analysis of accuracy among algorithms.

6   Conclusions

In this project, a system for the detection of emergency events, related to the categories of mobility, fires and health was developed analyzing the Twitter data flow. The following results were obtained:

- In order to treat our problem as a supervised classification problem we created a supervised dataset. This dataset contains Tweets from geographical locations near the campus.

– Three Machine Learning algorithms for the text classification were evaluated and the algorithm showing the best performance was the Vector Support Machines algorithm.

As future work, we plan to test more advanced word vectorization models such as the word2vec algorithm, which is a neural network-based implementation that learns representations of word vector distributions based on the Bag of Words and skip-gram architectures. Within the system for sending alerts, a module will be developed to learn when alerts should be sent, taking information on how many tweets have been generated for a topic, the location and verification of the account.

References

1. SF Flex Academy: Emergency response plan (2011)
2. Universidad de Alicante: UA Smart University (2020). https://web.ua.es/es/smart/el-proyecto.html. Accessed 24 July 2020
3. Karl, S.: An adaptable k-nearest neighbors algorithm for MMSE image interpolation. IEEE Trans. Image Process. 18, 1976–1987 (2009)
4. Mai, M., Leung, C.K., Choi, J.M.C., Kwan, L.K.R.: Big data analytics of twitter data and its application for physician assistants: who is talking about your profession in Twitter? In: Alhajj, R., Moshirpour, M., Far, B. (eds.) Data Management and Analysis. SBD, vol. 65, pp. 17–32. Springer, Cham (2020). https://doi.org/10.1007/978-3-030-32587-9_2
5. Murty, M.N., Raghava, R.: Support Vector Machines and Perceptrons. SpringerBriefs in Computer Science. Springer, Cham (2016). https://doi.org/10.1007/978-3-319-41063-0
6. Osuna, E.: Support vector machines: training and applications. Massachusetts Institute of Technology (1997)
7. Rennie, J.: Tackling the poor assumptions of Naive Bayes text classifiers. In: ICML (2003)
8. Sarkar, D.: Text Analytics with Python: A Practical Real-World Approach to Gaining Actionable Insights from Your Data. Apress, New York (2016)
9. Statista: Number of monthly active Twitter users worldwide from 1st quarter 2010 to 1st quarter 2019 (2019). https://www.statista.com/statistics/282087/number-of-monthly-active-twitter-users/. Accessed 24 July 2020
10. Zanghirati, G.: A parallel solver for large quadratic programs in training support vector machines. Parallel Comput. 29, 535–551 (2003)
11. Zhang, H.: The optimality of Naive Bayes. American Association for Artificial Intelligence (2004)
12. Zhenghui, M.: K-Nearest-Neighbours with a novel similarity measure for intrusion detection. In: 13th UK Workshop on Computational Intelligence (UKCI) (2013)

Evaluation of Similarity Measures in a Benchmark for Spanish Paraphrasing Detection
19th Mexican International Conference on Artificial Intelligence

Helena Gómez-Adorno[1（✉）], Gemma Bel-Enguix[2], Gerardo Sierra[2], Juan-Manuel Torres-Moreno[3,4], Renata Martinez[5], and Pedro Serrano[5]

[1] Instituto de Investigaciones en Matemáticas Aplicadas y en Sistemas, Universidad Nacional Autónoma de México, Mexico City, Mexico
helena.gomez@iimas.unam.mx

[2] Instituto de Ingeniería, Universidad Nacional Autónoma de México, Mexico City, Mexico
{gbele,gsierram}@iingen.unam.mx

[3] LIA-Université d'Avignon, Avignon, France
juan-manuel.torres@univ-avignon.fr

[4] Polytechnique Montréal, Montreal, Canada

[5] Facultad de Ciencias, Universidad Nacional Autónoma de México, Mexico City, Mexico
{renata_mtzg,pedrotlc}@ciencias.unam.mx

Abstract. In this paper, we present a similarity-based approach towards paraphrase detection in Spanish. We evaluate various models for semantic similarity computation using a gold-standard paraphrase corpus. It contains one original document and paraphrased documents on different levels (low and high), and reference documents on the same topic or same vocabulary. It allows to assess the similarity between a pair of texts or individual sentences. We found that some of the similarity metrics have a larger difference when comparing paraphrased sentences than others. Finally, we obtained a threshold for each of the similarity metrics with the aim of determining a classification boundary to decide if two sentences are paraphrased.

Keywords: Text similarity · Paraphrasing detection · Corpus linguistics

1 Introduction

Paraphrases are frequently defined as a series of expressions, linguistic forms, or alternative verbalizations, which convey the same information as an original expression within a language [13]. Other authors include within the paraphrases those expressions that have approximately the same meaning or equivalent content.

© Springer Nature Switzerland AG 2020
L. Martínez-Villaseñor et al. (Eds.): MICAI 2020, LNAI 12469, pp. 214–223, 2020.
https://doi.org/10.1007/978-3-030-60887-3_19

In recent years, paraphrasing has generated great interest in the computational linguistics community. Both, the ability to automatically change a text for a similar one and the ability to detect when the same thing is being said in two different texts are highly relevant to many Natural Language Processing (NLP) applications.

The task of paraphrase detection (PD) aims at identifying whether a set of sentences (typically a pair) are semantically equivalent. [4] mention that, although the identification of paraphrases is defined in semantic terms, it is traditionally approached with statistical classifiers based on the lexicon, n-grams and, syntactic matching.

Thus, solving this task requires several levels of analysis, among them syntactic and semantic issues. The construction of paraphrases includes mechanisms such as changing words by synonyms, rearranging sentences, dividing them, and breaking them down into multiple sentences in a different order. This is why those lexical and syntactic classifiers are not enough to solve this task.

The purpose of this paper is the assessment of similarity measures between pairs of sentences towards paraphrase detection. For that purpose, we used a gold-standard comparable corpus containing source documents and similar texts. The last group consists of: (a) paraphrases of the source, (b) texts that deal with the same topic, (c) texts that share a given amount of lexical units although not sharing the topic.

The paper is organized as follows. In Sect. 2 we present the state of the art in the area. Section 3 is devoted to introducing the Sushi Corpus. Section 4 explains the similarity measures that are used in this paper in order to evaluate them in the corpus. In Sect. 5, we provide an analysis of the results. Section 6 introduces the reference dataset. Finally, the paper ends with conclusions and future work in Sect. 7.

2 Related Work

According to [1], two documents can be similar without having a direct relationship to each other, but sharing a similarity with a third text. For example, different news articles derived from a specific source provided by an agency are similar. In the same way, students' homework on a particular subject can have high similarity. Many previous works on paraphrase detection use text similarity-based methods [8], while others use a supervised approach depending on machine learning and deep learning methods [7].

In [12], a textual German corpus for similarity detection is presented. Several simple measures were evaluated for both, the complete documents and individual sentence pairs. A paraphrase and semantic similarity approach was introduced in [5] using a machine learning algorithm trained on lexical, semantic, syntactic, semantic, and pragmatic features; the method is then evaluated on noisy user-generated short-text data such as Twitter.

In this work, we explore Text Similarity (TS) metrics for paraphrase detection. There are various approaches to similarity, such as vector space models (term-based), text alignment (linguistic knowledge-based), and n-gram overlapping (string-based). Substantial work has been carried out for paraphrase detection with TS metrics on various monolingual corpora of paraphrases (especially in English). Some of the most popular corpora to evaluate PD are the METER Corpus [3], the Microsoft Research Paraphrase Corpus [6], and the PAN Plagiarism Corpus [11].

3   The Sushi Corpus

For the assessment of similarity measures, we use a corpus made up of original texts and their paraphrase. This corpus is designed to assess the similarity between a pair of texts and to evaluate different similarity measures, both for whole documents or for individual sentences. The corpus is built around the subject of a Spanish blog article related to *Sushi* [2]. Several volunteers (undergraduate, graduate, and Ph.D. students) were asked to intentionally reformulate or paraphrase this article. The paraphrase of the article was carried out on two levels, according to the rules given in [12]:

- Low level: Only lexical variation.
- High level: Lexical, syntactic, textual or discursive organization variation and fusion or separation of sentences.

Two more types of texts were added to the corpus:

- Texts on the same theme and source as the original article, related to sushi.
- Texts on different theme as the original article but with overlapping vocabulary were gathered. That is, texts not related to sushi, but with exactly the same vocabulary as the original one. Some volunteers wrote a free text using the same content words as the original.

Therefore, the corpus consists of the following types of texts:

1. Original text (OR): Sushi article.
2. Paraphrased texts:
   (a) Low level (PL)
   (b) High level (PH)
3. Texts with the same theme as the original text (PNO)
4. Texts with different theme as the original text with overlapping vocabulary (SNO).

The corpus statistics are presented in Table 1. As it can be observed, the average number of words in the documents for each type of texts is similar. It can also be observed the lexical overlap between the original document and the other types of documents (PL, PH, PNO and SNO). As expected, the largest overlap is between the original document and the low level paraphrases documents, however, it is notable that there is a higher overlap between the *no sushi* documents and the original document than the *no paraphrase* documents and the original document.

4   Similarity Measures for Paraphrase Detection

Textual similarity metrics can be classified as in geometrical model metrics, the set theory model metrics, and lexical metrics. According to the earlier classification, in this section we describe the similarity metrics used in this work.

Table 1. Corpus statistics

Type of text	Quantity	Avg. no. of words	Overlap with original
Original	1	362.00	362
Low level	7	291.00	293
High level	5	303.20	221
No paraphrase	4	241.00	120
No Sushi	6	299.33	154

4.1   Metrics Based on the Geometrical Model

Documents written in natural language can be represented using vectors from an euclidean space. Suppose that we have two documents D_1 and D_2. We create a dictionary of size n, from the terms that form each document. So, the associated space to both documents has n dimension. The ith component from the vector that represents the document D_j, indicates the frequency in which the ith term from the dictionary appears in the document.

If D_1 and D_2 are two documents represented by vectors $\vec{x_1}$ and $\vec{x_2}$, respectively. The following metrics are defined as:

– Cosine similarity:

$$sim_{cos}(D_1, D_2) = \frac{\langle \vec{x_1}, \vec{x_2} \rangle}{\|x_1\|\|x_2\|} \qquad (1)$$

– Word2vec (w2v) based similarity: Word2vec is a model for representing words as a vector [9]. It captures the relation among words from the training on a large corpus and codifies the words into a vector of features. In this work we calculate the similarity of sentences using the word vectors obtained from word2vec in the Cosine Similarity equation described above.

4.2   Metrics Based on the Set Theory Model

In the Set theory model, the similarity between D_1 and D_2 is measured by a function that compares the elements shared by D_1 and D_2, taking into consideration the elements that are only in D_1 and the elements that are only in D_2:

$$sim(D_1, D_2) = f(D_1 \cap D_2, D_1 \triangle D_2) \qquad (2)$$

– Dice quotient

The Dice quotient takes the number of words shared by both strings and divides it by the total sum of the words of the texts D_1 and D_2:

$$sim_{Dice}(D_1, D_2) = \frac{2 \cdot |D_1 \cap D_2|}{|D_1 \vartriangle D_2|} \tag{3}$$

– Jaccard quotient

This coefficient is obtained by dividing the intersection of the terms by their union:

$$sim_{Jaccard}(D_1, D_2) = \frac{|D_1 \cap D_2|}{|D_1 \cup D_2|} \tag{4}$$

– n-grams

The n-grams are subsequences of elements (characters or words) taken from the sentence pairs of the compared texts. The computation of similarity by n-grams between the sentences pairs S_1 and S_2, consists in dividing the number of n-grams that share both sentences by the total number of n-grams:

$$sim_{n\text{-}gram}(D_1, D_2) = \frac{|ng(D_1) \cap ng(D_2)|}{|ng(D_2)|} \tag{5}$$

ng(D) are the n-grams in D.

4.3   Lexical Metrics

– Levenshtein

The result of this metric is the minimum number of operations required to transform one word into another. The editing operations are: an insert, deletion or substitution of a character. In this measure as long as the value is higher, the similarity value is very low.

– Jaro-Winkler

This metric is applied to words, it uses the number of characters that both words share, taking into account the characters that are in the same position and those that are transposed, with these data the metric is defined as:

$$sim_{J\text{-}W}(t_1, t_2) = \begin{cases} 0, & m = 0 \\ \frac{m}{3|t_1|} + \frac{m}{3|t_2|} + \frac{m-t}{3m}, & m \neq 0 \end{cases} \tag{6}$$

t_1 and t_2 are words and is the number of characters that match their position in both words, and t is the number of characters that match and is in different positions.

4.4   Metrics Based on Textual Energy

The analysis method uses textual energy [10]. It considers that a document could be coded as a set of small magnets called spins with two possible values: +1 (word present) and −1 (word absent). The method is language independent. It codifies the relationship between word and phrases as a graph. Every word is influenced by the others, directly or indirectly, since they all belong to the same system. Therefore, the words in the document interact giving sense to other words and coherence to the whole text.

Textual Energy connects, at the same time, sentences having common words because it includes the graph of intersection, as well as sentences which share the same neighbourhood without to necessarily share the same vocabulary.

Textual energy can be deduced through two graphs: one of document words and phrases, and the other of words for words. In a graph of words and phrases, the edges are the phrases and the arcs are the words in common. The other graph allows calculating the interactions of the words between them. Technically, the paths of length 2 are used in the corresponding graph to deduce the textual energy. With the document encoded as vector S of P phrases and M words, the textual energy is calculated by:

$$E = (SxS^T)^2 \tag{7}$$

where S^T is the transposed matrix of S. E is a matrix of $P \times P$, that indicates the similarity of a given phrase, with respect to all the others.

The textual energy let identify the relationship between phrases, even no sharing common words but through neighbour words. The following measures are calculated (All measures are normalized between 0 and 1):

- REC: measures the lexical similarity and is calculated by the coverage of the vocabulary.
- τ: measures the semantic similarity and is calculated by E.
- F1: is the lineal combination of Rec and τ.

5   Similarity Analysis Results on the Sushi Corpus

All similarity metrics were calculated as a function $sim(D_1, D_2)$, D_1 and D_2 are the texts for which their similarity is to be measured.

The result of applying the function to two documents is a matrix of size $|D_1| * |D_2|$, where $|D_1|$ is the number of sentences in D_1 and $|D_j|$ is the number of sentences pairs in the text D_j. The function takes the jth sentence of text D_1 and calculates its similarity with each of the sentences of text D_2. The rows in the matrix correspond to sentences of text D_1 and the columns to the sentences of text D_2, whereby the element of the matrix at position (i, j) indicates the level of similarity between row i^{th} of D_1 and row j^{th} of D_2.

The similarity averages presented in Table 2 were obtained by comparing each pair of sentences obtained from the combination of the Original document (D_1) and the rest of the documents of the corpus (D_2), using the metrics described in the previous section, and calculating the relevant arithmetic mean. The average similarity results are categorized according to the document type of the comparison, i.e., low paraphrasis, high paraphrasis, No paraphrasis, No sushi. The columns labeled as low and high paraphrasis represent the average similarity values for each metric for paraphrased sentence pairs. The columns labeled as baseline correspond to the average metric presented for non paraphrased sentence pairs in low paraphrasis (PL) and high paraphrasis (PH) texts. Analogously, the remaining two columns, represent non paraphrased texts (PNO), and non related texts with similar vocabulary (SNO), respectively. These were obtained to establish a better notion of a metric that can be indicative of a paraphrased sentence.

Table 2. Average similarity scores of the pairwise comparison between the original document and the other types of documents

Metrics	Low baseline	Low paraphrasis	High baseline	High paraphrasis	No paraphrasis	No Sushi
Cosine	0.665	0.817	0.677	0.798	0.678	0.679
W2V	0.705	0.895	0.680	0.872	0.686	0.680
Dice	0.052	0.171	0.037	0.083	0.027	0.018
Jaccard	0.427	0.697	0.433	0.539	0.491	0.426
N-grams	0.080	0.381	0.022	0.294	0.007	0.005
Levenshtein	133.679	91.437	142.571	140.857	131.821	137.929
Jaro-Winkler	0.631	0.764	0.642	0.676	0.647	0.623
Rec	0.034	0.356	0.026	0.210	0.030	0.022
Tau	0.669	0.888	0.634	0.893	0.613	0.506
F1	0.062	0.488	0.049	0.304	0.054	0.039

Figure 1 shows the same similarity results as in Table 2 in a graphical way. It can be observed that the F1 and Rec similarities present a grater difference when the metric is calculated on the paraphrased sentences (PL and PH), than when they are calculated on the non paraphrased sentences (PL baseline, PH baseline, PNO, and SNO).

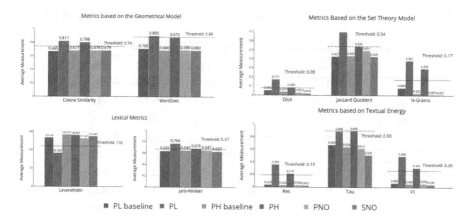

Fig. 1. Average similarity scores including the calculated threshold for each metric

6 Paraphrase Detection Using Similarity Metrics

For the paraphrase detection evaluation, we build a reference dataset based on the Sushi corpus. It consists of every possible sentence pair between the original text and the highly paraphrased sentences, low paraphrased texts, and non paraphrased texts with their corresponding tags. This yielded a dataset conformed by 6896 instances, 314 of which were confirmed paraphrased sentences from the original text.

Subsequently, the same measures were calculated for the described dataset. Given that the paraphrase detection problem presents a binary classification problem with arbitrary non negative real values (continuous output), the boundary must be established by a threshold value in order to minimize the number of false hits and maximize the performance of the similarity based detection system when applied to the dataset.

The values obtained by applying the measures to the dataset allowed us to perform a receiver operating characteristic analysis (ROC) in order to optimize the threshold assigned to each metric. Table 3 presents the evaluation metrics when the classification is performed using the similarity metrics with the corresponding threshold, i.e., if a sentence pair achieves more tan 0.74 of similarity score using the cosine metric then the sentence pairs are paraphrased. Paraphrasing detection based on similarity score achieve accuracy results greater than 0.9 in general.

Table 3. Paraphrase detection performance using each similarity metric individually

Metrics	Threshold	Accuracy	Precision	Recall	F1 score
Cosine	0.74	0.986	0.773	0.850	0.806
W2V	0.88	0.981	0.760	0.679	0.715
Dice	0.08	0.965	0.709	0.732	0.631
Jaccard	0.54	0.983	0.713	0.900	0.793
Levenshtein	110	0.980	0.743	0.693	0.713
Jaro-Winkler	0.67	0.986	0.767	0.907	0.829
N-grams	0.17	0.967	0.751	0.789	0.668
Rec	0.19	0.989	**0.853**	0.853	**0.843**
Tau	0.90	0.971	0.610	0.564	0.582
F1	0.26	0.980	0.685	0.864	0.759

7   Conclusions

We evaluated several semantic similarity models and evaluate their use for para-phrase detection in a Spanish benchmark. First, we described each of the simi-larity metrics. Then we evaluated the metrics on our reference corpus, analyzing their values on paraphrased and non sentences pairs. We found that the F1 and Rec metrics have larger difference when calculated on paraphrased and non paraphrased sentences pairs than the other measures.

Finally, we performed a classification of the sentences pairs using a threshold calculated for each metric. We show that it is possible to detect paraphrases with 84.3% of F1-score using the Rec similarity metric.

As future research we plan to extend the Spanish paraphrase corpus and validate the results on similarity based paraphrase detection. We aim to inves-tigate the performance of context aware based word embeddings for detecting paraphrasis in Spanish.

Acknowledgments. This work has been partially supported by PAPIIT projects IA401219, TA100520, AG400119 and CONACYT project A1-S-27780.

References

1. Bendersky, M., Croft, W.B.: Finding text reuse on the web. In: Proceedings of the Second ACM International Conference on Web Search and Data Mining, pp. 262–271 (2009)
2. Castro, B., Sierra, G., Torres-Moreno, J.M., Da Cunha, I.: El discurso y la semántica como recursos para la detección de similitud textual. In: Proceedings of the III RST Meeting (8th Brazilian Symposium in Information and Human Language Technology, STIL 2011). Brazilian Computer Society, Cuiabá (2011)
3. Clough, P., Gaizauskas, R., Piao, S.S., Wilks, Y.: METER: measuring text reuse. In: Proceedings of the 40th Annual Meeting of the Association for Computational Linguistics, pp. 152–159. ACL (2002)

4. Das, D., Smith, N.A.: Paraphrase identification as probabilistic quasi-synchronous recognition. In: Proceedings of the Joint Conference of the 47th Annual Meeting of the ACL and the 4th IJCNLP of the AFNLP: Volume 1, pp. 468–476. ACL (2009)

5. Dey, K., Shrivastava, R., Kaushik, S.: A paraphrase and semantic similarity detection system for user generated short-text content on microblogs. In: Proceedings of COLING 2016, the 26th International Conference on Computational Linguistics: Technical Papers, pp. 2880–2890 (2016)

6. Dolan, W., Quirk, C., Brockett, C., Dolan, B.: Unsupervised construction of large paraphrase corpora: exploiting massively parallel news sources. In: Proceedings of the 20th International Conference on Computational Linguistics (COLING) (2004)

7. El Desouki, M.I., Gomaa, W.H., Abdalhakim, H.: A hybrid model for paraphrase detection combines pros of text similarity with deep learning. Int. J. Comput. Appl. **975**, 8887 (2019)

8. Fernando, S., Stevenson, M.: A semantic similarity approach to paraphrase detection. In: Proceedings of the 11th Annual Research Colloquium of the UK Special Interest Group for Computational Linguistics, pp. 45–52 (2008)

9. Mikolov, T., Sutskever, I., Chen, K., Corrado, G., Dean, J.: Distributed representations of words and phrases and their compositionality (2013)

10. Molina, A., Torres-Moreno, J.M., SanJuan, E., Sierra, G., Rojas-Mora, J.: Analysis and transformation of textual energy distribution. In: 2013 12th Mexican International Conference on Artificial Intelligence, pp. 203–208. IEEE (2013)

11. Potthast, M., Stein, B., Eiselt, A., Cedeño, A.B., Rosso, P.: Overview of the 1st international competition on plagiarism detection. In: 3rd PAN Workshop. Uncovering Plagiarism, Authorship and Social Software Misuse, p. 1 (2009)

12. Torres-Moreno, J.M., Sierra, G., Peinl, P.: A German corpus for similarity detection tasks. Int. J. Comput. Linguist. Appl. **5**(2), 9–24 (2014)

13. Zhou, L., Lin, C.Y., Munteanu, D.S., Hovy, E.: ParaEval: using paraphrases to evaluate summaries automatically. In: Proceedings of the main conference on Human Language Technology Conference of the North American Chapter of the Association of Computational Linguistics, pp. 447–454. Association for Computational Linguistics (2006)

Similarity-Based Correlation Functions
for Binary Data

Ildar Z. Batyrshin[1]([⊠]) [iD], Ivan Ramirez-Mejia[1], Ilnur I. Batyrshin[2],
and Valery Solovyev[2] [iD]

[1] Instituto Politécnico Nacional, Centro de Investigación En Computación, CDMX, Mexico
batyr1@gmail.com, ramirez.alvarez.ipn@gmail.com
[2] Kazan Federal University, Kazan, Russia
batyrshin@gmail.com, maki.solovyev@mail.ru

Abstract. The purpose of this study is to survey the correlation and association coefficients introduced previously on the set of binary n-tuples and to determine coefficients satisfying the properties of correlation functions. These functions were recently introduced on the sets with involutive operation as functions generalizing classical correlation coefficients: Pearson's product-moment correlation, Spearmen's and Kendall' rank correlation coefficients, Yule's Q and Hamann's association coefficients, etc. It is shown that several, but not all, known correlation and association coefficients defined on the set of binary n-tuples, satisfy the properties of correlation functions. For these association coefficients, there were established similarity measures on the set of binary data that can be used for the generation of these association coefficients. A new parametric family of correlation functions for binary data is proposed. As a particular case, it contains Hamann's association coefficient.

Keywords: Binary data · Correlation coefficient · Association coefficient · Correlation function · Similarity measure · Negation of Binary n-Tuples

1 Introduction

Last decades it was drastically increased the number of works on similarity, correlation, association, and interestingness measures on different domains [1, 4, 5, 7–10, 12–17]. The study of such measures on the domain of binary n-tuples has a very long history. The first works on this topic appeared more than one hundred years ago [11, 18, 19] when Yule's Q and Yule's W association coefficients, and Jaccard's similarity measure were introduced. Now there are proposed several tens of such measures [8, 17]. Binary n-tuples $x = (x_1, \ldots, x_n)$, where, $x_i \in \{0, 1\}$, $i = 1, \ldots, n$, appear in many applications. In pattern recognition tasks, such n-tuple (vector) can represent the values of n binary or dichotomous attributes measured for some object x, and some measure of similarity between binary n-tuples can be used in the classification of objects on classes of similar objects. In statistics, a binary n-tuple can represent n measurements of one dichotomous variable, and some coefficient of correlation or association applied to two n-tuples can be used as a measure of the relationship between corresponding two variables.

L. Martínez-Villaseñor et al. (Eds.): MICAI 2020, LNAI 12469, pp. 224–233, 2020.
https://doi.org/10.1007/978-3-030-60887-3_20

Although the measures of similarity and correlation can be used for different purposes, in recent years, they are considered together for several reasons. First, the methods of their construction are similar and usually based on the representation of relationships between two n-tuples as 2×2 table. Second, in some applications, the correlation measures used as similarity measures. Third, in data mining, the measures of similarity, correlation, and association are considered together with other measures as measures of interestingness in the analysis of the relationship between data [16].

Recently it was introduced an axiomatic definition of correlation functions (association measures) as functions defined on a set with involution (negation) operation and satisfying several properties based on the properties of Pearson's product-moment correlation coefficient [2, 3]. Several methods of construction of correlation functions from similarity functions (similarity measures) satisfying suitable properties have been proposed [2, 3, 5, 6]. It was shown that many known correlation coefficients can be constructed as particular cases of these methods [2–6].

The paper has the following goals. First, to extend the definition of correlation functions on the binary domain and to determine what of the known correlation and association coefficients defined for binary data satisfy the properties of correlation functions. Second, to establish similarity measures that can be used for constructing these correlation functions using methods proposed in [2–4]. Third, based on these methods, to propose a new parametric family of correlation functions for binary data, including as a particular case Hamann's association coefficient.

The paper has the following structure. Section 2 gives an introduction to binary similarity, correlation, and association measures. Section 3 considers the basic definitions and properties of similarity and correlation functions and gives the method of construction of correlation functions from similarity functions. Section 4 determines what of the known binary association coefficients are correlation functions, and shows from what similarity measures these association coefficients can be constructed. Section 5 introduces a new parametric family of correlation functions on the binary domain. Section 6 contains conclusions and describes the directions of future work.

2 Binary Similarity Measures and Association Coefficients

For two binary n-tuples $x = (x_1, \ldots, x_n)$, $y = (y_1, \ldots, y_n)$, considered as n measurements of two dichotomous (binary) variables, denote:

- a, the number of measurements when $x_i = y_i = 1$;
- b, the number of measurements when $x_i = 1, y_i = 0$;
- c, the number of measurements when $x_i = 0, y_i = 1$;
- d, the number of measurements s when $x_i = y_i = 0$,

where $i = 1, \ldots, n$.

The numbers a and d are called the numbers of *positive* and *negative matches*, respectively. n-tuples can also be considered as vectors with n binary attributes. The numbers a, b, c, d, usually represented by 2×2 table, see Table 1.

Table 1. 2×2 table

		Y	$\overline{Y}$	
		1	0	
X	1	a	b	$a+b$
$\overline{X}$	0	c	d	$c+d$
		$a+c$	$b+d$	n

For each n-tuple $x = (x_1, \ldots, x_n)$, it is defined its negation: $\bar{x} = (1 - x_1, \ldots, 1 - x_n)$. This negation is involutive, i.e., for any binary n-tuple x it is fulfilled: $\bar{\bar{x}} = x$.

Consider examples of similarity measures and association coefficients [8, 17]:

Jaccard's similarity measure:

$$S_J = \frac{a}{a+b+c},$$

Simple Matching similarity measure:

$$S_{SM} = \frac{a+d}{a+b+c+d},$$

Hamann's association coefficient:

$$A_H = \frac{(a+d)-(b+c)}{a+b+c+d},$$

Yule's Q association coefficient:

$$A_{YQ} = \frac{ad-bc}{ad+bc}.$$

At this moment, it is known more than 40 similarity and association measures for binary data with different properties [8, 17]. Similarity measures usually used in classification and pattern recognition tasks. Association and correlation coefficients usually used for the analysis of relationships between variables. To compare similarity and association measures, analyze relationships between them, and possible transformation of one measure into another, in [4] it was proposed to consider these measures as functions defined on a set with involution (negation) operation. In the following section, we consider the basic definitions and results of such an approach applied to the set of binary n-tuples.

3 Similarity and Correlation Functions

Similarity measures and association coefficients considered in the previous section calculate for any two binary n-tuples $x = (x_1, \ldots, x_n)$ and $y = (y_1, \ldots, y_n)$, a similarity or an association value; hence they can be considered as functions of two arguments x and y. Usually, similarity measures take values between 0 and 1, and association coefficients take values between -1 and 1, for this reason, we will consider similarity measures as functions $S(x, y)$ taking values in the interval $[0,1]$ and association coefficients as functions $A(x, y)$ taking values in the interval $[-1,1]$. We will require that the "rational" similarity measures for any n-tuples x and y satisfy the following two properties [4]:

$$S(x, y) = S(y, x), \qquad \text{(symmetry)}$$
$$S(x, x) = 1. \qquad \text{(reflexivity)}$$

Such similarity measures are called *similarity functions*. The first property means that the value of similarity between x and y does not depend on the ordering of these arguments in the function S. The second property means that any n-tuple x similar to itself with the maximum value 1. Similarity functions formally coincide with fuzzy relations [4, 20]. We will also denote the similarity measures considered in the previous section and taking values in the interval $[0,1]$ as functions of four parameters $s(a, b, c, d)$. We can rewrite for them the symmetry and reflexivity properties as follows:

$$s(a, b, c, d) = s(a, c, b, d), \text{ (symmetry)}$$
$$s(a, 0, 0, d) = 1. \qquad \text{(reflexivity)}$$

The first property means that if in the formula of a similarity measure, for example, in Jaccard similarity measure, we replace b by c and c by b, then the value of the formula does not change. The second property means that when $b = 0$ and $c = 0$, the value of a reflexive similarity measure should be equal to 1. One can easily check that both properties are fulfilled for Jaccard and Simple Matching similarity measures.

A similarity function S is called *non-contradictive* or *consistent* if for all binary n-tuples x it is fulfilled:

$$S(x, \bar{x}) = 0. \quad \text{(consistency)}$$

This property is fulfilled for all reasonable similarity measures because any n-tuple x and its negation $\bar{x}$ have different values 1 or 0 in all components, i.e. $x_i \neq (\bar{x})_i$ for all $i = 1, \ldots, n$. This property can be presented in such form:

$$s(0, b, c, 0) = 0.$$

Dually to the similarity function, one can introduce a dissimilarity function $D(x, y)$ taking values in the interval $[0,1]$ and satisfying the properties:

$$D(x, y) = D(y, x), \qquad \text{(symmetry)}$$
$$D(x, x) = 0. \qquad \text{(irreflexivity)}$$

Similarity and dissimilarity functions are called *complementary* if

$$S(x, y) = 1 - D(y, x), \qquad D(x, y) = 1 - S(y, x).$$

A similarity function S is called *co-symmetric* if the following property is fulfilled for all binary n-tuples x and y:

$$S(\bar{x}, \bar{y}) = S(x, y), \quad \text{(co - symmetry)}$$

$$s(d, c, b, a) = s(a, b, c, d). \quad \text{(co - symmetry)}$$

Proposition 1 [2, 4]. A similarity function $S(x, y)$ is co-symmetric if and only if it satisfies the following property:

$$S(x, \bar{y}) = S(\bar{x}, y). \quad \text{(co - symmetry II)}$$

One can check that Jaccard's similarity measure is not co-symmetric, but Simple Matching similarity measure satisfies this property.

In [2–4], it was defined the correlation function on a set with involution operation as a function satisfying several simple properties, and it was shown that the correlation function could be constructed using similarity functions. It was shown [2–6] that Pearson's product-moment correlation, Spearman's rank correlation, Kendall's rank correlation, Yule's Q, and Hamann's association coefficients satisfy these properties and can be constructed from suitable similarity measures.

Below we extend the definition of correlation functions on the set of binary n-tuples.

Definition 1. An association coefficient taking values in the interval $[-1,1]$ will be called *correlation function (association measure)* and denoted by $A(x, y)$, if for all binary n-tuples $x = (x_1, \ldots, x_n)$ and $y = (y_1, \ldots, y_n)$ it satisfies the following properties [2–4]:

$$A(x, y) = A(y, x), \quad \text{(symmetry)} \tag{1}$$

$$A(x, x) = 1, \quad \text{(reflexivity)} \tag{2}$$

$$A(x, \bar{y}) = -A(x, y). \quad \text{(inverse relationship)} \tag{3}$$

An association coefficient that takes positive and negative values in the interval $[-1,1]$, satisfies the properties of symmetry and reflexivity and does not satisfy the inverse relationship property will be called a *weak correlation function*.

The first two properties of correlation functions are similar to the properties of similarity functions:

$$A(a, b, c, d) = A(a, c, b, d), \quad \text{(symmetry)}$$
$$A(a, 0, 0, d) = 1. \quad \text{(reflexivity)}$$

The last property can be written for association functions $A(a, b, c, d)$ as follows:

$$A(b, a, d, c) = -A(a, b, c, d). \quad \text{(inverse relationship)}$$

For correlation functions for all binary n-tuples x and y the following properties are fulfilled:

$$A(x, \bar{x}) = -1,$$
$$A(\bar{x}, \bar{y}) = A(x, y). \quad \text{(co - symmetry)}$$

Theorem 1 [2–4]. If S is a co-symmetric and consistent similarity function on the set of binary n-tuples, then the function:

$$A(x, y) = S(x, y) - S(x, \bar{y}), \tag{4}$$

is a correlation function.

4 Binary Correlation Functions

In [4–6], it was shown that Hamann's and Yule's Q association coefficients are correlation functions. Here we consider these association coefficients and show that some other but not all of the known association coefficients for binary data are also correlation functions. For association coefficients satisfying (1)–(3), we establish their relationship with similarity measures that can be used in (4) for constructing these association coefficients.

Consider Hamann's association coefficient:

$$A_H(x, y) = \frac{(a + d) - (b + c)}{a + b + c + d}.$$

One can check that it satisfies the properties (1)–(3) of the correlation function. Consider Simple Matching similarity measure:

$$S_{SM}(x, y) = \frac{(a + d)}{a + b + c + d}.$$

It is the co-symmetric and consistent similarity function. Using:

$$S_{SM}(x, \bar{y}) = \frac{(b + c)}{a + b + c + d},$$

in (4) we obtain Hamann's association coefficient:

$$A(x, y) = S_{SM}(x, y) - S_{SM}(x, \bar{y}) = \frac{(a + d)}{a + b + c + d} - \frac{(b + c)}{a + b + c + d}$$

$$= \frac{(a + d) - (b + c)}{a + b + c + d} = A_H(x, y).$$

Consider Yule's Q association coefficient [8, 17, 18]:

$$A_{YQ}(x, y) = \frac{ad - bc}{ad + bc}.$$

One can check that it is a correlation function. The function:

$$S_{YQ}(x, y) = \frac{ad}{ad + bc},$$

satisfies the properties of co-symmetric and consistent similarity functions. Using:

$$S_{YQ}(x, \bar{y}) = \frac{bc}{ad + bc},$$

from (4) we obtain

$$A(x, y) = S_{YQ}(x, y) - S_{YQ}(x, \bar{y})$$
$$= \frac{ad}{ad + bc} - \frac{bc}{ad + bc} = \frac{ad - bc}{ad + bc} = A_{YQ}(x, y).$$

Let us consider some other known association coefficients [8, 17] taking positive and negative values, determine what of them are correlation functions, and from what similarity measures they can be obtained using Theorem 1.

Consider Yule's W association coefficient [8, 17, 19]:

$$A_{YW}(x, y) = \frac{\sqrt{ad} - \sqrt{bc}}{\sqrt{ad} + \sqrt{bc}}.$$

One can check that it takes values in $[-1, 1]$ and satisfies the properties (1)–(3). Hence it is a correlation function. The function

$$S_{YW}(x, y) = \frac{\sqrt{ad}}{\sqrt{ad} + \sqrt{bc}},$$

satisfies the properties of co-symmetric and consistent similarity functions. Calculate:

$$S_{YW}(x, \bar{y}) = \frac{\sqrt{bc}}{\sqrt{ad} + \sqrt{bc}},$$

and from (4) obtain Yule's W:

$$A(x, y) = S_{YW}(x, y) - S_{YW}(x, \bar{y})$$
$$= \frac{\sqrt{ad}}{\sqrt{ad} + \sqrt{bc}} - \frac{\sqrt{bc}}{\sqrt{ad} + \sqrt{bc}} = \frac{\sqrt{ad} - \sqrt{bc}}{\sqrt{ad} + \sqrt{bc}} = A_{YW}(x, y)$$

Consider *phi coefficient* that is *Pearson's product-moment correlation coefficient* applied to binary variables [8, 17]:

$$A_\rho(x, y) = \frac{ad - bc}{\sqrt{(a + b)(a + c)(b + d)(c + d)}}.$$

One can check that it is a correlation function. Consider the *Sokal and Sneath similarity measure* [8, 17]:

$$S_\rho(x, y) = \frac{ad}{\sqrt{(a + b)(a + c)(b + d)(c + d)}}.$$

It is a co-symmetric and consistent similarity function. We have:

$$S_\rho(x, \bar{y}) = \frac{bc}{\sqrt{(a + b)(a + c)(b + d)(c + d)}}.$$

Using (4), we obtain phi coefficient.

We discover that not all association coefficients satisfy the properties of a correlation function. For example, the following association coefficients [8]:

$$A(x, y) = \frac{ad - bc}{(a + b + c + d)^2},$$

$$A(x, y) = \frac{4(ad - bc)}{(a + d)^2 + (b + c)^2},$$

$$A(x, y) = \frac{n^2(na - (a + b)(a + c))}{(a + b)(a + c)(b + d)(c + d)},$$

are not reflexive; hence they are not correlation functions. Another association coefficient [8]:

$$A(x, y) = \frac{\sqrt{ad} + a - (b + c)}{\sqrt{ad} + a + b + c},$$

does not satisfy the property (3); hence it is not a correlation function.

5 New Parametric Family of Binary Correlation Functions

Gover and Legendre [17] introduced the parametric family of similarity measures $(\theta > 0)$:

$$S_\theta(x, y) = \frac{a + d}{(a + d) + \theta(b + c)}.$$

One can check that this similarity measure is a co-symmetric and consistent similarity function; hence using Theorem 1, we can construct a correlation function. We have:

$$S_\theta(x, \bar{y}) = \frac{b + c}{(b + c) + \theta(a + d)},$$

and from (4) we obtain a new correlation function:

$$A_\theta(x, y) = \frac{a + d}{(a + d) + \theta(b + c)} - \frac{b + c}{(b + c) + \theta(a + d)}.$$

Taking into account that $a + b + c + d = n$, after equivalent mathematical transformations, we can represent the new correlation function in the following form:

$$A_\theta(x, y) = \frac{\theta n[(a + d) - (b + c)]}{\theta n^2 + (\theta - 1)^2 (a + d)(b + c)}. \tag{5}$$

The new correlation function is the parametric generalization of Hamann's association coefficient, which is obtained from (5) when $\theta = 1$.

6   Conclusion and Future Work

The paper determined the class of association coefficients for binary data satisfying the properties of correlation functions (association measures) similar to the properties of Pearson's product-moment correlation coefficient. Based on the theoretical results obtained recently for correlation functions, the paper established relationships between similarity measures and association coefficients belonging to the class of correlation functions. It was shown that some known association coefficients do not belong to this class. Such measures require additional investigation. The paper proposed a new parametric family of correlation functions for binary data containing Hamann's association coefficient as a particular case.

In the future, it is planned to analyze the properties of all association measures known for binary data and to introduce new correlation functions for these data. Such correlation functions can be used as measures of interestingness and measures of relationship in data mining and intelligent data analysis.

Acknowledgements. The work is supported by the program of the development of the Scientific-Educational Mathematical Center of Volga Federal District № 075-02-2020-1478.

References

1. Batagelj, V., Bren, M.: Comparing resemblance measures. J. Classif. **12**, 73–90 (1995)
2. Batyrshin, I.: Association measures and aggregation functions. In: Castro, F., Gelbukh, A., González, M. (eds.) MICAI 2013. LNCS (LNAI), vol. 8266, pp. 194–203. Springer, Heidelberg (2013). https://doi.org/10.1007/978-3-642-45111-9_17
3. Batyrshin, I.Z.: On definition and construction of association measures. J. Intell. Fuzzy Syst. **29**, 2319–2326 (2015)
4. Batyrshin, I.: Towards a general theory of similarity and association measures: similarity, dissimilarity and correlation functions. J. Intell. Fuzzy Syst. **36**(4), 2977–3004 (2019)
5. Batyrshin, I.Z.: Data science: similarity, dissimilarity and correlation functions. In: Osipov, G.S., Panov, A.I., Yakovlev, K.S. (eds.) Artificial Intelligence. LNCS (LNAI), vol. 11866, pp. 13–28. Springer, Cham (2019). https://doi.org/10.1007/978-3-030-33274-7_2
6. Batyrshin, I.Z.: Constructing correlation coefficients from similarity and dissimilarity functions. Acta Polytech. Hung. **16**(10), 191–204 (2019)
7. Chen, P.Y., Popovich, P.M.: Correlation: Parametric and nonparametric measures. Sage, Thousand Oaks, CA (2002)
8. Choi, S.S., Cha, S.H., Charles, C.T.: A survey of binary similarity and distance measures. J. Syst. Cybern. Inform. **8**, 43–48 (2010)
9. Clifford, H.T., Stephenson, W.: An Introduction to Numerical Classification. Academic Press, New York (1975)
10. Gibbons, J.D., Chakraborti, S.: Nonparametric Statistical Inference, 4th edn. Dekker, New York (2003)
11. Jaccard, P.: The distribution of the flora in the Alpine zone. New Phytol. **11**, 37–50 (1912)
12. Janson, S., Vegelius, J.: Measures of ecological association. Oecologia **49**, 371–376 (1981)
13. Kendall, M.G.: Rank Correlation Methods, 4th edn. Griffin, London (1970)
14. Legendre P., Legendre, L.F.: Numerical Ecology, 2nd edn. Elsevier, English Ed. (1998)

15. Lesot, M-J., Rifqi, M., Benhadda, H.: Similarity measures for binary and numerical data: a survey. Int. J. Knowl. Eng. Soft Data Paradigms, **1**, 63–84 (2009)
16. Tan, P.N., Kumar V., Srivastava, J.: Selecting the right interestingness measure for association patterns. In: Proceedings of the Eighth ACM SIGKDD International Conference on Knowledge Discovery and Data Mining, pp. 32–41 (2002)
17. Warrens, M.J.: Similarity measures for 2 × 2 tables. J. Intell. Fuzzy Syst. **36**(4), 3005–3018 (2019)
18. Yule, G.U.: On the association of attributes in statistics. Philos. Trans. R. Soc. A **194**, 257–319 (1900)
19. Yule, G.U.: On the methods of measuring the association between two attributes. J. R. Stat. Soc. **75**(6), 579–652 (1912)
20. Zadeh, L.A.: Similarity relations and fuzzy orderings. Inf. Sci. **3**, 177–200 (1971)

Best Paper Award, Third Place

Dissimilarity-Based Correlation of Movements and Events on Circular Scales of Space and Time

Ildar Batyrshin[1]([⊠]) [iD], Nailya Kubysheva[2] [iD], and Valery Tarassov[3]

[1] Instituto Politécnico Nacional, Centro de Investigación en Computación, Mexico City, Mexico
batyr1@gmail.com
[2] Kazan Federal University, Kazan, Russia
aibolit70@mail.ru
[3] Bauman Moscow State Technical University, Moscow, Russia
vbulbov@yahoo.com

Abstract. Circular scales appear in many applications related to a comparative analysis of the timing of events, wind directions, animals and vehicle movement directions, etc. The paper introduces a new non-statistical correlation function on circular scales based on a recently proposed approach to constructing correlation functions (association measures) using (dis)similarity measures on the set with an involutive operation. An involutive negation and a dissimilarity function satisfying required properties on a circular set are introduced and used for constructing the new correlation function. This correlation function can measure correlation both between two grades of circular scale and between sets of measurements of two circular variables.

Keywords: Circular scale · Correlation coefficient · Correlation function · Similarity function · Negation on circular scale

1 Introduction

The paper introduces the dissimilarity-based correlation function on circular scales. Such scales appear in many applications related to the comparative analysis of time events, animals and vehicle movements, wind directions in weather analysis and forecasting, etc. [12–19]. It was proposed several correlation coefficients on a circular scale. Traditionally these coefficients based on a statistical approach to the calculation of correlation between circular variables given by sets of measurements of these variables [12–16, 18]. Recently it was introduced a general approach to the construction of a correlation functions (association measures) on sets with involution (reflection) operation using (dis)similarity functions [2–4, 9, 10]. Depending on the domain, an involution operation can denote a negation, complement, multiplication on -1, and another operation mapping elements of the domain into "opposite elements." In this approach, the correlation between the two objects is positive if they are "similar" and negative if they are "opposite." An "object" can denote an element of a set, a variable given by the set of measurements, a time series, a vector, etc. As usual, when the generalization of some

L. Martínez-Villaseñor et al. (Eds.): MICAI 2020, LNAI 12469, pp. 237–246, 2020.
https://doi.org/10.1007/978-3-030-60887-3_21

theory includes the old one, the new approach to the construction of correlation coefficients includes as particular cases the classical correlation and association coefficients (Pearson's, Spearman's, Kendall's, Yule's Q, etc.) [2, 3, 8–10]. Based on (dis)similarity-based approach, the correlation functions (association measures) have been introduced on the set [0,1] of probability and membership values, on the set of fuzzy sets, on the set of subintervals of [0,1], for bipolar rating profiles, and for binary data [4–7, 11].

Based on this new approach to the construction of correlation functions, the paper introduces the reflection (negation) operation involutively mapping the grades of circular scale into "opposite" grades. Further, the dissimilarity function satisfying some necessary properties on the set of the grades of the circular scale is introduced. Finally, using this dissimilarity function, a new correlation function on a circular scale is constructed. Some examples of calculation of the new dissimilarity and correlation functions are given.

The paper has the following structure. Section 2 gives a short introduction to circular scales. Section 3 gives basic definitions of (dis)similarity and correlation functions and describes the method of construction of correlation function from these functions. Section 4 introduces an involutive negation, new dissimilarity, and correlation functions on circular scales and provides some examples. Sections 5 and 6 contain short discussions of how the new correlation functions can be used for calculation of correlation between variables given by sets of measurements on circular scales, and how circular scales can be considered as relative and dynamic scales in the analysis of movements of vehicles. Section 7 presents conclusions and discussion of future work.

2   Circular Scales

Consider some examples of circular scales. A 24-h clock called a military time contains 24 grades $(1, 2, \ldots, 12, 13, \ldots, 24)$ where 24 is also considered as 0. Such a scale can have more grades if it includes minutes. It is also used circular (round or circle) calendars with 12 months or a circular calendar with seven days of the week. Another example gives the scale of angular measurements from 0 °C to 360 °C. In such a circular scale, 360 °C can be considered as 0 °C. Angles can be measured in radians. There are also used several types of the compass rose circular scales: 8-point compass rose (windrose) with the grades (N, NO, O, SO, S, SW, W, NW), see Fig. 1 [17], 16-point or 32-point compass rose, an ancient 12-wind rose, etc.

In the following section, we consider basic definitions of dissimilarity and correlation functions that will be introduced later on circular scales.

3   Similarity and Correlation Functions

Correlation functions (association measures) were introduced in [2–4] on a set with involutive operation as functions satisfying some basic properties of Pearson's product-moment correlation coefficient. It was shown [2, 3, 8–11] that most of the known correlation and association coefficients satisfy the properties of correlation functions and can be constructed by methods proposed in [2–4, 9, 10] using some similarity or dissimilarity functions. Formally, (dis)similarity functions are fuzzy relations, and many properties of fuzzy relations can be considered for these functions [1, 8, 10, 20]. In this

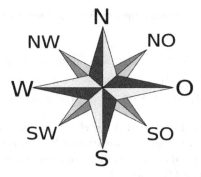

Fig. 1. 8-Point Compass Rose, adopted from [17].

section, we consider some related definitions and methods that will be used further for the construction of the correlation function on a circular scale.

Let Ω be a nonempty set with involutive operation $N(x)$ called *reflection* or *negation* such that for any element x in Ω the reflection of x also belongs to Ω, and double reflection of x equals to x:

$$N(N(x)) = x. \quad \text{(involutivity)}$$

Such involutive operation usually considered for many types of data: the complement of sets or fuzzy sets, the negation in binary or fuzzy logic, the multiplication on -1 on the set of real numbers, etc. If for some x in Ω it is fulfilled

$$N(x) = x,$$

then such element is called a *fixed point* of the reflection N. The set of all fixed points of N in Ω denoted by $FP(N, \Omega)$ or FP. For example, on the set of real numbers, the operation of multiplication on -1 is an involutive operation: $N(x) = (-1) \cdot x = -x$, due to $N(N(x)) = N(-x) = -(-x) = x$. This operation has a fixed point $x = 0$ because $N(0) = -0 = 0$. But the set $F \backslash \{0\}$ of real numbers without 0 has not fixed points with respect to this involutive operation. Note that not all sets have fixed points of a reflection operation defined on the set. For example, the negation operation in binary logic defined on the set of truth values $\Omega = \{0, 1\}$ has not fixed points.

Definition 1 [3, 8]. Let Ω be a set with a reflection operation N, and V be a subset of $\Omega \backslash FP(N)$ closed under operation N, i.e., for any x in V its reflection $N(x)$ belongs to V. The *correlation function (association measure)* on V is a function $A(x, y)$ such that for any x and y in V it takes values in $[-1,1]$, and satisfies the following properties:

$$A1.\, A(x, y) = A(y, x), \qquad \text{(symmetry)}$$
$$A2.\, A(x, x) = 1, \qquad \text{(reflexivity)}$$
$$A3.\, A(x, N(y)) = -A(x, y). \quad \text{(inverse relationship)}$$

It is shown that correlation functions satisfy the following properties fulfilled for all x and y in V:

$$A(x, N(x)) = -1, \qquad \text{(opposite elements)}$$
$$A(N(x), N(y)) = A(x, y), \qquad \text{(co-symmetry)}$$
$$A(x, N(y)) = A(N(x), y). \qquad \text{(co-symmetry - II)}$$

As it follows from the definition, the correlation between the elements of Ω and possible fixed points x_{Fp} of the reflection operation N on Ω is not defined. Depending on the possible applications of correlation functions, these correlations can be defined as follows: $A(x, x_{FP}) = 0$ or $A(x, x_{FP}) = 1$.

Correlation functions can be obtained from similarity and dissimilarity functions [3, 8] defined as follows.

A function $S(x, y)$ is called a *similarity function* on Ω if for all x, y in Ω it takes values in $[0,1]$ and satisfies the following properties:

$$S(x, y) = S(y, x), \qquad \text{(symmetry)}$$
$$S(x, x) = 1. \qquad \text{(reflexivity)}$$

A function $D(x, y)$ is called a *dissimilarity function* on Ω if for all x,y in Ω it takes values in $[0,1]$ and satisfies the following properties:

$$D(x, y) = D(y, x), \qquad \text{(symmetry)}$$
$$D(x, x) = 0. \qquad \text{(irreflexivity)}$$

Similarity and dissimilarity functions are dual concepts. They can be easily obtained one from another as follows:

$$S(x, y) = 1 - D(y, x), \quad D(x, y) = 1 - S(y, x).$$

Such similarity and dissimilarity functions are called *complementary*. Generally, they will be referred to as *(dis)similarity functions,* and they can be considered together [8], but for a specific domain, depending on the method of their construction, it is sufficient to consider only one of these functions. On circular scales, we will obtain dissimilarity function from a distance; hence we will consider further a dissimilarity function.

A non-negative real-valued function $d(x, y)$ of elements of Ω will be referred to as a *distance* if for all x, y in Ω it satisfies the following properties:

$$d(x, y) = d(y, x), \qquad \text{(symmetry)}$$
$$d(x, x) = 0. \qquad \text{(irreflexivity)}$$

If for some positive real value M for all x, y in Ω it is fulfilled $d(x, y) \leq M$, then the function

$$D(x, y) = \frac{d(x, y)}{M},$$

will be a dissimilarity function taking values in the interval $[0,1]$.

Let N be a reflection operation on a set Ω and $FP(N, \Omega)$ be a set of all fixed points of N on Ω. The set $FP(N, \Omega)$ can be empty. Let V be a subset of $\Omega \backslash FP(N)$ closed under operation N. A dissimilarity function D is called *non-contradictive* or *consistent* on V if for x in V it is fulfilled:

$$D(x, N(x)) = 1. \quad \text{(consistency)}$$

This property means that the dissimilarity between opposite elements in maximal.

A dissimilarity function D is called *co-symmetric* on V if for all x, y in V it is fulfilled:

$$D(N(x), N(y)) = D(x, y). \quad \text{(co-symmetry)}$$

This property means symmetry of dissimilarity functions with respect to opposite elements (complements, is we talk about sets).

Theorem 1 [3, 8]. Let Ω be a set with a reflection operation N, and V be a subset of $\Omega \backslash FP(N)$ closed under operation N. If D is a co-symmetric and consistent dissimilarity function on V, then the function:

$$A(x, y) = D(x, N(y)) - D(x, y), \tag{1}$$

is a correlation function on V.

Originally in Theorem 1, an association measure (correlation function) was constructed using similarity functions but replacing similarity functions by complementary dissimilarity functions we obtain (1).

4 Dissimilarity and Correlation Functions on Circular Scales

Let $(c_1, \ldots, c_n)$ be a sequence of the grades of the n-point circular scale. We will suppose here that n is even, i.e., $n = 2m$ for some positive integer m. Instead of names of the grades of the circular scale, we will use the indexes $I_{2m} = (1, \ldots, 2m)$ of these grades. Generally, what point of the scale will obtain the index 1 is not important because we will calculate differences between them, that are invariant under rotation of indexing when the circular order of indexes coincides with the circular order of the points on the scale. For example, for the 8-point compass rose shown on Fig. 1, with the grades (**NO, O, SO, S, SW, W, NW, N**), the indexing $I_8 = (1, 2, \ldots, 7, 8)$ can correspond to the points **NO, O, ..., NW, N**, respectively. In the examples below, we will use this indexing of the points of the circular scale. For simplicity of interpretation of operations, we write this indexing of 8-point compass rose as follows:

$$1(\mathbf{NO}), 2(\mathbf{O}), 3(\mathbf{SO}), 4(\mathbf{S}), 5(\mathbf{SW}), 6(\mathbf{W}), 7(\mathbf{NW}), 8(\mathbf{N}). \tag{2}$$

But if one assigns the index 1 to **O**, then **SO** will have index 2, and so on, and the final index 8 will be assigned to **NO**. Below we will operate with the indexes of a circular scale, for this reason, the set $I_{2m} = (1, \ldots, 2m)$ sometimes also will be referred to as a circular scale.

Let us introduce a negation N on the circular scale with indexes $I_{2m} = (1, \ldots, 2m)$ as follows:

$$N(k) = \begin{cases} k + m, & if \quad k \leq m \\ k - m, & if \quad k > m \end{cases}.$$ (3)

It is important to note that the set $I_{2m} = (1, \ldots, 2m)$ has an even number of grades. For this reason, the negation (3) has not fixed points, and we can apply Theorem 1 for the set $V = \Omega$ and for all x, y in Ω.

The negation (3) maps any point of the circular scale into the "opposite" point of the scale. For example, for circular scale from Fig. 1 with indexing (2) we have $m = 4$, and we obtain for $N(\mathbf{NO})$: $N(1) = 1 + 4 = 5$, i.e., $N(\mathbf{NO}) = \mathbf{SW}$, and we obtain for $N(\mathbf{W})$: $N(6) = 6 - 4 = 2$, i.e., $N(\mathbf{W}) = \mathbf{O}$.

Let us introduce the distance on the circular scale $I_{2m} = (1, \ldots, 2m)$ as follows:

$$d(k, j) = min\{|k - j|, 2m - |k - j|\}.$$ (4)

The distance on the circular scale equals to the minimum of "directional" distances between 2 points k and j calculated in clockwise and counterclockwise directions. For example, to calculate the distance $d(1, 7)$ between two points $1(\mathbf{NO})$ and $7(\mathbf{NW})$ we calculate, first, the directional distance in the clockwise direction starting from 1: $|k - j| = |1 - 7| = 6$. Since the sum of distances in clockwise and counterclockwise directions equal to $2m = 8$, we calculate the distance in the counterclockwise direction as follows: $2m - |k - j| = 8 - 6 = 2$. The minimal distance between $1(\mathbf{NO})$ and $7(\mathbf{NW})$ will be equal to $d(1, 7) = min\{|1 - 7|, 8 - |1 - 7|\} = min\{6, 2\} = 2$.

Since the distance (4) has the maximal value $M = m$, we obtain the dissimilarity function on a circular scale as follows:

$$D(k, j) = \frac{1}{m} d(k, j) = \frac{1}{m} \left[min\{|k - j|, 2m - |k - j|\} \right].$$ (5)

It is easy to see that D is symmetric and irreflexive. Also, we can prove the fulfillment of the following properties of dissimilarity function (5).

Proposition 1. The dissimilarity function (5) is consistent, i.e., for all $k = 1, \ldots, 2m$ it is fulfilled:

$$D(k, N(k)) = 1.$$ (6)

Proposition 2. The dissimilarity function (5) is co-symmetric, i.e., for all $k = 1, \ldots, 2m$ it is fulfilled:

$$D(N(k), N(j)) = D(k, j).$$ (7)

From the definition of dissimilarity functions it follows that a dissimilarity function D is co-symmetric if the corresponding distance d is co-symmetric, i.e., for all $k, j = 1, \ldots, 2m$ it is fulfilled:

$$d(N(k), N(j)) = d(k, j).$$ (8)

From Proposition 1, Proposition 2, (6)–(8), and Theorem 1 it follows that the following function will be a correlation function on a circular scale:

$$A(k,j) = D(k, N(j)) - D(k,j) = \qquad (9)$$

$$= \frac{1}{m}\left[min\{|k - N(j)|, 2m - |k - N(j)|\} - min\{|k - j|, 2m - |k - j|\}\right]. \qquad (10)$$

As we can see from (9) the *correlation between k and j is positive if* distance $D(k,j)$ is less than $D(k, N(j))$, i.e. *k is closer to j than to its negation $N(j)$. In inverse case, when k is closer to the negation $N(j)$ than to j, the correlation between k and j is negative.*

Example 1 Consider 8-point Compass Rose from Fig. 1 with indexing (2). We have $m = 4$, $2m = 8$. Compare correlation values for different pairs of points from this circular scale. For example, grades 1(**NO**) and 4(**S**) have indexes $k = 1$ and $j = 4$ respectively. From (3) and $m = 4$ we have $N(4) = 4 + m = 8$, i.e., $N(\mathbf{S}) = \mathbf{N}$. Using (9) and (10) calculate correlation $A(\mathbf{NO}, \mathbf{S})$ step by step:

$$A(1, 4) = D(1, N(4)) - D(1, 4),$$

$$D(1, N(4)) = D(1, 8) = \frac{1}{4}[min\{|1 - 8|, 8 - |1 - 8|\}] = \frac{1}{4}[min\{7, 1\}] = \frac{1}{4}.$$

$$D(1, 4) = \frac{1}{4}[min\{|1 - 4|, 8 - |1 - 4|\}] = \frac{1}{4}[min\{3, 5\}] = \frac{3}{4}.$$

$$A(1, 4) = \frac{1}{4} - \frac{3}{4} = -\frac{2}{4} = -\frac{1}{2}, \text{ i.e., } A(\mathbf{NO}, \mathbf{S}) = -0.5.$$

As we can see from the example, the correlation between **NO** and **S** is negative because **NO** is closer to the negation of **S**, $N(\mathbf{S}) = \mathbf{N}$, than to **S**, see Fig. 1. Table 1 presents correlation values between all directions (points) of 8-Point Compass Rose.

Table 1. Correlations of directions in 8-point Compass Rose, see Fig. 1.

		1	2	3	4	5	6	7	8
		NO	O	SO	S	SW	W	NW	N
1	NO	1	0.5	0	−0.5	−1	−0.5	0	0.5
2	O	0.5	1	0.5	0	−0.5	−1	−0.5	0
3	SO	0	0.5	1	0.5	0	−0.5	−1	−0.5
4	S	−0.5	0	0.5	1	0.5	0	−0.5	−1
5	SW	−1	−0.5	0	0.5	1	0.5	0	−0.5
6	W	−0.5	−1	−0.5	0	0.5	1	0.5	0
7	NW	0	−0.5	−1	−0.5	0	0.5	1	0.5
8	N	0.5	0	−0.5	−1	−0.5	0	0.5	1

Another interpretation of the obtained result can be the following. Suppose that the circular scale from Fig. 1 is used for measuring the direction of the movement of some vehicles. Then, comparing the movement of one vehicle in the direction **NO** with the directions of the movement of other vehicles we can say that the directions **N** and **O** are "near" to **NO** and positively correlated with **NO**, but the directions **W, SW** and **S** are "opposite" to **NO** and negatively correlated with it. In such a case, the correlation function can measure the sign and the strength of the relationship between directions. In our example, we obtained a negative correlation between **NO** and **S** because they move in "opposite" directions.

As a simple example, suppose that some vehicle moves in direction **W**. Then its movement is positively correlated with the movements of other vehicles moving in the "similar" directions **NW, W** and **SW** and it is negatively correlated with the movements of vehicles moving in the "opposite" directions **NO, O** and **SO**, (see bold font numbers in Table 1).

5   Correlation of Measurements in Circular Scales

Using the correlations of points of a circular scale, one can calculate the correlation between sets of measurements in circular scales. If these measurements are ordered in time, we can say about the time series of circular values. Consider two n-tuples $x = (x_1, \ldots, x_n)$ and $y = (y_1, \ldots, y_n)$ such that x_i, y_i for all $i = 1, \ldots, n$ belong to the same circular scale $I_{2m} = (1, \ldots, 2m)$. As an example, one can consider the directions of movements of two vehicles x_i, y_i at different moments of time $i = 1, \ldots, n$. If these measurements have been done during the movements of these vehicles at the same moments of time and ordered in time, these n-tuples will give time series of circular measurements or trajectories. It can be proposed several methods of correlation between sets of circular measurements or trajectories.

Since we can calculate correlations between n-tuples element-by-element, then the total correlation between them can be calculated as an average correlation:

$$A(x, y) = \frac{1}{n} \sum_{i=1}^{n} A(x_i, y_i).$$

It is clear that the correlation $A(x, y)$ takes values in the interval $[-1,1]$.

If we talk about the movements of two vehicles x and y then $A(x, y)$ will be negative if they generally move in "opposite" directions, and this correlation will be positive if they generally move in the "similar" direction.

In an analysis of circular time correlations of events, the correlation between two sets of measurements can be negative, for example, if the events from one set usually happen in the morning and events from another set happen usually in the evening. For example, John is usually jogging in the morning, and Bob is jogging in the evenings, then these events for John and Bob will be negatively correlated. For example, a high positive correlation of the hours of visiting the store by some groups of customers can give rise to an analysis of the demands of these groups of customers. The time correlations can be useful in the analysis of the weather conditions or contamination levels during the day or the year, in the analysis of the behavior of animals or birds, etc.

6 Relative and Dynamic Circular Scales

The usage of the same scale, for example, for analysis of the correlation between the movements of different agents (vehicles, people, animals) can be useful in many tasks, but in some situations, it is more convenient to use for different agents different scales. For example, if we will calculate the correlation between the movements of two ships moving to the North on the opposite sides of the Earth, then in the circular scale of one ship, these two ships will move in opposite directions towards each other and can be considered as negatively correlated. From another point of view, we can consider that their movements as positively correlated because they move in the same direction.

A similar situation appears when we want to calculate the correlation between the movement of two vehicles in the city when we consider their movements as positively correlated if both move to the Center of the city. But if they move from the North and from the South of the city, they will move in opposite directions. Hence if we will use one circular scale for both vehicles, then their movements will be negatively correlated.

To consider such situations, one can introduce relative circular scales, and for each vehicle to measure the angle between the "goal of the movement" (Center of the city) and the real direction of the movement of the vehicle. To measure such angles, one can use the angular circular scale $I_{360} = (1, \ldots, 360)$ where direction 360 (or equally 0) will correspond to the direction to the goal of the movement. In this case, vehicles can move in opposite directions, one from the North and another from the South of the city, but the correlation between their movements will be positive if both are moving to the common goal, to the Center of the city.

Another situation appears when the goals of vehicles are different and dynamically changed. Such situations appear for "predator and prey" or "leader and follower" agents. In these cases, the goals of movements for different agents can be different and will depend on the movements of other agents. Hence the correlation between movements of different agents can be calculated in a different manner.

7 Conclusion and Future Work

The paper introduced a new correlation function on a circular scale. It does not require that the circular scale should be angular as it is usually required in previous works on correlations on circular scales. Another advantage of the proposed method is that it gives a possibility to measure the correlation between two points of the scale and does not require the sets of measurements of circular variables. In future, the proposed approach will be extended on circular scales with an odd number of points. Also, the comparison of the proposed circular correlation coefficient with the existing circular correlations will be made.

Acknowledgements. The investigation is partially supported by projects IPN SIP 20200853 and RFBR No 20-07-00770.

References

1. Averkin, A.N., et al.: Fuzzy Sets in Models of Control and Artificial Intelligence, edited by Pospelov, D.A. Nauka, Moscow (1986). (in Russian)
2. Batyrshin, I.: Association measures and aggregation functions. In: Castro, Félix, Gelbukh, Alexander, González, Miguel (eds.) MICAI 2013. LNCS (LNAI), vol. 8266, pp. 194–203. Springer, Heidelberg (2013). https://doi.org/10.1007/978-3-642-45111-9_17
3. Batyrshin, I.Z.: On definition and construction of association measures. J. Intell. Fuzzy Syst. **29**, 2319–2326 (2015)
4. Batyrshin, I.Z.: Association measures on [0,1]. J. Intell. Fuzzy Syst. **29**(3), 1011–1020 (2015)
5. Batyrshin, I., Villa-Vargas, L.A., Solovyev, V.: Association measures on the set of subintervals of [0,1]. In: NAFIPS - WConSC 2015, pp. 1–3. IEEE (2015)
6. Batyrshin, I.: Association measures on sets with involution and similarity measure. In: Zadeh, L., Abbasov, A., Yager, R., Shahbazova, S., Reformat, M. (eds.) Recent Developments and New Direction in Soft-Computing Foundations and Applications. Studies in Fuzziness and Soft Computing, vol. 342, pp. 221–237. Springer, Cham (2016). https://doi.org/10.1007/978-3-319-32229-2_16
7. Monroy-Tenorio, F., et al.: Correlation measures for bipolar rating profiles. In: Melin, P., Castillo, O., Kacprzyk, J., Reformat, M., Melek, W. (eds.) Fuzzy Logic in Intelligent System Design. NAFIPS 2017. Advances in Intelligent Systems and Computing, vol. 648, pp. 22–32. Springer, Cham (2018). https://doi.org/10.1007/978-3-319-67137-6_3
8. Batyrshin, I.: Towards a general theory of similarity and association measures: similarity, dissimilarity and correlation functions. J. Intell. Fuzzy Syst. **36**(4), 2977–3004 (2019)
9. Batyrshin, I.Z.: Constructing correlation coefficients from similarity and dissimilarity functions. Acta Polytech. Hung. **16**(10), 191–204 (2019)
10. Batyrshin, I.Z.: Data science: similarity, dissimilarity and correlation functions. In: Osipov, G.S., et al. (eds.) Artificial Intelligence. Lecture Notes in Computer Science (LNAI), vol. 11866, pp. 13–28. Springer, Cham (2019). https://doi.org/10.1007/978-3-030-33274-7_2
11. Batyrshin, I.Z., Ramirez-Mejia, I., Batyrshin, I.I., Solovyev, V.: Similarity-based correlation functions for binary data. In: Martínez-Villaseñor, L., et al. (eds.) Advances in Computational Intelligence. LNAI, vol. 12469, p. 514. Springer, Cham (2020). https://doi.org/10.1007/978-3-030-60887-3_20
12. Fisher, N.I., Lee, A.J.: A correlation coefficient for circular data. Biometrika **70**(2), 327–332 (1983)
13. Jammalamadaka, S.R., Sengupta, A.: Topics in Circular Statistics, vol. 5. World Scientific (2001)
14. Johnson, R.A., Wehrly, T.: Measures and models for angular correlation and angular–linear correlation. J. R. Stat. Soc. B **39**(2), 222–229 (1977)
15. Lee, A.: Circular data. Wiley Interdisc. Rev. Comput. Stat. **2**(4), 477–486 (2010)
16. Mardia, K.V.: Statistics of directional data (with discussion). J. R. Stat. Soc. B **37**, 349–393 (1975)
17. Open - 8 Point Compass Rose @clipartmax.com, https://www.clipartmax.com/middle/m2i8Z5G6H7b1K9Z5_open-8-point-compass-rose/, last accessed 2020/08/05
18. Pewsey, A., García-Portugués, E.: Recent advances in directional statistics. arXiv preprint arXiv:2005.06889 (2020)
19. Tarassov V.B.: Development of fuzzy logics: from universal logic tools to natural pragmatics and non-standard scales. Proc. 9th Intern. Conf. ICSCCW, Budapest, 908–915 (2017)
20. Zadeh, L.A.: Similarity relations and fuzzy orderings. Inf. Sci. **3**, 177–200 (1971)

Data Augmentation with Transformers for Text Classification

José Medardo Tapia-Téllez(iD) and Hugo Jair Escalante$^{(\boxtimes)}$(iD)

Instituto Nacional de Astrofísica Optica y Electronica, Puebla, Mexico
{tapiatellez,hugojair}@inaoep.mx

Abstract. The current deep learning revolution has established to transformer based architectures as the state of the art in several natural language processing tasks. However, it is not clear whether such models can be also used for enhancing other aspects of the learning pipeline in the NLP context. This paper presents a study in such a direction, in particular, we explore the suitability of transformer models as a data augmentation mechanism for text classification. We introduce four ways of using transformer models for augmenting data in text classification. Each of these variants take the outputs of a transformer model, feed with training documents, and use such outputs as additional training data. The proposed strategies are evaluated in benchmark data using CNN and LSTM based classifiers. Experimental results are promising: improvements over a model training on the plain documents are consistent.

Keywords: Text classification · Data Augmentation · Transformers

1 Introduction

Data Augmentation (DA) is the process of obtaining/generating additional data for training machine learning models. Through the DA process, one is able to reduce the risk of overfitting and increase the robustness of the machine learning models when not enough data is available. DA has emerged in the context of deep learning because these models require of large amounts of data in order to learn adequately. These techniques are commonly used in computer vision where considerable improvements in performance are reported, see e.g., [5,13]. Despite this success, DA has not been thoroughly explored in the context of Natural Language Processing (NLP), where there are plenty of domains in which collecting manually labeled documents is complicated.

In this paper we explore DA in the context of NLP, specifically for text classification. We rely on the success that transformer-based models (e.g., Bert [6] and GPT [12]) have reported in different NLP task and use them to generate synthetic documents that are in turn used to augment the initial training set associated to text classification tasks. Transformers are powerful models implementing self attention mechanisms that allow them to capture long range dependencies among words in a sequence. Outstanding results in a wide variety of

© Springer Nature Switzerland AG 2020
L. Martínez-Villaseñor et al. (Eds.): MICAI 2020, LNAI 12469, pp. 247–259, 2020.
https://doi.org/10.1007/978-3-030-60887-3_22

tasks have been reported with these methods when used as both (pretrained) feature extractors and end-to-end learners [6,8,12]. Since transformers are in essence language models they can be sampled conditioned on certain inputs and be used as text generators. We use such feature of transformers and use it as a data augmentation mechanism. In a nutshell, we sample pretrained transformers conditioned on training documents of a text classification task, and use the outputs as augmented training instances. Four variants for generating artificial samples are proposed and evaluated in benchmark data. Experimental results show the usefulness of the augmented samples and motivate further research on data augmentation for NLP.

The contributions of this paper are as follows:

- We explore the suitability of using transformers as data augmenters in the context of text classification.
- We propose four variants to augment a training set of documents with the outputs of transformers conditioned on the documents.
- We show the augmentation process is promising, motivating further research.

The remainder of this paper is organized as follows. The next section reviews related work on data augmentation for NLP. Next Sect. 3 describes the proposed methodology for data augmentation. Then Sect. 4 presents an experimental evaluation of the augmentation procedures. Finally, Sect. 5 outlines conclusions and future work directions.

2   Related Work

This section briefly reviews related work on data augmentation for NLP tasks. The idea of augmenting the available training documents for NLP tasks is not new, see for instance [1,7,11]. However, early approaches for data augmentation (either term or document expansion methods) mainly dealt with the task of identifying words associated to the content of documents that could be added to it. These methods mostly relied on thesaurus, semantic nets like WordNet [1,7] or co-occurrences [4,11]. In this way, a document could include more related terms eventually addressing issues associated to synonymy and polysemy. Differently to early expansion strategies, *modern* data augmentation aims at generating artificial instances (instead of extending the content available instances) based on the available ones. In this sense, data augmentation is closer to oversampling (e.g., see [3]) than to classical expansion methodologies.

Recent data augmentation efforts for NLP have adopted quite diverse methodologies, in the following we summarize the main paradigms. Yang et al. generate artificial (word-embedding) representations for documents by using the most similar word embeddings to the words appearing in the initial document [15]. In this way, the artificial representations resemble documents with meaning related to the original ones. One should note, however, that no document is actually generated, but only the embedding-based representations. Wei et al. introduce four Easy Data Augmentation (EDA) strategies for generating artificial documents: synonym replacement (choosing randomly n words from a sentence and replace them by a synonym); random insertion (inserting a random synonym of a random word in a random position in the sentence); random swap (randomly choose two words in the sentence and swap their positions); and random deletion (randomly remove each word in the sentence with probability p.). EDA improves performance on both, convolutional and recurrent neural networks, and is particularly helpful for smaller data sets. Kobayashi et al. propose a method for augmenting label sentences [9], their method is called contextual augmentation. They stochastically replace words with other words that are predicted by a bi-directional language model at the word positions. It is important to emphasize that they feedback the language model with a label-conditional architecture, which allows the model to augment sentences without breaking the label-compatibility. Their results for six different classification tasks show improvements when using a convolutional neural network.

Parallel to our work, Kumar et al. proposed a method that used transformers for data augmentation in NLP [10]. The authors feed transformer models with documents and label information and they synthesize new samples. Although they use transformers, the way new instances are generated is different: they feed the transformer models with a combination of the source document and its label, but they ask the model to predict the most probable sequence following the fed input. This way of synthetization was first proposed in [2] obtaining satisfactory results. Compared to these related efforts, our proposal adopts a different approach to generate synthetic documents: we mask words and ask the model to predict the missing words, the predicted words are used to generate new documents. We also developed a data augmentation procedure based on sentences, this resembles to some extend the method in [10].

3 Data Augmentation with Transformers

This section describes the proposed methodology for data augmentation in text classification. We first briefly introduce transformers, then we describe the augmentation process and the four considered variants.

3.1 Transformers

A Transformer is a deep network architecture for processing sequential data and that is based entirely on *attention* mechanisms [14]. This type of models do not aim to explicitly model sequential information as in classical recurrent neural nets (RNNs), instead they implicitly capture such information via self attention layers. An attention mechanism indicates which parts of a document (words, sentences) should have more weight for the modeling processes. A self attention module, thus, learns which parts of an input document should receive higher weight for the layers in upper levels. Transformers learn in an end-to-end way multiple self attention mechanisms in parallel and rely on an encoder-decoder architecture to solve a variety of NLP tasks.

Although there are several transformer based models out there, in this work we rely on two of the most effective and popular ones, namely: BERT and GPT-2. BERT (*Bidirectional Encoder Representations from Transformers*) is a language model based on transformers [6]. It implements bidirectional self attention mechanisms and has been trained by using term masking strategies. GPT-2, on the other hand, implements an unidirectional language model trained under a predict-next-word objective [12]. Both transformer models were trained using huge corpora and under a variety of settings. Pretrained models are available out there so that anyone can use them as starting point for their research. Using such pretrained models for solving a variety of NLP tasks is straightforward. However, the benefits of these methods for data augmentation have not been explored deeply so far (see [2,10]), we hope our study helps to better understand the capabilities and limitations of transformers for data augmentation.

3.2 Generation of Synthetic Documents

The goal of this work is to explore the benefits of using transformers for data augmentation in the context of text classification. As previously mentioned, we propose four variants to generate synthetic documents. In all of them, the idea is to feed a transformer (either BERT or GPT-2) with training documents of the classification task at hand and use different strategies to generate artificial documents. Each of the synthetically generated documents will be assigned the same label as the source document we use it for the generation process. Synthetic and original training documents are both used for training a classification model. In the remainder of this section we detail the four variants, where the first three correspond to masking words using BERT, and the remaining one aims at expanding sentences using GPT-2.

Single Masking Augmentation. The first method corresponds to Single Masking Augmentation. Where a specific sentence is first tokenized. Then a random word in the sentence is selected and masked, this allows for BERT not to take into account this word and be able to predict new words based on the other words in the sentence and the position of the masked item. We then regroup the sentence but now with the masked item, this masked sentence is inserted into BERT where a series of tokens (words) for unmasking the masked word are predicted. Based on a predefined number of sentences to be augmented per sentence in the training set, we produce new sentences with the respective tokens in the masked position. For example, if we want to generate three new sentences, then the first three tokens would be used to create them. Figure 1 graphically depicts this variant, in the left plot we show the procedure and in the right plot we show sample generated sentences.

Double Masking Augmentation. The second method corresponds to Double Masking Augmentation. The idea is similar as in Single Masking Augmentation but now we mask two random words of the original sentence. To do this we apply the Single Masking Augmentation process in series. It is important to mention that the obtained sentences from the Single Masking Augmentation

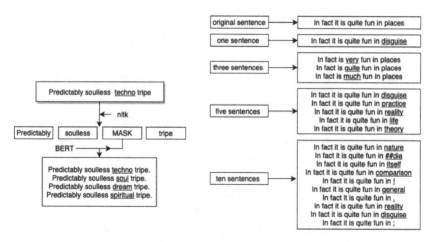

Fig. 1. Left: Single masking augmentation method. Words in a sentence are masked (one at a time) and artificial sentences are produced by asking BERT to unmask the masked words. Right: Examples for one, three, five and ten sentences created from the original sentence using Single Masking Augmentation

process are now masked in a second and different random position see the top of Fig. 2 for an illustration. Finally a token is provided based on the order of the respective sentences. The final results are sentences with two words changed based on BERT. As with Single Masking Augmentation we apply this method in the production of one, three, five and ten sentences as shown in the bottom of Fig. 2.

Triple Masking Augmentation. The third method corresponds to Triple Masking Augmentation. The idea is the same as in the previous two methods, but we now mask three words of the sentence. As we can see in Fig. 3, the Single Masking Augmentation procedure is applied three times in series, thus creating the set of new sentences with three different words in them. As in the previous two methods, we run experiments for the generation of one, three, five and ten sentences as shown in Fig. 4

Augmented Sentence. The fourth proposed variant for data augmentation consists in augmenting sentences with GPT-2. The procedure is pretty straightforward: as input we take a sentence that is introduced into the GPT-2 model, then we ask the model to predict an augmented sentence with up to fifty words as show in Fig. 5. One should note that this procedure is very similar to that one described in [2, 10].

4   Experimental Results

This section presents an experimental evaluation of the data augmentation strategies described in Sect. 3. We first describe the experimental settings and then we present the results obtained by each of the developed strategies.

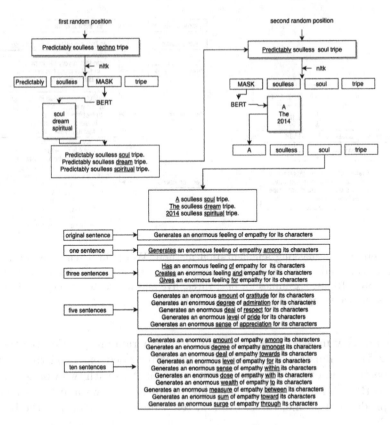

Fig. 2. Top: double masking augmentation: we apply Single masking augmentation in series to generate sentences with two modified words. Bottom: examples for one, three, five and ten sentences created from the original sentence using Double Masking Augmentation.

4.1　Experimental Settings

For the experiments we used four benchmark datasets for text classification, namely: SST-2 (Stanford Sentiment Treebak), IMDB (Sentiment-related movie reviews), Spam, and Sentence Type (sentences classified as command, question and statement). Each of the considered datasets have 600 instances, for the experimental evaluation we used 80% of instances for training and the remainder for testing. In a preliminary stage we used 20% of the training data as validation set for hyperparameter tuning.

As classification models we used straighforward state of the art models based on deep learning: a Convolutional Neural Network (CNN) and a Long Short-Term Memory Network (LSTM-RNN). The CNN and LSTM-RNN were implemented based on the implementations of Wei et al. [16]. The CNN has the following layers: embedding, convolutional one dimension (activation = relu), dropout, global max pooling one dimension, dense (relu), dense (sigmoid); it

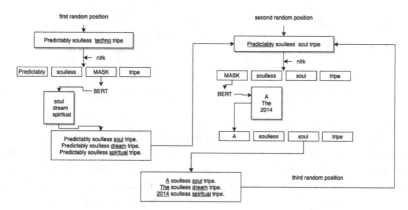

Fig. 3. Triple masking augmentation: we apply Single masking augmentation three times in series to generate sentences with two modified words.

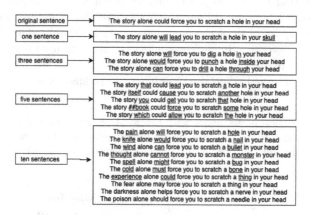

Fig. 4. Examples for one, three, five and ten sentences created from the original sentence using Triple Masking Augmentation

was trained with a binary cross entropy loss and an Adam optimizer, and it was trained for ten epochs. The LSTM-RNN has the following layers: embedding, LSTM, FC1, activation (relu), dropout, output layer, activation (sigmoid); it was trained also with a binary cross entropy and an RMSprop optimizer and trained for ten epochs.

4.2   Results

Each of the four strategies described above were evaluated using the datasets and classifiers just mentioned, where we used test set accuracy as the leading evaluation measure. The augmentation methods using BERT (i.e., single, double, and triple masking) were tested for an augmentation of one, three, five and up to ten sentences, this in order to determine whether there was a relationship

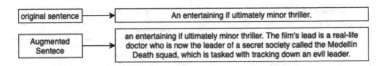

Fig. 5. Example of an augmented sentence through the Augmented Sentence method

between the number of augmented sentences and accuracy. The augmentation method using GPT-2 (i.e., augmented sentence) always produced a sentence of the specified length and our goal was used it as reference, as this method resembles state of the art augmentation methods introduced elsewhere [2,10].

Table 1 shows the results obtained with the Single Masking Augmentation strategy for different numbers of augmented sentences. Based on the results from Table 1, we can conclude, that on average, the original training data plus the data generated by the proposed augmentation method improves accuracy with respect to the baseline (Plain, which refers to the model trained only with the original data) for both of the considered classification models. We can also conclude that, on average, Single Masking with ten augmented sentences has the best accuracy with the LSTM-RNN, this was not the case for CNN, where the best result was obtained when three sentences were augmented. Table 2 shows

Table 1. Classification performance with Single Masking for one, three, five and ten sentences augmented. Plain indicates the performance obtained by each of the models without performing any augmentation. The best result in each row is shown in **bold**.

		Plain	Plain + Single Masking			
			1	**3**	**5**	**10**
Spam	RNN	0.79	0.86	0.84	0.86	**0.92**
	CNN	0.59	**0.68**	0.67	0.65	0.63
IMDB	RNN	0.71	0.72	**0.77**	0.75	0.71
	CNN	**0.79**	0.76	0.77	0.72	0.74
Sentence Type	RNN	0.35	0.42	0.46	0.53	**0.56**
	CNN	0.40	0.43	**0.44**	0.40	0.40
SST2	RNN	0.59	0.63	**0.68**	0.66	0.67
	CNN	0.68	0.63	0.66	0.65	**0.71**
Average	RNN	0.61	0.65	0.68	0.70	**0.71**
	CNN	0.61	0.62	**0.63**	0.60	0.62

the results obtained with the Double Masking Augmentation Strategy. From this table it can be seen that in average this strategy is able to improve the performance of the baseline. Where we can see that augmenting ten sentences

Table 2. Results for DA with Double Masking for one, three, five and ten sentences augmented

		Plain	Plain + Double Masking			
			1	3	5	10
Spam	RNN	0.79	0.82	0.77	0.78	**0.86**
	CNN	0.59	0.65	0.57	0.66	**0.74**
IMDB	RNN	0.71	0.71	0.69	**0.76**	0.70
	CNN	**0.79**	0.76	0.76	0.74	0.70
Sentence Type	RNN	0.35	0.37	0.44	0.35	**0.50**
	CNN	0.40	0.43	0.40	0.35	**0.50**
SST2	RNN	0.59	0.64	0.66	**0.69**	0.66
	CNN	0.68	0.68	0.65	**0.70**	0.70
Average	RNN	0.61	0.63	0.64	0.64	**0.68**
	CNN	0.61	0.63	0.59	0.61	**0.66**

gave better results for both models, LSTM-RNN and CNN. This result seems to corroborate the hypothesis that the more amount of augmented sentences yields better classification performance.

Table 3 shows the results obtained with the Triple Masking Augmentation strategy. Based on the results from Table 3 it can be seen that this strategy also improves the performance of the baseline for both considered classification models. Augmenting ten sentences resulted in better classification performance for the LSTM-RNN and one sentence worked better for the CNN.

Table 3. Results for DA with Triple Masking for one, three, five and ten sentences

		Plain	Plain + Triple Masking			
			1	3	5	10
Spam	RNN	0.79	0.74	0.77	0.67	**0.85**
	CNN	0.59	**0.69**	0.65	0.62	0.57
IMDB	RNN	0.71	0.72	0.70	0.70	**0.77**
	CNN	**0.79**	0.75	0.76	0.70	0.76
Sentence Type	RNN	0.35	0.43	0.38	0.47	**0.51**
	CNN	0.40	0.42	0.35	0.36	**0.50**
SST2	RNN	0.59	0.64	0.69	**0.70**	0.63
	CNN	0.68	0.70	**0.71**	0.70	0.63
Average	RNN	0.61	0.63	0.63	0.63	**0.69**
	CNN	0.61	**0.64**	0.62	0.59	0.61

Last but not least, the results for the Augmented Sentence strategy are presented in Table 4. From this table it can be observed that there were improvements mostly for the LSTM-RNN model, whereas the augmentation did not seem to improve the performance of the CNN classifier.

Table 4. Results obtained for the Augmented sentence method.

		Plain	Plain + Augmentation
Spam	RNN	**0.79**	0.75
	CNN	0.59	**0.61**
IMDB	RNN	0.71	**0.75**
	CNN	**0.79**	0.76
Sentence Type	RNN	0.35	**0.46**
	CNN	0.40	**0.41**
SST2	RNN	0.59	**0.62**
	CNN	**0.68**	0.67
Average	RNN	0.61	**0.64**
	CNN	**0.61**	**0.61**

Finally we show in Table 5 a summary of results that shows the average performance across datasets obtained by the different variants. It is clear from this table that only in two out of the 24 configurations we tested, the usage of data augmentation decreased the performance of the baseline model, and in most cases an improved was reported. Also, it can be seen that better results were obtained with the masking strategies than with the sentence augmentation method, which can be seen as a reference of the state of the art augmentation techniques [2, 10].

Table 5. Summary of the results for the four method along with their average and standard deviation.

		Plain	Plain + Single Masking				Plain + Double Masking				Plain + Triple Masking				Plain + A. S.
			1	3	5	10	1	3	5	10	1	3	5	10	
Avg.	RNN	0.61	0.65	0.68	0.70	**0.71**	0.63	0.64	0.64	0.68	0.63	0.63	0.63	0.69	0.64
	CNN	0.61	0.62	0.63	0.60	0.62	0.63	0.59	0.61	**0.66**	0.64	0.62	0.59	0.61	0.61
Tot. avg.		0.61	0.63	0.65	0.65	0.66	0.63	0.61	0.62	**0.67**	0.63	0.62	0.61	0.65	0.62
Std. Dev.		0.0	0.02	0.03	**0.07**	0.06	0.0	0.03	0.02	0.01	0.01	0.01	0.02	0.05	0.02

In order to get a better insight on how the proposed methods work when using different amounts of training data, we subsampled the training set and evaluated the performance of the augmentation for the RNN-LSTM classifier. For this experiment we considered only the Single Masking method. Our hypothesis was that the smaller the size of the training set, the larger the impact of the different augmentation strategies in the classification performance. Figure 6 shows the result of the experiment for the four datasets.

From Fig. 6, we cannot see the behavior we were expecting. However, still it can be seen that the augmentation strategy outperforms the baseline when half size of the training set (and on) data was considered. The fact we could not obtain larger improvement margins with fewer documents could be due to the fact that training data was very limited, also, one should note we trained the model for 10 epochs only.

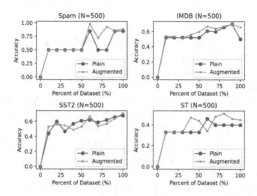

Fig. 6. Performance on text classification tasks with and without Augmented Data for Single Masking in three sentences augmented

5   Conclusions

We proposed four data augmentation strategies for text classification based on Transformers (BERT and GPT-2), namely: Single Masking, Double Masking, Triple Masking, and Augmented Sentence. The latter resembling a state of the art solution. The four evaluated methods, on average, improved the classification performance of two classification models a CNN and an LSTM-RNN. Regarding the masking methods, Single Masking obtained the best performance, however, Double and Triple Masking methods also improved the performance of the baseline. In general, better performance was observed when more sentences were added.

Experimental results with different amounts of training data were also reported. These results seem to indicate two things: first of all, we need more data in order to perform more experiments and to validate through graphs the results from the tables, and second, the curves related to the augmented data classification are, in general, on top of the curves related to the plain data, thus indicating that the augmentation methods do improve accuracy.

As future work, we would first like to work with more data, with it we could produce better graphs and a second round of average results for the methods. Second, we would like to implement a class-label-related as part of our model, this in order to conserve the label within the text augmentation process. And finally, we would like to run experiments on combinations of the methods developed in this work. This would surely bring light over which could be the best tool in order to enhance our training data.

Acknowledgements. This work was partially supported by CONACyT under project grant A1-S-26314, *Integración de Visión y Lenguaje mediante Representaciones Multimodales Aprendidas para Clasificación y Recuperación de Imágenes y Videos.*

References

1. Agirre, E., Arregi, X., Otegi, A.: Document expansion based on WordNet for robust IR. In: Coling 2010: Posters, pp. 9–17 (2010)
2. Anaby-Tavor, A., et al.: Not enough data? Deep learning to the rescue! arXiv preprint 1911.03118 (2019)
3. Bowyer, K.W., Chawla, N.V., Hall, L.O., Kegelmeyer, W.P.: SMOTE: synthetic minority over-sampling technique. J. Artif. Int. Res. **16**(1), 321–357 (2002)
4. Cabrera, J.M., Escalante, H.J., Montes-y-Gómez, M.: Distributional term representations for short-text categorization. In: Gelbukh, A. (ed.) CICLing 2013, Part II. LNCS, vol. 7817, pp. 335–346. Springer, Heidelberg (2013). https://doi.org/10.1007/978-3-642-37256-8_28
5. Cubuk, E.D., Zoph, B., Mane, D., Vasudevan, V., Le, Q.V.: Autoaugment: learning augmentation policies from data. ArXiv preprint 1805.09501 (2018)
6. Devlin, J., Chang, M., Lee, K., Toutanova, K.: BERT: pre-training of deep bidirectional transformers for language understanding. In: Proceedings of the 2019 Conference of the NAACL, pp. 4171–4186, June 2019
7. Gong, Z., Cheang, C.W., Hou U, L.: Multi-term web query expansion using WordNet. In: Bressan, S., Küng, J., Wagner, R. (eds.) DEXA 2006. LNCS, vol. 4080, pp. 379–388. Springer, Heidelberg (2006). https://doi.org/10.1007/11827405_37
8. Howard, J., Ruder, S.: Universal language model fine-tuning for text classification. ArXiv preprint 1801.06146 (2018)
9. Kobayashi, S.: Contextual augmentation: data augmentation by words with paradigmatic relations. In: Proceedings of the 2018 Conference of NAACL, pp. 452–457 (2018)
10. Kumar, V., Choudhary, A., Cho, E.: Data augmentation using pre-trained transformer models. ArXiv preprint 2003.02245 (2020)
11. Lavelli, A., Sebastiani, F., Zanoli, R.: Distributional term representations: an experimental comparison. In: Proceedings of the 13th ACM International Conference on Information and Knowledge Management, pp. 615–624. ACM (2004)
12. Radford, A., Wu, J., Child, R., Luan, D., Amodei, D., Sutskever, I.: Language models are unsupervised multitask learners. Technical report (2019)
13. Shorten, C., Khoshgoftaar, T.M.: A survey on image data augmentation for deep learning. J. Big Data **6**, 60 (2019). https://doi.org/10.1186/s40537-019-0197-0

14. Vaswani, A.: Attention is all you need. Adv. Neural Inf. Process. Syst. **30**, 5998–6008 (2017)
15. Wang, W.Y., Yang, D.: That's so annoying!!!: A lexical and frame-semantic embedding based data augmentation approach to automatic categorization of annoying behaviors using #petpeeve tweets. In: Proceedings of the 2015 Conference on Empirical Methods in Natural Language Processing, pp. 2557–2563 (2015)
16. Wei, J., Zou, K.: EDA: easy data augmentation techniques for boosting performance on text classification tasks. In: Proceedings of 2019 Conference on Empirical Methods in Natural Language Processing, pp. 6382–6388. Association for Computational Linguistics (2019)

Information Retrieval-Based Question Answering System on Foods and Recipes

Riyanka Manna[1], Dipankar Das[1], and Alexander Gelbukh[2(✉)]

[1] Department of Computer Science and Engineering,
Jadavpur University, Kolkata, India
riyankamanna16@gmail.com, dipankar.dipnil2005@gmail.com
[2] CIC, Instituto Politécnico Nacional, Mexico City, Mexico
gelbukh@gelbukh.com
http://www.gelbukh.com

Abstract. Question Answering (QA) is an emerging domain of research that retrieves a textual segment from the set of documents in response to user's queries. To recommend the answer in response to cooking recipe related questions is just an early stage of research and requires the significant refinement. In this paper, we have developed a question answering system on cooking recipes by using Natural Language Processing (NLP) and Information Retrieval (IR) technique. In recent years, with the rapid growth of information, the IR system has more importance in question answering domain. Users can also face difficulties to find expected answers from a huge amount of information. QA solves the information-overloading problem and IR returns the precise answers to the users. Answers from search engines are not only the results for a user's query but these collective words should justify the questions. We have a standard dataset on recipes and foods from famous cities in India which is collected from various Indian recipe websites. We have used Apache Lucene for information retrieval and we have prepared the gold standard dataset for the question answering system on cooking recipes.

Keywords: Natural language processing · Information retrieval · Question Answering · Apache Lucene · Question types

1 Introduction

Question Answering (QA) system is one of the important research areas in Natural Language Processing domain. The prime objective of QA systems is to retrieve the answer from the large collection of data based on the keywords contained in a question. QA system basically provides a system that enables meaningful and useful capabilities to the high-end questioner. India has different cities with different cultures, and the cities have different varieties of cuisines. In India, every state has its own famous and special foods. We have collected data of the famous foods of various Indian cities. Different recipe related information like

© Springer Nature Switzerland AG 2020
L. Martínez-Villaseñor et al. (Eds.): MICAI 2020, LNAI 12469, pp. 260–270, 2020.
https://doi.org/10.1007/978-3-030-60887-3_23

ingredients, cooking methods, preparation time, cooking time is collected from the recipes. The data has been crawled from various Indian recipe websites and has been indexed through Apache Lucene.

For this experiment, we have collected various real questions from Yahoo Answers! like *'What are the famous foods in Kolkata?'*, *'How to make paneer bhurji?'*. We have classified user questions into different categories like factoid, descriptive etc. type of questions. First, we have processed the questions to query words from the questions by removing stop words. Query words are processed through Apache Nutch, and retrieve the search results and returned the answers to the users. The user can search city wise recipes which gives a complete overview of a city's food specialty. A menu card on the foods of the different cities can be derived; it will help travel buffs to plan their tour accordingly.

During this experiment, we had met some challenges like integrating the modules, extracting answers from users' questions from the dataset, classifying question types and indexing. To integrate the modules, we have chosen an easy and effective technique for our work. The retrieval of significant knowledge (e.g., components, ingredients, processing etc.) from a recipe will be very helpful to the common users as well as the specific users such as chefs, house-wives, nutritionists.

2 Related Works

Question Answering is an area that operates on top of search engines (Google, Yahoo, etc.) in order to provide users with more accurate and actual responses where search engine outputs remain relevant, but difficult to understand, and sometimes incoherent. This system deals with the development and evaluation of subsystems that aimed correct answers given by a QA system. The automatic system would be useful for improving QA system performance, helping humans in the assessment of QA systems output. QA system on foods and recipes is done by Ling Xia [7] on Chinese cuisine where the cooking domain questions were classified into four categories such as ingredients, method, raw material dish suggestions, essential food preparation and the problem review worked out over these cases. Within this research module, questions have been initially prepared by the user for lexical analysis, domain vernacular formation and stop words removal. Afterward, questions are broadly categorized and according to the question type concise answers were given.

To investigate the question recommendation and answer retrieval [8], the QA system uses a statistical language model in order to identify spectrum of user expectations and to build the customer list of questions. It also analyzes the candidate answers, measures the similarity between the question and the answer using various syntactic and semantic characteristics, and then obtains the possible reply list, allowing the user to pick the best response easier. Moreover, the IR-based QA system [6], user questions are evaluated, system answers are extracted from a reversed database, and a personalized score feature is implemented. The objective of the designed QA system was to obtain the answers to questions which scrutinize German governmental services. With the aid of recovery strategies, task trees and legislative solutions, the program effectively manages unanswered issues.

The paper [5] describes the function of information retrieval and information extraction in QA systems. In this paper the authors compare the similarities and differences between the main building modules, and then describe the possible ways of merging the three technologies and performing trade-offs for various application domains. In the paper [2], the different sources such as, structured databases, non-structured, free text and pre-compiled semantic base information have been presented. For researchers in this area, this can be helpful and depending on the issue, they can select the system. They have also identified and corrected problems or suggest new QA systems. Different approaches of Question Answering [1] i.e. Linguistic approach, Statistical approach and Pattern Matching approach was discussed as well as the competitive study was analyzed. Based on these approaches, a different QA system was developed. According to [3] an efficient method was proposed for extracting answers from retrieved documents from a large collection of text. A question answer framework has been proposed, utilizing various methods of question classification to obtain support from category information.

The paper [4] provides a comparative overview of the Question system. The survey suggests a general question answering architecture and different levels of question processing. Different representations like POS tagging, expected answer type classification, semantic roles, discourse analysis was discussed. This part should contain sufficient detail so that all procedures can be repeated. It can be divided into subsections if several methods are described.

3   Dataset

To verify the significance of the proposed approach, we have collected data from various Indian recipe websites. We have made the xml documents for each city and added these xml files to the search engine. We have considered the recipe name, time to cook, level of difficulty, ingredients, method. As per our system requirement we have converted the data to our own format, so that it could be easily indexed by a search module. An example of a data entry is shown in Table 1.

3.1   Dataset Statistics

We have collected recipes of the 20 Indian cities from the source of sanjeevkapoor[1], ndtv[2], punjabi-recipes[3], tarladalal[4] etc. In each of the cities several field names or tags are added such as *City, Recipe, Ingredients, TimeTocook, HowToCook, WhenToServe, LevelOfCooking.* These tags give us various details

---

[1] https://www.sanjeevkapoor.com/.
[2] https://food.ndtv.com/.
[3] http://punjabi-recipes.com/.
[4] https://www.tarladalal.com/.

Table 1. Data format for this experiment.

```
<doc>
<field name="city">Kolkata</field>
<field name="recipe">Chicken Biriyani</field>
<field name="ingredients">1 cup whole milk yogurt Kosher salt 2 teaspoons Kash-
miri chili powder (see Cook's Note) 1/2 teaspoon ground turmeric 2 1/2 pounds
bone-in skinless chicken thighs 1 1/2 cups vegetable oil 2 large red onions, thinly
sliced 2 cups basmati rice 8 whole cloves 7 green cardamom pods 3 bay leaves 1/4
teaspoon whole black peppercorns 2 large pinches saffron 1/4 cup whole milk 1/2 cin-
namon stick 1 teaspoon cumin seeds 1 tablespoon finely grated ginger 5 cloves garlic,
finely grated 2 medium tomatoes, chopped into 1/2-inch pieces 1/2 cup cilantro leaves
1/2 cup mint leaves</field>
<field name="TimeToCook">Yield: 4, Prep time: 10 MINUTES, Cook time: 35
MINUTES, Total time: 45 MINUTE</field>
<field name="HowToCook">Heat the oil over medium-high heat in a large nonstick
skillet or frying pan. Once the oil is shimmering, add the chicken pieces and let them
cook, undisturbed, for 3-5 minutes until golden brown. Turn the chicken pieces and
add the onion, jalapeno, ginger, salt, garam masala, cumin, turmeric, and salt. Saute
for 3 minutes, or until the onions have softened. Add the garlic, tomatoes, and raisins
to the pan. Stir well, then add the rice and broth. Allow the liquid to come to a boil,
then cover the pan and turn the heat down to medium-low. Let the rice steam for 15
minutes. Turn off the heat and fluff the rice with a fork. Re-cover the pan, and allow
the rice to continue to steam for another 10 minutes. Garnish with cilantro leaves
and almond slices. Serve the Biryani straight out of the pan, accompanied by lime
wedges for squeezing.</field>
<field name="WhenToServe">Lunch, Dinner</field>
<field name="LevelOfCooking">Medium</field>
</doc>
```

of the recipe. Each of the cities contain at least 5–6 recipes. Here we can group only two types of queries, such as easy and complex ones. Statistics of the categories is shown in Table 2.

Table 3 shows the number of WH-type questions from the prepared question bank.

4   QA System Architecture

Our QA system comprises the four modules. The System architecture is shown in Fig. 1. These four main modules are as follows:

A. Apache Lucene module
B. Query Processing module
C. Document Processing module
D. Answer Processing module.

Table 2. Statistics of various categories.

SL. No.	Category	Count
1	City	20
2	Recipe	100
3	Ingredients	100
4	TimeToCook	100
5	HowToCook	100
6	WhenToServe	100
7	LevelOfCooking	100

Table 3. Question Category.

SL. No.	Category
1	What
2	How
3	Why
4	When
5	Which
6	Other

4.1   Apache Lucene Module

Apache Lucene[5] is a Java library which builds a index that is easily searchable for retrieval. Apache Lucene has two important aspects: first one is the way how the data is to be stored and the way how to search that indexed data. For indexing the content, it uses the inverted index technique. For searching, the Apache Lucene offered the multiple ways for instance Lucene API and query parser. In which, the query parser translates an input expression into an arbitrarily complex query and work in conjunction with analyzer to analyze that what user want to looking for which called as query object. After that, the query object fed into the index searcher to retrieves the relevant answer with respect to user entered query. The third and most important aspect of Apache Lucene in the task of information retrieval is Apache Nutch. The Apache Nutch is a highly extensible and open source web crawler software which entirely coded in the java programming language. The Nutch perform the crawling by which retrieval system scan websites and collect details about each page. The Nutch have three main components, namely, the crawler, the segment directory and the indexer, each of which is discussed below with respect to our proposed work.

At the beginning, the URLs of the collected information with respect to cooking recipe's are stored into a text file where each URL associate with each

---

[5] http://nutch.apache.org/.

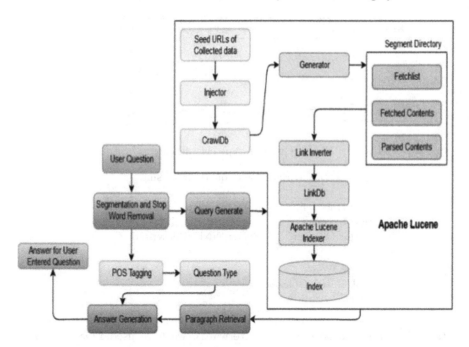

Fig. 1. System architecture.

cooking related document. The injector module of Apache Nutch transformed collected URLs to CrawlDb and at the time transformation, each pattern of the URLs is verified. CrawlDb is a crawl database which stores every URLs together with the metadata that it knows about. Afterwards, the generator module generates Fetch-list from CrawlDb and fed into the Segment Directory. The segment directory is a basically a folder containing all the data related to one fetching batch. To fetched the content of URLs, the fetcher module extracts their content and stored back into Segment Directory. Afterwards, the Link inverter inverts links of the documents to give preference to inlinks over outlinks.

Our QA system architecture is shown in Fig. 1.

4.2 Query Processing Module

The goal of question-answering system is to extract the segments of information from the question. The question sets the keyword to be used by the IR system in data representation. The purpose of the query processing module is to examine the query and process for a defined natural language database input by generating representations of the necessary information in certain formats.

Processing Questions Using NLTK. We have used NLTK tool kit of python to design the QA system. NLTK includes a necessary library file which helps to perform our works. At first step, we have used word tokenizer to divide the

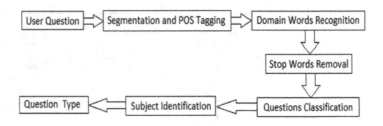

Fig. 2. Question type classification.

Table 4. Example of query processing.

User Question:	POS Tagging:
How to cook	how/WRB to/TO cook/VB
Butter Chicken?	Butter/NNP Chicken/NNP
	?/

Query Words:
cook Butter Chicken

sentence into the tokens or words. Afterwards, we have used POS Tagger to obtain the query words which is shown in Fig. 2.

POS Tagger. We have a list of questions which yield query words for the search engine. Each of these questions have a 'Wh' word and one or more query word. Next, we get the query words using NLTK POS tagger to identify the particular words. i.e. Noun or Pronoun or verb etc. The 'Wh' word is used to determine the type of the answer. The proper nouns among the query words are used for searching the data. Other words like verb, adverbs, noun, adjective, cardinal numbers are later used for answer preparation. When there is more than one query word then we search as a single phrase. For the query words we have used NNP, NN tags and removed other as stop words.

An example of query processing is shown in Table 4.

4.3   Document Processing Module

This module contains revised queries which are fed into the IR system, and retrieves the bunch of documents in a ranked order. The key role of this module is to obtain relevant information from one or more data systems and obtained documents are sorted and organized. It is intended to produce a series of candidate sentences containing responses. The function of classifying or acknowledging the sort of question is to identify the type and the labeled the entity of the response.

Apache Nutch[6] is an open source search enginebased on Lucene, upgraded and modified to ease operations. For our convenience, we run Nutch and used it

---

[6] http://nutch.apache.org/.

Table 5. Example of answer processing.

SL. No.	Question	Template	Count
1	How to cook Butter Chicken?	Butter Chicken is prepared by	13
2	What are required utensils are to cook "Punjabi dhaba style daal"?	The required utensils are	26
3	What are the Ingredients of Chilli Paneer?	Ingredients of Chilli Paneer are	3
4	How long to Preparation MalaiKofta?	Preparation time for MalaiKofta is	8

through its web interface and create a core. A Nutch Core is a running instance of a Lucene index that contains all the Nutch configuration files required to use it. After creating a core, we have added documents to the core and perform indexing. Nutch has a web interface that can be used to add documents. We add the city food documents in xml format. We have searched using the url. After the three text files i.e. the WH word file, (VB, NNS, NN, JJ, CD, RB) word file, NNP-word file are read which was earlier written in the previous module. Then the query terms from the NNP- word file are sent to the URL of the Nutch web interface to retrieve the search results. For final outcomes, the retrieve results are stored in csv format.

4.4 Answer Processing Module

After the bunch of words from the WH word file and (VB, NNS, NN, JJ, CD, RB) word(s) file is read then it is checked with the Answer Processing module for providing precise answers from the specific question. If the conditions are matched then precise answer is displayed but if it does not happen then the bunch of words from both the WH word file and (VB, NNS, NN, JJ, CD, RB) word file is checked again before which the (VB, NNS, NN, JJ, CD, RB) from the question sentence is tokenized further for more accurate and precise result. In order to give precise answers in a complete sentence, we have a set of template answers for every type question. Here the answers for both factoid and non-factoid questions could be obtained. The template is present in the front part of the answers whereas in the rear end data is extracted from the dataset which gives more specific answers in a concise manner.

An example of answer processing is shown in Table 5. Here, the answers for both factoid and non-factoid questions could be obtained. The template is present in the front part of the answers whereas in the rear end data is extracted from the dataset which give more specific answers in a concise manner.

5   Experimental Setup

Integration of these modules has been a crucial part of this project. To simplify our problem, we have implemented the modules by apache lucene to retrieve the results. Then these results are further used for answer presentation.

The submodules of our QA system are shown in Fig. 3.

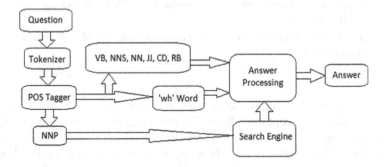

Fig. 3. Submodules of system.

6   Result Analysis

In this system we have tested the several types of questions. There are some questions which are difficult to answer like "How much oil needed to cook Butter Chicken?", in these cases, we have generated search query by removing stop words from the question and then retrieve search results from documents. We have used a test set of questions and questions have a specific type.

Experimental Results. By using our response generating process, we have tested our approach to 50 categorized questions, and its defined into three cases: correct answer, incorrect answer, and no answer. QA System efficiency is measured with scope and precision. The following are described in respect to scope and precision:

$$\text{Precision} = \frac{\text{number of correctly classified questions}}{\text{number of obtained classified}},$$

$$\text{Recall} = \frac{\text{number of correctly classified questions}}{\text{number of questions}}.$$

Evaluation Results We have categorized all the 50 different questions into two categories: easy type and complex type depending on the complexity of answer processing. Easy type question - are those questions whose answers are retrieved directly using a XML tag and also factoid question e.g. "How to cook butter chicken?" For giving answer to this question, we have retrieved only "method" tag of butter chicken from dataset. Complex type question—are those questions

Table 6. Experimental results.

QT	Q	CA	ICA	NA	COV (%)	ACC (%)
Easy type	21	18	0	3	85.71	85.71
Complex type	29	17	0	12	58.62	58.62
Total	50	35	0	15	70	70

Table 7. Results per question type.

Question type	Count	Mean of MMR (%)
What	18	79.16
How	12	82.30
Why	1	0
When	1	0
Which	4	55
Other	14	38.46

whose answers are retrieved not directly from a XML tag. In this case, we have retrieved some specific answer from a XML tag. e.g. *"How much oil is required for 1lb cauliflower to make aloo gobhi masala?"* In this case we have to retrieve a specific answer from XML tag "ingredients", not the total ingredients. In the question sets contained 21 easy type and 29 complex type questions and we have checked each question in our system and prepare the results given below. The result of evaluation is shown in Table 5. Where, QT: the type of question; Q: the number of questions; CA: the number of correct answers; ICA: the number of incorrect answers; NA: the number of no answer; Cov: coverage; Acc: accuracy.

Mean Reciprocal Rank. The medial range reciprocal is a statistical test to assess any method that generates a range of suggested answers to a collection of question. The subsequent grade of the answer is the multiplicative opposite of the very first best answer grade. The reciprocal outcomes for a sample question are considered as the final reciprocal outcomes. So, Mean Reciprocal Rank (MRR) is as $(31 + 1/2 + 1/4 + 1/5 + 1/5)/50 = 64.30\%$.

Experimental results are shown in Table 6, and the results per question type are shown in Table 7.

7   Discussion

The xml format for the search engine needs a specific format. Initially acquired data were arranged according to predefined format. This eases the process of adding documents otherwise there are some techniques which are a bit complex. We have opted for Apache Lucence as search engine and has many options on how it can be used. Most importantly Apache Lucence has this web interface which comes easy for our work. It reduces the functionalities of the answer processing module. In this system, we assign the task of collecting the correct result among the search results to the answer processing module. In the search engine if we have more than one query word then normal query shows all the possible results related to the query words. As mentioned earlier more than one query word are sent as a phrase to the search engine. In the first module, the POS tagger not able to identify proper nouns until they start with capital letter. So, we have made the proper nouns correctly updated into the question set.

Acknowledgement. The authors would like to express gratitude to the Department of Computer Science & Engineering, Jadavpur University for providing infrastructural facilities and support.

References

1. Barskar, R., Ahmed, G.F., Barskar, N.: An approach for extracting exact answers to question answering (QA) system for English sentences. Proc. Eng. **30**, 1187–1194 (2012). https://doi.org/10.1016/j.proeng.2012.01.979. International Conference on Communication Technology and System Design 2011
2. Bouziane, A., Bouchiha, D., Doumi, N., Malki, M.: Question answering systems: survey and trends. Proc. Comput. Sci. **73**, 366–375 (2015). https://doi.org/10.1016/j.procs.2015.12.005. International Conference on Advanced Wireless Information and Communication Technologies (AWICT 2015)
3. Dwivedi, S.K., Singh, V.: Research and reviews in question answering system. Proc. Technol. **10**, 417–424 (2013). https://doi.org/10.1016/j.protcy.2013.12.378. First International Conference on Computational Intelligence: Modeling Techniques and Applications (CIMTA) 2013
4. Kolomiyets, O., Moens, M.F.: A survey on question answering technology from an information retrieval perspective. Inf. Sci. **181**(24), 5412–5434 (2011). https://doi.org/10.1016/j.ins.2011.07.047
5. Moldovan, D., Surdeanu, M.: On the role of information retrieval and information extraction in question answering systems. In: Pazienza, M.T. (ed.) Information Extraction in the Web Era. LNCS (LNAI), vol. 2700, pp. 129–147. Springer, Heidelberg (2003). https://doi.org/10.1007/978-3-540-45092-4_6
6. Ostendorff, M., Düver, J., Ploch, D., Lommatzsch, A.: An interactive e-government question answering system. In: LWDA, September 2016
7. Xia, L.: Answer planning based answer generation for cooking question answering system. J. Chem. Pharm. Res. **6**, 474–480 (2014)
8. Xianfeng, Y., Pengfei, L.: Question recommendation and answer extraction in question answering community. Int. J. Database Theory Appl. **9**, 35–44 (2016)

Sentiment Detection in Economics Texts

Olumide E. Ojo[1], Alexander Gelbukh[1], Hiram Calvo[1(✉)],
Olaronke O. Adebanji[2], and Grigori Sidorov[1]

[1] Natural Language and Text Processing Laboratory, Center for Computing
Research, Instituto Politécnico Nacional, Mexico City, Mexico
olumideoea@gmail.com, gelbukh@gelbukh.com, {hcalvo,sidorov}@cic.ipn.mx
[2] Statistics Department, University of Ilorin, Ilorin, Nigeria
olaronke.oluwayemisi@gmail.com

Abstract. Deriving intelligence from text is important as it can provide
valuable information on how events influence public opinion. In this work,
a classification task was done in order to obtain the sentiment behind the
polarity of an economic text using machine learning and deep learning
methods. We analyzed the text for keywords that can be categorized into
positive, negative and neutral reviews and found more insights. In the
final result of classifying three groups (positive, negative and neutral),
the models were unable to perform up to 80% accuracy, where only one
variant has the accuracy of 80% as the best on the test dataset.

Keywords: Machine learning · Deep learning · Sentiment analysis ·
Economic text

1 Introduction

In the world today, there is increasing digitalization with the web becoming a
critical domain. The web allows data processing and knowledge exchange, and
affects various aspects of life including marketing, advertisement, education,
governance, etc. The networking through social media facilitates the creation
and exchange of information as an integral part of the web. Web-based services
through social media allow individuals to create a domain profile and supports
successful communication with a list of other network users. The advent of data
collection, which includes diverse views and opinions, is gaining more impact,
becoming an attraction for researchers and generating significant computational
challenges. Efficient mining of information from such data on a wide scale helps
to discover useful knowledge of vital importance in various fields including the
social-economic, health and government sectors. The growing importance of sentiment analysis being applied in nearly every business and social domain depends
on web-based human activities such as reviews, discussion forums, blogs, microblogs, etc. Computers can recognize, analyze and understand the feelings behind

This work has been possible thanks to the support of the Government of Mexico via
CONACYT, SNI, grant A1-S-47854; and Instituto Politécnico Nacional (IPN), grants
SIP 2083, SIP 20200811 and SIP 20200859, IPN-COFAA, IPN-EDI, and IPN-BEIFI.

L. Martínez-Villaseñor et al. (Eds.): MICAI 2020, LNAI 12469, pp. 271–281, 2020.
https://doi.org/10.1007/978-3-030-60887-3_24

text with resources required to detect trends and extract information from data, thereby enhancing operations of government and private agencies, identifying potential threats, reducing crimes, and improving public services.

Social media brings people, companies, and organizations together so that they can generate ideas and share information with others. Sentiment analysis uses natural language processing (NLP) techniques to automate the detection or classification of sentiments from text and has multiple application in different domain. Although the feeling of a word depends on the context in which it is used, sentiment analysis using machine learning is one of the most useful and well-studied tasks in NLP which can be used for a domain's automatic rating. This sort of an examination of feelings can also be regarded as a classification task. The importance of a review in product sales, economic decisions and others cannot be underestimated. As machine learning knowledge rapidly grows, it would be much easier to go through thousands of reviews if a model is used to polarize those reviews and learn from them. With the study on data, various approaches and techniques can be used to extract thoughts, emotions or opinions with resources being able to detect positive, negative, and neutral polarity.

This project's main goal is to observe the performance of machine learning and deep learning methods on the economic text. We used the dataset of Malo et al. [1] provided by Reuters which contains subjective sentences from economic review summaries and the target concept is to extracts the words behind the polarity of the opinions and to calculate classification accuracy from the opinion polarity experiment using machine learning. In their literature, they used the overall phrase-structure information and domain-specific use of language to detect polarity-lexicons that can be used to investigate how financial sentiments relate to future company performance. It is important to understand what influences the markets as a proxy for the global economies, especially the financial markets, as well as how they respond and how powerful that influence is. We had chosen machine and deep learning model on the text because it leads to conceptual characterization so they can be constant to the local change in the input data. The era of automatically extracting opinions is here and it is impossible to keep up with the analysis of new knowledge using lexicon-based methods [22].

Two major approaches are adopted in machine learning: supervised machine learning [10,12] and unsupervised methods [7–9,11]. Supervised approaches involve a training set of texts with manually defined polarity values, and they learn the features (e.g. words) that correspond with the value from these examples. Gelbukh and Kolesnikova [10,12] developed techniques enabling the automated sorting of word combinations into pre-established groups, for techniques relating to automated collocation classification. Gambino and Calvo [3] considered the use of text-learned emotions to classify polarities using NLP techniques. Supervised techniques have the downside of requiring high quality training data for each type of text and language. On the other hand, unsupervised systems are more versatile across various types of texts and domains, and can be implemented more easily once the lexical and semantic resources have been created. Previous work uses various neural models [5,14,18] to gain the understanding of how to predict text, including convolution models [2,13,15] and the implementation of attention models [16]. Successful classifiers have been applied to financial data [21,23].

Machine learning algorithms create a classifier automatically by learning the category characteristics from a collection of classified documents, and then using the classifier to classify documents into predefined categories. Before we can apply our machine learning algorithms to the text data we must process the text data. CountVectorizer implements the bag-of-word representation, which is a transformer. Fitting the CountVectorizer is tokenization of the training data and vocabulary creation. Data features were rescaled using the term frequency – inverse frequency of document (tf-idf) process. Features can be simple word count, characters, or even syntactic n-grams [4]. We focus on the supervised classification techniques so we can compare results. The techniques and algorithms we propose for the clustering and classification of the text are Decision tree classifier, Gradient boosting classifier, Bernoulli Naive Bayes algorithm, Linear Support Vector Machine (SVM) algorithm [6], and the Logistic Regression Model. Pre-trained models, which include word embeddings [12,17] are also useful for classifying text and other NLP tasks which may prevent a new model from being trained from scratch. The models will be pre-trained on the data to extract the keywords behind the sentiments of the economic text and the accuracy of the model will be measured.

2   Related Work

A substantial amount of work has been conducted in recent years in the area of sentiment analysis [2,3,5,8,14,17]. The social media, which is now accessible and embraced as a universal means of communication, has flourished thereby making the world a global village. More people now rely on it for breaking news and other facts, and more informed decisions are being made. Sentiment analysis is deriving intelligence from content generated on social media and the objective analysis of the feelings shared by users is of significant importance. The recent developments in computational natural language processing and machine learning have greatly benefited sensitivity analysis. Most of the recent research focuses on reviewing blog-writing opinions and informal social media texts. Just a few studies have examined how to model the financial and economics sentiments.

Machine learning approaches and deep learning networks have shown good success in detecting sentiments [2,3,10,12,13]. In Malo et al. [1], authors used the combination of categorical grammar, annotation, lexicon acquisition and semantic networks to examine the text feelings and to identify polarity-lexicons. To classify the sentence and economic reviews and measure their accuracy, we have proposed a batch of machine learning methods with semantic analysis.

3   Approach

There have been various opinion studies on economic text but the bone of contention remains how the feeling can be measured correctly and in a timely manner. The problem with rule-based approach is that the knowledge content is seriously dependent upon a trained expert in terms of prediction and pattern recognition. There has been growing interest in using algorithms that learn to understand language to resolve the learning capacity of this process, without being explicitly programmed. We identified six common approaches used in recent studies [19–21], namely logistic regression, support vector machines, naive Bayes, gradient boosting classifier, random forest, k-nearest neighbors, and decision tree classifier. Machine learning models for sentiment analysis can help us extract knowledge from areas that are becoming increasingly relevant and try to make more effective use of the data. These models have been extensively tested, and provide accurate results when working with various dataset types. Although, large data sets help to create better accuracy in machine learning models, the data collected is not large and unbalanced, i.e. the number of positive and negative samples are unequal.

3.1   Features

Certain features were extracted as the sentence author opines them. Each of the sentences is annotated with language information as well as polarity at the sentence level. We experimented with features such as the bag-of-words, word level n-grams, count vectorizer, and TF-IDF vectors.

N-grams. This is just any combination of adjacent or n-tuple of words or characters that can be found in the text. The fundamental point of n-grams is that they capture the structure of the language from a statistical point of view. There are various n-gram sequences which include a 1-gram (or unigram), 2-gram (or bigram) and 3-gram (or trigram).

Bag-of-Words. Representation of the bag-of-words which is commonly used to represent text for machine learning only counts how often each word appears in a text. Extracting features from text, it generates a vocabulary of all the specific terms with the output as a single word-count vector for each document.

Count Vectorizer. Count Vectorizer allows not only tokenization of text documents and the construction of a vocabulary of known words, but also the encoding of new documents using that vocabulary. It is a highly versatile module for text representation of features which provides the ability to pre-process the text data before creating the vector representation. A machine learning algorithm can then directly use the encoded vectors.

TF-IDF. This is an acronym for "Term Frequency – Inverse Document Frequency", the components of each word's resulting scores. Term Frequency summarizes the frequency with which a given word appears in a document and Inverse Document Frequency down-scale words which appear a lot throughout documents. TFIDF can tokenize documents, practice vocabulary and reverse weighting of document frequencies and encrypt new documents.

$$tfidf_{i,d} = tf_{i,d} \cdot idf_i \tag{1}$$

3.2 Models

To generate feature vectors for the data, the following classifiers were experimented as a supervised machine learning model. We measured the efficiency of classification, and found the measure of accuracy.

Logistic Regression Model. To predict an outcome that comes in a categorical form, the logistic regression is used. It is usually suited by highest probability. Logistic Regression not only provides a measure of the significance of a predictor but also of its correlation direction (positive or negative).

Support Vector Machine. This algorithm, which can be used for problems with classification or regression, uses a technique called the kernel trick to transform data and then finds an optimal boundary between potential outputs based on those transformations. The algorithm produces a line or hyperplane that divides the data into groups and works well for many practical problems.

Naive Bayes Classifier. Naive Bayes is a type of classifier using the Bayes Theorem. As a general statement, we can state Baye's theorem as follows

$$P(\theta|\mathbf{D}) = P(\theta)\frac{P(\mathbf{D}|\theta)}{P(\mathbf{D})} \tag{2}$$

The data are represented by $\mathbf{D}$ and parameters are represented by θ. It predicts the probabilities of membership for each class, such as the probability that a given record or data point belongs to a certain class. The highest-probability class is known as the most likely class.

Random Forest. The random forest algorithm, an ensemble of many decision trees, operates by creating an uncorrelated forest of trees at training time and outputting the class mode or mean forecast of the individual trees. The prediction by committee is more accurate than that of any individual tree.

Gradient Boosting Machine. This learning technique for regression and classification problems, is in the form of a combination of weak prediction models, typically decision trees, to create a strong predictive model. It uses our base model's loss function as a proxy to minimize the overall model error.

Decision Tree Classifier. The decision tree algorithm learns from the data in the form of a tree structure to build a classification or regression model. The internal nodes in this structure are labeled by features. It is an approach where the data is continually divided by a certain parameter and labelled by tests on the weight of the functions (smaller subsets) in the data while a related decision tree is built at the same time.

4   Experiments

Basically, we used different models to predict on the test data after fitting the training data to the model and testing data. We translated the data into numeric form and split the data into sets for training and testing. The training set was used to train the algorithm, while the test set was used to evaluate the machine learning model's output. Machine learning algorithms were used to learn from the training data, passing features and labels into the training as parameters. The models are intended to predict the results, and the accuracy, precision, and f1-score were obtained using bag representations of unigrams, bigrams, and trigrams as features within the model. A vocabulary was developed to keep a list of the word vectors and turn the text array into a TF-IDF function matrix. The model has been used to re-scale the meaning of essential words in the text making them comparable to each other. The classification and evaluation of the given instances into coarse-grained groups such as positive, neutral, and negative was performed. A graphical distribution of the sentiment polarity in the datasets can be seen in Fig. 1.

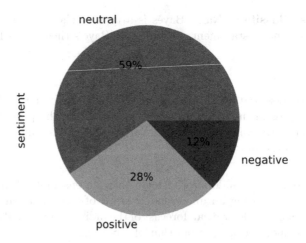

Fig. 1. Percentage of sentiment polarity in the datasets

4.1   Results and Discussion

In this study, we introduced machine learning techniques to analyze the feelings in the data for better and faster decision making, and we were able to compare the output of those techniques, thus adding to the state-of-the-art literature on tasks of sentiment analysis. This dataset includes human-labeled financial news from a retail investor's perspective with economic applications. These algorithms were applied in the datasets to predict the text's sentiment polarity and classify it according to the polarity. The performance of these methods was measured using the overall accuracy measurement.

Table 1. Accuracy values of the Models

Model	n-grams	Precision	Recall	f1-score	Accuracy %
Logistic Regression	Unigram	0.77	0.76	0.74	76.39%
	Bigram	0.75	0.72	0.68	72.16%
	Trigram	0.72	0.68	061	67.53%
Support Vector Machine	Unigram	0.77	0.77	0.76	77.11%
	Bigram	0.45	0.75	0.73	74.85%
	TrigramTrigram	0.72	0.71	0.68	71.44%
Naïve Bayes	Unigram	0.72	0.71	067	71.34%
	Bigram	0.56	0.64	0.54	64.02%
	Trigram	0.54	0.61	0.49	61.44%
Gradient Boosting	Unigram	0.35	0.59	0.44	59.38%
	Bigram	0.35	0.59	0.44	59.38%
	Trigram	0.35	0.59	0.44	59.38%
Decision Tree	Unigram	0.68	0.69	0.68	68.56%
	Bigram	0.67	0.69	0.67	68.87%
	Trigram	0.67	0.67	0.32	67.32%
Random Forest	Unigram	0.77	0.74	0.71	74.12%
	Bigram	0.740.74	0.71	0.67	71.44%
	Trigram	0.71	0.66	0.59	66.49%
K-Nearest Neighbors	Unigram	0.67	0.68	0.67	67.73%
	Bigram	0.74	0.61	0.48	61.03%
	Trigram	0.74	0.60	0.460.46	60.21%

Fig. 2. Word cloud view of the keywords pertaining to positive, negative, and neutral sentiment in the datasets

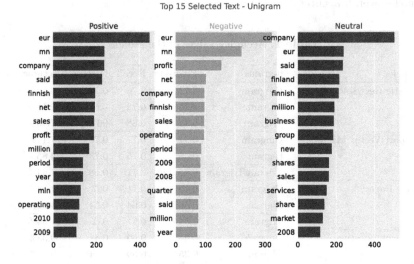

Fig. 3. Top 15 selected unigrams in the text

The word clouds (see Fig. 2) offered an indication of the words that could affect the polarity of the text. We extracted the n-gram functions such as unigram (see Fig. 3), bigram (see Fig. 4), and trigram (see Fig. 5) to examine the selected texts. With regard to the n-gram features, there was a strong correlation between the bi-gram and tri-gram as they gave meaningful results. Finally we present the machine learning accuracy score for all models used in Table 1.

The Support Vector Machine did the best of all the models while the Gradient Boosting Classifier did the least. The efficiency of a data mining classification algorithm is greatly impacted by the quality and quantity of the data.

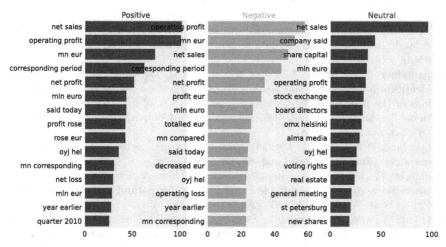

Fig. 4. Top 15 selected bigrams in the text

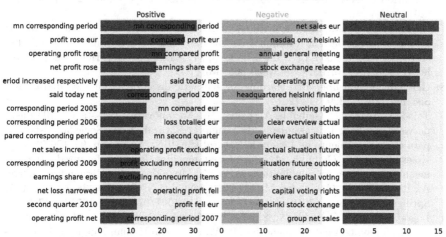

Fig. 5. Top 15 selected trigrams in the text

5   Conclusion

In this study, we successfully introduced various machine learning approaches for sentiment analysis in economics text involving word embedding learning, classifications, opinion extraction, and lexicon learning. Insufficient amount of data affected the performance of these models. We intend to explore more ways to improve the accuracy of classification of sentiments in the future.

References

1. Malo, P., Sinha, A., Korhonen, P., Wallenius, J., Takala, P.: Good debt or bad debt: detecting semantic orientations in economic texts. J. Assoc. Inf. Sci. Technol. **65**(4), 782–796 (2014)
2. Aroyehun, S.T., Angel, J., Pérez Alvarez, D.A., Gelbukh, A.: Complex word identification: convolutional neural network vs. feature engineering. In: Proceedings of the Thirteenth Workshop on Innovative Use of NLP for Building Educational Applications, pp. 322–327. Association for Computational Linguistics, New Orleans, Louisiana (2018)
3. Gambino, O.J., Calvo, H.: Predicting emotional reactions to news articles in social networks. Comput. Speech Lang. **58**, 280–303 (2019)
4. Sidorov, G., Velasquez, F., Stamatatos, E., Gelbukh, A., Chanona-Hernandez, L.: Syntactic n-grams as machine learning features for natural language processing. Expert Syst. Appl. **41**, 853–860 (2014)
5. Gomez-Adorno, H., Markov, I., Sidorov, G., Posadas-Duran, J., Sanchez-Perez, M.A., Chanona-Hernandez, L.: Improving feature representation based on a neural network for author profiling in social media texts. In: Computational Intelligence and Neuroscience (2016)
6. Hernández-Castañeda, Á., Calvo, H., Gelbukh, A., Flores, J.J.G.: Cross-domain deception detection using supportvector networks. Soft Comput. **21**(3), 585–595 (2017)
7. Calvo, H., Gelbukh, A., Kilgarriff, A.: Distributional thesaurus versus WordNet: a comparison of backoff techniques for unsupervised PP attachment. In: Gelbukh, A. (ed.) CICLing 2005. LNCS, vol. 3406, pp. 177–188. Springer, Heidelberg (2005). https://doi.org/10.1007/978-3-540-30586-6_17
8. Gelbukh, A.: Unsupervised learning for syntactic disambiguation. Computación y Sistemas **18**(2), 329–344 (2014)
9. Tejada-Cárcamo J., Gelbukh, A., Calvo, H.: An innovative two-stage WSD unsupervised method. Procesamiento de Lenguaje Natural, N 40. Sociedad Española para el Procesamiento de Lenguaje Natural (SEPLN), Spain, pp. 99–106 (2008)
10. Kolesnikova, O., Gelbukh, A.: Supervised machine learning for predicting the meaning of verb-noun combinations in Spanish. In: Sidorov, G., Hernández Aguirre, A., Reyes García, C.A. (eds.) MICAI 2010. LNCS (LNAI), vol. 6438, pp. 196–207. Springer, Heidelberg (2010). https://doi.org/10.1007/978-3-642-16773-7_17
11. Lugo-Garcia, T., Gelbukh, A., Sidorov, G.: Unsupervised learning of word combinations for syntactic disambiguation. In: Workshop on Human Language Technologies at the ENC-2004, 5th Mexican International Conference on Computer Science. Avances en la Ciencia de la Computación, pp. 311–318 (2004)
12. Kolesnikova, O., Gelbukh, A.: A study of lexical function detection with word2vec and supervised machine learning. J. Intell. Fuzzy Syst. **39**, 1993–2001 (2020)
13. Poria, S., Cambria, E., Gelbukh, A.: Aspect extraction for opinion mining with a deep convolutional neural network. Knowl. Based Syst. **108**, 42–49 (2016)
14. Aroyehun, S.T., Gelbukh, A.: Aggression detection in social media: using deep neural networks, data augmentation, and pseudo labeling. In: Proceedings of the First Workshop on Trolling, Aggression and Cyberbullying (TRAC-1), Santa Fe, USA (2018)
15. Gómez-Adorno, H., Fuentes-Alba, R., Markov, I., Sidorov, G., Gelbukh, A.: A convolutional neural network approach for gender and language variety identification. J. Intell. Fuzzy Syst. **36**(5), 4845–4855 (2019). ISSN 1064-1246

16. Galassi, A., Lippi, M., Torroni, P.: Attention, please! a critical review of neural attention models in natural language processing. arXiv preprint arXiv:1902.02181 (2019)
17. Aroyehun, S.T., Gelbukh, A.: Detection of adverse drug reaction in tweets using a combination of heterogeneous word embeddings. In: Proceedings of the Fourth Social Media Mining for Health Applications (#SMM4H) Workshop & Shared Task, ACL. Florence, Italy, Association for Computational Linguistics, pp. 133–135 (2019)
18. Radford, A., Narasimhan, K. Salimans, T., Sutskever, I.: Improving language understanding by generative pre-training (2018). https://s3-us-west-2.amazonaws.com/openai-assets/research-covers/language-unsupervised/language_understanding_paper.pdf
19. Pranckevicius, T., Marcinkevicius, V.: Comparison of Naive Bayes, random forest, decision tree, support vector machines, and logistic regression classifiers for text reviews classification. Balt J. Mod. Comput. 5(2), 221–32 (2017)
20. Athanasiou, V., Maragoudakis, M.: A novel, gradient boosting framework for sentiment analysis in languages where NLP resources are not plentiful: a case study for modern Greek. Algorithms 10, 34 (2017)
21. Serrano-Silva, Y.O., Villuendas-Rey, Y., Yáñez-Márquez, C.: Automatic feature weighting for improving financial Decision Support Systems. Decis. Support Syst. 107, 78–87 (2018)
22. Lytras, M.D., Mathkour, H.I., Abdalla, H., Yáñez-Márquez, C., Ordóñez de Pablos, P.: The social media in academia and education: research r-evolutions and a paradox: advanced next generation social learning innovation. J. Univ. Comput. Sci. (J.UCS) 20(15), 1987–1994 (2014)
23. Villuendas-Rey, Y., Rey-Benguría, C.F., Ferreira-Santiago, A., Camacho-Nieto, O., Yáñez-Márquez, C.: The naïve associative classifier (NAC): a novel, simple, transparent, and accurate classification model evaluated on financial data. Neurocomputing 265, 105–115 (2017)

Depression Detection in Social Media Using a Psychoanalytical Technique for Feature Extraction and a Cognitive Based Classifier

Seyed Habib Hosseini-Saravani, Sara Besharati, Hiram Calvo,
and Alexander Gelbukh[(⊠)]

Center for Computing Research, Instituto Politécnico Nacional, Av. JD Bátiz e/MO de
Mendizábal s/n, Nva. Ind. Vallejo, 07738 Mexico City, Mexico
hosseinihaamed@gmail.com, besharatisara62@gmail.com,
hcalvo@cic.ipn.mx, gelbukh@gelbukh.com

Abstract. Depression detection in social media is a multidisciplinary area where psychological and psychoanalytical findings can help machine learning and natural language processing techniques to detect symptoms of depression in the users of social media. In this research, using an inventory that has made systematic observations and records of the characteristic attitudes and symptoms of depressed patients, we develop a bipolar feature vector that contains features from both depressed and non-depressed classes. The inventory we use for feature extraction is composed of 21 categories of symptoms and attitudes, which are primarily clinically derived in the course of the psychoanalytic psychotherapy of depressed patients, and systematic observations and records of their characteristic attitudes and symptoms. Also, getting insight from a cognitive idea, we develop a classifier based on multinomial Naïve Bayes training algorithm with some modification. The model we develop in this research is successful in classifying the users of social media into depressed and non-depressed groups, achieving the F1 score 82.75%.

Keywords: Depression detection · Social media · Natural Language Processing · Psychoanalysis · Rational Speech Act · Naïve Bayes

1 Introduction

Drastic changes in human lifestyle over the past century has led to an increase in the number of people suffering from depression, which is believed to be a "disease of modernity" [1]. It is predicted that, by 2030, one of the three leading causes of illness will be depression [2], and the epidemic of child and adolescent depression is now one the most important concerns in academia. Depression negatively affects how a person

This work was done with support of the Government of Mexico via CONACYT, SNI, CONACYT grant A1-S-47854 and grants SIP 20200811, SIP 20200859 of the Instituto Politécnico Nacional (IPN), IPN-COFAA and IPN-EDI.

© Springer Nature Switzerland AG 2020
L. Martínez-Villaseñor et al. (Eds.): MICAI 2020, LNAI 12469, pp. 282–292, 2020.
https://doi.org/10.1007/978-3-030-60887-3_25

feels, thinks and acts and can bring a variety of emotional and physical problems that can decrease the person's ability to function in the society. Therefore, it is highly beneficial if depression can be detected before it hurts both the individuals and the society.

Depression symptoms show themselves in different categories of human behavior and activities and can vary from mild to severe [3]. One of the most important sources that helps in detecting symptoms of depression in individuals is the language they use. In fact, language is the manifestation of how a person feels and thinks. Cognitive and linguistic analyses [4] have shown that people with depression use language differently. For example, they frequently use adjective and adverbs conveying negative emotions, such as "lonely", "sad", or "miserable", and they have a tendency to use first person singular more than second or third person, which shows that they are more focused on themselves, and less connected with others. In addition, the style in which the people with depression use language show that they frequently use absolutist words, such as "always", "nothing" or "completely" [4, 5].

Social media, as one of the most important sources of individuals' thoughts, opinions and feelings [6], can carry very important information about the users [21, 22]. In the area of psychoanalytical and psychological studies, relying on social media as a behavioral health assessment tool even has many advantages over self-report methodology in behavioral surveys because social media measurement of behavior captures social activity and language expression in a naturalistic setting—in contrast with behavioral surveys, where responses are prompted by the experimenter and typically comprise recollection of health facts [6]. As a result, social media has been an important source of data for researchers [6, 7], and also there have been many works on depression detection in social media using Natural Language Processing (NLP) [8] techniques, which can help us mine the data in the social media in order to detect the signs of depression in the users of social media.

In this research, getting insight from cognitive and psychoanalytical findings, we apply modified multinomial Naïve Bayes algorithms to the area of depression detection in social media. The contributions of this research are mainly in two areas: (a) feature extraction, (b) learning method. For feature extraction, using an inventory [3] that has made systematic observations and records of the characteristic attitudes and symptoms of depressed patients, we develop a bipolar feature vector that contains features from both depressed and non-depressed classes. Then we use these features to train a classifier whose learning is based on multinomial Naïve Bayes [9] algorithm. However, we have made a modification in the training phases of our classifier in order for it to increase the importance of some of the features in certain conditions. In our model, features of each of the target classes are of more importance when they are observed in the training data of that certain class. The idea behind this modification is based on Rational Speech Act (RSA) theory, [10] which is a Bayesian based cognitive theory whose main idea is that not all features of a thing or a person have the same value for human brain when during the recognition of a thing or a person by human brain. Based on RSA theory, this characteristic of human brain can help humans have a rational inference of what others say.

In the following paragraphs, we will discuss previous work in Sect. 2, explain the methodology in detail in Sect. 3, discuss the experimental results in Sect. 4, and summarize our findings and talk about future work in Sect. 5.

2 Related Work

De Choudhury et al. [11, 12] used crowdsourcing methodology to build a large corpus of postings on Twitter that have been shared by individuals diagnosed with clinical depression. They developed a model (an SVM classifier) trained on this corpus to determine if posts could indicate depression. Their model leveraged signals of social activity, emotion, and language manifested on Twitter. For feature extraction, they proposed several features to characterize the postings in their dataset. The features could be categorized into two types: post-centric and user-centric features. Post-centric features—emotion, time, linguistic style—captured properties in the post, while the user-centric features— engagement, ego-network—characterized the behavior of the post's author. Their models could predict if a post is depression-indicative, with accuracy of more than 70% and precision of 0.82. This work demonstrated how sets of behavioral markers manifested in social media can be harnessed to predict depression-indicative postings, and thereby understand large-scale depression tendencies in populations.

Evaluating depressive symptoms using the Center for Epidemiological Studies-Depression (CES-D) scale, Sungkyu et al. [13] developed a Web application to identify depressive symptom–related features from users of Facebook as a popular social networking platform. They provided tips and facts about depression to participants and measured their responses using EmotionDiary, the Facebook application that they had developed. To identify the Facebook features related to depression, correlation analyses were performed between CES-D and participants' responses to tips and facts or Facebook social features. Last, they interviewed depressed participants (CES-D $\geq$ 25) to assess their depressive symptoms by a psychiatrist. The results of that paper showed that Facebook activities had predictive power in distinguishing depressed and nondepressed individuals.

Tsugawa et al. [14] evaluated the effectiveness of using a user's social media activities for estimating degree of depression. They used the results of a web-based questionnaire for measuring degree of depression of Twitter users. For feature extraction for estimating the presence of active depression, they extracted several features from the activity histories of Twitter users. That paper showed that (a) features obtained from user activities can be used to predict depression of users with an accuracy of 69%, (b) topics of tweets estimated with a topic model are useful features, (c) approximately two months of observation data are necessary for recognizing depression, and longer observation periods do not contribute to improving the accuracy of estimation for current depression; sometimes, longer periods worsen the accuracy.

In another research, Shen et al. [15] constructed well-labeled depression and non-depression dataset on Twitter, and extract six depression-related feature groups covering not only the clinical depression criteria, but also online behaviors on social media. They proposed a multimodal depressive dictionary learning model to detect the depressed users on Twitter, and analyzed a large-scale dataset on Twitter to reveal the underlying online behaviors between depressed and non-depressed users. Their proposed model—Multimodal Depressive Dictionary Learning (MDL)—achieved the best performance with 85% in F1-Measure. In another similar research [16] Shen et al. studied a problem of enhancing detection in a certain target domain with ample Twitter data as the source domain. They first systematically analyzed the depression-related feature patterns across domains and summarized two major detection challenges, namely "isomerism" and "divergency". We further propose a cross-domain Deep Neural Network model with Feature Adaptive Transformation & Combination strategy (DNN-FATC) that transfers the relevant information across heterogeneous domains.

Based on the CLEF/eRisk 2017 pilot task, which is focused on early risk detection of depression, Stankevich et al. [17] performed a research using CLEF/eRsik dataset [18] which consists of text examples collected from messages of Reddit users. They classified users into two groups: risk case of depression and non-risk case, considering different feature sets for depression detection task among Reddit users by text messages processing. For feature extraction, they used bag-of-words, embedding and bigram models. They used support vector machines (SVM) for classification, and the best model in that research was tf·idf with morphology set as features, which achieved best results on the test data with 63% F1-score. Embedding features in that research always obtained better recall score than tf·idf but the accuracy and the precision scores were lower.

3 Methodology

The inventory we used in this research for feature extraction [3] is designed to measure the behavioral manifestations of depression and is composed of 21 categories of symptoms and attitudes that are primarily clinically derived in the course of the psychoanalytic psychotherapy of depressed patients and systematic observations and records of their characteristic attitudes and symptoms. These categories are listed in Table 1.

Table 1. The categories of symptoms and attitudes

1. Mood	12. Social Withdrawal
2. Pessimism	13. Indecisiveness
3. Sense of failure	14. Body Image
4. Lack of satisfaction	15. Work Inhibition
5. Guilty feeling	16. Sleep Disturbance
6. Sense of punishment	17. Fatigability
7. Self-Hate	18. Loss of Appetite
8. Self Accusations	19. Weight Loss
9. Self Punitive Wishes	20. Somatic Preoccupation
10. Crying Spells	21. Loss of Libido
11. Irritability	

Each category consists of a graded series of 4 to 5 self-evaluative statements that describe a specific behavioral manifestation of depression and are ranked from 0 to 3 to reflect the range of severity of the symptom from neutral to maximal severity. The statements of three of the categories are shown in Table 2 (to see the complete statements visit [3]).

Table 2. Statement of the categories of symptoms and attitudes

Category	Statements
Sense of Failure	0 - I do not feel like a failure 1 - I feel I have failed more than the average person 2a - I feel I have accomplished very little that is worthwhile or that means any¬ thing 2b - As I look back on my life all I can see is a lot of failures 3 - I feel I am a complete failure as a person (parent, husband, wife)
Self Hate	0 - I don't feel disappointed in myself la - I am disappointed in myself lb - I don't like myself 2 - I am disgusted with myself 3 - I hate myself
Social Withdrawal	0 - I have not lost interest in other people 1 - I am less interested in other people now than I used to be 2 - I have lost most of my interest in other people and have little feeling for them 3 - I have lost all my interest in other people and don't care about them at all

To extract the features, we used the keywords of the statements in each category. First, we removed the stop words from the statements and obtained a list of words including the main verbs and the adjectives that we expect to be frequently used by the people who suffer from depression. The main idea behind our technique is that either the words from this list of words or their synonyms are frequent in the language of people with symptoms of depression; therefore, using NLTK library in python, we found the synonyms of the words in our list and added them to our list and called this list the "depressed class features" list.

On the other hand, psychoanalytical studies tell us that the people who are not suffering from depression not only do not use the words in "depressed class features" list frequently but also might use the words with the opposite meaning. For example, a depressed person might show the signs of depression through producing utterances like "I hate other people"; however, a person who is not depressed is more willing to use sentences like "I love my friends". Therefore, to distinguish depressed people from non-depressed ones, the antonyms of the words in "depressed class features" list were also important because the occurrence of these words could reduce the probability that a person is depressed. As a result, we created a second list including the antonyms of the words in "depressed class features" list and called it the "non-depressed class features" list.

Finally, our ultimate feature list was the combination of "depressed class features" and "non-depressed class features" lists. Based on this feature extraction method, which we call a bipolar feature extraction, we developed a classifier with two target classes. In addition, we develop another model based on word frequency as the feature extraction in order to compare the results of our bipolar feature extraction with this model.

Using the features we obtained, we apply a Bayesian classifier whose training is based on multinomial Naïve Bayes algorithm to classify each of the users in the test data into either depressed or non-depressed class.

3.1 Training and Test Data Set

We use a subset of the CLEF/eRsik dataset [18], which was provided as the part of the pilot task of early risk detection of depression and consists of text examples collected from messages of Reddit users. The dataset consists of 15% of positive examples (risk case of depression) and 85% of negative examples (non-risk case), and each example of the dataset contains all text messages—between 10 to 2000 messages—of the user for a certain period of time with different time intervals. The dataset that we had access to consists of 340 examples (40 positive samples and 300 negative samples). To make a balance between the two classes, we used 25 positive examples and 50 negative examples as the training data, and allocated 15 positive and 25 negative examples for the test data set. Note that all 40 positive examples were used in this work.

3.2 Classification Method

To classify the users, we use a classifier based on Naïve Bayes algorithm, which is a kind of a frequently used supervised learning method that examines all its training input and applies Bayes theorem with the "Naïve" assumption of conditional independence between features given the value of the class variable [19]. Equation 1 below shows Bayes theorem, where C stands for class variable and x_1 through x_n are dependent feature vectors:

$$P(C \mid x_1, \cdots x_n) = \frac{P(C)P(x_1, \ldots, x_n \mid C)}{p(x_1, \ldots, x_n)}.$$ (1)

There are different kinds of Naïve Bayes classifiers based on their training and classification algorithms and their *attitude* toward the data distribution. The classifier we used is based on a multinomial Naïve Bayes algorithm which is implemented to the data that is multinomially distributed and is one of the Bayes variants that is usually used in text classification [19]. The distribution is parametrized by vectors $\theta y = (\theta y_1, \ldots, \theta y_n)$ for each class y, where *n* is the number of features—the size of the vocabulary—and θy_i is the probability $P(x_i \mid y)$ of feature *i* appearing in a sample belonging to class y. The parameter $\hat{\theta} y_i$ is estimated by a smoothed version of maximum likelihood, i.e. relative frequency counting:

$$\hat{\theta} y_i = \frac{N_{y_i} + \alpha}{N_y + \alpha n}.$$ (2)

In the equation above, $Ny_i = \sum_{x \in T} x_i$ is the number of times feature i appears in a sample of class y in the training set T, and $N_y = \sum_{i=1}^{n} N_{y_i}$ is the total count of all features for class y. The smoothing priors $\alpha \geq 0$ accounts for features not present in the learning samples and prevents zero probabilities in further computations. Setting $\alpha = 1$ is called Laplace smoothing, while $\alpha < 1$ is called Lidstone smoothing [20].

Our Modified Classifier

The classifier we developed in this research is a modification of Naïve Bayes algorithm; however, because of the feature extraction method we used in this research, the conditional value of some of the features in the feature vector gets doubled in certain conditions. Based on the bipolar nature of the features we extract for the two classes in this research—depressed and non-depressed people—, we modified our classifier in a way that the words in the depressed class feature list and the words in the non-depressed class feature list get a double importance when they appear in classes depressed and non-depressed respectively. Therefore, in this model:

$$\hat{\theta}_{y_i} = \begin{cases} \frac{N_{y_i} + \alpha}{N_y + \alpha n} \times 2 & \text{if } N_{y_i} \text{ only in feature list of class } y \\ \frac{N_{y_i} + \alpha}{N_y + \alpha n} & \text{else.} \end{cases} \tag{3}$$

4 Experimental Results

To have the ratio of correctly predicted positive observations to the total predicted positive observations, and also the ratio of correctly predicted positive observations to the all observations in actual class, we used the metrics precision and recall respectively. In addition, to take both false positives and false negatives into account, we used F1 Score, which is the weighted average of precision and recall. Equations 4, 5, and 6 show the formulae for the calculation of precision, recall, and F1 Score respectively:

$$\text{Precision} = \frac{TP}{TP + FP} \tag{4}$$

$$\text{Recall} = \frac{TP}{TP + FN} \tag{5}$$

$$\text{F1 Score} = \frac{2 \times (\text{Recall} \times \text{Precision})}{(\text{Recall} + \text{Precision})} \tag{6}$$

where TP (True Positive), TN (True Negative), FP (False Positive), and FN (False Negative) refer to the correctly predicted depressed users, correctly predicted non-depressed users, incorrectly predicted depressed users, and incorrectly predicted non-depressed respectively. Tables 3, and 4 show the confusion matrices of the modified Bayes model and basic Bayes model, both with bipolar feature vector, and Tables 5 and 6 show the confusion matrices of modified and basic Bayes models, both with 1000 most frequent words as the feature vector.

Table 3. Confusion matrix of the modified Bayes model with bipolar feature vector

	Positive	Negative
Positive	12	3
Negative	2	23

Table 4. Confusion matrix of the basic Bayes model with bipolar feature vector

	Positive	Negative
Positive	14	1
Negative	6	19

Table 5. Confusion matrix of modified Bayes model with 1000 most frequent words as the feature vector

	Positive	Negative
Positive	10	5
Negative	1	24

Table 6. Confusion matrix of basic Bayes model with 1000 most frequent words as the feature vector

	Positive	Negative
Positive	9	6
Negative	1	24

As Table 7 shows, considering F1 score, the models that used bipolar feature vectors have a much better performance than the ones with term frequency as feature extraction method. Also, it can be seen in Fig. 1 that the models developed by the basic Bayes classifiers were biased to one class, but our modified Bayes classifiers were able to make a balance between the two classes; consequently, they improved the F1 score of the models based on basic Bayes classifiers. However, the performance of our modified model was much more significant for the models with bipolar feature extraction.

Table 7. Comparison of the results obtained from our models

	Precision %	Recall %	F1 Score %
Modified – bipolar	80.00	85.71	82.75
Basic – bipolar	93.33	70.00	79.99
Modified – frequency	66.66	90.90	76.91
Basic – frequency	60.00	90.00	72.00

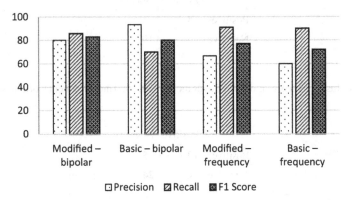

Fig. 1. Comparison of the results of different models developed in this research

5   Conclusions and Future Work

We introduced a new method for feature extraction using findings in the realms of Psychoanalysis and Psychology. We called our feature extraction method a *bipolar feature extraction* since the final feature vector we obtained from this method consists of two groups of features that are clearly opposite to each other in meaning. In addition, getting insight from a cognitive idea, we developed a modified Bayesian classifier, which improved the performance of the basic Bayesian classifiers, especially when it was used with a bipolar feature vector. The results obtained from this research showed an achievement in classifying social media users into depressed and non-depressed classes when we used our bipolar feature vector instead of word frequency. Also, the modified classifier we developed was successful in improving the performance of our models reaching the F1 score 82.75%.

For future work, we intend to get insight from findings in cognitive and psychoanalytical studies to predict the severity of depression in users of social media.

References

1. Hidaka, B.H.: Depression as a disease of modernity: explanations for increasing prevalence. J. Affect. Disord. **140**(3), 205–214 (2012)

2. Mathers, C.D., Loncar, D.: Projections of global mortality and burden of disease from 2002 to 2030. PLoS Med. **3**(11), e442 (2006)
3. Beck, A.T., Ward, C.H., Mendelson, M., Mock, J., Erbaugh, J.: An inventory for measuring depression. Arch. Gen. Psychiatry **4**(6), 561–571 (1961)
4. Al-Mosaiwi, M., Johnstone, T.: In an absolute state: elevated use of absolutist words is a marker specific to anxiety, depression, and suicidal ideation. Clin. Psychol. Sci. **6**(4), 529–542 (2018)
5. Al-Mosaiwi, M.: People with depression use language differently – here is how to spot it. In: The Conversation. https://theconversation.com/people-with-depression-use-language-differently-heres-how-to-spot-it-90877(2020). Accessed 24 Jul 2020
6. Paul, M.J., Dredze, M.: You are what you tweet: analyzing twitter for public health. In: Fifth International AAAI Conference on Weblogs and Social Media (2011)
7. Sadeque, F., Xu, D., Bethard, S.: Measuring the latency of depression detection in social media. In: Proceedings of the Eleventh ACM International Conference on Web Search and Data Mining, pp. 495–503 (2018)
8. Manning, C,. Schutze, H.: Foundations of Statistical Natural Language Processing. MIT Press, Cambridge (1999)
9. Kibriya, A.M., Frank, E., Pfahringer, B., Holmes, G.: Multinomial Naive Bayes for text categorization revisited. In: Webb, G.I., Yu, X. (eds.) Australasian Joint Conference on Artificial Intelligence. Lecture Notes in Computer Science, vol. 3339, pp. 488–499. Springer, Berlin, Heidelberg (2004). https://doi.org/10.1007/978-3-540-30549-1_43
10. Monroe, W., Potts, C.: Learning in the rational speech acts model. arXiv preprint arXiv:1510.06807 (2015)
11. De Choudhury, M, Gamon, M., Counts, S., Horvitz, E.: Predicting depression via social media. In: Seventh International AAAI Conference on Weblogs and Social Media (2013)
12. De Choudhury, M., Counts, S., Horvitz, E.: Social media as a measurement tool of depression in populations. In: Proceedings of the 5th Annual ACM Web Science Conference, pp. 47–56 (2013)
13. Park, S., Lee, S.W., Kwak, J., Cha, M., Jeong, B.: Activities on facebook reveal the depressive state of users. J. Med. Internet Res. **15**(10), e217 (2013)
14. Tsugawa, S., et al.: Recognizing depression from twitter activity. In: Proceedings of the 33rd Annual ACM Conference on Human Factors in Computing Systems, pp. 3187–3196 (2015)
15. Shen, G., et al.: Depression detection via harvesting social media: a multimodal dictionary learning solution. In: IJCAI, pp. 3838–3844 (2017)
16. Shen, T., et al.: Cross-domain depression detection via harvesting social media. In: International Joint Conferences on Artificial Intelligence (2018)
17. Stankevich, M., Isakov, V., Devyatkin, D., Smirnov, I. Feature engineering for depression detection in social media. In: ICPRAM, pp. 426–431 (2018)
18. Losada, D.E., Crestani, F.: A test collection for research on depression and language use. In: Fuhr, N., et al. (eds.) Experimental IR Meets Multilinguality, Multimodality, and Interaction. Lecture Notes in Computer Science, vol. 9822, pp. 28–39. Springer, Cham (2016). https://doi.org/10.1007/978-3-319-44564-9_3
19. scikit-learn. 1.9. Naïve Bayes. scikit-learn. https://scikit-learn.org/stable/modules/Naïve_bayes.html. Accessed 5 May 2020
20. Franco-Penya, H.H., Sanchez, L.M.: Tuning Bayes Baseline for dialect detection. In: Proceedings of the Third Workshop on NLP for Similar Languages, Varieties and Dialects (VarDial3), pp. 227–234 (2016)

21. Lytras, M.D., et al.: Guest editorial: the social media in academia and education: research R-evolutions and a paradox: advanced next generation social learning innovation. J. Univ. Comput. Sci. (J.UCS) 20(15), 1987–1994 (2014)

22. Moreno-Moreno, P., Yáñez-Márquez, C.: The new informatics technologies in education debate. In: Lytras, M.D., et al. (eds.) WSKS 2008. CCIS, vol. 19, pp. 291–296. Springer, Berlin, Heidelberg (2008). https://doi.org/10.1007/978-3-540-87783-7_37

Does Supervised Learning of Sentence Candidates Produce the Best Extractive Summaries?

Sandra J. Gutiérrez Hinojosa[1], Hiram Calvo[1(⊠)],
Marco A. Moreno-Armendáriz[1], and Carlos Duchanoy[2]

[1] Center for Computing Research, Instituto Politécnico Nacional,
Mexico City, Mexico
sandyguh04@gmail.com, {hcalvo,mam_armendariz}@cic.ipn.mx
[2] Cátedra CONACyT, Center for Computing Research, Instituto Politécnico
Nacional, Av. JD Bátiz e/ M.O. de Mendizábal, 07738 Mexico City, Mexico
duchduchanoy@cic.ipn.mx

Abstract. In this work multi-document, extractive summaries have
been obtained using supervised learning algorithms in a well-known
dataset (DUC 2002); the methodology has three steps: the pre-processing
step, which filters irrelevant words and reduces vocabulary using stem-
ming; the representation step, which transforms sentences into vectors;
and the classification step which selects sentences for the summary. Noting that the last step is crucial because it determines the relevance of
each sentence according to the information included in the embeddings.
We found that the classifiers performance is not related to the summary
quality mainly classifier's goal is not aligned to summarizer's goal, as
classifier is based on selecting whole sentences, while summarization is
evaluated by n-grams, for example ROUGE-n, and therefore it is relevant while comparing performances between different works in the state
of the art.

1 Research Problem

The main goal of a summary is to find the main ideas in a document reducing
the original documents size; algorithms created for solving this task have a relevant application given the exponential growth of textual information online, and
the need to find the main ideas of documents in a shorter time. In order to perform this task automatically there are two different approaches: extracting the
main sentences from the documents or paraphrasing the main ideas. Abstractive
methods are highly complex, as they need to simulate human cognitive process
for generating summaries [6]. Therefore, research community is focusing more

This work has been possible thanks to the support of the Mexican government through
the FOINS program of Consejo Nacional de Ciencia y Tecnología (CONACYT) under
grants Problemas Nacionales, 5241, Cátedras CONACYT 556; the SIP-IPN research
grants SIP 2083, SIP 20200640, and SIP 20200811; IPN-COFAA and IPN-EDI.

L. Martínez-Villaseñor et al. (Eds.): MICAI 2020, LNAI 12469, pp. 293–296, 2020.
https://doi.org/10.1007/978-3-030-60887-3_26

on extractive summaries, trying to achieve more coherent and meaningful summaries [5,9,11].

In this work, multi-document, extractive summaries have been obtained using supervised learning algorithms. The supervised learning has three phases: training, validation, and testing [1]; each one requires labeled data, i.e., examples of documents and the sentences which belong to the summary [10]. In this paper three interesting questions arise: (1) is a supervised algorithm capable of learning characteristics of a sentence in order to classify it as a candidate or not for a summary?, (2) in which way the classifier performance is affected when a classifier is trained in a certain corpus, and then tested in another? and (3) is the classifier performance related to the summary quality?

2   Motivation

The task of automatic summaries has been studied in multiple works with the supervised approach [3,4,13]. However, in some other works [2,7] the task is considered solved due to the performance of the classifiers, without considering the quality of the summary that these classifiers provide. This has led us to the task of studying the relationship between the summary quality and classifier performance.

3   Technical Contribution

We used pre-trained sentence embeddings, for further details in the embedding settings refer to [8]. In this supervised learning setting, 80% of the data was used for training and the remaining, for testing. Also, we used k-fold cross validation with k = 5 in the training phase and oversampling technique to handle unbalanced data. Finally, we used cross-validation method and five classifiers: Gaussian Naïve Bayes (GNB), Bernoulli Naïve Bayes (BNB), Multi-layer Perceptron with hyperbolic tangent (MLP-tanh), logistic sigmoid (MLP-log) and rectified linear unit (MLP-relu), where each multilayer perceptron has four layers with 100, 50, 20 and 10 neurons, respectively.

The performance measures were recall, precision, accuracy and F1 score for the classifiers, which contain information about the number of misleading classifications on each class, while ROUGE-n measure, based on n-gram overlapping, was used as intrinsic quality of the summaries.

The following tables show performance of five classifiers and the summary quality of each one. We selected the classifier with best accuracy performance in the validation phase and then verified its performance measuring the summary quality (See Tables 1 and 2).

We find that the relation between classifier performance and summary quality is not clear, may be because the classifiers work with sentences while the quality of the summaries is evaluated by groups of words, this result answers the third research question, but there is still room for improvements, for example

Table 1. Summary quality and classifier performances with unbalanced data

Summary quality				Classifier performance		
Classifier	ROUGE-1	ROUGE-2	F1 score	Training	Validation	Testing
GNB	0.2789	0.0831	0.3386	0.6980	0.6977	0.6995
BNB	0.4297	**0.1737**	**0.4389**	0.8432	0.8410	0.8424
MLP-tanh	0.3854	0.1160	0.3887	0.9928	0.9297	0.9240
MLP-log	**0.4342**	0.1530	0.4095	0.9592	**0.9592**	**0.9522**
MLP-relu	0.3834	0.1497	0.3963	**0.9963**	0.9444	0.9399

Table 2. Summary quality and classifier performances with balanced data

Summary quality				Classifier performance		
Classifier	ROUGE-1	ROUGE-2	F1 score	Training	Validation	Testing
GNB	0.3524	0.0793	0.3721	0.7863	0.7767	0.7624
BNB	**0.4685**	**0.1887**	**0.4385**	0.8205	0.7890	0.7553
MLP-tanh	0.3278	0.0633	0.3231	0.9763	0.8085	**0.8014**
MLP-log	0.3586	0.0967	0.3718	0.5009	0.5018	0.4929
MLP-relu	0.3753	0.1153	0.3905	**0.9960**	**0.8253**	0.7943

experimenting with other corpora (DUC 2001). Nonetheless, embeddings capture relevant characteristics for summaries in DUC 2002, this result answers the first research question and this work is different from previous works because we experiment with other classifiers.

With respect to the second research question we are working in it, using trained classifier in DUC 2001 to solve extractive summary task. It is important to note that sentences must be labeled in order to apply supervised approach, we are following a methodology proposed in previous work [12].

The uncorrelated results between classifier performance and summary quality suggest us to evaluate experiments with supervised approach using n-grams instead of sentence embeddings, as future work. We believe that using universal sentence encoder as inputs to the classifiers is another interesting direction for further research. Finally, an additional research effort should be directed toward proposing another evaluation metric besides the ROUGE metric.

References

1. Acevedo-Mosqueda, M.E., Yáñez-Márquez, C., López-Yáñez, I.: Alpha-beta bidirectional associative memories: theory and applications. Neural Process. Lett. **26**(1), 1–40 (2007)
2. Azhari, M., Jaya Kumar, Y.: Improving text summarization using neuro-fuzzy approach. J. Inf. Telecommun. **1**(4), 367–379 (2017)

3. Cao, Z., Li, W., Li, S., Wei, F.: Improving multi-document summarization via text classification. In: Thirty-First AAAI Conference on Artificial Intelligence (2017)
4. Cao, Z., Wei, F., Li, S., Li, W., Zhou, M., Houfeng, W.: Learning summary prior representation for extractive summarization. In: Proceedings of the 53rd Annual Meeting of the Association for Computational Linguistics and the 7th International Joint Conference on Natural Language Processing (Volume 2: Short Papers). pp. 829–833 (2015)
5. Gambhir, M., Gupta, V.: Recent automatic text summarization techniques: a survey. Artif. Intell. Rev. 47(1), 1–66 (2016). https://doi.org/10.1007/s10462-016-9475-9
6. Gupta, V., Lehal, G.S.: A survey of text summarization extractive techniques. J. Emerg. Technol. Web Intell. 2(3), 258–268 (2010)
7. Hardy, H., Shimizu, N., Strzalkowski, T., Ting, L., Zhang, X., Wise, G.B.: Cross-document summarization by concept classification. In: Proceedings of the 25th annual international ACM SIGIR conference on Research and development in information retrieval. pp. 121–128. ACM (2002)
8. Lau, J.H., Baldwin, T.: An empirical evaluation of doc2vec with practical insights into document embedding generation. arXiv preprint arXiv:1607.05368 (2016)
9. López-Yáñez, I., Yáñez-Márquez, C., Camacho-Nieto, O., Aldape-Pérez, M., Argüelles-Cruz, A.J.: Collaborative learning in postgraduate level courses. Comput. Human Behav. 51, 938–944 (2015)
10. Mohri M., R.A., Talwalkar, A.: Foundations of machine learning. MIT Press (2018)
11. Moreno-Moreno, P., Yanez-Marquez, C., Moreno-Franco, O.A.: The new informatics technologies in education debate. Int. J. Technol. Enhanc. Learn. 1(4), 327–341 (2009)
12. Nallapati, R., Zhai, F., Zhou, B.: Summarunner: A recurrent neural network based sequence model for extractive summarization of documents. In: Thirty-First AAAI Conference on Artificial Intelligence (2017)
13. Tohalino, J.V., Amancio, D.R.: Extractive multi-document summarization using multilayer networks. Phys. A 503, 526–539 (2018)

Authorship Link Retrieval Between Documents

Hiram Calvo[2]([✉]), Consuelo Varinia García-Mendoza[1], Esteban Andrés
Ruiz-Chávez[1], and Omar Juárez Gambino[1]

[1] Instituto Politécnico Nacional, Escuela Superior de Cómputo, Mexico City, Mexico
andress.rch@gmail.com, {cvgarcia,jjuarezg}@ipn.mx
[2] Instituto Politécnico Nacional, Centro de Investigación en Computación,
Av. JD Bátiz e/ MO de Mendizábal s/n, 07738 Mexico City, Mexico
hcalvo@cic.ipn.mx

Abstract. In this paper we propose a method for automatic author clustering called Document Authoring Link Retriever, DALIR. Documents are represented using Doc2Vec, experimenting with several parameters; afterwards, vectors are clustered (or *linked* together) using K-means and Hierarchical Agglomerative Clustering. We experimented with different vector representation sizes, different fixed number of clusters, and clustering methods. We evaluated our method on the author clustering task of PAN @ CLEF 2017. We used the BCubed F-score evaluation scheme of this task, being able to overcome some of the reported results from the first places of this challenge, although our method requires to manually establish a number of clusters a priori.

Keywords: Style analysis · Author profiling · Clustering · Computational linguistics.

1 Introduction

An important goal for Computational Linguistics is attributing the authorship of a text, and thus reduce plagiarism, which is virtually impossible to manually solve due to the large amounts of documents that already exist and the large amount of time it takes for a person to perform this task [18]. For this purpose, there are techniques for analyzing style and models of similarity in texts. Journalism and law are areas where actually knowing the author of a document may involve saving a life [6].

With the advance of research in technology, and particularly on Computational Linguistics, now a computer can simulate some human linguistic ability, as well as how to identify writing styles. Thanks to Natural Language Processing, which provides techniques for implementation of tools that help, for this

This work was done with support of the Government of Mexico via CONACYT, SNI, and Instituto Politécnico Nacional (IPN) grants SIP 2083, SIP 20200811, SIP 20201252, and SIP 20201362, IPN-COFAA and IPN-EDI.

© Springer Nature Switzerland AG 2020
L. Martínez-Villaseñor et al. (Eds.): MICAI 2020, LNAI 12469, pp. 297–305, 2020.
https://doi.org/10.1007/978-3-030-60887-3_27

particular case it is possible to try to associate a text with its respective author. In this work we present a tool that helps the analysis of texts to group them according to their potential author.

Attributing authorship is a problem in which there is a set of authors together with samples of texts written by them. When a set of samples is analyzed, a model is built on the writing style of each of the authors, then the challenge is to identify the author of a text whose author is unknown by comparing the writing style present in the anonymous text and the styles of the authors who belong to the set [19].

Attribution of closed authorship is when the author of the anonymous text belongs to an available set of authors. On the contrary, when the author does not belong or is not part of the authors within the set, it is called attribution of open authorship.

Author Clustering is a task in which, within a collection of documents or texts, the main objective is to group documents written by the same author so that each group corresponds to a different author. This task can also be seen as the establishment of authorship links between documents [20]. Authorship attribution is used to group each of the documents with their respective authors.

PAN is a series of scientific events and shared tasks on forensic text and digital text stylometry, fostering forensic investigation of digital texts by organizing shared task evaluations. Tasks are computer events that invite researchers and professionals to work on solving a specific problem of interest [20]. CLEF stands for Conference and Labs of the Evaluation Forum and is a self-organized body whose main mission is to promote research, innovation and the development of information access systems [1]. Each year a series of new tasks are published which seek to be solved among the research community, including social media and education [15] and share that solution [5].

2   State of the Art

In this section some works related to style analysis and author clustering are mentioned.

Researchers at the National University of San Luis in Argentina performed intrinsic plagiarism detection based on global histograms by analyzing an author's writing style and identifying style changes in the histograms [7].

Mexican researchers won the PAN @ CLEF 2017 contest, presenting a work that groups authors and classifies authorships by performing a grouping analysis of characteristics and using the log-entropy and tf-idf models [8].

In the same contest, another work was presented which groups authors by means of a simple similarity measure that works with the probability distribution of the sequence of characters in a document [3].

Another work participating in the contest [10] proposes methods for the task of identifying the author, dividing into grouping of authors and detection of style gaps. Clustering based on locality-sensitive hashing of vectors of real values is used, which are mixtures of stylometric characteristics and a bag of

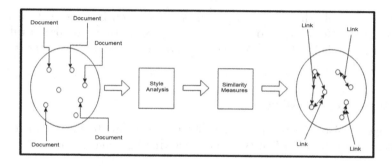

Fig. 1. DALIR model architecture

n-grams, and a statistical approach based on some different tf-idf characteristics that characterize the documents. By applying the Wilcoxon Signed Rank test to these characteristics, style inferences are determined [10].

Performance in PAN @ CLEF 2017 is measured through BCubed F, BCubed Recall and BCubed Precision [4]. A summary of scores can be found in Table 1.

Table 1. Summary of performance of PAN @ CLEF 2017, author clustering participants, ranked by BCubed F.

Participants	Performance		
	B^3 F	B^3 Recall	B^3 Precision
Gómez-Adorno et al.	0.573	0.639	0.607
García et al.	0.565	0.518	0.692
Kocher & Savoy	0.552	0.517	0.677
Halvani & Graner	0.549	0.589	0.569
Alberts	0.528	0.599	0.550
BASELINE-PAN16	0.487	0.331	0.987
Karas et al.	0.466	0.580	0.439

3   DALIR Model Architecture

Figure 1 shows the general architecture of the Document Authoring Link Retriever (DALIR) model. The left circle shows the documents to group together and the inputs to the model. The style analysis and similarity measures boxes indicate the methodology that the model follows in order to group the input documents. The circle on the right represents the output of the model and contains documents grouped by model according to the similarities between them.

3.1   Document Representation

In order to being able to compare two different documents, we use Doc2Vec [11]. Doc2Vec is a form of neural embeddings associated with a group of words

[2]. In this module the content of the files is read and is assigned to a vector representation. However, in order to do this, text must be preprocessed first. Then, with the help of Doc2Vec, vectors are generated as characteristics for each of the documents.

3.2 Text Preprocessing

This stage consists of several functions for preprocessing texts contained in the documents prior to using Doc2Vec. The following functions are used:

- **Stop words deletion.** The text is compared with the stop word list of the language as defined by NLTK [13]. Words contained in this list are removed from the original text.
- **Lowercase conversion.** Each token is transformed into a lowercase letters.
- **Lemmatization.** Each token is converted into its representative stem.

It is important to perform this step since there are words that do not add significant meaning to the context of documents such as articles or prepositions, as well as transforming words into their simplest form (stems) to find their representation in an easier way during the following step.

3.3 Doc2Vec Vector Generation

After the text has been normalized, each document is given an identifier or label, in this case it was a number. A function numbered the documents from 0 to $n-1$ being $n-1$ the total of documents and added them to a list. The list of tagged documents was entered into Doc2Vec in this way a trained vector model was generated with the documents.

Doc2Vec is a model that is considered to be an implementation of word2vec. It consists of generating characteristic vectors of paragraphs or documents regardless of their length. His training uses the architectures of word2vec, since it requires the representation of words in vectors to generate the vector resulting from a paragraph or document [12]. Doc2Vec has two algorithms to generate vectors: Distributed Memory (DM) and Distributed Bag of Words (DBOW). The first uses the vectors of the words and averages them with the current vector of the next paragraph, and for the next word uses the vector resulting from the previous average as input. The second algorithm ignores the context before the word and forces the vector to be predicted according to the vectors of the words. Illustration 5 graphically shows this.

The Doc2Vec [12] parameters for generation of the vectors are by default to exception of Window, Min_count, Workers and DBOW_words whose values can be seen in a Table 2. Window considers the number of adjacent words (previous and after) the word objective. Min_cout ignores words with the frequency indicated. Workers is the number of processor cores, in order to speed up training. The computer where this was performed process contains 4 cores in its processor, therefore *workers* was set to 4. DBOW_words elaborates the vectors of the

Table 2. Doc2Vec parameters

Parameter	Value
Window	10
Min_cout	1
Workers	4
DBOW_words	1

words with the skip-gram model, which according to [16] provides better results than bag of words.

The vectors obtained as output in the module style analysis are the input for the similarity measures module. In this module the comparison and clustering of vectors obtained in the style analysis module is performed, giving as a result the number of cluster to which each document corresponds to in the form of a list.

3.4 Vector Clustering

The grouping of the vectors obtained in the training was done with the K-means [9] and Hierarchical Agglomerative Clustering (HAC) [17] algorithms, which are some of the most commonly used clustering methods [14]. These algorithms determine the number of cluster to which each document belongs.

K-means clustering consists of automatically partitioning a data set into groups of k. It proceeds by selecting the data pool centers and then iteratively refining them as follows [25]: Each instance is assigned to its closest group center; then each cluster center is updated to be the average of its constituent instances. The algorithm converges when there are no more changes to the assignment of instances to clusters [21].

A hierarchical grouping consists of generating groups according to the data to be grouped, analyzes the input data and is joining to form a new group or dividing to form two new groups, depending on the case. There are two types of hierarchical grouping: (a) agglomerative: These are those that start with a number of groups equal to the input data and compare them with another of the data and gradually unite the most similar and form new groups. (b) Dissociative: To the contrary of agglomerative clustering methods, these group all input data into the same group and divide the groups to form new smaller ones. In this work we use a Hierarchical Agglomerative Clustering (HAC) method.

3.5 Evaluation

To perform the evaluation, the outputs of the clusters were written in .json files according to the format indicated in the PAN @ CLEF competition 2017. This format is necessary for the results to be entered to the competition's evaluation software, which uses the BCubed performance of standard measures Precision, Recall and F-Measure, as detailed in [4].

4 Results

We experimented with all parameters (clusters, algorithm to train Doc2Vec, vector's dimension, clustering algorithm). Ranges of clusters were from 1 to 7; parameters for Doc2Vec were DBOW or DM, and tested dimensions were 200, 300, 500; clustering methods were K-means or HAC. The best results obtained with the training corpus and the test corpus of the PAN @ CLEF17 are shown in Tables 3 and 4. These results are the averages of the evaluations provided by the PAN @ CLEF17 automatic evaluator. The training results are shown in Table 3 and the test results in Table 4. In both cases the language, cluster number, Doc2Vec algorithm, vector length, clustering algorithm and the result of F1-score, recall and accuracy metrics are given.

Table 3. Results with the PAN @ CLEF17 training corpus

Language	Clusters	Doc2Vec	Vector	Clustering	B^3 F	B^3 Recall	B^3 Precision
Std. English	4	DM	300	K-Means	0.5791	0.6776	0.5359
Std. Dutch	6	DM	300	K-Means	0.5597	0.5802	0.5593
Std. Greek	6	DBOW	300	K-Means	0.5406	0.5502	0.5449
All std	6	DBOW	200	HAC	0.5579	0.5501	0.5986
All non-std	6	DBOW	200	K-Means	0.5447	0.5389	0.5869

Table 4. Results with the PAN @ CLEF17 test corpus

Language	Clusters	Doc2Vec	Vector	Grouping	B^3 F	B^3 Recall	B^3 Precision
Std. English	4	DBOW	500	HAC	0.6038	0.6383	0.5992
Std. Dutch	5	DBOW	200	K-Means	0.5862	0.6254	0.5697
Std. Greek	4	DBOW	200	HAC	0.5438	0.5688	0.5499
All std	5	DBOW	500	HAC	0.5618	0.5562	0.5846
All non-std	5	DBOW	200	HAC	0.5531	0.5473	0.5843

Table 5 shows our results compared with the best results of PAN @ CLEF 2017 [8]. Best results are shown in bold. From this table, it is possible to see that in general we attain better results in BCube F measure, mostly due to recall. However, we are able to overcome Precision only for the Dutch language.

5 Analysis and Conclusions

This paper described a method that solves a specific task related to the grouping of authors and proposed by the PAN @ CLEF17 contest with the help of libraries such as NLTK, scikit-learn, and gensim with which the preprocessing of the text, the word embeddings and data grouping are carried out.

Table 5. Comparison with the best results of PAN @ CLEF 2017.

	B^3 F	B^3 Recall	B^3 Precision
English	0.5913	0.6175	**0.6483**
Us	**0.6181**	**0.6383**	0.5992
Dutch	0.5765	**0.7204**	0.5508
Us	**0.5962**	0.6254	**0.5697**
Greek	0.5517	0.5743	**0.6222**
Us	**0.5742**	**0.5878**	0.5613
All	0.5733	**0.6379**	**0.6069**
Us	**0.5962**	0.6172	0.5767

Regarding the results obtained in the experimentation of this model, we can observe that some of them are similar to those obtained by the first places of the participants (*cf.* Tables 1 and 5), although not excelling in the first place for all languages, according to the evaluation measures (BCubed F, Recall and Precision). Some of the obtained results can be comparable and positioned in a position of the contest as results of the proposed task, among the first three places. One of the main problems we identified is the preprocessing of the text before the generation of vectors, as well as similarity measures for clustering.

An important point to notice is that each language uses a different configuration, and a different number of clusters, obtained experimentally. The results between 4, 5 and 6 clusters represent better results due to the fact that the average groups that should be formed this one among these same amounts, which is why those same values are more prominent than those of 3 and 7 clusters. The lesser the number of clusters, the lesser the recall, but higher precision; and conversely if the number of clusters increases, say forming a single cluster, a recall of 100% is obtained, and conversely, for n groups, where n = amount of documents to be clustered, precision goes up to 100%. When the text is preprocessed (standardized) we are able to obtain better results; this is because the method only needs to work with the words that give context and meaning to the texts, allowing the construction of better distributed vectors in vectorial space.

With regard to the ways of generating vectors there is not a very noticeable difference between DM and DBOW, the same goes for the K-Means and HAC grouping algorithms. Models that work with words to create their feature vectors do not heavily rely on the modification of the parameters for creating vectors, that is, there is no significant improvement if the parameter values, such as dimensions of the vector are changed. What is more important, are the texts that are used for obtaining the vector representation. This can represent a significant improvement in the grouping of authors. As a future work, we plan to experiment with different sources for generating vector representation.

References

1. The CLEF Initiative (Conference and Labs of the Evaluation Forum) - Homepage. http://www.clef-initiative.eu/. Accessed Aug 2020
2. Acevedo-Mosqueda, M.E., Yáñez-Márquez, C., López-Yáñez, I.: Alpha-beta bidirectional associative memories: theory and applications. Neural Process. Lett. **26**(1), 1–40 (2007)
3. Alberts, H.: Author clustering with the aid of a simple distance measure. In: CLEF (Working Notes) (2017)
4. Amigó, E., Gonzalo, J., Artiles, J., Verdejo, F.: A comparison of extrinsic clustering evaluation metrics based on formal constraints. Inf. Retr. **12**(4), 461–486 (2009)
5. Brennan, M.R., Greenstadt, R.: Practical attacks against authorship recognition techniques. In: IAAI (2009)
6. Elayidom, M.S., Jose, C., Puthussery, A., Sasi, N.K.: Text classification for authorship attribution analysis. arXiv preprint arXiv:1310.4909 (2013)
7. Funez, D.G., Errecalde, M.L.: Detección de plagio intrínseco basado en histogramas. In: XVIII Congreso Argentino de Ciencias de la Computación (2012)
8. Gómez-Adorno, H., Aleman, Y., Ayala, D.V., Sanchez-Perez, M.A., Pinto, D., Sidorov, G.: Author clustering using hierarchical clustering analysis. In: CLEF (Working Notes) (2017)
9. Jain, A.K.: Data clustering: 50 years beyond k-means. Pattern Recognit. Lett. **31**(8), 651–666 (2010)
10. Karas, D., Spiewak, M., Sobecki, P.: Opi-jsa at clef 2017: Author clustering and style breach detection. In: CLEF (Working Notes) (2017)
11. Lau, J.H., Baldwin, T.: An empirical evaluation of doc2vec with practical insights into document embedding generation. arXiv preprint arXiv:1607.05368 (2016)
12. Le, Q., Mikolov, T.: Distributed representations of sentences and documents. In: International Conference on Machine Learning. pp. 1188–1196 (2014)
13. Loper, E., Bird, S.: Nltk: the natural language toolkit. arXiv preprint cs/0205028 (2002)
14. López-Yáñez, I., Sheremetov, L., Yáñez-Márquez, C.: A novel associative model for time series data mining. Pattern Recognit. Lett. **41**, 23–33 (2014)
15. Lytras, M.D., Mathkour, H., Abdalla, H.I., Yáñez-Márquez, C., De Pablos, P.O.: The social media in academia and educationresearch r-evolutions and a paradox: advanced next generation social learning innovation. J. UCS **20**(15), 1987–1994 (2014)
16. Mikolov, T., Chen, K., Corrado, G., Dean, J.: Efficient estimation of word representations in vector space. arXiv preprint arXiv:1301.3781 (2013)
17. Müllner, D., et al.: fastcluster: Fast hierarchical, agglomerative clustering routines for r and python. J. Stat. Softw. **53**(9), 1–18 (2013)
18. Pérez Afonso, J.: Detección intrínseca de plagio. Ph.D. thesis, Universitat Politècnica de Valencia (2014)
19. Posadas, J.: Detección automática de plagio usando información sintáctica. Ph.D. thesis, Center for Computing Research, Instituto Politécnico Nacional (2016)

20. Tschuggnall, M., et al.: Overview of the author identification task at pan-2017: Style breach detection and author clustering. In: CLEF (Working Notes) (2017)
21. Wagstaff, K., Cardie, C., Rogers, S., Schrödl, S., et al.: Constrained k-means clustering with background knowledge. Icml. **1**, 577–584 (2001)

Image Processing and Pattern Recognition

A Proposal of an Empirical Methodology to Approximate an Electroencephalographic Signal with Appropriate Representatives of the Fourier Transformation

José Alfredo Zavaleta-Viveros[1]([✉]) [ID], Porfirio Toledo[1] [ID],
Martha Lorena Avendaño-Garrido[1] [ID], and Jesús Enrique
Escalante-Martínez[2] [ID]

[1] Facultad de Matemáticas, Universidad Veracruzana, Xalapa, Veracruz, Mexico
`pepezavaleta20@gmail.com`, {`ptoledo,maravendano`}`@uv.mx`
[2] Facultad de Ingeniería Mecánica y Eléctrica, Universidad Veracruzana, Poza Rica,
Veracruz, Mexico
`jeescalante@uv.mx`

Abstract. This paper proposes a methodology to approximate an electroencephalographic (EEG) signal. To make it, the signal is decomposed using the Fast Fourier Transform (FFT). The methodology consist of an algorithm that select representatives of each of the brainwaves frequency bands classification, ensuring the selection of at least one representative of each band. The approximation is constructed with a small number of amplitude and frequency components that are obtained from the FFT. Once the representatives were selected, the approximation of the signal is expressed as a sum of periodic functions. The quality of this approximation is measured using three values: the Hausdorff distance between approximated and observed signals, the percentage of signal points outside of fitting region built around the approximation signal, and the variance of these points. Furthermore, a numerical example is developed where the implementation of the methodology is shown and the results are discussed.

Keywords: Fast fourier transform · EEG Signal · Approximation methodology · Signal fitting · Empirical mode decomposition

MCS: 65D10 · 65T50 · 68U10 · 68W25

1 Introduction

Approximation or curve fitting is a method used for study sundry phenomena consists of the iterative construction of a curve that achieves a good approximation to a given data set or function. In general, this process is useful because it allows us to interpret, approximate, manage, and study a data set easily.

© Springer Nature Switzerland AG 2020
L. Martínez-Villaseñor et al. (Eds.): MICAI 2020, LNAI 12469, pp. 309–324, 2020.
https://doi.org/10.1007/978-3-030-60887-3_28

In this sense, the objective of this paper is to propose a methodology to achieve an approximation of a signal from an EEG of an adult male rat. To fulfill this purpose, we resort to the FFT since this allows us to decompose the original signal into its amplitude and frequency components to write it later as a sum of cosines. We propose an approach that does not take all the values of the FFT, but rather, some representative candidates that will be selected considering taking at least one representative from each of the classifications in brainwaves frequency bands. The aim is to ensure a good fit with a small number of terms ensuring the quality of the approximation by calculating three values. We consider important to analyze these signals since could have a subsequent application in the analysis of epileptic crises.

Decomposition proposals similar to this have been studied before, like the case of the Empirical Mode Decomposition (EMD) proposed by N. Huang in [1], or improvements, such as the Weighted Sliding Empirical Mode Decomposition (WSEMD) in [2], or the Variational Mode Decomposition (VMD) featured in [3], which address more complex modifications to the original EMD. One of the advantages of EMD is that it does not assume that the signal to be decomposed is periodic, neither does it assume that it decomposes into a sum of cosines as in our case. To achieve this, EMD makes use of the Hilbert-Huang Transform, decomposing the functions into functions of intrinsic mode (IMF). However, it has limitations, such as sensitivity to noise and sampling, in addition to being a more complex and computationally expensive analysis, unlike the FFT which is computationally efficient [4]. In [5] an EMD method based on the FFT is proposed, however, it also uses the Hilbert transform and in [6] can be found a method that is intended to be faster and easier using also the FFT within the EMD, saving resources in terms of calculation and time, however, it is only limited to decomposition and takes a different way to ours. In this paper, we propose a simpler and faster method than the EMD, by using only the FFT, Fourier series, an algorithm for selecting candidates and quality values that are easy to calculate which guarantee a good fit to the original signal.

2   Experimental Data

2.1   Electroencephalographic Signal

The EEG signal data that we consider in this work was obtained from an adult male rat, which was implanted through surgery, a pair of nail-type electrodes at the level of the frontal cortex (AP = 1.5 mm, ML = ±3 mm, for Bregma) and another pair in the parietal cortex (AP = −3.5 mm, ML = ±3 mm, for Bregma).

Once the electrodes were implanted, the EEG record was obtained in both regions, the signal was obtained with a Grass electroencephalograph and the ADQCH4 program at a frequency 300 Hz and a sensitivity of 7 μA. The measurement of the record lasted approximately one hour and the values of the signal amplitude were taken every 3.3 ms, high-pass filters 70 Hz were applied. For our analysis we just consider one minute of such record which consisting of 18000 data.

During the recording the rat was free and their natural behavior can not be controlled, thus, the signal presents sudden and abrupt variations that could be due to any action of the animal; for example, a blink, a natural movement, etc. Due to this, we will divide the signal into six parts and average them to obtain a more uniform and representative behavior, obtaining a signal with a length of 10 s and 3000 data. This signal is designated by the letter S.

2.2 Brainwaves Frequency Bands Classification

The brain of any living being works through an electrical flow, which is propagated by electrical signals that are produced and exchanged in neurons through action potentials. These signals are also known as brainwaves or brain oscillations, can be measured and observed through the EEG, are constituted of different types of waves, and their classification depends on the frequency values they have, there are six classification bands: infra-low delta waves (0 to 0.5 Hz) and delta waves (0.5 to 4 Hz), these two appear mostly in activities such as deep sleep and basic biological needs of the living being; theta waves (4 to 8 Hz) that are related to activities such as creativity, intuition, meditation; alpha waves (8 to 13 Hz) related to motor coordination and relaxation activities; beta waves (13 to 30 Hz), present in activities such as listening or solving a problem; and gamma waves (30 to 70 Hz) related to the representation and construction of objects these occur mostly in demanding activities, or that require attention to several things at once. All of the above according to [7–10]. It is important to consider this classification because the decomposition of the signal has components with different frequency values, and therefore waves of different types of band. Brain oscillatory activity is characterized by presenting all these bands together, to a greater or lesser extent at the same time, so signals that have all types of frequency bands are considered, particulary in the approximation.

3   The Fourier Transform

To carry out the analysis of the EEG signal we use the Fourier Transform (FT). This allows us to "decompose" a function $g(t) : \mathbb{R} \rightarrow \mathbb{R}$, which is assumed to be periodic in its components, providing all the values of the amplitude that constitute it and their corresponding frequencies, with which later we can express it by sums of cosine functions. Taking these values, it is possible to approximate the Fourier series corresponding to the function $g(t)$, according to [4]. Let $x :=$ $\{x_j | x_j = g(j), \ 0 \le j \le L-1\}$, a set of L values. The function $g(t)$ is considered as a signal whose domain is discrete in time, and their image are the values of signal amplitude which vary for each time value in the domain. Applying the FT we obtain a series of complex values $X = \{X_0, X_1, ..., X_{L-1}\}$, in our case we will use the FFT to obtain these and approximate the FT [11].

By getting the FFT we obtain the set L of complex values of the form $X_j =$ $P_j + iQ_j, \ 0 < j < L-1$. These values play the role of a "indicator" of the amplitude corresponding to the frequency in the original signal. In general the FT takes a function with time as domain and amplitude as codomain, to transform it into a function with frequency as domain and amplitude as codomain, that is, a function $\mathcal{F} : f \rightarrow A$, where A is the set of all amplitudes. This allows us decompose the signal $g(t)$ into each of its frequency and amplitude components obtaining a frequency spectrum. Then we can describe the function using the expression [4], for $n = 0, 1, ..., L - 1$,

$$S(m) = \sum_{n=0}^{L-1} A_n \cos\left(\frac{2\pi nm}{L} + \phi_n\right),$$ (1)

where the amplitude value A_n is given by

$$A_n = \sqrt{P_n^2 + Q_n^2},$$

and the phase ϕ_n by

$$\phi_m = \tan^{-1}\left(\frac{Q_n}{P_n}\right).$$ (2)

The set of amplitude values given by the FFT is $A = \{A_0, A_1, ..., A_{L-1}\}$. Considering all the components from (1) we have an approximation of $g(t)$ given by the Discrete Fourier Transform. Our goal is to achieve an approximation of $g(t)$ without using all the components of $S(m)$ but describe accurately the original signal using a small number of terms, guaranteeing precision of the approach by quality values, which will be described in Sect. 5.

4   Approximation Method

The proposed method consists of selecting representative values of the decomposition obtained by applying the FFT. With those values, the EEG signal is approximated securing its quality. This will allow study of the original signal through representation as a finite sum of cosines as shown in (1), whose handling is relatively simple. To guarantee the quality of the approximation we use three values, two of them we will obtain considering a fitting region that is built around the approximation. The percentage of points of the original signal that is outside the region and the variance of those points are considered. On the other hand, we will calculate the Hausdorff distance between the original signal and the approximation, to have a measure between these sets of points.

4.1   Use of the Fast Fourier Transform

Given the signal S, we obtain each of the amplitude values of each component of the signal X_j using the FFT. We obtain a set will consist of L amplitude values, that is, the same number of elements as S and we determine their corresponding frequencies considering

$$F = \frac{1}{T},\tag{3}$$

where T is the period and F is the sampling frequency of the signal. The value of the frequencies is designated with f_i, where $1 \leq i \leq L$, and it is calculated as follows

$$f_i = \frac{F \cdot i}{L}.\tag{4}$$

When the FFT is applied a spectrum of symmetrical values is obtained, where the values of the amplitudes are repeated. Thus we will only consider the first half of the spectrum of the FFT for the election of representants, that is, will be taken $N = \frac{L}{2}$ data. Figure 1 shows the spectrum of the FFT of S up 70 Hz, whose detailed analysis will be presented later.

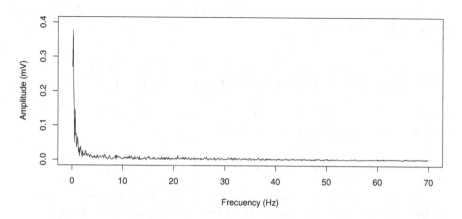

Fig. 1. Fourier Transform of the EEG signal S of the rat 70 Hz.

4.2 Aproximation of S, Choosing the Representantives of Frequency and Amplitude

With the amplitude and frequency values choosed and calculated we can approximate the signal S using (1) as a function that depends on t, then we obtain

$$\mathcal{S}(t) = \frac{A_0}{2} + \sum_{n=1}^{L-1} A_n \cos\left(2\pi f_n t + \phi_n\right).\tag{5}$$

To understand how the approximation is constituted, see that, in the spectrum shown in Fig. 1, the components with the greatest amplitude are those with the lowest frequencies; mostly are waves that go from 0 to 1 Hz, that is, infralow delta and delta waves. As these components are stronger, they affect the approximation to a greater extent and provide more information on the shape of S than the weak ones, that is, the most important components are those with a greater amplitude. Those with a smaller amplitude will only cause small variations on the biggest waves, but we must take into account the classification of waves presented in Subsect. 2.2 then representants of small values are considered too. The proposed algorithm considers representatives of all frequency bands as presented below.

1. A value k is chosen with which we will divide the number of elements N, since,

$$1 \leq \frac{N}{k^h}, \quad h = 1, 2, ... \tag{6}$$

With the value of h we will define $\mathcal{N} = k^h$, and for the rest of the algorithm consider only $\mathcal{N}$ elements, this to obtain a divisible amount of data.

2. With the value $\mathcal{N}$ we generate a partition O_k of the frequency interval of the FFT in $h + 1$ intervals,

$$O_k = \left\{ \left(0, \frac{F}{L}\right], \left(\frac{F}{L}, \frac{F\mathcal{N}}{k^{h-1}L}\right], ..., \left(\frac{F\mathcal{N}}{k^2 L}, \frac{F\mathcal{N}}{kL}\right], \left(\frac{F\mathcal{N}}{kL}, \frac{F\mathcal{N}}{L}\right] \right\}. \tag{7}$$

3. From each interval of the partition O_k we will take the maximum value of the corresponding amplitudes, that is, from each $r = h, ..., 1$

$$A_{h+1-r} = \max \left\{ \mathcal{F}(f) : f \in \left(\frac{F\mathcal{N}}{k^r L}, \frac{F\mathcal{N}}{k^{r-1}L}\right] \right\}. \tag{8}$$

The above will generate h amplitude values selected representatively, we consider A_1 from $\left(0, \frac{F}{L}\right]$, and A_0 as the DC of the FFT, so finally, we have $h + 1$ values of amplitude took as shown in Fig. 2.

4. Once we have the amplitudes A_r, their corresponding frequencies f_{i_r} are selected considering the position i it has in the set of data given by the r subscript of A_r, the values of the phase ϕ_{i_r} are calculated with the Eq. (2), taking respectively P_{i_r} and Q_{i_r} for each A_r too. Then all these values are used to find the approximation $S(t)$ with the expression (5).

5   Quality Values of the Approximation $S(t)$

To guarantee a quality measure of our approximation $S(t)$ three values are used with which we can measure how good is the proposed approximation. The Hausdorff distance was considered to measure the distance between the signal and the approximation, for two sets X and Y, this distance is defined by

$$d_H(X, Y) = \max \left\{ \sup_{x \in X} \inf_{y \in Y} |x - y|, \sup_{y \in Y} \inf_{x \in X} |x - y| \right\}.$$

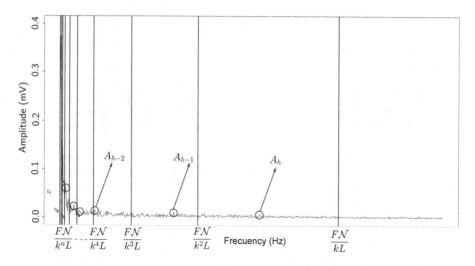

Fig. 2. The strongest value of the amplitude corresponding to each interval is taken.

The functionality of this value is that the maximum distance of one set to the closest point of the other is calculated, for our purpose a small value of d_H suggests that the approximation and the original signal are closer.

The other two values that we will use to measure the quality of the approximation are obtained by building a fitting region R around the approximate signal. We will consider the percentage of points of the signal S that are outside of R and its variance. To build R we will choose a constant $\epsilon > 0$, such that

$$\epsilon \le \frac{\max(S) - \min(S)}{2}. \tag{9}$$

This restriction is established because if we consider a larger ϵ, all the values of S would be within the fitting region, which is not our interest. The variance of the signal points that are outside of R will tell us how scattered the set is. This consideration is very important, since, in the approximation, it is preferable to have two points outside of R but close to this region, instead of one point outside but far away. This ensures that our approximation will be closer to the points that are outside, so the variance value and the percentage value complement each other. In Fig. 3 we can see this idea illustrated, note that in a) there are two points y_i and y_j outside R, these points are at distances d_i and d_j respectively, while in b) there is only one point y_k outside R, whose distance is equal to the sum of the two in a); however, the variance in a) will be less than in b). In this way we obtain better approximations when something similar occurs to the situation presented in a), so the variance will provide us with a way of measuring how far the points that are outside of R are separated. We consider the set $C = \{P : P = x \pm \epsilon, x \in S(t)\}$, that contains the approximation $S(t)$.

From C we construct a region R, the border of C has the same shape as the approximate signal then its shape is irregular and will have small variations

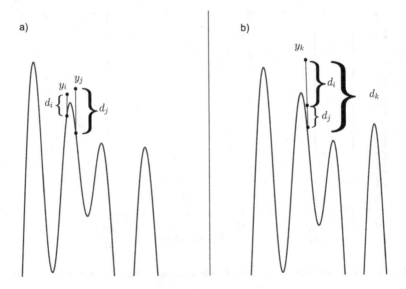

Fig. 3. In a) there are two points y_i and y_j outside the bands, distances d_i and d_j respectively are away from the band, while in b) if there is only one point y_k outside the band, whose distance is equal to the sum of the two in a), the variance in a) will be less than in b).

because the components with high frequencies, then by applying a smoothing to C we obtain a more uniform region R. This will cause some points that are close to C to leave this set, therefore, smoothing helps us avoid these variations in the border of the fitting region.

Through these three quality values, we seek to guarantee a good fit in the approach to the signal. Next, we will show a numerical implementation of the algorithm to the EEG signal taken from the rat and we will compare these presented values.

6   Numeric Implementation

6.1   Fourier Transform

The first thing we will do is calculate the FFT of S, with $L = 3000$, obtaining $\mathcal{F}(f_i)$ that will consist of 3000 amplitude values, that is, we will get a set $\{A_i\}$ with $1 \leq i \leq 3000$, and determine the corresponding frequencies as explained in Sect. 4, also, by considering what was mentioned in Sect. 2, the frequency F in the experiments 300 Hz, therefore from (3), $F = \frac{1}{T} = 300$ Hz, where T is the period in seconds. So with the 3000 data of $\mathcal{F}(f_i)$ we obtain a signal of approximately 10 s in length. From (4), the value of the corresponding frequencies will be given by the expression $f_i = \frac{i}{10}$, for $0 \leq i \leq 3000$.

We have that $L = 3000$, considering the symmetry of the FFT we will use only half of the data, that is, $N = \frac{L}{2} = 1500$. Besides, frequencies less than or equal 70 Hz will only be considered since, as mentioned in Sect. 2, high pass filters 70 Hz were applied to eliminate possible interference from the equipment. In Fig. 1 we show the FFT of the signal S.

All the operative, numerical and graphic processes shown were performed using the R programming language with the IDE RStudio.

6.2 Application of the Algorithm

We will proceed to apply the selection algorithm presented in Sect. 4 to construct the signal approximation, we will consider for the implementation the fixed value of $k = 2$. For this case the condition (6) is met up to $h = 10$, then $\mathcal{N} = 2^{10}$ and so we will have 11 intervals of the form (7),

$$O_2 = \left\{ \left(0, \frac{1}{10}\right], \left(\frac{1}{10}, \frac{2}{10}\right], \left(\frac{2}{10}, \frac{2^2}{10}\right], ..., \left(\frac{2^8}{10}, \frac{2^9}{10}\right], \left(\frac{2^9}{10}, \frac{2^{10}}{10}\right] \right\}.$$

However, the interval $\left(\frac{2^9}{10}, \frac{2^{10}}{10}\right]$ it will not be taken into account because the frequency corresponding to it falls beyond 70 Hz. Then this partition has 10 intervals and it can be appreciated in Fig. 4.

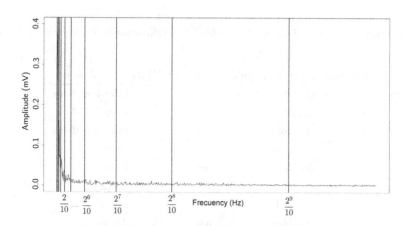

Fig. 4. Partition of the frequency domain for the case $k = 2$.

Once this is done, we proceed to choose the representative amplitudes following the algorithm presented in Sect. 4.2, taking each interval of the partition O_2, obtaining 10 amplitude values. These will be given by (8), with $r = 10, 9, ..., 2$, as follows,

$$A_{11-r} = \max \left\{ \mathcal{F}(f) : f \in \left(\frac{2^{10}}{10(2^r)}, \frac{2^{10}}{10(2^{r-1})} \right] \right\}.$$

Note that for the interval $(0, \frac{1}{10}]$, the value of the amplitude will always be the first of the set of amplitudes, regardless of the value of k. With these values, we will construct the approximations $\mathcal{S}_r(t)$, which we will call iterations, where r denotes the number of iteration.

$$\mathcal{S}_1(t) = \frac{A_0}{2} + A_1 \left[\cos \left(2\pi f_{i_1} t + \phi_1 \right) \right],$$

$$\mathcal{S}_2(t) = \frac{A_0}{2} + \sum_{n=1}^{2} A_n \left[\cos \left(2\pi f_{i_n} t + \phi_n \right) \right],$$

$$\vdots$$

$$\mathcal{S}_{10}(t) = \frac{A_0}{2} + \sum_{n=1}^{10} A_n \left[\cos \left(2\pi f_{i_n} t + \phi_n \right) \right].$$

Table 1. Values of every iteration for $\mathcal{S}_r(t)$ and classification in brainwaves frequency bands.

Iteration	Amplitude (mV)	Frequency (Hz)	Type of wave
1	0.33904	0.1	Infra-low delta
2	0.26996	0.2	Infra-low delta
3	0.37465	0.3	Infra-low delta
4	0.15038	0.5	Delta
5	0.07565	0.9	Delta
6	0.03788	1.8	Delta
7	0.01601	3.3	Theta
8	0.01506	6.6	Alpha
9	0.01246	20.9	Beta
10	0.00903	26.5	Gamma

Considering the proposed algorithm, let us see in the Table 1 that we have frequencies corresponding to all the bands of each classification, from infra-low delta to gamma, which makes the approximation proposal have relevance in a biological sense due to the presence of every kind of wave.

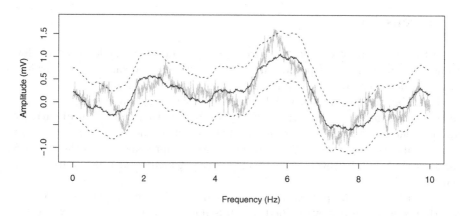

Fig. 5. A plot of the approximation for 10 iterations (black) and the original data (grey), the fitting region is denoted by dashed lines.

6.3 Quality Values Calculation

For each of the iterations, we will calculate the three quality values described in Sect. 5. First, in order to construct the fitting region R we must establish the value of ϵ, then we consider (9) as $\epsilon = \frac{1}{5}[\max(S) - \min(S)]$.

The numerical information obtained by the algorithm is presented in Table 2 and the graph of its approximation in Fig. 5, where the smoothed adjustment region is also shown after applying the approximation algorithm when $n = 10$.

Table 2. Comparison of the quality values for $n = 1, 2, ..., 10$, the Hausdorff distance, percentage of points outside of the fitting region, and variance of such points.

Iteration	Hausdorff Distance	Percentage	Variance
1	1.13320	22.03333	0.02792
2	0.94994	18.36666	0.00982
3	0.63275	4.86666	0.00219
4	0.54103	3.93333	0.00232
5	0.60785	2.63333	0.00169
6	0.59693	2.8	0.00189
7	0.58329	2.83333	0.00193
8	0.56855	2.8	0.00190
9	0.55938	2.8	0.00190
10	0.55114	2.8	0.00190

7   Discussion

Note that from iteration six, the three quality values vary little and have low values, the percentage of points outside is practically the same as its variance and Hausdorff distance, with very slight variations of order 10^{-2}.

It can be seen that values with higher frequencies only causes small variations in the approximation signal, their affectation is such that the values of variance and percentage do not change even taking decimals of order 10^{-5}. Also note that taking only six iterations in $\mathcal{S}_r(t)$ ensures a good approximation to the signal S for the case $k = 2$, however, this proposal does not include all types of frequency bands.

Figure 6 shows the percentage of points out of R in each iteration. We can see that step three is a watershed in adjusting the curve since it is where the most drastic variation is obtained.

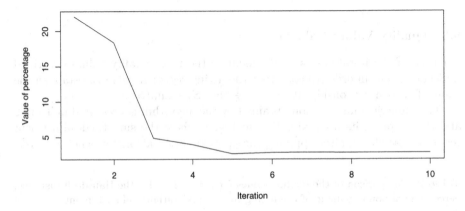

Fig. 6. Values corresponding to the percentage of points outside of the fitting region for each n.

Figure 7 presents the values of the variance of the points outside R. Again, the variation that occurs in the third step is remarkable. Likewise, the value of the Hausdorff distance, which can be seen in Fig. 8, suggests a behavior similar to the previous ones with the difference of having a very low value in the fourth iteration.

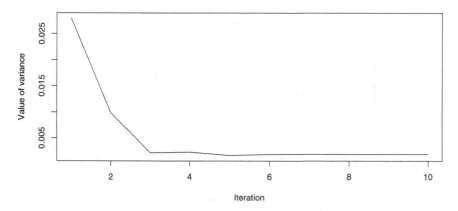

Fig. 7. Values corresponding to the variance of the points outside of the fitting region for each n

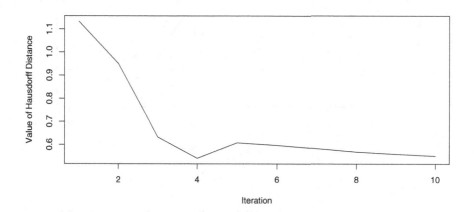

Fig. 8. Values corresponding to the Hausdorff distance for each n.

The approximation algorithm for $k = 2$ shows that the values of the percentage outside the region and the variance change abruptly in the first three or four iterations, while the Hausdorff Distance continues decreasing and at iteration six attenuates its variation. The above can be seen graphically in Fig. 9, where in the top the first iteration is presented, in the middle the third and in the bottom the seventh iteration; note that iterations seven and ten (see in Fig. 5) are very similar.

The previous analysis was developed for values of $k \geq 3$ too, for which similar conclusions were obtained, however, the quality values are worsening to the case $k = 2$.

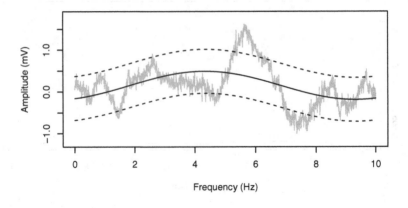

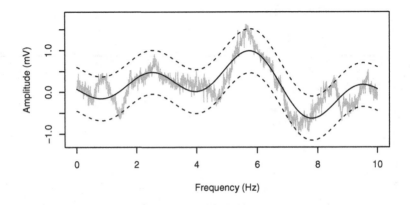

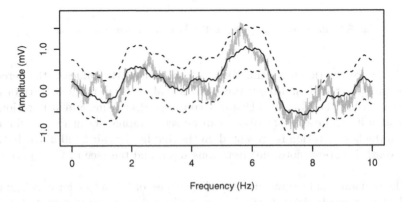

Fig. 9. Plots of some iterations. The iteration one in the top, iteration three in the middle, and iteration seven in the bottom.

8   Conclusions

A methodology that consists of an algorithm for the selection of amplitude and frequency representatives provided by the decomposition obtained with the FFT was proposed and developed. Using Eq. (5) we approximate an EEG signal obtained according to Sect. 2. The classification into frequency bands of the brain signals was considered, thus, the proposed algorithm makes a selection of some of the best representatives of each type of brain signal to obtain a good fit in and this adjustment was ensured by three quality values.

A numerical example of the algorithm was performed too, calculating the approximation and the quality values. A fitting region was built around the proposed approximation, and with which, the percentage of points of the original signal that were outside the region was calculated, as well as the variance of such points; furthermore, the Hausdorff distance between the original signal and the approximation was considered. Finally, using these three values, it was concluded that the sixth iteration presents a good numerical approximation, but does not include a representative of all the classifications of brainwaves frequency bands.

We believe that our research, particularly the development of the proposed method, have a strong didactic potential, that could serve as an introduction to deeper studies on signals and their mathematical analysis.

Acknowledgment. The authors thank the researcher M. L. López Meraz from the Centro de Investigaciones Cerebrales of the Universidad Veracruzana for providing the data for this work. In addition, the first author is grateful for the support of CONACYT through scholarship 741711 to carry out PhD studies at the Universidad Veracruzana.

References

1. Huang, N.E., et al.: The empirical mode decomposition and the Hilbert spectrum for nonlinear and non-stationary time series analysis. Proc. Roy. Soc. London. Ser. A: Math. Phys. Eng. Sci. **454**(1971), 903–995 (1998). https://doi.org/10.1098/rspa.1998.0193
2. Faltermeier, R., Zeiler, A., Tomé, A.M., Brawanski, A., Lang, E.W.: Weighted sliding empirical mode decomposition. Adv. Adapt. Data Anal. **03**(04), 509–526 (2011). https://doi.org/10.1142/S1793536911000891
3. Dragomiretskiy, K., Zosso, D.: Variational mode decomposition. IEEE Trans. Signal Process. **62**(3), 531–544 (2014). https://doi.org/10.1109/TSP.2013.2288675
4. James, J.F.: A Student's Guide to Fourier Transforms, 2nd edn. Cambridge University Press, Cambridge (2002)
5. Chen, Z.S., Rhee, S.H., Liu, G.L.: Empirical mode decomposition based on Fourier transform and band-pass filter. Int. J. Nav. Archit. Ocean Eng. **11**(2), 939–951 (2019). https://doi.org/10.1016/j.ijnaoe.2019.04.004
6. Myakinin, O.O., Zakharov, V.P., Bratchenko, I.A., Kornilin, D.V., Artemyev, D.N., Khramov, A.G.: The Empirical Mode Decomposition algorithm via Fast Fourier Transform. **9217**, 1–6 (2014). https://doi.org/10.1117/12.2061808

7. Schmidt, H., Petkov, G., Richardson, M.P., Terry, J.R.: Dynamics on networks: the role of local dynamics and global networks on the emergence of hypersynchronous neural activity. PLoS Comput. Biol. **10**(11), 1–16 (2014). https://doi.org/10.1371/journal.pcbi.1003947

8. Botcharova, M.: Modelling and analysis of amplitude, phase and synchrony in human brain activity patterns. Ph.D. thesis, University College London, London, United Kingdom (2014)

9. Zainuddin, B.S., Hussain, Z., Isa, I.S.: Alpha and beta EEG brainwave signal classification technique: A conceptual study. In: 2014 IEEE 10th International Colloquium on Signal Processing and its Applications. pp. 233–237. IEEE (2014). https://doi.org/10.1109/CSPA.2014.6805755

10. van Putten, M.J.A.M., Tjepkema-Cloostermans, M.C., Hofmeijer, J.: Infraslow EEG activity modulates cortical excitability in postanoxic encephalopathy. J. Neurophysiol. **113**(9), 3256–3267 (2015). https://doi.org/10.1152/jn.00714.2014

11. Hasan, A., Al-Amin, M.M., Owaziuddin, M.M.: Applications of fourier series in electric circuit and digital multimedia visualization signal process of communication system, vol. 4, pp. 72–80. American Institute of Sciencie (2019)

Rotten Fruit Detection Using a One Stage Object Detector

K. Perez-Daniel[✉][iD], A. Fierro-Radilla[iD], and J. P. Peñaloza-Cobos[iD]

Universidad Panamericana, Mexico City 03920, Mexico
{kperezd,afierro,0212667}@up.edu.mx

Abstract. Digital images and computer sciences have become two powerful tools in several areas, such as astronomy, medicine, forensics, etc. In the last years, computer sciences are getting involved in agricultural and food science to decide based on estimated or actual parameters named features. Rottenness is the state of decomposing or decaying the quality of the fruit, which not only affects the taste and appearance but also modifies its nutritional composition, causing the presence of mycotoxins dangerous for humans. Nowadays, rottenness detection is carried out using human inspection or using Ultraviolet light to highlight spots of rottenness represented as fluorescence. Recent computer vision approaches address this problem using hyperspectral imaging systems. In this paper, we propose to use a one-stage object detector inspired by RetinaNet to detect whether a fruit is fresh or rotten. One of the main stages of RetinaNet is based on computing a multi-scale convolutional feature pyramid network on top of a backbone. Therefore, in this work, we analyze the performance of RetinaNet using different artificial neural networks as backbone to determine the highest accuracy for fruit and rottenness detection. The experiments were done using a dataset composed of 13599 images divided by 6 classes, 3 fresh fruits, and 3 rotten fruits. The performance evaluation considers the mean average precision in the detection and the inference time of tested backbone models.

Keywords: Rotten fruit detection · One stage object detector · Backbone · RetinaNet

1 Introduction

Determining the quality of food is fundamental for planning distribution, pricing, agricultural evaluation, determining the most accurate preservation method, among others. Talking about the particular case of fruits, identifying the quality of fruits may be a laborious, expensive, and time-consuming task. However, avoiding quality inspection leads to extensive losses, not only because of the presence of external defects decreases the price of food, but also because of eating rotten fruits may lead to some diseases.

Supported by organization Universidad Panamericana.

L. Martínez-Villaseñor et al. (Eds.): MICAI 2020, LNAI 12469, pp. 325–336, 2020.
https://doi.org/10.1007/978-3-030-60887-3_29

Examples of fruit defects are bruises, rain damage, abrasions, and rottenness. Rottenness is the state of decomposing or decaying the quality of the fruit, which not only makes the fruit look worse but also affect the taste and appearance, causing economic losses. Currently, most of the rottenness detection methods are carried out manually using Ultraviolet (UV) light, which highlights potential rotten areas [8]; however, this method is harmful to the workers. For this reason, computer vision-based techniques have been recently proposed to evaluate food quality. In this sense, Zhu et al. [23] proposed a non-invasive method for fruit rottenness detection based on hyperspectral imagining to detect bruises on apples. Nevertheless, using hyperspectral imaging requires specialized devices to capture hyperspectral information. Then, processing this multi-channel visual information affects the computational processing time of detection algorithms.

In the past two decades, several techniques have been used to avoid unsafe and subjective methods, such as electrical impedance, X-rays, thermal, and hyper-spectral imaging [23]. Computer sciences are getting more popular in agricultural and food science, for example, for food separating according to several criteria such as color, size, and texture [5]. However, traditional computer vision systems still lack algorithms capable of detecting external defects.

Most of the proposed fruit defect detection approaches are done using low level features, such as color, shape and texture [2,7,8,18,22,23]. However these algorithms are not capable of obtaining enough information for efficient and generalized rottenness detection. Another way of doing fruit defect detection is by using convolutional neural networks as in [5,23], which obtain non-linear representations of images, making possible the detection of fresh and rotten fruits. Nevertheless, many of the previously proposed rottenness detectors are dedicated to a single fruit class. Currently, there are several one-stage and two-stage detectors that have been successfully used for object detection and recognition [9,15,17,19,20]. However, to the best of our knowledge, this kind of detectors have not being used yet for fruit detection and classification. Therefore, in this paper, we propose using the One Stage Object Detector RetinaNet [15] for the detection of fresh and rotten fruits.

For experimentation, we used the dataset "Fruits fresh and rotten for classification" [14], which is composed of 13,600 images of apples, oranges, and bananas, divided and 6 classes, 3 rotten and 3 fresh classes.

The rest of the paper is organized as follows: Sect. 2 exposes previous works on the detection of rottenness. Section 3 presents a brief overview of RetinaNet. In Sect. 4, is explained our proposed method. Section 5, presents the experimental results. Finally, the conclusions and future work are discussed in Sect. 6.

2   Related Works

In the citrus industry, the losses due to the fungi Penicillium are dramatic, so it is essential to make detection to determine which fruit is defected. In [8], authors used Artificial Neural Networks (ANN) and Decision Trees, obtaining 98% accuracy for fruit defect detection.

Slight bruises on apples are very common in harvesting and storage, which is caused by impact, compression, vibration, or abrasion; so, in [23], it is proposed the use of hyperspectral images from which authors obtained spatial and spectral information from apple images. Then, the hyperspectral information is processed using an Extreme Learning Machine (ELM), Partial Least Squares Discriminant Analysis algorithm (PLS-DA), and Classification and Regression Tree (CART) models, achieving a maximum classification rate of 95.97%.

In [18], the authors proposed the use of Gray Level Co-occurrence Matrix (GLCM) for feature extraction and then used the k-nearest neighbors (k-NN) algorithm for rotten fruit classification achieving a maximum of 96.3% accuracy. On the other hand, authors in [1] proposed determining the maturity stages of avocados using the Principal Component Analysis (PCA) and the k-NN classifier and using Lab color space to determine color features. Similarly, authors in [4] propose an evaluation of the ripening stages of apples according to the CIELab color features and physicochemical feature parameters. PCA was used to find the correlation between feature spaces.

Authors in [22] used low-level feature for image representation and then implemented a multi-class kernel support vector machine (kSVM) for fruit type classification, reporting 88.2% of accuracy.

Some other published works use image segmentation [2,7] instead of hand-crafted image features classification wherein [2] authors reported a review of different methods and showed that the best segmentation algorithm is Linear Regression. In the same analysis, the authors show different feature extractor algorithms whose performance depends on the type of fruits. However, in general, they report that the best classification algorithm is multi-class SVM and ANN. The main drawback of the works mentioned earlier is that classification accuracy not only depends on the classification technique itself but also on the feature extractor algorithm.

Other approach uses inductive characterization and a set of low level features, such as color features, texture features and geometric features, for food products classification, attaining a 93% of correct classification, in average [3].

New computer vision techniques use convolutional neural networks (CNN) for detection since CNNs are able to obtain feature maps at different abstraction levels. In this sense, the authors in [5], proposed using Deep Residual Neural Network (ResNet) for external defect detection on tomatoes achieving a precision of 94.6%. Recently, Fan et al. [6] report a CNN for apple detection, achieving a speed of 5 fruits per second, and 96.5% accuracy. Although these results are promising, using CNNs in real time implementations requires high computational resources, which make impractical its application in devices with limited computational capabilities, such as CCTV cameras, raspberry, mobile phones, etc. Then, it is crucial not only to develop solutions with high accuracy but also able to be implemented in real-time scenarios. For this reason, we propose using the one-stage detector RetinaNet for fresh and rotten fruit detection. In this sense one of the main advantages of RetinaNet is its accuracy and efficacy in real-time detection applications [11].

3   RetinaNet Overview

Object detection methodologies based on Artificial Neural Networks can be classified on the number of stages they require for detection. The two-stage detectors rely on the Region Proposal Network (RPN) to identify the candidate areas in the image to search for the target object. Finally, the object is detected using bounding box regression and classification on the most likely candidates. On the other hand, one-stage detectors, instead of using an RPN to identify the candidate areas, propose those areas. In general one-stage detectors are faster and less accurate than two-stage detectors [13]. However, in specific applications, one-stage detectors can outperform two-stage detectors.

RetinaNet [15] is a one-stage detector composed of following main parts:

1. Backbone. The backbone network is crucial for computing the convolutional feature map over the entire input image. Typically, RetinaNet uses the ResNet [10] architecture as a backbone considering a Fully Convolutional Network (FCN) structure to get the image features. From this fully convolutional structure are generated proportionally sized feature maps at multiple scales. Consecutive layers generate feature maps representing the same scale. On the other hand, feature maps of deeper layers represent smaller scales. Therefore, a Feature Pyramid Net (FPN) is created and used as a feature detector. In this sense, the FPN can be seen as a multi-scale feature encoder, where the scale-decreased network is referred as *backbone*.
2. The classification subnet considers an FCN processed on each level of the generated FPN to share information across all levels in the pyramid. This stage is responsible for determining the existence and class number of an object if any.
3. The regression and the classification subnets are processed in parallel on each level of the FPN. The main difference between the classification and the regression subnets is that while the classification subnet determines the class of an object, the regression subnet defines the relative offset between an anchor box and its corresponding ground-truth box.

The performance of RetinaNet is highly influenced by the FPN architecture. The feature maps generated by the FPN are semantically and spatially consistent. Higher-level feature maps of the pyramid are designed to detect larger objects, while feature maps on the bottom of the pyramid are better at detecting small objects. Feature maps in the FPN of RetinaNet can be used independently to make predictions, and as stated above, the multi-scale feature maps depend on the ANN architecture used as backbone. For this reason, in this paper, we analyze which is the most suitable backbone for the detection and classification of fresh and rotten fruits.

4   Proposed Methodology

Figure 1 shows an overview of the proposed methodology for rotten and fruit detection and classification. The proposed dataset is curated, labeled and augmented. This visual information is processed by RetinaNet to obtain the learning model for inference. The output of inference provides the class and location of fresh and rotten fruits.

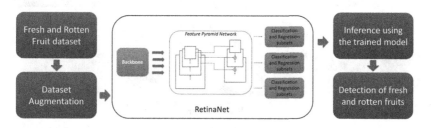

Fig. 1. Overview of the proposed process for detection of fresh and rotten fruits.

4.1   Dataset Description

Considering the need for automatic detection of fresh and rotten fruit in real-life scenarios, we propose using an RGB-based database, instead of using traditional hyperspectral imaging for food quality detection. The main drawback of using hyperspectral imaging is its high-dimensionality, which not only slows down the training process but also limits the on-line detection. Additionally, using RGB information facilitates the implementation of detection algorithms in real-life scenarios.

We propose using a modified version of the RGB-database presented in [14]. The original dataset consists of 13599 RGB images divided into 6 categories, 3 of fresh fruits and 3 of rotten fruits, as described in Table 1. This dataset considers images with a wide variety of viewing conditions and backgrounds, which is augmented using rotation every 15° from 15° to 75°, and other kinds of image processing operations, such as vertical flip, translation and salt, and pepper noise addition.

Figure 2 shows some samples of the images in the dataset used in this approach. From this figure, we can observe the challenges presented to define the boundaries between some instances, especially in the case of fresh and rotten banana, where the degree of occlusion is high. For this reason, instances with a high degree of occlusion are not considered in the database. Although this dataset is large enough for training, it does not include compressed images. Including compressed images is not only useful as a data augmentation technique but also provides compressed samples which may be useful implementation in real-life scenarios, since most of the CCTV RGB-cameras store compressed recording

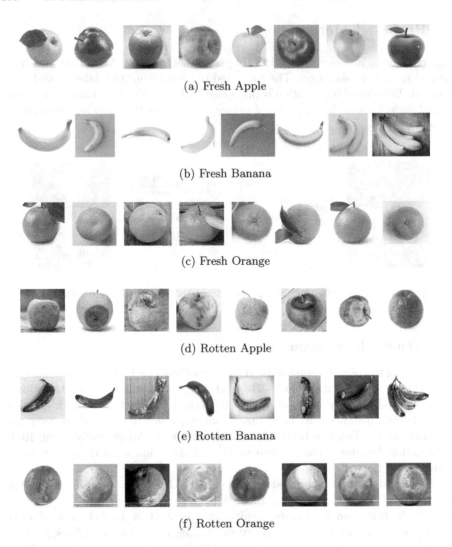

(a) Fresh Apple

(b) Fresh Banana

(c) Fresh Orange

(d) Rotten Apple

(e) Rotten Banana

(f) Rotten Orange

Fig. 2. Training dataset

using the standard H.264/AVC [12]. Table 1 describes the number of instances per class for the original dataset presented in [14], as well as the modified version which does not consider cases with a high degree of occlusion, neither cases with more than one instance per image. Additionally, some hundred samples of compressed instances were included.

Using this modified dataset with images of 380×380 pixels, we propose using the RetinaNet [15] deep learning architecture for training purposes. To evaluate the performance of Retinanet, in detecting fresh and rotten fruits, we use 80% of the images for training, and the remaining 20% is used for validation. The labeled dataset is randomly shuffled and split into the training and validation sets.

4.2 Training Process

In this paper, we consider the one-stage detector called RetinaNet [15] for the detection and classification of fresh and rotten fruits (apple, banana, and oranges).

Table 1. Database description for the original dataset in [14] and the modified version used in this approach

Dataset	Fresh fruits			Rotten Fruits		
	Apples	Banana	Oranges	Apples	Banana	Oranges
[14]	2088	1962	1854	2943	2754	1998
Modified	2276	1803	1889	2152	2398	2113

The training stage used in this research considers a comparison in the performance of RetinaNet using different ANNs as the backbone. In this sense, ResNet-50, and ResNet-101 were pre-trained using the MS-COCO [16] dataset, while VGG19 was pre-trained on the ImageNet [21] dataset. RetinaNet models were trained using the following configuration parameters: Batch size of 16, 680 steps per epoch, 100 epochs, and Adam as a stochastic optimizer. RetinaNet was implemented on the Google Colaboratory platform with a single 12GB Nvidia Tesla K80 GPU enabled. The implementation was done using Keras and Tensorflow as frameworks.

5 Experimental Results

In this section, we present the evaluation results, in the training and validation stages. The training and validation stages were done in the Google Colaboratory platform using the same configuration specified in the previous section. The dataset was hosted on Google Drive to improve computational efficiency. For an scale-invariant evaluation in classification and detection, the validation dataset was randomly scaled up and down 50%.

As stated in Sect. 3, RetinaNet relies on the backbone network to obtain the feature maps. These feature maps are used to learn those features which are dominant and scale-invariant. In this paper, we evaluate the performance of RetinaNet using ResNet-50, ResNet-101, and VGG as a backbone.

Figure 3 shows the performance results obtained during the training process of RetinaNet using different backbones. This figure presents the classification loss, the regression loss, and the total loss obtained per epoch, as well as the learning rate achieved. From this figure, we can observe that using ResNet-50 as backbone provides the better results during training than other ANNs. In general, ResNet-50 outperforms RestNet-101 and VGG-19, since it converge faster than others, in the classification and regression subnets.

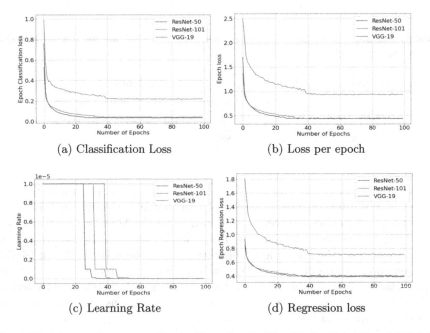

(a) Classification Loss

(b) Loss per epoch

(c) Learning Rate

(d) Regression loss

Fig. 3. Training performance of RetinaNet using ResNet-50, ResNet-101 and VGG-19 based models as backbone

Table 2 presents the obtained results of RetinaNet using ResNet-50, ResNet-101, and VGG as backbone during the validation process. Results are presented in terms of the mAP (mean Average Precision), the size of the training model, and the average inference time of RetinaNet running on the aforementioned Google Colaboratory platform. From this table, we can see that the best results are attained using ResNet-50 as backbone. This is, using ResNet-50 as backbone outperforms other ANNs not only in terms of the mAP but also in terms of the model size and the inference time.

Table 2. Performance comparison of RetinaNet Models using different backbones.

Backbone	Avg. mAP	Size of the trained model (MB)	Avg. Inference Time (sec)
ResNet-50	0.9885	139	0.4523
ResNet-101	0.9478	212	7.4195
VGG-19	0.7921	153	5.4213

Table 3 presents the per class accuracy obtained using each ANN as backbone in RetinaNet for fresh and rotten fruit detection. This table shows that ResNet-50 outperforms other backbones for this application, attaining an average accuracy of 0.9885 out of 1. The accuracy obtained using RetinaNet with ResNet-50 as the backbone is above 0.98 in all class cases, as shown in Table 3.

Table 3. Per class mAP using different ANNs as backbone in RetinaNet for fresh and rotten detection.

Backbone	Fresh fruits			Rotten Fruits		
	Apples	Banana	Oranges	Apples	Banana	Oranges
ResNet-50	0.9870	0.9895	0.9892	0.9868	0.9925	0.9861
ResNet-101	0.9587	0.9381	0.9405	0.9439	0.9552	0.9504
VGG-19	0.7757	0.8092	0.8672	0.7421	0.8255	0.7311

Figure 4 shows some visual examples of fresh and rotten fruit detection in the validation subset, which also considers noisy, rotated, scaled, and unaltered samples. In this figure, we can observe some samples per class during detection, as well as its confidence score. Most of the classes present high confidence scores, above 0.9. However, specially in the case of rotten orange, although the object is well classified and detected, the confidence score is low compared to other classes.

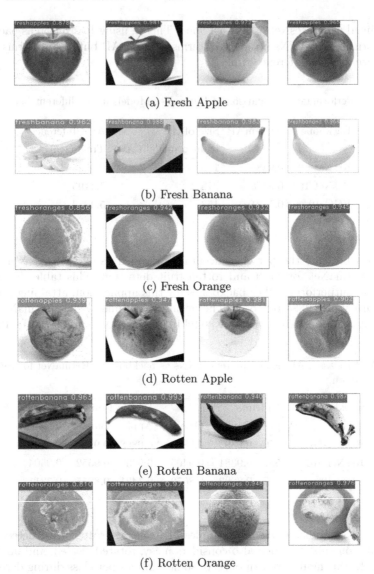

(a) Fresh Apple

(b) Fresh Banana

(c) Fresh Orange

(d) Rotten Apple

(e) Rotten Banana

(f) Rotten Orange

Fig. 4. Examples of fresh and rotten fruit detection and classification using the proposed approach.

6   Conclusions and Future Work

In this paper, we presented a RetinaNet based approach for fresh and rotten fruit detection and classification. To determine the most accurate backbone for RetinaNet, we evaluate ResNet-50, ResNet-101, and VGG-19. From this evaluation we conclude the best results are obtained using ResNet-50 as backbone for fresh and rotten fruit detection, attaining an accuracy above 0.98 per class, in terms of mAP. The dataset considered in this paper includes data augmentation by means of rotation, flipping, translation, noise and compression using the standard H.264 to represent fresh and rotten apples, banana and oranges. According to obtained results, RetinaNet outperforms previous approaches not only in terms of mAP reached but also in terms of the number of classes considered for detection and classification. As future work, we propose including more samples per class and more classes as well as implementing the model in hardware constrained devices for real-time detection in real-life scenarios.

References

1. Arzate-Vázquez, I., et al.: Image processing applied to classification of avocado variety hass (persea americana mill) during the ripening process. Food Bioprocess Technol. **4**(7), 1307–1313 (2011)
2. Bhargava, A., Bansal, A.: Fruits and vegetables quality evaluation using computer vision: a review. Journal of King Saud University - Computer and Information Sciences, pp. 1–15 (2018)
3. Calvo, H., Moreno-Armendáriz, M.A., Godoy-Calderón, S.: A practical framework for automatic food products classification using computer vision and inductive characterization. Neurocomputing **175**, 911–923 (2016)
4. Cárdenas-Pérez, S., et al.: Evaluation of the ripening stages of apple (golden delicious) by means of computer vision system. Biosyst. Eng. **159**, 46–58 (2017)
5. da Costa, A.Z., Figueroa, H.E.H., Fracarolli, J.A.: Computer vision based detection of external defects on tomatoes using deep learning. Biosyst. Eng. **190**, 131–144 (2020)
6. Fan, S., et al.: On line detection of defective apples using computer vision system combined with deep learning methods. J. Food Eng. **286**, 110102 (2020)
7. Goel, L., Raman, S., Dora, S.S., Bhutani, A., Aditya, A.S., Mehta, A.: Hybrid computational intelligence algorithms and their applications to detect food quality. Artif. Intell. Rev. **53**(2), 1415–1440 (2019). https://doi.org/10.1007/s10462-019-09705-8
8. Gómez-Sanchis, J., Martín-Guerrero, J.D., Soria-Olivas, E., Martínez-Sober, M., Magdalena-Benedito, R., Blasco, J.: Detecting rottenness caused by penicillium genus fungi in citrus fruits using machine learning techniques. Expert Syst. Appl. **39**(1), 780–785 (2012)
9. He, K., Gkioxari, G., Dollár, P., Girshick, R.: Mask R-CNN. In: 2017 IEEE International Conference on Computer Vision (ICCV), pp. 2980–2988 (2017)
10. He, K., Zhang, X., Ren, S., Sun, J.: Deep residual learning for image recognition. In: 2016 IEEE Conference on Computer Vision and Pattern Recognition (CVPR), pp. 770–778 (2016)
11. Hoang, T.M., Nguyen, P.H., Truong, N.Q., Lee, Y.W., Park, K.R.: Deep retinanet-based detection and classification of road markings by visible light camera sensors. Sensors (Basel, Switz.) **19**, 281 (2019)
12. ITU: H.264 : Advanced video coding for generic audiovisual services (2018). urlhttps://www.itu.int/rec/T-REC-H.264-201906-I/en
13. Jiao, L., et al.: A survey of deep learning-based object detection. IEEE Access **7**, 128837–128868 (2019)
14. Kalluri, S.R.: Fruits: fresh and rotten for classification Dataset (2018). urlhttps://www.kaggle.com/sriramr/fruits-fresh-and-rotten-for-classification

15. Lin, T., Goyal, P., Girshick, R., He, K., Dollár, P.: Focal loss for dense object detection. IEEE Trans. Pattern Anal. Mach. Intell. **42**(2), 318–327 (2020)
16. Lin, T.-Y., et al.: Microsoft COCO: Common objects in context. In: Fleet, D., Pajdla, T., Schiele, B., Tuytelaars, T. (eds.) ECCV 2014. LNCS, vol. 8693, pp. 740–755. Springer, Cham (2014). https://doi.org/10.1007/978-3-319-10602-1_48
17. Liu, W., et al.: SSD: Single shot multibox detector. In: ECCV (2016)
18. Nosseir, A., Ahmed, S.E.A.: Automatic classification for fruits' types and identification of rotten ones using k-nn and svm. Int. J. Online Biomed. Eng. **15**(03), 47–61 (2019)
19. Redmon, J., Divvala, S., Girshick, R., Farhadi, A.: You only look once: Unified, real-time object detection. In: 2016 IEEE Conference on Computer Vision and Pattern Recognition (CVPR), pp. 779–788 (2016)
20. Ren, S., He, K., Girshick, R., Sun, J.: Faster R-CNN: towards real-time object detection with region proposal networks. IEEE Trans. Pattern Anal. Mach. Intell. **39**(6), 1137–1149 (2017)
21. Russakovsky, O., et al.: ImageNet large scale visual recognition challenge. Int. J. Comput. Vis. **115**(3), 211–252 (2015). https://doi.org/10.1007/s11263-015-0816-y
22. Zhang, Y., Wu, L.: Classification of fruits using computer vision and a multiclass support vector machine. Sensors (Basel, Switz.) **12**, 12489–12505 (2012)
23. Zhu, X., Li, G.: Rapid detection and visualization of slight bruise on apples using hyperspectral imaging. Int. J. Food Prop. **22**(1), 1709–1719 (2019)

MSUP: Model for the Searching of Unidentified People Comparing Binary Vectors Using Jaccard and Dice

José-Sergio Ruiz-Castilla[✉], Farid García-Lamont, José Ángel Regalado-García, Adán Vidal-Peralta, and Carlos Rafael Hernández-Magos

Centro Universitario UAEM Texcoco, Universidad Autónoma del Estado de México, Texcoco, Estado de México, México
jsergioruizc@gmail.com, fglamont@yahoo.com.mx,
angel.95rg@gmail.com, adanvidal16@hotmail.com,
crhdzmag15@gmail.com

Abstract. The last years in Mexico were reported thousands of missing people. Almost every day were found peoples dead in some place. The authorities open a folder investigation, but most times is not possible an identification effective of the person. In other hand, the familiars report the missing of any member to Public Ministry (PM) or to National Search Commission (NSC). So, the authorities of PM or NSC document the personal data. However, there is not a connection between instances governmental. In this work, we propose a platform and algorithms to find missing people. These algorithms are: The first module is the Characterization of Unidentified People (CUP); the second module is the Characterization of Wanted People (CWP) and finally the Searching for Missing People (SMP). SMP will focus on an algorithm with similarity metrics with the ability to find one or more "similar" people found on the platform. This last module will determine which similar people were found, as well as the dependence on the government where they are physically. To implement the solution, it is necessary to: Establish a vector of characteristics obtained from the CUP module and the CWP module to apply algorithms based on similarity metrics until "matching" and evaluate the proposed algorithms to obtain the best result. For the solution we propose can benefit the institutions that have unidentified corpses under your responsibility. If a human remains is properly characterized, it increases the possibility of identifying and claiming. Therefore, it can reduce or avoid the problem of excess of thousands unclaimed corpses.

Keywords: Similarity metrics · Missing people · People search · Match · Matching

1 Introduction

1.1 Scene in Mexico

When a dead person is found, he usually lacks identification. The elements to identify a person are their physical characteristics. The physical characteristics can be: approximate

L. Martínez-Villaseñor et al. (Eds.): MICAI 2020, LNAI 12469, pp. 337–349, 2020.
https://doi.org/10.1007/978-3-030-60887-3_30

age, weight, height, gender, tattoos, moles, scars, among others. Also, they can be: shoes, clothes, watch, jewelry, etc.

The authorities must lift the corpse under a protocol. The protocol consists of recording the details of the corpse and the environment. An autopsy is performed and said body is protected. Meanwhile, Family members search for the missing person through photographs and physical features.

When a family member disappears, the search begins in public places and at hospitals. The search continues in the morgue or places where corpses are sheltered. However, when a corpse lacks identification, its location is difficult.

This research work seeks to facilitate the search and location of missing and found dead without identifying. The MSUP is proposed (Fig. 2). The MSUP must be implemented in a Web system for governmental instances mainly.

The forensic analysis is very important for identification of corpses. Dorado and Sánchez mentioned in their book (*"What the dead tell"*, *"Lo que cuentan los muertos"*) as the forensic get all characteristics from corpse or bones. The characteristics are recorded in several forms. The characteristics of the bones and prostheses are useful when a corpse is burned, decomposed or dismembered. The characteristics usually are: approximated age, gender, height, denture, among others [1].

One problem in Mexico is that during the forensic process the data is not completed on the corresponding forms. During autopsy, all data related to the human body must be documented. The main objective is to know the cause of death. However, it is a great opportunity to collect traits for possible identification. It is not a problem only in Mexico, in the work of *Chattopadhyay et al.* documents that Calcutta, India, poses as a disaster the autopsy process in unidentified people. 89% of people studied between 15 to 60 years of age. 7.4% are not recognized before 7 days. Most people were found on roads, paths and rivers. The number of unidentified deaths in the city of Calcutta is quite alarming. It would not be incorrect to describe it as a *"disaster in disguise"* [6].

1.2 Search for Missing People

The *"Registro nacional de datos de personas extraviadas o desaparecidas (RNPED), National data registry of lost or missing persons"*, publishes that until 2018 the "Federal Law Statistics" there are 1,171 people missing, while in the "Common Law Statistics" there are, 36,266 people missing or not located. The data recorded are: Date and time of disappearance, country where it disappeared, place of disappearance, nationality, stature, complexion, gender, age, characteristics, ethnicity, disability and place where the disappearance was recorded [2].

The RNPED use the form the Fig. 1 to search to a lost or disappeared person. However, when the person is dead and unidentified is not possible to locate.

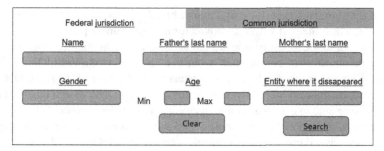

Fig. 1. Form for search people using the website from *RNPED.*

According to the RNPED data, people missing every year from 2014 to 2018 are shown in Table 1. Of which, 26,938 are men and 9,327 are women. In addition, more than 50% are between 15 and 25 years old.

Table 1. Disappeared people from 2014 to 2018 in México [4].

Date	Federal Law	Common Law	Total	Variation
Oct 2014	554	23271	23605	ND
Jun 2015	443	25293	25736	2131
Apr 2015	557	25398	25955	219
Jul 2015	662	25917	26580	625
Oct 2015	916	26670	27586	1006
Jun 2016	946	27215	28161	575
Apr 2016	1027	27162	28189	28
Jul 2016	1044	27428	28472	283
Oct 2016	966	28937	29903	1431
Jun 2017	1030	29912	30942	1039
Total	**8145**	**267203**	**275129**	

2   Related Publications

2.1   People Search

Since 2009 Grandsman proposed to extract blood to children. The above, because the next problem. During the military dictatorship in Argentina (1976–1983), up to 30,000 people disappeared, including an estimated 500 newborn infants and young children who were handed over to military families to be raised as their own. *"Las Abuelas de Plaza de Mayo" (The Grandmothers of the Plaza de Mayo)* is a human rights organization that formed in order to identify their missing grandchildren and reunite them with their biological families. However, the extraction of blood was an illegal practice but can be been an effective strategy if where approved for the government [3].

The cases of unidentified people are not necessarily for homicide. After a disaster, there may be unidentified people such as the 9/11 attacks on the World Trade Center, Hurricane Katrina, or the Southeast Asian tsunami. In the case of 9/11, the DNA was used because the majority of relatives contributed a sample and, on the other hand, the authorities were able to obtain a sample from each unidentified person. [7].

In the work of Andreev et al. exposed a growth of cases of unidentified people in Russia. In this case, these are mostly people who die by drinking alcohol excessively. To identify are used the passport or other documents in the pockets. The 13.5% are unidentified people, but with a growing trend [9].

Bell proposes as a resource to identify people, the dental information. This resource is very effective, however when you do not have the record you lose the possibility. On the other hand, sometimes the bodies are found burned and much dental information is lost. In summary, it is insufficient and other additional characteristics are required. [10]

2.2 Person's Characteristics

The characteristics of a person are related to gender, age, height, weight, skin color, among others. However, there are other more specific ones such as scars, tattoos, moles, etc. Finally, there are other forensic types. Forensic characteristics are the most useful to identify a person. The characteristics of the denture, prosthesis, fractures, absence of limbs.

3 Similarity Metrics

3.1 Binary Similarity

Ie publication Seung-Seok Choi et al., we found 76 formulas for studying binary similarity and distance metrics. In the set of formulas there are formulas that omit (d) that refers to the similarity (0−0). For example: Jaccard, Dice, Czecanowski, 3w-Jaccard, Nei & Li, Sokal & Sneak-I among others [8].

There are formulas for binary coefficient which not use la variable d. This formulas considered for this propose are the next [5] [8].

Jaccard
$$S1(X1 \ X1 = \frac{a}{a+b+c+d} \tag{1}$$

Dice
$$S1(X1 \ X1) = \frac{2a}{2a+b+c} \tag{2}$$

3W-Jaccard
$$S1(X1 \ X1) = \frac{3a}{3a+b+c} \tag{2}$$

We only using solutions with (a), (b), and (c), because the binary matrix has 93% of zeros. Figure 7. With the metrics with (d) the % of similitude was above 90%, the above made very difficult find the best similitude. In this case, Dice and 3 W-Jaccard it assign two or three times the value of (a) but also they omit (d) [5, 8, 11].

4   Method

4.1   MSUP Model

We propose a model for searching the unidentified people. The model not use personal data but the characteristics. We use views for the capture the characteristics. Each matrix or vector has 900 0 s and 1 s. However, each binary vector is stored as a text file independent. The files are read and compared each other. See Fig. 2.

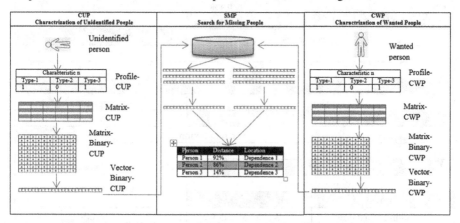

Fig. 2. MSUP: Model for Searching for unidentified people.

The MSUP model is integrated for three modules: The first is *CUP (Characterization of Unidentified People)*, the second is *CWP (Characterization for Wanted People)* and the third is *SMP (Search for Missing People)*.

The *module CUP* has the function of get the *Profile-CUP* of characteristics of unidentified person. The characteristics are related of her skin, body, hair, clothes, etc. Each characteristic will a binary data as zero or one $(0-1)$. The data binary are recorded in a *Matrix-CUP*. After, The *Matrix-CUP* is converted in at a *Binary-Vector-CUP*; finally, the *Binary-Vector-CUP* is stored in a *Dataset-CUP* in a *Server*.

Fig. 3. Photos as source of characteristics.

Multiples photo was taken of volunteer men and women, Fig. 3. These persons were characterized as disappeared persons. With these photos was filled the matrix of characteristics. Accord of matrix of Fig. 4.

The characteristics include of body and clothing and accessories. Other characteristics: the teeth, inlaid teeth, bone prostheses, among others. See the Fig. 4. We obtained about 900 binary elements, shown as 0 s and 1 s.

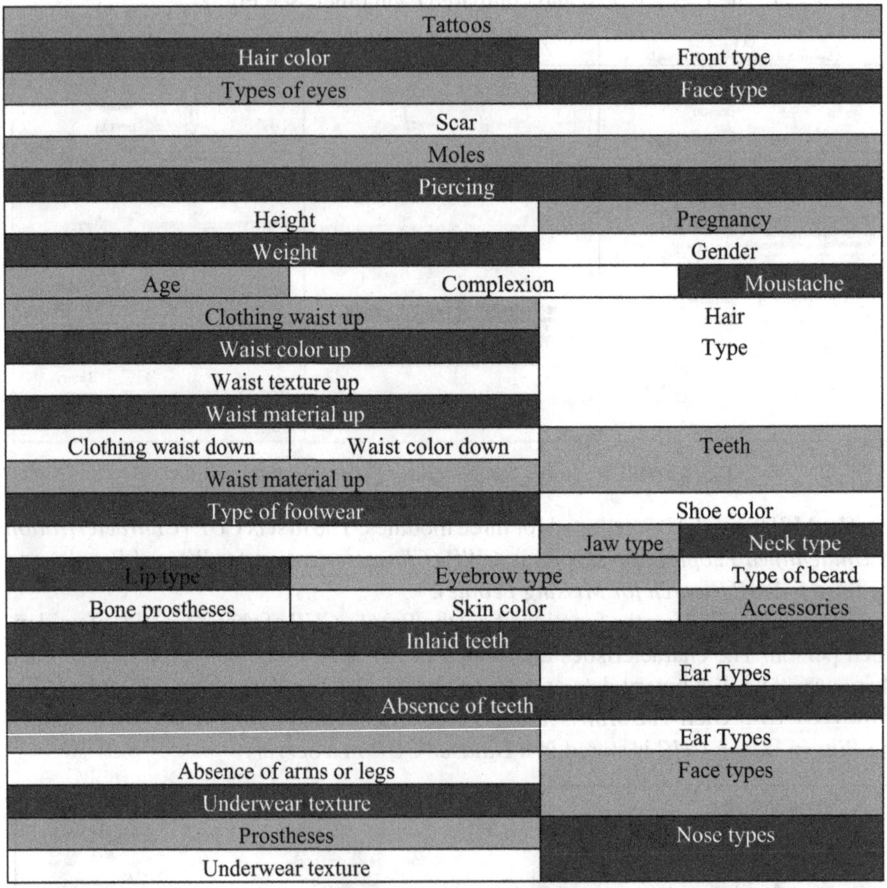

Tattoos				
Hair color			Front type	
Types of eyes			Face type	
Scar				
Moles				
Piercing				
Height			Pregnancy	
Weight			Gender	
Age		Complexion		Moustache
Clothing waist up			Hair	
Waist color up			Type	
Waist texture up				
Waist material up				
Clothing waist down		Waist color down	Teeth	
Waist material up				
Type of footwear			Shoe color	
		Jaw type	Neck type	
Lip type		Eyebrow type	Type of beard	
Bone prostheses		Skin color	Accessories	
Inlaid teeth				
			Ear Types	
Absence of teeth				
			Ear Types	
Absence of arms or legs			Face types	
Underwear texture				
Prostheses			Nose types	
Underwear texture				

Fig. 4. Set of types of characteristics.

The characteristics included have multiples answers. For each characteristic was created a vector generally for the types. By example, for each tooth was recorded: absence or presence, inlaid teeth end prostheses. In the case of hair: were recorded: presence or absence, color, dyeing color, straight or curly, with extensions, among others.

The process for the characterization is possible using an interface with a set the forms. Each form contains a question about a characteristic. The form has "n" answers, all with zeros. The user only chooses an option, and record 1 as answer. By example, the form shows ages ranges, the user choose the option accord the approximated age of unidentified person. After, the vector is recorded in the *Binary-Vector-CUP*. Each characteristic is processed with the same way. See the Fig. 5.

Approximated height							
0- 50 cm	51- 70 cm	71- 90 cm	91- 110 cm	11- 130 cm	131- 150 cm	151- 170 cm	171- 190 cm
0	0	0	0	0	0	1	0

Note: The user record 1 on range of 151-170 cm because the unidentified person is approximated 160 cm. Only the vector with 0s and 1s is recorded in the *Binary-Matrix-CUP*.

Fig. 5. *Binary-Vector* of approximated age.

This method was use for each characteristic. The 0 represents the absence while the 1 represents the presence of the characteristic. This process was used to normalize the data as binary. A vector was created for each characteristic.

The body has been divided into sections. Sections are marked to indicate the existence of a mole, scar, or tattoo during characterization. For now, the tattoo on the body for example a skull is not included.

The *CWP module* has a function of get the *Profile-CWP* of characteristics of wanted persons. The characteristics are of skin, body, clothes, moles, tattoos, etc. The characteristics are recorded as zeros and ones in a *Binary-Vector-CWP*, after in a *Binary-Vector-CWP*.

The user that characterizes a wanted person does the following task. The user obtains the characteristics of the person wanted. Then, mark with 1 each characteristic as in Fig. 2. Each binary vector will be added to the *Binary-Matrix-CWP*.

The last module is *SMP* for do match. The "Matching" makes comparison of two strings of characters or 0 s and 1 s. When both strings are equals the distance is 1.0 0r 100%. If there are coincidences then there is a percentage between 0 and 99%. When we obtain a set of records with a percentage each, just choose the highest percentage.

4.2 Characterization of the Missing Person

The characterization is made from a photo or characteristics listed from a member familiar. The physical characteristics and clothes. As well, accessories as like as watch, rings, earrings, and piercings among others. In the Fig. 6 we can see to a) the person's photo and at b) the characterization binary.

The first algorithm is detailed in the Code 1. This code find and open each file for compare two strings end calculate the percentage of similarity. In this case each file have the characterization of a person. The comparing is made with the string of the wanted person and the each string of peoples unidentified registered

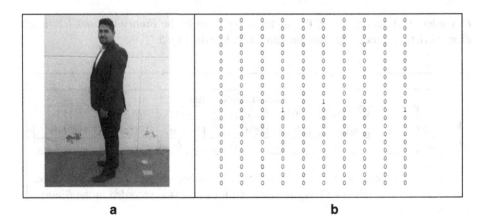

a   b

Fig. 6. a) The person's photo and at b) The characterization binary.

```
Find the path
Open the folder files
Identification of the folder file according to gender
Set the path
Repeat
Read the file as a characters string
Calculate the similarity of the two strings
Calculate percentage of similarity
Store result
End of repeat
Show the results
```

The second algorithm set the gender. There is a folder for each gender. El gender 1 is male, 2 is female, 3 is transgender and 4 is intersexual. This makes it easier the search because only search in the corresponding folder. See the Code 2.

```
Open the file
Read the file
Generate a binary string
If index 476 = 1 then Gender is Male
If index 477 = 1 then Gender is Female
If index 478 = 1 then Gender is Transgender
If index 479 = 1 then gender is Intersexual
Return Gender
```

The third algorithm allow calculate the grade of similitude. The algorithm make a comparison for find a similitude percentage. For this calculate we use metrics of similitude. See to Code 3.

```
Initialize a = b = c = d = 0
If length (String1) = length(String2)
```

```
Repeat from Index = 0 to length(String2)
If(String[Index]) = 1 and String2[Index] = 1 then a = a+1
If(String[Index]) = 1 and String2[Index] = 0 then b = b+1
If(String[Index]) = 0 and String2[Index] = 1 then c = c+1
If(String[Index]) = 0 and String2[Index] = 0 then d = d+1
Print a,b,c,d
SS3 = a / (a + b + c) //Jaccard
SS2 = (2*a) / ((2*a) + b + c) //Dice
SS1 = (3*a) / ((3*a) + b + c) //3w-Jaccard
Return SS1, SS2, SS3
```

The Table 2 sample had 41 characterizations of unidentified people, including men and women. Then, two characterizations were made of a man and a woman from the same sample group, but as a wanted person. Matrices and binary vectors were obtained, once the characterizations had been carried out. Finally, we apply the metrics through the corresponding algorithms.

Table 2. Table of binary coefficient. [5, 8, 11]

	1 (presence)	0 (absence)	Sum
1 (presence)	1,1 (a)	1.0 (b)	a + b
0 (absence)	0,1 (c)	0,0 (d)	c + d
	a + c	b + d	A + b + c + d

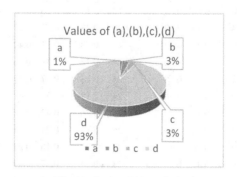

Fig. 7. Ratio of 0's and 1's

5   Results

To test the methodology, two cases were studied. The case 1 and 2 yielded the results of Table 3. Are shown the profile of 41 people. Two profiles of unidentified people were included. The results are percentages of similarity. In this case, the higher values indicate more similar to the wanted people.

Table 3. Results of Case 1 and Case 2 with Jaccard, Dice and 3W-Jaccard.

Case 1				Case 2			
	Jaccard	Dice	3 W-Jaccard		Jaccard	Dice	3W-Jaccard
1	0.13	0.22	0.30	1	0.20	0.33	0.42
2	0.25	0.41	0.51	2	0.26	0.42	0.52
3	0.26	0.42	0.52	3	0.17	0.29	0.38
4	0.14	0.24	0.32	4	0.52	0.69	0.77
5	0.17	0.29	0.38	5	0.21	0.35	0.44
6	0.17	0.29	0.38	6	0.24	0.39	0.49
7	0.24	0.38	0.48	7	0.38	0.55	0.65
8	0.18	0.30	0.39	8	0.22	0.37	0.46
9	0.13	0.23	0.31	9	0.06	0.12	0.17
10	0.30	0.46	0.56	10	0.18	0.31	0.40
11	0.30	0.46	0.56	11	0.13	0.22	0.30
12	0.29	0.44	0.55	12	0.06	0.11	0.16
13	0.27	0.43	0.53	13	0.11	0.19	0.26
14	0.25	0.39	0.49	14	0.13	0.23	0.31
15	0.25	0.39	0.49	15	0.12	0.22	0.29
16	0.25	0.41	0.51	16	0.13	0.22	0.30
17	0.67	0.80	0.86	17	0.18	0.31	0.40
18	0.20	0.34	0.43	18	0.19	0.32	0.41
19	0.34	0.51	0.61	19	0.15	0.26	0.34
20	0.24	0.39	0.49	20	0.23	0.37	0.47
21	0.28	0.44	0.54	21	0.18	0.31	0.40
22	0.30	0.46	0.56	22	0.22	0.36	0.46
23	0.41	0.58	0.68	23	0.23	0.37	0.47
24	0.30	0.46	0.56	24	0.19	0.33	0.42
25	0.16	0.27	0.36	25	0.26	0.42	0.52
26	0.21	0.34	0.44	26	0.14	0.24	0.33
27	0.25	0.40	0.50	27	0.09	0.16	0.23
28	0.26	0.42	0.52	28	0.23	0.37	0.47
29	0.28	0.43	0.53	29	0.14	0.25	0.33
30	0.24	0.38	0.48	30	0.16	0.28	0.37
31	0.20	0.33	0.42	31	0.27	0.42	0.52
32	0.31	0.47	0.57	32	0.21	0.34	0.44
33	0.36	0.53	0.62	33	0.18	0.31	0.40

(*continued*)

Table 3. (*continued*)

Case 1				Case 2			
	Jaccard	Dice	3 W-Jaccard		Jaccard	Dice	3W-Jaccard
34	0.27	0.43	0.53	34	0.18	0.30	0.39
35	0.35	0.52	0.62	35	0.10	0.19	0.26
36	0.28	0.43	0.53	36	0.13	0.22	0.30
37	0.33	0.50	0.60	37	0.08	0.14	0.20
38	0.21	0.34	0.44	38	0.25	0.41	0.51
39	0.26	0.42	0.52	39	0.15	0.26	0.34
40	0.14	0.24	0.33	40	0.33	0.50	0.60
41	0.23	0.38	0.48	41	0.20	0.33	0.43

Figure 8 shows case 1. As we can see, the unidentified person 17 corresponds more to the characteristics of the person sought. Also, we can see other cases such as 23, 33, 35 and 37 with higher %, however, it is clear that 17 corresponds to the person sought.

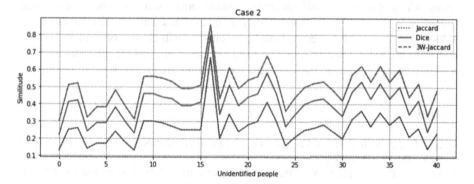

Fig. 8. Graph of case 1, the record 17 is major.

In Case 2, were obtained the results of Table 3. As we can see, record 4 shows the highest %. Similarly, 17 of 41 results. Other high values such as 7, 31, 38 and 40 were found. However, the record 4 corresponds to the person wanted.

In Fig. 9. We can see that register 4 is the highest. Therefore it corresponds to the person wanted. We can see other cases such as 7 or 40. However, the value of register 4 makes it very evident that he is the person wanted.

In this case, of the three techniques used, we can see that the 3 W-Jaccard technique is the most effective, as it shows higher values for all cases. On the other hand, we can see the consistency between the three techniques. It is possible that the values are not 100% due to the characterization of the people. The characterization is done by users through the interfaces and it is possible that they introduce some erroneous data. The above, because most of the characteristics are qualitative.

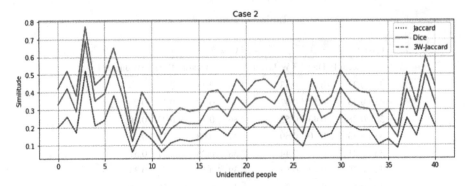

Fig. 9. Graph of Case 2. In this case the record 4 is the major.

6   Conclusions

We can conclude that it is possible to generate the profiles of unidentified people and the profiles of wanted people. With the profiles you can generate matrices and binary vectors in a standard way. Once the profiles are generated they can be stored in text files as strings of 0s and 1s. We can apply similarity metrics and find the closest match. Similar records can be found. In that case, the highest percentages are candidates. Once the most similar record or records have been found, it will be necessary to go to make an identification of the person sought to the government instance. In this case, there are government agencies where the unidentified candidates are to complete the search.

References

1. Dorado E., Sánchez, J.A.: Lo que cuentan los muertos (*What the dead tell*). In: Paidos (ed.) Spain, ISBN: 978-84-998-436-6 (2015)
2. RNPED "Registro Nacional de datos de personas extraviadas o desaparecidas", (NRLMP National Data Registry of Lost or Missing Persons) (2019). www.rndep.segob.gob.mx
3. Grandsman, A.: J. Lat. Am. Carib. Anthropol. **14**(1), 162–184 (2009). https://doi.org/10.1111/j.1935-4940.2009.0001043.x. ISSN 1935-4940. Bay the American Anthropological Association. All rights reserved
4. Centro Diocesano para los Derechos Humanos Fray Juan de Larios. Diagnóstico del Registro Nacional de Datos de Personas Extraviadas o Desaparecidas (RNPED). Saltillo, Coah, México (2017). www.frayjuandelarios.org
5. Rodríguez, S.M.E.: Coeficientes de asociación, Ed. Plaza y Valdez. Ciudad de México (2001)
6. Chattopadhyay, S., Shee, B., Sukul, B.: Unidentified bodies in autopsy–A disaster in disguise. Egypt. J. Forensic Sci. **3**(4), 112–115 (2013). https://doi.org/10.1016/j.ejfs.2013.05.003
7. Ritter N.: Missing Persons and Unidentified Remains: The Nation's Silent Mass Disaster, National Institute of Justice, vol. 256, Washington, DC (2007)
8. Choi, S.-S., Cha, S.-H., Tappert, C.C.: A survey of binary similarity and distance measures. J. Syst. Cybern. Inf. **8**, 43–48 (2010)
9. Andreev, E., Pridemore, W.A., Shkolnikov, V.M., Antonova, O.I.: An investigation of the growing number of deaths of unidentified people in Russia. Eur. J. Pub. Health **18**(3), 252–257 (2008). https://doi.org/10.1093/eurpub/ckm124

10. Bell, G.L.: Dentistry's role in the resolution of missing and unidentified person cases. Dent. Clin. North Am. **45**(2), 293–308 (2001)

11. Batyrshin Ildar Z, Kubysheva Nailya, Solovyev Valery, Villa-Vargas Luis A. Visualization of Similarity Measures for Binary Data and 2×2 Tables, Computación y Sistemas, ISSN: 2007-9737, vol 20, 3, pp. 345-353 (2016). https://doi.org/10.13053/cys-20-3-2457

Comparison Methods for Fuzzy C-Means Initialization Applied to Image Segmentation

Virna V. Vela-Rincón[1]([⊠]), Celia Ramos-Palencia[1], Dante Mújica-Vargas[1], Jean Marie Vianney Kinani[2], and Eduardo Ramos-Díaz[2]

[1] Tecnlógico Nacional de México/CENIDET, Cuernavaca, Morelos, Mexico
viryvela@cenidet.edu.mx
[2] Instituto Politécnico Nacional-UPIIH, San Agustín Tlaxiaca, Hidalgo, Mexico

Abstract. Fuzzy C-means (FCM) is one of the most used clustering algorithms, some research seeks to achieve a better quality in the results. It is well known that an adequate selection of the initial centroids will achieve a better clustering result. In this paper, a comparison of some initialization methods applied to the FCM algorithm is made, the experimental results suggest that the K-means++ initialization method is the most suitable for image segmentation, since it produces a better initialization condition, in addition to improve convergence times.

Keywords: Initialization methods · Fuzzy C-means · Image segmentation · K-means++

1 Introduction

Clustering breaks down a given set of objects into subgroups or clusters based on similarity. Its objective is to divide the data set in such a way that the objects that belong to the same cluster are as similar as possible, while the objects that belong to different clusters are as different as possible [1]. Clustering can be applied to image segmentation, one of the most popular methods is the Fuzzy C-means algorithm, which allows the gradual belonging in a closed interval $\mu \in [0, 1]$ of the data with respect to the clusters or regions of interest.

Fuzzy C-means cannot guarantee a unique clustering result because the initial centroids are chosen at random, this yields a higher number of iterations and the clustering result unstable. Many investigations try to find the best way to initialize this centroids, such is the case in [2] that extracts the most vivid and distinguishable colors called **dominant colors** from a given set of colors, this are obtained by computing the similarities between color points in the set and well-known reference colors. These similarities are used to calculate the degree of belonging of each color point to the reference colors. Based on the degrees of membership, we identify the dominant colors and then select to guide the selection of the initial centroids. [3] presents a initialization scheme called Hierarchical

© Springer Nature Switzerland AG 2020
L. Martínez-Villaseñor et al. (Eds.): MICAI 2020, LNAI 12469, pp. 350–362, 2020.
https://doi.org/10.1007/978-3-030-60887-3_31

Approach (HA), integrates the splitting and merging techniques to obtain the initialization condition for FCM algorithm. Initially, the splitting technique is applied to split the color image into multiple homogeneous regions. Then, the merging technique is employed to obtain the reasonable cluster number for any kind of input images. In addition, the initial cluster centers for FCM algorithm are also obtained. [4] introduces the Fuzzy C-means++ algorithm which utilizes the seeding mechanism of the K-means++ algorithm and improves the effectiveness and speed of Fuzzy C-means. By careful seeding that disperses the initial cluster centers through the data space, the resulting Fuzzy C-means++ approach samples starting cluster representatives during the initialization phase. The cluster representatives are well spread in the input space, resulting in both faster convergence times and higher quality solutions. Finally an improved fuzzy clustering algorithm based on Spark is proposed in [5], that integrates the L2 norm and uses the K-means++ algorithm improved by the Canopy algorithm [6] to initialize the cluster center; their experimental results show that performs well in clustering accuracy and computational performance.

An experimental comparison between three batch clustering algorithms is performed in [7], the Expectation – Maximization (EM) algorithm significantly outperforms the other methods, so, they proceed to investigate the effect of various initialization methods on the final solution produced by this algorithm. The initialization methods considered are (1) parameters sampled from a previous non-informative, (2) random perturbations of the marginal distribution of the data, and (3) the result of the hierarchical agglomerative grouping. All the methods proposed in the literature for centroids initialization present positive results in different areas, there is some papers that compares different initialization methods for the K-means clustering algorithm [8–10] however, a comparison of these has not been carried out to determine which allows to generate a better clustering result in the Fuzzy C-means algorithm. The aim of the paper is to compare different methods to initialize clustering algorithm, specifically Fuzzy C-means to improve the data clustering or image segmentation. The rest of the paper is organized as follows: Sect. 2 presents background information. Section 3 will describe the initialization methods. Section 4 will compare the clustering and segmentation results produced by FCM algorithm using considered initialization methods. Finally, conclusions and future work in Sect. 5.

2   Fuzzy C-Means Algorithm

The Fuzzy C-means algorithm assigns each pixel to the nearest cluster, it allows a gradual membership in a closed interval of data $\mu \in [0, 1]$, with respect to the cluster or regions of interest. This flexibility allows to express the membership of a data to all the cluster or regions simultaneously. The problem of dividing a set of data into different clusters is a the task of minimizing the square distances between the data and the centers of the clusters. Formally, a fuzzy clustering model of a given data set X in c clusters is defined to be optimal when it minimizes the following objective function J_f [11]:

$$J_f(X; U_f, C) = \sum_{i=1}^{n} \sum_{j=1}^{c} \mu_{ij}^m d_{ij}^2 , \tag{1}$$

where U_f represents the membership matrix and C a vector with the centers of the cluster, μ_{ij} is the membership of pixel x_i in the jth cluster, d_{ij}^2 is a norm metric, and the parameter m controls the fuzziness of the resulting partition, and $m = 2$ is regularly used. The objective function J_f is minimized by two steps. First the degrees of membership are optimized by setting the parameters of the clusters, then the prototypes of the clusters are optimized by setting the degrees of membership. The equations resulting from the two iterative steps form the Fuzzy C-Means clustering algorithm [11].

$$u_{ij} = \frac{1}{\sum_{k=1}^{c} \left(\frac{d_{ij}^2}{d_{ik}^2}\right)^{\frac{2}{m-1}}} = \frac{d_{ij}^{-\frac{2}{m-1}}}{\sum_{k=1}^{c} d_{ik}^{-\frac{2}{m-1}}} , \tag{2}$$

$$c_j = \frac{\sum_{i=1}^{n} \mu_{ij}^m x_i}{\sum_{i=1}^{n} \mu_{ij}^m} , \tag{3}$$

3   Seeding Initialization Methods

Seeding is a important method for clustering initialization, which generally is an procedure to select seeds from a data set, each being as the initial center of a cluster. Then four techniques were studied: K-means++, cluster, sample and uniform random seeding.

Uniform Seeding. This technique selects the number of centers uniformly at random from the range of data set, it is the usual way in which centroids are initialized in the FCM algorithm.

Sample Seeding. This method selects a number of points from the data set at random, which are the initial centroids [12]. The algorithm is very simple, and it is showed on Algorithm 1.

Algorithm 1. *Sample seeding Algorithm*

Input: Data X, number of centers k
Output: Set of center C
 1: $C \leftarrow \{c_1, c_2, \ldots, c_k\}$, c_i is selected randomly from X

Cluster Seeding. Cluster seeding performs a preliminary clustering phase in a random 10% sub sample of the data only if the number of points in the sub sample is greater than the number of centers. This preliminary phase initializes itself using sample seeding. When the number of points in the 10% random sub sample is less than the number of center, then selects randomly the number of centers of the data set [12]. The algorithm is very simple, and it is showed on Algorithm 2.

Algorithm 2. *Cluster seeding Algorithm*

Input: Data $X = \{x_1, x_2, \ldots, x_n\}$, number of centers k
Output: Set of cluster C
1: $S \leftarrow \{x_1, x_2, \ldots, x_d\}$, is selected randomly from X with $d < n$
2: $C \leftarrow \{c_1\}$, c_i is selected randomly from S
3: **while** $C \setminus C' \neq v$ **do**
4:     **for all** $u \in S$ **do**
5:         $C' \leftarrow C$
6:         Assign u to group C_j corresponding to its nearest centroid c_j
7:         Compute a new set of centroids to update C
8:     **end for**
9: **end while**

K-Means++ Seeding. The K-means++ is one of the most used strategies for seeding that distribute the initial centers across the space covered by dataset elements. The K-means++ first choose an initial center from points, then compute the square distances between all points and select the next random centroid with probability $\frac{d_{min}(u)^2}{\sum_{v \in X} d_{min}(v)^2}$ [13]. This procedure can be defined as in Algorithm 3.

Algorithm 3. *K-means++ seeding Algorithm*

Input: Data X, number of centers k
Output: Set of center C
1: $C \leftarrow \{c_1\}$, c_i is selected randomly from X
2: **while** $|C| < k$ **do**
3:     **for all** $u \in X$ **do**
4:         Select a new center c_i from X with probability $\frac{d_{min}(u)^2}{\sum_{v \in X} d_{min}(v)^2}$
5:         $C \leftarrow C \cup \{c_i\}$
6:     **end for**
7: **end while**

---

4   Experiments and Results

In this section, we first describe the databases where the initialization methods were tested, secondly, how the experimentation is carried out, then the metrics considered to measure quality, and finally the results of the evaluation of the quality of clustering and image segmentation.

4.1   Databases

The four initialization methods of FCM was tested on 300 images selected of the following databases: Sky database, a collection of images for sky segmentation [14], Weizmann Segmentation Evaluation Database images [15], ISIC 2019 database, contains dermoscopic images [16], Berkeley Segmentation Data Set 500 (BSDS500) [17] and Microsoft Research Cambridge Object Recognition Image Database [18], just to depict, in Fig. 1 some of the images of these databases are shown, in which, we put the name of the image and the number of regions established to segment each image is suggested through parameter c.

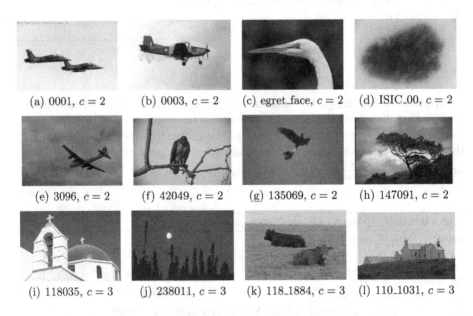

(a) 0001, $c = 2$ (b) 0003, $c = 2$ (c) egret_face, $c = 2$ (d) ISIC_00, $c = 2$

(e) 3096, $c = 2$ (f) 42049, $c = 2$ (g) 135069, $c = 2$ (h) 147091, $c = 2$

(i) 118035, $c = 3$ (j) 238011, $c = 3$ (k) 118_1884, $c = 3$ (l) 110_1031, $c = 3$

Fig. 1. Real images for the experimentation.

4.2   Experiments

The purpose of the considered experiment is to segment the test images with each of initialization methods to FCM to find the better performance. The comparison of these initialization methods was evaluated to some metrics to measure the quality of the clustering and others to measure the quality of the Image Segmentation. Moreover, the number of iterations and runtime are also taken into account, to observe which initialization method is faster. The four methods were implemented in the MATLAB R2018b environment, in addition to number of cluster c, we considered the parameters $m = 2$, and $\varepsilon = 0.0001$.

4.3 Quality Metrics

We describe the four considered metrics to measure the quality of the clustering:

- Sum of Squared Errors, calculates the sum squared distance of all objects to the centroids of their respective clusters, this is usually is using to validate the initialization methods [19].

$$SSE = \sum_{k=1}^{K} \sum_{\forall x_i \in G_k} (x_i - c_k)^2 \tag{4}$$

  Where G_k is the $k - th$ cluster and c_k is the centroid of the group.
- Partition Coefficient (PC) measures the amount of overlapping between cluster [20]

$$PC = \frac{1}{N} \sum_{i=1}^{c} \sum_{j=1}^{N} (\mu_{ij})^2 \tag{5}$$

  where μ_{ij} is the membership of data point j in cluster i. The optimal number of cluster is at the maximum value.
- Adjusted Rand index, calculate a similarity measure between two clusterings by considering all pairs of samples and counting pairs that are assigned in the same or different clusters in the predicted and true clusterings [21,22].

$$ARI = \frac{RI - E[RI]}{max(RI) - E[RI]} \tag{6}$$

Where RI is the non-adjusted version of the Rand Index and is described by the following formulation:

$$RI = \frac{a + b}{C_2^n} \tag{7}$$

This score takes values between 0 and 1, and higher values indicate better clusters in terms of ARI than lower ones.

On the other hand, the metrics used to measure the quality of the Image Segmentation were [23]:

- Accuracy, to measure the quality of the clustering.

$$Accuracy = \frac{TP + TN}{TP + TN + FP + FN} \tag{8}$$

- DICE similarity coefficient (DSC), quantify the overlap between segmentation results with the ground truth.

$$DSC = 2 \bullet \frac{\text{Area}(X \cap Y)}{\text{Area}(X) + \text{Area}(Y)} \tag{9}$$

- Intersection over union (IOU), shows if the classes are well classified, take values in $[0, 1]$, with a value of 1 indicate a perfect segmentation.

$$IOU = \frac{TP}{TP + FP + FN} \tag{10}$$

Where TP are the true positives, TN the true negatives, FP the false positives and FN the false negatives. X and Y represent the ground truth and segmented images, respectively.

4.4   Evaluation on Clustering Quality

This paper aims to compare different methods to initialize clustering algorithm, specifically Fuzzy C-means. One of the experiments to determine which of the four methods has the better performance, is compare the number of iterations and the runtime of each one. Therefore, Table 1 shows the execution average performance of the four initialization methods, demonstrated that cluster seeding has the shortest execution performance, with less number of iterations, 11; this is because it performs a clustering phase on a sub sample of 10% of the data, that is, it reduces the images to 10% to find the initial centroids, while the other methods use the original image. In second place is K-means ++, with 14 iterations, approximately 0.09 s slower than cluster seeding.

Table 1. Execution average performance.

Method	Iterations	Runtime(s)
Uniform	15	0.453
Sample	17	0.518
Cluster	**11**	**0.349**
K-means++	14	0.443

Furthermore, to observe the evolution of the four methods based on the number of iterations, the minimization of the quadratic error was calculated through its objective function and the result was plotted in terms of the number of iterations, as shown in Fig. 2 for sample image 238011. In order to standardize the graphs, the iterations were plotted from 0 to 20, it is important to emphasize that uniform seeding requires more than 20 iterations to converge, specifically 39, while K-means++ only needs 4 iterations. As can be seen in the figure, apparently there is not a change significantly the quantification of the objective function from the eighth iteration, with this in mind, is this possible that these iterations allow improve fine detail in segmentation

Finally, the average value of each metric considered to measure the clustering quality is presented in Table 2, undoubtedly, there is a significant reduction of final SSE compared to initial, and the K-means++ seeding has the best average score. It should be noted that this method also has the best PC score, while cluster seeding has the second place. On the other hand, cluster seeding has the best ARI average score, followed by K-means++ seeding. To summarize, these quantitative results suggest that K-means++ seeding has the best performance in terms of clustering quality for two of the three metrics considered, the superiority is due to the fact that this technique selects the initial centroids by means of a probability function, and the others do so randomly or with subsamples.

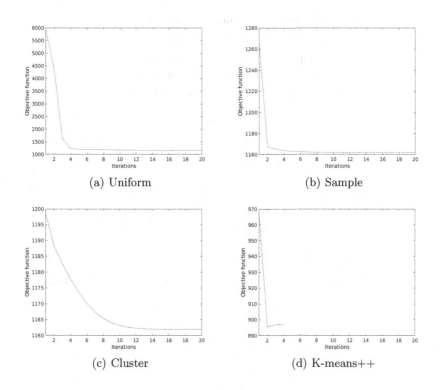

Fig. 2. Objective function vs iteration in image 238011.

Table 2. Evaluation average on clustering quality.

Metrics	Initial SSE	Final SSE	PC	ARI
Uniform	2.4E+06	4.176E+03	0.902	0.706
Sample	7.9E+09	4.286E+03	0.893	0.705
Cluster	**1.0E+06**	4.235E+03	0.907	**0.763**
K-means++	5.9E+08	**4.154E+03**	**0.910**	0.723

4.5   Quantitative Evaluation on Segmentation Results

The quantitative average performance of the experiments is presented in Table 3, specifically the K-means++ method obtains a better result in terms of segmentation in two of the metrics considered (Accuracy and DICE coefficient).

Table 3. Evaluation average on clustering quality.

Metrics	Accuracy	Dice	IOU
Uniform	0.774	0.725	0.767
Sample	0.790	0.789	0.757
Cluster	0.825	0.809	**0.813**
K-means++	**0.836**	**0.832**	0.800

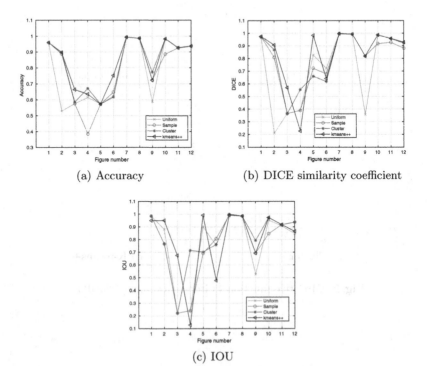

(a) Accuracy   (b) DICE similarity coefficient

(c) IOU

Fig. 3. Graphical results of the segmentation performance.

Fig. 4. Samples of segmentation results.

While the best average score for IOU is cluster seeding. These results show that K-means++ has better results in two of the metrics used due to the selection of its initial centroids, which unlike the other methods uses a probability function and is not random. To visualize the results, Fig. 3, just for depict, shows the behavior of these metrics in only 12 test images.

4.6 Qualitative Evaluation on Segmentation Results

Due to space limitation, here we only show quantitative results for three sample image segmented through Fuzzy C-means using the four initialization methods, uniform, sample, cluster and K-means++, are illustrated in Fig. 4. It is noteworthy that segmented regions obtained by initialization with K-means++ produces better segmentation results compared to the other initialization methods. In other words, K-means++ to successfully segment the images in the clusters determined by the parameter c, while the other initialization methods have trouble assigning some pixels to the correct cluster. This is corroborated by comparing it with the ground truth, in addition to reinforcing the quantitative results presented previously.

5   Conclusions

In this paper, a comparison of different initialization methods was made to improve the sensitiveness of FCM algorithm when choosing the initial cluster centers as they are important on the segmentation quality. Stated another way, they can achieve a better segmentation result, because of a better initialization condition provided to the FCM algorithm. The experimental results conclude that the K-means++ initialization method produce a better initialization condition for FCM algorithm than sample and uniform initialization methods, by successfully reducing the execution time and producing more homogeneous regions in the segmentation results. This is reinforced with the quantitative results, demonstrated that K-means++ has a superior result in the considered metrics.

As future work, we intend to incorporate other features, such as texture, to improve segmentation. In addition, testing the initialization methods in different color spaces, even extending it to intuitionistic or hesitant fuzzy set theory.

Acknowledgment. The authors of this work express their gratitude to CONACYT, as well as to the Tecnológico Nacional de México/CENIDET for financing through the project "Controlador Difuso para ajuste de coeficientes de rigidez de un modelo deformable para simulación en tiempo real de los tejidos del hígado humano".

References

1. De Oliveira, J.V., Pedrycz, W.: Advances in Fuzzy Clustering and its Applications. Wiley, Hoboken (2007)
2. Kim, D.W., Lee, K.H., Lee, D.: A novel initialization scheme for the fuzzy c-means algorithm for color clustering. Pattern Recogn. Lett. **25**, 227–237 (2004)
3. Tan, K.S., Lim, W.H., Isa, N.A.M.: Novel initialization scheme for fuzzy c-means algorithm on color image segmentation. Appl. Soft Comput. **13**, 1832–1852 (2013)
4. Stetco, A., Zeng, X.J., Keane, J.: Fuzzy c-means++: fuzzy c-means with effective seeding initialization. Expert Syst. Appl. **42**, 7541–7548 (2015)
5. Ma, Y., Cheng, W.: Optimization and parallelization of fuzzy clustering algorithm based on the improved kmeans++ clustering. MS&E **768**, 072106 (2020)
6. McCallum, A., Nigam, K., Ungar, L.H.: Efficient clustering of high-dimensional data sets with application to reference matching. In: Proceedings of the Sixth ACM SIGKDD International Conference on Knowledge Discovery and Data Mining, pp. 169–178 (2000)
7. Meila, M., Heckerman, D.: An experimental comparison of several clustering and initialization methods. arXiv preprint arXiv:1301.7401 (2013)
8. Khan, S.S., Ahmad, A.: Cluster center initialization algorithm for k-means clustering. Pattern Recogn. Lett. **25**, 1293–1302 (2004)
9. Steinley, D., Brusco, M.J.: Initializing k-means batch clustering: a critical evaluation of several techniques. J. Classif. **24**, 99–121 (2007)
10. Celebi, M.E., Kingravi, H.A., Vela, P.A.: A comparative study of efficient initialization methods for the k-means clustering algorithm. Expert Syst. Appl. **40**, 200–210 (2013)
11. Bezdek, J.C., Keller, J., Krisnapuram, R., Pal, N.: Fuzzy Models and Algorithms for Pattern Recognition and Image Processing, vol. 4. Springer, Heidelberg (1999). https://doi.org/10.1007/b106267
12. Gan, G., Ma, C., Wu, J.: Data Clustering: Theory, Algorithms, and Applications. SIAM, USA (2007)
13. Arthur, D., Vassilvitskii, S.: K-means++: the advantages of careful seeding. Technical report, Stanford (2006)
14. Alexandre, E.B.: IFT-SLIC: geração de superpixels com base em agrupamento iterativo linear simples e transformada imagem-floresta. Ph.D. thesis, Universidade de São Paulo (2017)
15. Winn, J., Criminisi, A., Minka, T.: Object categorization by learned universal visual dictionary. In: Tenth IEEE International Conference on Computer Vision (ICCV 2005) Volume 1, vol. 2, pp. 1800–1807. IEEE (2005)
16. Combalia, M., et al.: Bcn20000: dermoscopic lesions in the wild. arXiv preprint arXiv:1908.02288 (2019)
17. Martin, D., Fowlkes, C., Tal, D., Malik, J.: A database of human segmented natural images and its application to evaluating segmentation algorithms and measuring ecological statistics. In: Proceedings of 8th International Conference on Computer Vision, vol. 2, pp. 416–423 (2001)
18. Criminisi, A., et al.: Microsoft research Cambridge object recognition image database (2004)
19. Ortiz-Bejar, J., Tellez, E.S., Graff, M., Ortiz-Bejar, J., Jacobo, J.C., Zamora-Mendez, A.: Performance analysis of k-means seeding algorithms. In: 2019 IEEE International Autumn Meeting on Power, Electronics and Computing (ROPEC), pp. 1–6. IEEE (2019)

20. Bezdek, J.: Pattern recognition with fuzzy objective function algorithms (1981)
21. Hubert, L., Arabie, P.: Comparing partitions. J. Classif. **2**, 193–218 (1985)
22. Santos, J.M., Embrechts, M.: On the use of the adjusted rand index as a metric for evaluating supervised classification. In: Alippi, C., Polycarpou, M., Panayiotou, C., Ellinas, G. (eds.) ICANN 2009. LNCS, vol. 5769, pp. 175–184. Springer, Heidelberg (2009). https://doi.org/10.1007/978-3-642-04277-5_18
23. Ling, C., Diyana, W.M., Zaki, W., Hussain, A., Hamid, H.A.: Semi-automated vertebral segmentation of human spine in MRI images. In: 2016 International Conference on Advances in Electrical, Electronic and Systems Engineering (ICAEES), pp. 120–124. IEEE (2016)

SPEdu: A Toolbox for Processing Digitized Historical Documents

Fabio Gomes Rocha[1,2] and Guillermo Rodriguez[3(✉)]

[1] Universidade Tiradentes, Aracaju, Sergipe, Brazil
[2] Instituto de Tecnologia e Pesquisa - ITP, Aracaju, Sergipe, Brazil
fabio_gomes@email.unibr.br, gomesrocha@gmail.com
[3] ISISTAN (UNICEN-CONICET) Research Institute, Tandil,
Buenos Aires, Argentina
guillermo.rodriguez@isistan.unicen.edu.ar,
https://www.itp.org.br/pesquisa/laboratorios/LACIA

Abstract. Historical-educational documentary sources have gained considerable attention in educational contexts. However, some sources suffer from serious problems such as inadequate infrastructure, poor preservation, and lack of qualified personnel. In addition, a large part of documents is not digitilized, making research difficult. As a consequence, there is a need for transcription, digitalization, and cataloging sources of information for the analysis of large volumes of data. To deal with this issue, we present SPEdu, a tool to digitalize sources of information demanded by research on the History of Education. The workflow of SPEdu is divided into three steps. Firstly, SPEdu adquires images from an information source. Secondly, the tool prepocesses the images and extracts features from them. Finally, a supervised machine learning module was built to classify images between text and non-text. To assess the viability of SPEdu, we used the Official Gazette of the State of Sergipe. Regarding the third step, we evaluated the performance of classification algorithms, such as J48, Logistic Regression, Multi-layered Perceptron (MLP), Naive Bayes, Random Forest, and Random Tree. Results have revealed that Random Forest outperformed remaining techniques with an average rate of 95% of accuracy.

Keywords: History of education · Supervised machine learning · Information retrieval · Image processing · Digitalization of documents

1 Introduction

Printed information sources represent a complex challenge for information extraction [1]. This is because these sources contain several page *layouts* with multiple articles, in which they are designed to allow people to define their own reading order. Paragraphs and images are distributed over several pages in an unpredictable way, making the extraction of data from printed information sources (e.g. gazettes or newspapers) a daunting task. According to [2], the processing of data from gazettes is a task with a high degree of difficulty, since they

© Springer Nature Switzerland AG 2020
L. Martínez-Villaseñor et al. (Eds.): MICAI 2020, LNAI 12469, pp. 363–375, 2020.
https://doi.org/10.1007/978-3-030-60887-3_32

are printed on low quality paper, which tends to change color over time. Such changes generate noise that increases with the time of existence of the document.

In this sense, Rajeswari and Magapu state that in the digital age, the conversion of printed documents to electronic form has become a necessity for the availability of information [3]. Nonetheless, the authors also state that in order to allow documents to be found, it is necessary that their metadata be extracted through optical character recognition (OCR) tools. OCR is a technology that allows for converting different types of documents such as *Portable Document Format* (PDF) files, images captured in digital cameras and documents scanned into an editable and searchable format [3].

To extract data through OCR, Vasilopoulos and Kavallieratou state that it is necessary to employ methods that combine document layout analysis with text detection [4]. The first step in the process occurs with the digitization of the document. This can be done by means of photographs or scanning of sources. However, the scanning process aims to turn the printed or handwritten document into a digital document, but this fails to allow the document to be searched automatically. In this way, the searcher will remain dependent on transcription, cataloguing and indexing of documents.

Kaur and Jindal indicate that in order to be able to recognize and detect the texts, it is necessary to pre-process the images, removing noises [5]. Since historical documents are complex sources, after the preprocessing it is necessary to analyze the page layout. This will allow proper recognition of the texts. Layout analysis then aims to recognize the distinction between regions that are textual and those that are not, then making it possible to extract the texts [5].

To deal with layout analysis, we present SPEdu, a tool to digitalize historical sources of information in Portuguese. The workflow of SPEdu is divided into three steps. Firstly, SPEdu adquires images from an information source. Secondly, the tool prepocesses the images and extracts features from them. Finally, a classification model is trained to determine whether an image is text or not.

To evaluate SPEdu, we used as a case-study the gazette "Diário Oficial do Estado de Sergipe", published in Brazil, which has more than 100 years of paper edition. This gazette contemplates the political history of the region, since all governmental acts must be published in it. Examples are the opening of schools and possessions of government secretaries, which can subsidize researchers in the field of education. After training classification algorithms, RandomForest obtained best results in terms of accuracy.

The remainder of the paper is organized as follows. Section 2 describes the related work. Section 3 introduces our approach for image processing and describes feature extraction techniques. Section 4 evaluates classification strategies and discusses the results. Finally, Sect. 5 concludes the article and identifies future research lines.

2   Related Work

Several authors have been studying techniques that allow for layout analysis, using machine learning in printed sources, mainly newspapers. Zeni and Weldermarian [1] performed layout analysis in an experiment with one hundred newspapers, using the decision tree algorithm and obtaining an average accuracy of 84%. Bukhari et al. [6] used the Multilayer Perceptron classifier (MLP) for text and non-text separation. The work, however, failed to obtain acceptable precision. Palfray et al. [7] used forty-two images from the Rouen Journal, in old French, for an experiment, obtaining a precision of 85.84%.

Another work that uses historical sources is that of Hebertet et al. [8], with the PlaIR system, performing an experiment in two books, in French, obtaining as a result the accuracy between 77.07 and 87.61%. Pramanik and Bag [9] obtained an overall accuracy of 88.74%, the MLP obtained accuracy of 88.74%, the Support Vector Machine (SVM) obtained an accuracy of 86.45% and the Random Forest an accuracy of 86.17% for document classification in Bengla (one of the Indian languages). Researchers Chathuranga and Ranathunga [10] proposed an algorithm for content extraction from old newspapers. In the test with forty-four images, they obtained an accuracy of 69.79%. In the work of Vasilopoulos [11], using seventy-four scanned pages of newspapers, in Arabic, with a layout analysis algorithm prepared by the researcher, an accuracy of 90.4% was obtained.

3   Image Processing and Feature Extraction Techniques

To work with historical documents, a cycle involving acquisition, pre-processing and image classification is adopted, as shown in Fig. 1. When using scanners, considerable care is taken in the handling of the document, preventing it from being damaged.

The images used in this work were captured in a hand scanner, portable NIP-A4 model, by Marc Nipponic. This device was chosen due to the fact that old gazettes are highly fragile. Thus, it would not be possible to use a desk scanner for the task. Mobile phones and cameras were also used for the photographs of the gazettes. However, in the detection process, the quality obtained with those devices was lower than the one obtained with the scanner.

An image can present several problems, such as noise, low contrast, inadequate inclination, and elements that make it difficult to recognize characteristics, among others. Thus, in order to carry out the detection and recognition of texts, the second step of the proposed model was carried out: the conversion of the image into gray tones. With this, the color image, which has three color channels, is converted into an image with a single grayscale channel, reducing the amount of information regarding the image. The result of this procedure is presented in Fig. 2.

The grayscale image is generated after a conversion performed from the weighted sum of the color channels, considering the capacity of the human eye to absorb the light emitted by each color. Thus, the representation of the image through luminance produces a grayscale image, where each color (RGB) is associated with a luminance value, given by the Eq. 1:

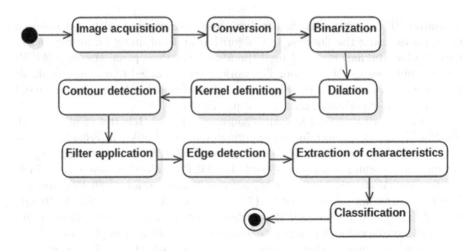

Fig. 1. Acquisition, pre-processing and classification cycle.

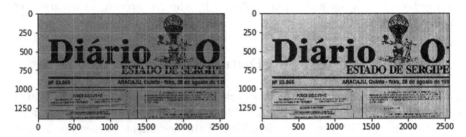

Fig. 2. Grayscale image conversion.

Equation 1 - Conversion of color image to grayscale

$$Gray\,Image = (red\,image * 0.299) + (green\,image * 0.587) + (blue\,image * 0.114)$$

$$(1)$$

Source: Based on Mello, Santos and Oliveira [12]).

The gray image is the result of the luminance value, which was found by means of Eq. 1, based on the color components R (red), G (green) and B (blue). As a result, the generated image employs levels of gray ranging from 0 (absolute black) to 255 (absolute white) [12]. As the image had three color channels and was converted to a channel with only grayscale, the final size of the image is reduced, implying the amount of information available for the computer to process.

The next step in image processing is binarization (threshold) [3], which can be categorized as global and local [13]. Binarization is a technique that allows for segmenting the objects of the background. This process transforms the image that is in shades of gray into a binary image, which has only two colors - black (represented by 0) and white (represented by 1), removing the other colors. There is also a form of threshold based on a single limit, called "Otsu binarization"

[14]. An alternative is the adaptive threshold proposed by Sauvola et al. [15]. In this work, we used binarization to remove the yellowish background, which makes difficult the analysis and classification of the objects in the image. This process is presented in Fig. 3.

Fig. 3. Grey to black and white image conversion. (Color figure online)

However, in order to mark the text it is necessary to apply the inversion of the binarization, where the white will be black and the black will be white, as shown in Fig. 4. This will allow for marking each white board on the black background. It is important to remember that the original picture was kept, being processed a copy of the image, automatically generated by the tool.

Fig. 4. Inverted image binarization.

Next, a morphological operation was performed, which is a transformation applied to a binary or gray scale image. Operations can be performed to expand or reduce the size of objects in the image, close gaps, among other things. For text detection, the morphological operation is used to gather the texts into a single group. In this way, texts can be detected in blocks. There are several morphological operations. In this work, the dilation operation was adopted. For this, a structuring element such a matrix was defined.

According to [16], the dilation operation allows *pixels to* expand, resulting in the union of gaps. In the operation performed in this work, the size of the optimal matrix was 51×51. In the case of texts, the tendency is to unify in a block, as shown in Fig. 5.

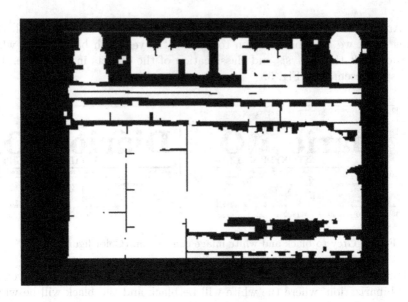

Fig. 5. Dilated image.

In order to reduce the noise generated by the image, after applying the dilation, the median filter was used. A filter is a way of processing the *pixels* of an image, generating a new output image through a *kernel*. The *kernel* is a small matrix, which can be used to blur the image, improve sharpness, and detect edges, among other features. Basically, the *kernel* is scrolled over the image, from left to right and from top to bottom, allowing mathematical operations to be performed on the original image and creating a new image as a result.

According to Frery in Melo [12], the median filter has as output one *pixel*, which, in turn, has as value the median of the observations. Although it has a similar effect to the median filter, the median filter has as a positive point the preservation of the edges and contours of an object. Considering this aspect, this filter was adopted to reduce some points of the image, keeping the edges intended to be detected. In the evaluation, the *kernel* size defined was 25. The result obtained is presented in Fig. 6.

To identify the objects, OpenCV contour detection was used, which is based on the algorithm proposed by Suzuki and Abe [17]. As a result, the detection located contours of objects that are different from the background. Thus, an outline is a representation of the boundary of a shape.

The *findCountours* function in the OpenCV library detects the outline in an image. In order to select only the external contour in a picture, and not the contour within the contour, cv2.RETR_EXTERNAL was used. The parameter cv2.CHAIN_APPROX_SIMPLE was used for reducing the number of redundant points, leaving only the essential points of the detection. Thus, the result is the detection of an outline on the objects that can be marked in an image, as shown in Fig. 7, being one of the steps to search for data in historical sources.

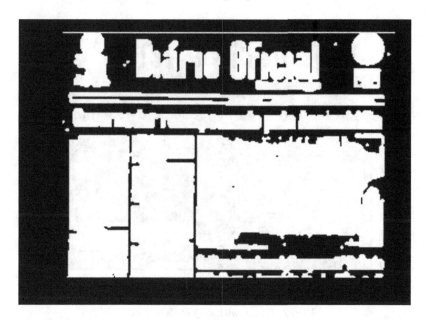

Fig. 6. Binary image after applying the median filter over the dilated image.

To conduct the third step of SPEdu, it is necessary to extract some features related to the detected outlines. Then, these features are used to build classification models to identify whether a extracted image from the second step is a text or a non-text. In this work, information was extracted from 100 issues of the Official Gazette of the State of Sergipe, totaling 1,100 pages, in order to characterize the detected object, test and train the supervised machine learning model. Thus, the following features from extracted items were identified:

(a) Ratio: is the result of the image width divided by the image height. Thus, if the value obtained as a result is less than 1, the object or figure has a height greater than the width. If the ratio is greater than 1, the object has a width greater than the height and, if it is equal to 1, the object is a square. In the work of [18], the proportion is used to reduce the research space. In our work, the proportion is one of the features extracted in the detection. According to Kumar, Sailaja and Begum [19], the proportion is the ratio between the width and height of an image, according to Eq. 2.

Equation 2 - Proportion of an image

$$Proportion = \frac{image\,width}{image\,height} \tag{2}$$

Source: Adapted formula from Kumar, Sailaja and Begum [19]).

Fig. 7. Image with object detection.

b) Extension: is the area of the shape divided by the bounding box area of the object. As a result, every extension is smaller than 1, since the object must always be smaller than the figure as a whole, that is, the extension is the area covered by something [19], according to the Eq. 3.

Equation 3 - Extension of an object

$$Extension = \frac{object\ area}{rectangle\ area} \tag{3}$$

Source: Adapted formula from Kumar, Sailaja and Begum (2019).

c) Convex hull: is used to check the curve for convexity defects and correct it [19]. Thus, given a set of points in Euclidean space, the convex hull is the smallest possible set that contains these points, as shown in Fig. 8.
Thus, the space in the image is smaller than the space used by the convex hull to delimit the object.
d) Solidity: is the quality or state of being firm or strong in relation to its structure [19], being the result of the division of the contour area by the area of the convex hull, according to Eq. 4.

Equation 4 - Solidity of an object

$$Solidity = \frac{boundary\ area}{convex\ hull\ area} \tag{4}$$

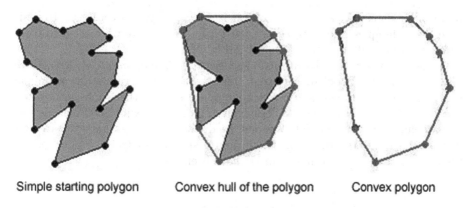

Simple starting polygon   Convex hull of the polygon   Convex polygon

Fig. 8. Application of the convex hull.

e) Area of the object: represents the number of pixels within the outline, i.e., the total of points on an object (Eq. 5).

Equation 5 - Area of an object

$$Area = object\ width * object\ height \qquad (5)$$

Finally, the width and height of the objects detected were also used in isolation.

4   Evaluating Classification Strategies

To classify the items, it is required to perform the layout analysis. Thus, based on the extracted data, an information base was created to train the classification model. This would allow SPEdu to automatically detect in subsequents images whether the items in a document were text, pictures or just noise. In this process, 5,000 pieces of information related to ten issues of the gazette surveyed, with the features listed above, were used.

In this work, the hypotheses are:

- H1: the object is an image;
- H2: the object is a text;
- H3: the object is a noise;

Each one of these items can accept only one hypothesis and rejects the others. This is a useful technique in layout analysis, aiming to validate if the detected items are text, images, etc.

To corroborate the aforementioned hyphoteses, the following machine learning techniques were tested: NaiveBayes, MLP, Logistic Regression, Tree, Random Tree and Random Forest. We aimed to determine which of these techniques best perform for detecting whether the object is a text, a figure, or a noise. To evaluate the techniques, the features extracted from the gazette were used, to train and validate the algorithms, as shown in Fig. 9.

In order to test the algorithms, the *"Waikato Environment for Knowledge Analysis"* (WEKA) tool was utilized. The "k-fold" cross validation technique was adopted to test the model, which is applied so that the whole database is used in training and testing [20]. Thus, given a database with 2000 records and being defined k = 10, the database will be divided into 10 subsets, where each one will have 200 records each, being used 9 subsets for training and one for testing, being rotated, until all subsets have been used [20].

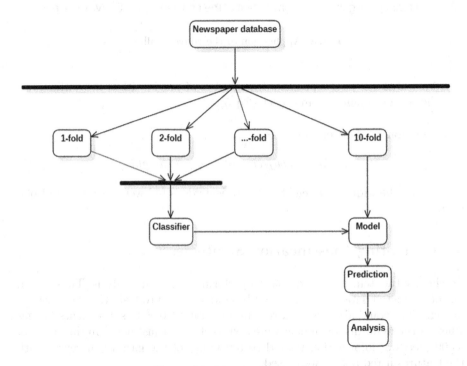

Fig. 9. Machine learning cycle.

In this way, the database of features extracted from the gazette was divided into *folds*. Thirty tests were then performed with each algorithm, using values from 1 to 30 as the variation. This allowed the analysis of the average of the following machine learning techniques: J48 (Tree), *Logistic* (Logistic Regression), Multi-layered Perceptron (MLP), *Naive Bayes, Random Forest and Random Tree*, resulting in 180 tests. As a result, the *Random Forest* algorithm presented the highest average of accuracy, as can be seen in Fig. 10.

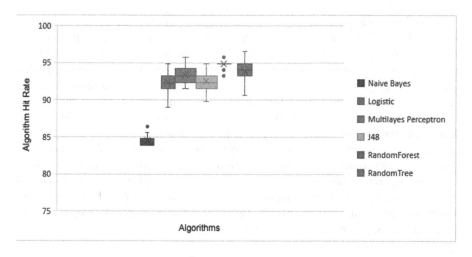

Fig. 10. Comparison of machine learning techniques in terms of accuracy.

Table 1 shows the average of accuracy for each of the tested algorithms. The *Random Forest* algorithm is said to have performed better, with an average 95% for classifying items such as text, images, and noise.

Table 1. Accuracy of machine learning algorithms

Algorithms	NaiveBayes	Logistic	MLP	J48	RandomForest	RandomTree
Average	85	92	93	93	95	94

Based on average performance, *RandomForest* obtained the best result in layout analysis. The *RandomForest* method emerged in 1995 in Kan's work presented at the *International Conference on Document Analysis and Recognition*. This method is a combination of tree classifiers, which is, according to [21], a knowledge representation model that uses nodes to represent decisions. According to [22], the tree is a predictive method that, by means of rules, uses a path in a tree format, which is used for classification. Thus, a *Random Forest* combines several tree type classifiers, being a more robust algorithm in relation to noise, but still considered a light algorithm [23,24].

5   Conclusion and Future Work

In this article, we have presented SPEdu, a tool to digitalize historical sources of information. The tool extracts information from the digital version of a source, extracts features from the images, and finally, classifies the images as text and non-text. To assess the classification module of SPEdu, we analyzed several supervised learning techniques using the gazette "Diário Oficial do Estado de Sergipe (Brazil)" as case-study. For the analysis, features such as proportion, extension, convex hull, strength, and area of the object were extracted to built the classification models using J48 (Tree), *Logistic* (Logistic Regression), Multilayered Perceptron (MLP), *Naive Bayes, Random Forest* and *Random Tree*. To sum up, Random Forest outperformed the remaining algorithms in the task of classification of text and non-text, with an average accuracy of 95%. Moreover, Random Forest showed little discrepancy in the results, with few errors, proving to be a more efficient option compared to the other algorithms. We also demonstrated that our approach surpassed state-of-the-art comparable approaches.

As future work, we are planning to evaluate SPEdu with more historical sources of information. Furthermore, we intend to extend our tool to incorporate historical sources in English. Finally, we would like to explore Deep Learning algorithms to improve the classification task.

References

1. Zeni, M., Weldemariam, K.: Extracting information from newspaper archives in Africa. IBM J. Res. Dev. **61**(6), 12:1 (2017)
2. Jana, S., Das, N., Sarkar, R., Nasipuri, M.: Recognition system to separate text graphics from Indian newspaper. In: Kar, S., Maulik, U., Li, X. (eds.) FOTA 2016. SPMS, vol. 225, pp. 185–194. Springer, Singapore (2018). https://doi.org/10.1007/978-981-10-7814-9_14
3. Rajeswari, S., Magapu, S.B.: Development and customization of in-house developed OCR and its evaluation. Electron. Libr. (2018)
4. Vasilopoulos, N., Kavallieratou, E.: Complex layout analysis based on contour classification and morphological operations. Eng. Appl. Artif. Intell. **65**, 220–229 (2017)
5. Kaur, R.P., Jindal, M.K.: Headline and column segmentation in printed Gurumukhi script newspapers. In: Panigrahi, B.K., Trivedi, M.C., Mishra, K.K., Tiwari, S., Singh, P.K. (eds.) Smart Innovations in Communication and Computational Sciences. AISC, vol. 670, pp. 59–67. Springer, Singapore (2019). https://doi.org/10.1007/978-981-10-8971-8_6
6. Bukhari, S.S., Al Azawi, M.I.A., Shafait, F., Breuel, T.M.: Document image segmentation using discriminative learning over connected components. In: Proceedings of the 9th IAPR International Workshop on Document Analysis Systems, pp. 183–190 (2010)
7. Palfray, T., Hebert, D., Nicolas, S., Tranouez, P., Paquet, T.: Logical segmentation for article extraction in digitized old newspapers. In: Proceedings of the 2012 ACM Symposium on Document Engineering, pp. 129–132 (2012)
8. Hebert, D., Palfray, T., Nicolas, S., Tranouez, P., Paquet, T.: PIVAJ: displaying and augmenting digitized newspapers on the web experimental feedback from the "Journal de Rouen" collection. In: Proceedings of the First International Conference on Digital Access to Textual Cultural Heritage, pp. 173–178 (2014)
9. Pramanik, R., Bag, S.: Shape decomposition-based handwritten compound character recognition for Bangla OCR. J. Vis. Commun. Image Represent. **50**, 123–134 (2018)

10. Chathuranga, R.S., Ranathunga, L.: Procedural approach for content segmentation of old newspaper pages. In: 2017 IEEE International Conference on Industrial and Information Systems (ICIIS), pp. 1–6. IEEE (2017)

11. Vasilopoulos, N., Wasfi, Y., Kavallieratou, E.: Automatic text extraction from Arabic newspapers. In: Campilho, A., Karray, F., ter Haar Romeny, B. (eds.) ICIAR 2018. LNCS, vol. 10882, pp. 505–510. Springer, Cham (2018). https://doi.org/10.1007/978-3-319-93000-8_57

12. De Mello, C.A.B., de Oliveira, A.L.I., Dos Santos, W.P.: Digital Document Analysis and Processing. Nova Science Publishers, Hauppauge (2012)

13. Gllavata, J., Ewerth, R., Freisleben, B.: A robust algorithm for text detection in images. In: Proceedings of the 3rd International Symposium on Image and Signal Processing and Analysis, 2003, ISPA 2003, vol. 2, pp. 611–616. IEEE (2003)

14. Otsu, N.: A threshold selection method from gray-level histograms. IEEE Trans. Syst. Man Cybern. 9(1), 62–66 (1979)

15. Sauvola, J., Seppanen, T., Haapakoski, S., Pietikainen, M.: Adaptive document binarization. In: Proceedings of the Fourth International Conference on Document Analysis and Recognition, vol. 1, pp. 147–152. IEEE (1997)

16. Gonzalez, R.C., Woods, R.E.: Digital Image Processing. Pearson, London (2018)

17. Suzuki, S., et al.: Topological structural analysis of digitized binary images by border following. Comput. Vis. Graph. Image Process. 30(1), 32–46 (1985)

18. Quddus, A., Cheikh, F.A., Gabbouj, M.: Wavelet-based multi-level object retrieval in contour images. In: Proceedings of the International Workshop on Very Low Bit Rate Video Coding, pp. 1–5 (1999)

19. Ramesh Kumar, P., Sailaja, K.L., Mehatab Begum, S.: Human identification based on ear image contour and its properties. In: Pandian, D., Fernando, X., Baig, Z., Shi, F. (eds.) ISMAC 2018. LNCVB, vol. 30, pp. 1527–1536. Springer, Cham (2019). https://doi.org/10.1007/978-3-030-00665-5_143

20. Duchesne, P., Rémillard, B.: Statistical Modeling and Analysis for Complex Data Problems, vol. 1. Springer, Heidelberg (2005). https://doi.org/10.1007/b105993

21. Goldschmidt, R., Passos, E., Bezerra, E.: Data Mining. Elsevier, Brazil (2015)

22. Baeza-Yates, R., Ribeiro-Neto, B.: Recuperação de Informação-: Conceitos e Tecnologia das Máquinas de Busca. Bookman Editora (2013)

23. Gislason, P.O., Benediktsson, J.A., Sveinsson, J.R.: Random forests for land cover classification. Pattern Recogn. Lett. 27(4), 294–300 (2006)

24. Breiman, L.: Random forests. Mach. Learn. 45(1), 5–32 (2001)

Road Signs Segmentation Through Mobile Laser Scanner and Imagery

K. L. Flores-Rodríguez[1]([⊠]), J. J. González-Barbosa[1],
F. J. Ornelas-Rodríguez[1], J. B. Hurtado-Ramos[1], and P. A. Ramirez-Pedraza[2]

[1] Instituto Politécnico Nacional, CICATA-QRO, Qro., Santiago de Quertaro, Mexico
kfloresr1800@alumno.ipn.mx, {jgonzalezba,fornelasr,jbautistah}@ipn.mx
[2] Centro de Investigación en Óptica A. C., Cátedra CONACYT, Ciudad de Mxico,
Mexico
pedro.ramirez@conacyt.mx
http://www.cicataqro.ipn.mx
https://www.cio.mx/

Abstract. This work aims to present an urban segmentation to acquire road signs descriptions and annotations. The process implies geometrical characteristics from 3D points clouds (like dimensions, and shape), and visual characteristics from image data (like color wear, and damage) computation. We handle visual and spatial information of the road signs individually to fusion through GPS data in future work. The process for obtaining spatial information from 3D point clouds includes: (i) object segmentation through 3D point cloud density, (ii) use of the retro-reflective attribute of the material to differentiate possible road signs, (iii) plane orientation determination via singular value decomposition, (iv) 2D point cloud projection to geometric shape estimation. The process for getting visual information from images comprises: (i) color segmentation of the road signs in two-parts: border-color and inside-color, (ii) color identification using HSV color model (iii) geometric shape association via contour comparison, (iv) local features extraction and description from semantic data as numbers, characters, and drawings. We chose to work with low rise road signs because the sensors for mobile laser scanning has an elevation angle that delimits the acquisition. We select an experimentation ground truth from the KITTI data set to prove an adequate visual and spatial segmentation.

Keywords: 3D point clouds · Mobile laser scanning · Road signs · Visual features · Spatial feature · Local features

1 Introduction

Detection of urban furniture such as road signs using mobile laser scanning (MLS) combined with image information facilitates inventory, preservation, and maintenance tasks. The road signs condition evaluation is more suitable if it integrates image color data. Among some of the difficulties faced of the staff performing urban management tasks are: (i) insecurity in some places, (ii) unfavorable

© Springer Nature Switzerland AG 2020
L. Martínez-Villaseñor et al. (Eds.): MICAI 2020, LNAI 12469, pp. 376–389, 2020.
https://doi.org/10.1007/978-3-030-60887-3_33

weather, (iii) vehicular traffic and conglomeration avoiding task performing, (iv) signs hidden by vegetation, and (v) inaccessible sites to public passage.

There is notable progress in road signs detection through MLS and image data. The main of this progress is about improvement in road assistance and driving safety, [4,10,17] and [7]. The works like in [5,14,28,29] and [3], whose goal is to inspect and locate road signs to maps buildings. Most of them use the retro-reflective attribute of road signs to detect them. The use of visual features like color, 2D geometric shape, and 2D patterns from images is mainly for road signs appearance. Currently, the most popular classification method is deep neural networks (DNN). DNN allows getting a general detection of road signs without digging into the details. The areas of opportunity of these works are diverse if the goal is inventory, preservation, and maintenance tasks. Using the retro-reflective attribute of road signs is susceptible to dismiss worn signs. Lighting changes in urban environments influence the colors in images promoting confusion in the hue that must be detected. If the mobile laser scanner and image data acquisition is not appropriate geometric shape and pattern extraction can be affected; as a consequence, bad conditions road sings can be omitted.

This work presents a system with a segmentation procedure to enhance the inventory and examination of road signs based on LiDAR data, panoramic images ad GPS. The process implies compute geometrical characteristics from 3D points clouds (i. e. dimensions, and shape) and visual characteristics from images data (i. e. color wear, damage). We designed a meticulous segmentation process from both visual and spatial features to extract dimensions, material intensity, color, shape, and semantic data from the road signs condition.

We use different methods to get the geometrical features as dimensions, 3D shape, and orientation from the 3D point clouds. 3D point clouds is delimited respect to vehicle's field of view since they must perceive the relevant information of the road signs. The geometrical information computed from 3D point clouds summarizes the following steps: (i) object segmentation method to cluster 3D points (ii) differentiate between road signs and other objects using the retro-reflectivity attribute of the material; (iii) compute plane orientation via singular value decomposition (SVD) to do a 2D projection, and (iv) estimate geometric shape using 2D feature extraction process. In order to have more detail as qualitative features, spatial information requires visual information. The visual information stands in images from panoramic cameras and provides color, 2D geometric shape, and semantic information (numbers, characters, and drawings). Compute visual information from images is summarized in the following steps: (i) color segmentation using the hue, saturation, value color model (HSV) facing up lighting conditions; (ii) hue color identification of road signs through border-color and inside-color; (iii) 2D geometric association from the road signs to define the shape; (iv) semantic data extraction via numbers, characters, and drawings as local features extraction and description using the A-KAZE method.

In Sect. 2 is a brief review of works with a discussion about how other authors did the road signs segmentation and detection. Section 3 shows the methodology describing the tools for handle the data. Also, in Subsect. 3.1 we describe the segmentation process from the spatial information. And in Subsect. 3.2 is the segmentation process from the visual information. Further, in Sect. 4, we present the ground truth building from the KITTI database for experimentation and comparison purposes. Experiments and results are in Sect. 5. The last section in Sect. 6 is for the conclusion and discussion about the experiments and results. Also, we include some future proposals work.

2   State of the Art

There is a wide variety of approaches using MLS systems, [9,13,18,20,31]. The goal of these developments is to obtain photo-realistic maps for use in autonomous navigation and urban planning applications. The works in [6,22,26] and [3] detect various urban objects such as vehicles, trees, pedestrians, and road signs. Currently, the most popular method for classification is DNN. These methods allow the detection of urban objects in a general way without going into detail. In more specific tasks such as inventorying and maintenance of urban services, analysis of road signs in autonomous driving, the generation of maps, and the reconstruction of environments, it is necessary to consider more details of urban objects. For these detection tasks, researchers include MLS and cameras. Spatial features from MLS and visual features from cameras allow acquiring more information to make a more precise detection of objects.

Many works related to the detection of urban objects using 3D point clouds and color images begin with planes extraction for data reduction and subsequent object segmentation. Planes extraction remove the ground and buildings facades or structures which are considered irrelevant. Extraction methods such as *Support Vector Machine* (SVM), [27], *Principal Component Analisys* (PCA), [23], *Random Sample Consensus* (RANSAC), [12] use a core function to separate data, statistical processes, and mathematical approximation models. After planes extraction, the remaining points are clustered individually for later identification. The point clustering is known as object segmentation, and MeanShift [8] and K-Means [21] are the classic ones.

Urban object as the road signs have physical characteristics for safety and to avoid obstructing other activities. Road signs have specific standards to be easily and quickly identified by drivers and pedestrians. These standards include colors, shapes, and sizes for classifying into informative, restrictive, and preventive signs. Another feature of great importance is the painting of the signs that allows them to reflect the light from the headlights of vehicles at night or in areas with poor lighting. Most current research considers this feature for the rapid detection of road signs and includes image data for enhancing, [10,14,16,24,29,30].

Currently, the most popular classification method is DNN because it facilitates the semantic feature extraction. DNN use convolutional neural networks (CNN) based on supervised learning and contain several hierarchical specialized hidden layers. The first layers detect lines and curves; then, they specialize until they reach deeper layers that recognize complex shapes such as a face or the silhouette of an animal, [19]. Works like [1,2,4,7,17,25] and [32] use DNN to road signs classification. The results are favorable, but there is still some uncertainty in using such a powerful tool for a particular problem. Road signs detection is a problem with well-defined and known variables that could cause difficulty being occluded, altered, or damaged. We explore a segmentation process from both visual and spatial features that allows getting a reliable road signs description condition.

3   Methodology

In this section, we describe the urban segmentation process to acquire road signs descriptions and annotations. The process starts assuming the acquisition of data from a multi-sensor system. The multi-sensor system minimum requirements are a LiDAR sensor, a camera, and a GPS receiver. The Fig. 1 presents the block diagram of the segmentation process to obtain spatial and visual features of the road signs. The multi-sensor system delivers 3D point clouds through the LiDAR sensor, GPS data, and image data from cameras. The 3D point clouds allow to obtain global features such as dimensions, shapes, and location, once complete the following process:

1. Plane extraction: Planes like ground and buildings can be subtracted from point clouds to streamline the segmentation process for smaller objects.
2. Object segmentation: After plane subtractions, it is necessary to identify the remaining objects as individual elements for later recognition.
3. Intensity filter: The laser sensor provides a degree of intensity in a range of 0 to 1 for those points that encountered reflective material. Road sign paint is a reflective material so that vehicle lights can illuminate them properly. This feature makes it possible to differentiate the road signs from other objects in the segmentation.
4. Shape extraction: Road signs have specific shapes and sizes standards to be easily and quickly identified by drivers and pedestrians. These standards allow detailing what type of sign it is how informative, restrictive, and preventive signs.

On the other hand, imagery allows access to color information, 2D shape, and semantic data, with the following process:

1. Color extraction: Road signs designs are of brighter colors to easily express a message. This feature allows for the rapid identification of the sign type.
2. 2D shape extraction: Road signs have a specific geometric shape to differentiate them. It is useful to perform a contour approximation for the detection of shape in images.
3. Semantic information extraction: Some road signs may contain the same color and shape when trying to advertise or inform. However, there is information expressed in numbers, characters, or drawings that show different messages within these signs. It is convenient to extract these features using pattern recognition methods.

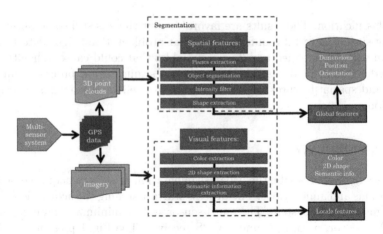

Fig. 1. Block diagram of the segmentation process for obtaining spatial features and visual features from road signs.

The road signs segmentation process imply to handle visual and spatial features individually to fusion through GPS data in the future. Spatial features refer to road signs global characteristics since they differ from the rest of the elements considering a reference frame in the environment. In contrast, the visual features refer to local characteristics because their frame of reference is the same sign. These characteristics allow getting descriptions and annotations of the different road signs.

3.1 Spatial Features Segmentation

Spatial feature segmentation seeks to extract physical properties from road signs in point clouds. The Fig. 2 presents the block diagram of the spatial feature segmentation process. The 3D point cloud segmentation starts with planes extraction considering vehicle field of view delimitation. The next step is object segmentation through 3D point cloud density. After segmentation, an intensity filter uses the material reflective (0.5–1) to differentiate possible road signs. The last step is the shape extraction getting the plane orientation via SVD technique and 2D point cloud projection for geometric shape estimation.

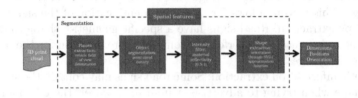

Fig. 2. Block diagram of the spatial feature segmentation process.

Planes Extraction. The first step of this process is the plane's extraction, such as ground and buildings. The last is true since the road signs must be close to the sidewalks and up to the ground. Also, the extraction of these elements allows working with fewer data. We design a ground removal based on the lowest useful point in the point cloud and its neighbors. After performing the road removal, we use a workspace delimitation based on the vehicle field of view. The delimitation is useful because of we fusion visual information taken from the front of the vehicle. Therefore, we define two geometric delimitation to reduce the point cloud data: (i) field of view delimitation, and (ii) ground removal.

Given a 3D point cloud $\mathbf{N}$, the process looks for those points which meet the following statement:

$$\mathbf{A} = \{ (x_i \; y_i \; z_i)^T \} = \{ \mathbf{A} \in \mathbf{N}, \Re^3 \mid x_i \geq 0 \; \& \; -\epsilon \leq y_i \leq \epsilon \} \qquad (1)$$

where ϵ is a threshold defined as average road width. The statements allows eliminating parts of buildings; the rest goes in the next segmentation step.

The removal process starts ensuring detection of the lowest z-point belonging to the ground. The algorithm 1 describes the ground removal process briefly. The lowest z-point is the one who has a minimum number of neighbors N_z within a maximum distance MaxThreshold. Selecting the lowest z-point that meets the criteria described above allows the extraction of most points from the ground. We know that on long slopes, the method will have a significant error.

Algorithm 1: *Road removal.*

Data: $\mathbf{N} = \{ n_i = (x_i, y_i, z_i)^T \in \Re^3 \}$ 3D point cloud
1 % Initialization:
2 $MinThreshold \leftarrow$ close neighbor distance
3 $MaxThreshold \leftarrow$ neighbor maximum distance
4 $N_z \leftarrow$ minimum neighbors in road
5 $n_{min1} \leftarrow \min_{z_i} \{ n_i \}$
6 $n_{min2} \leftarrow \min \{ |n_{min1} - n_i| \}$
7 **while** $|n_{min1} - n_{min2}| > MinThreshold$ **do**
8    $\{ n_i \} \leftarrow \{ n_i : n_i \in \mathbf{N}, \sim (n_i \in n_{min1}) \}$ % excluding n_{min1}
9
10   $n_{min1} \leftarrow \min_{z_i} \{ n_i \}$
11   $n_{min2} \leftarrow \min \{ |n_{min1} - n_i| \}$
12 **while** $Neighbors \geq N_z$ **do**
13   $EuclideanDistance = \{ sort(||n_{min1} - n_i||) \}$
14   **if** $EuclideanDistance_i \leq MaxThreshold$ **then**
15      $Neighbors{+}{+}$
16      $P_i \leftarrow$ next point in the point cloud
17   $n_{min1} \leftarrow P_i$

Result: $\mathbf{N} \leftarrow \{ n_i : n_i \in \mathbf{N}, \sim (n_i \in P_i) \}$ % 3D point cloud without road

---

Algorithm 2: *Density-based object segmentation.*

Data: $\mathbf{N} = \{ n_i = (x_i, y_i, z_i, Nd_i)^T \in \Re^3 \}$ 3D point cloud ($P_i = (x_i, y_i, z_i)$) with their density (Nd_i)
1 $MaxNeighborhood \leftarrow$ neighbor maximum density
2 **for** j **do**
3    $n_j = (P_j, Nd_j) \leftarrow$ point with maximum density $(P_j, Nd_j) \in \mathbf{N}$
4    $m_j \leftarrow \{ m_j \in \mathbf{N} \mid |P_j - P_i| < MaxNeighborhood \}$
5    $\mathbf{N} \leftarrow \{ n_i : n_i \in \mathbf{N}, \sim (n_i \in n_j) \}$
6

Result: $\{ m_{1...j} \}$ clustered objects

Where $||n_i - n_j||$ represent the euclidean distance in $X - Y$ plane and $|n_i - n_j|$ is the euclidean distance using three components.

Object Segmentation. After plane extraction, we group the remaining 3D points as individual objects through an object segmentation method based on 3D point cloud density. The point density in a 3D point cloud defines the number of measurements per area. A higher point density presents a lower point spacing. Therefore, each object in a 3D point cloud represents a point accumulation, so a high-density value. A density segmentation searches for neighborhoods where there is a high accumulation of points. The Algorithm 2 describes the object segmentation process briefly. First is the point cloud density computation Nd_i for each point P_i. For each maximum density point calculates the neighborhood within a particular distance using euclidean distance and extract the object N_i.

Intensity Detection. After the object segmentation, we use an intensity filter to road signs verification besides several objects, including parts of buildings, pillars, vehicles. The LiDAR sensor gets a retro-reflective value of each point [0-1], denoting 1 the most intense value. The road signs have this retro-reflective attribute for easily reflect the light from vehicles. However, there is a possibility that some signs have worn material. We apply an intensity filter to verify the points that probably belong to the road signs. Given the intensity value in the point cloud $N(I_i)$ verify:

$$I_i \in RoadSign \rightarrow N(I_i) \geq LowerValue \tag{2}$$

Being I_i an element in the point cloud N. If I_i is greater than $LowerValue$, it belongs to a road sign. The value of LowerValue may vary from 0.4 to 0.9.

Shape Extraction. In the last step, the process determines the plane orientation via SVD technique for the road signs. The goal is shape extraction by performing a 2D point cloud projection, and a geometric shape estimation. The process allows estimating the dimensions, position, and orientation from the road signs since the 3D point cloud reserves data from the real world. First, there is applied the SVD technique for dimension reduction. A matrix $\mathbf{M}$ can be decomposed into three matrices,

$$M = U\Sigma V^T \tag{3}$$

U stands for an orthonormal basis of eigenvectors. Σ encloses the eigenvalues. The Matrix V^T has the singular decomposing orthogonal axes vectors. We use V^T as rotation matrix for road sign re-projection and dimension reduction. After re-projection, we can use morphological transformation for contour extraction, (see 2D shape extraction section from visual features segmentation).

3.2   Visual Features Segmentation

The visual features segmentation seeks to extract measurable attributes from the images. The Fig. 3 presents the block diagram of the visual feature segmentation process. The first step is color extraction though "Hue, Saturation, Value" (HSV) model. We decide to divide the extraction in two (border-color and inside-color) considering that most of the signs have more than one color. After color extraction, there is the 2D shape extraction based on contour approximation. We use the geometric forms of the road signs to make a contour comparison and decide which form is closest. The last step is semantic information extraction from local feature extraction and descriptions of numbers, characters, and drawings. We use the A-KAZE invariant feature and descriptor method.

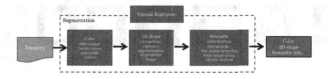

Fig. 3. Block diagram of the visual feature segmentation process.

Color Extraction. In real situations, colors are often not correctly perceived by cameras due to constant changes in lighting. We use the HSV color model, which defines color in terms of its components. The HSV model allows facing real situations such as brightness due to a large amount of light or opacity due to the absence of light. Generally, most of the signs have more than one color to express a message. Then, we divide the color extraction in two: border-color and inside-color. The selection of colors depends on the road signs to detect. HSV color allows applying a color mask in the images for seeking those pixels within color ranges. In Fig. 4, we show a color extraction example from the original in the left looking for a road sign, in the center is the border-color extraction, and in the right the inside-color extraction.

Fig. 4. Left: original image, center: border-color, right: inside-color. (Color figure online)

2D Shape Extraction. The 2D shape extraction use as input the border-color detection from the color extraction step. The process implies a morphological transformation that starts with binarization (seen Fig. 5). The method applies a large mask for data dilatation to increase the road sign extraction probability. Next, it uses a smaller size mask for data erosion to remove pixels considered noise. An opening next to closing transformation helps to refine the noise removal.

This type of morphological transformation does not ensure obtaining proper contours for different reasons. It is common for color extraction to lose data due to lighting conditions or possible nearby objects with similar colors. However,

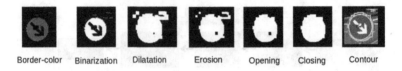

Border-color   Binarization   Dilatation   Erosion   Opening   Closing   Contour

Fig. 5. 2D shape extraction: morphological transformation. Border-color detection, color binarization, data dilatation, data erosion, data opening, and data closing.

a geometric shape association via contour comparison can give us good results. This contour comparison aims to match as much as possible the geometric function of the figures with the contours. Road sign shapes are generally simple geometric shapes such as circles, squares, triangles, and rectangles. In the Fig. 6, we show the comparison for a circular road sign. The process implies an overlay of the geometric shape regarding the area occupied by the contour. The goal is to look for the geometric shape that fills most of the silhouette.

Fig. 6. Geometric shape association via contour comparison. Road sign shapes are generally simple geometric shapes such as circles, squares, triangles, and rectangles.

Semantic Information Extraction. The last step is semantic information extraction from local feature and descriptions of numbers, characters, and drawings. We use the A-KAZE invariant feature and descriptor method, [11]. The method consists of three main tasks: the building of non-linear scale-space, the detection of features, and the description of features. The semantic information extraction starts once having the road signs colors and 2D shape. The features come from corners, curves, and edges of numbers, characters, and drawings in the road signs. In the Fig. 7, we show some invariant features from a road sign.

4   Ground Truth Building Procedure

In this work, we use the KITTI database to carry out preliminary experiments and compare results, [15]. The database was acquire a multi-sensor system composed of two digital cameras, a LiDAR sensor, a GPS receiver, and a GPS/IMU system. We selected the 2017 3D object detection and evaluation data. We empirically build a Ground Truth from the KITTI database choosing low rise road signs. Only ten different road signs remain because most of the road signs keep

Fig. 7. Invariant features extraction from a road sign, the features are from corners, curves, and edges of numbers, characters, and drawings.

in the blind point of the LiDAR sensor, (see Fig. 8). We sought to integrate a minimum of 40 images and point clouds per road sign. In the future, we will increase the database. We built the feature Table 1 describing the ten different road signs. The table includes: (i) reflectivity value: 0.5–0.9, (ii) shapes: circles, squares, inverted triangles, and rectangles, (iii) three dimensions: residential, medium-speed roads, and high-speed roads (plate thickness 2–3 mm), (iv) color: border-color and inside-color, and (v) description: name of the road signs. We make a manual color analysis of road signs samples to define the HSV minimum color value and maximum color value.

Fig. 8. Ten different road signs from KITTI data set: giving orders, and information.

5 Experiments and Results

The following section presents the results for the visual and spatial segmentation process. We used the ten road signs from the Ground Truth section for experimentation. We applied 20 data (image and point cloud) for training and 20 data for testing. The training process implies visual and spatial features extraction from the 20 data. The objects were extracted manually from 3D point clouds,

Table 1. The ten road signs feature table: reflectivity, shape, dimensions, colors, and descriptions.

No.	Reflectivity [0–1]	Shape	Dimensions mm (2–3 mm thickness)	Border-color	Inside-color	Description
1	0.5–0.9	Circle	⊘420, ⊘600, ⊘750	Blue	White	Pass right
2	0.5–0.9	Circle	⊘420, ⊘600, ⊘750	Red	White	No entry
3	0.5–0.9	Circle	⊘420, ⊘600, ⊘750	Red	Blue	No stopping
4	0.5–0.9	Circle	⊘420, ⊘600, ⊘750	Green	Yellow	Bus stop
5	0.5–0.9	Square	840 × 840	White	Red and black	Max zone 30
6	0.5–0.9	Inverted triangle	SL 630, SL 900, SL 1260	Red	White	Give way
7	0.5–0.9	Rectangle	420 × 630, 600 × 900	Blue	White	Calming zone start
8	0.5–0.9	Rectangle	420 × 630, 600 × 900	Blue	White and red	Calming zone end
9	0.5–0.9	Square	420 × 420, 600 × 600, 840 × 840	Blue	White	Parking
10	0.5–0.9	Diamond	420 × 420, 600 × 600, 840 × 840	White	Yellow	Priority

and the process segmentation evaluation is for retro-reflective and shape extraction. We obtained the road sign image manually too, and they are arranged together for reducing visual segmentation. The testing process refers to reproduce the segmentation for both visual and spatial features for 20 different data. We compare the results using a vector approximation method via euclidean distance. The results come from evaluating the count of true positives (TP) and false positives (FP), (see Table 2). The count allows us to carry out a precision evaluation using the next formula:

$$Precision = \frac{TP}{TP + FP} \qquad (4)$$

We obtained a precision result of 0.88. We analyzed the road signs with higher FP counts. The road signs with less TP count are because the border colors as green, white, and yellow are easily confused with trees, the buildings, and the sky.

Table 2. Segmentation results: the ten road signs, 20 training data, 20 testing data, true positives counts TP, and false positives counts FP. Precision assessment.

No.	Training	Testing	TP	FP	No.	Training	Testing	TP	FP
1	20	20	20	0	6	20	20	20	0
2	20	20	20	0	7	20	20	18	2
3	20	20	20	0	8	20	20	18	2
4	20	20	10	10	9	20	20	20	0
5	20	20	20	0	10	20	20	10	10
								TP = 176	FP = 24

6 Conclusion and Discussion

In this work, we presented an urban segmentation to acquire road signs descriptions and annotations thought laser scanning and cameras. We achieved the segmentation of road signs by obtaining spatial and visual information from 3D point clouds and imagery. The spatial information segmentation from 3D point clouds comprised: (i) object segmentation through 3D point cloud density, (ii) use of the retro-reflective property of the material to differentiate possible road signs, (iii) plane orientation determination via SVD, (iv) 2D point cloud projection to geometric shape estimation, (v). The visual information from images comprised: (i) color segmentation of the road signs in two-parts: border-color and inside-color, (ii) color identification using HSV color system (iii) geometric shape association via contour comparison, (iv) local features extraction and description from semantic data as numbers, characters, and drawings. We know that on long slopes, our ground removal method will have a significant error. But it is irrelevant for only reduce 3D point cloud. Most of the research use retro-reflective filter at the beginning; however, there is a possibility that some signs have worn material. The morphological transformation does not ensure obtaining proper contours for different reasons. It is common for color extraction to lose data due to lighting conditions or possible nearby objects with similar colors. We built a ground truth from the KITTI database to prove an adequate visual and spatial segmentation. We identified ten low rise road signs, each one with 20 training data (point cloud and image) and 20 testing data. We evaluated the precision of the segmentation thanks to the count of TP and FP. We obtained a precision of 0.88, and we identified possible segmentation errors. Some border-colors are easily confused with trees, buildings, and sky. Our ground truth is small because KITTI database does not have enough road signs, they focused on vehicles, pedestrians, and bicycle detection. For the road signs is necessary to place the LiDAR high enough. Besides, the closest signs have a more significant number of points and better detail. We recommend a minimum of 500 points per road sign 3D point cloud to accomplish road signs inventory and examination for maintenance. Future work includes expanding the database and conducting experiments for at least 25 road signs. We will increase the ground truth data (image and point cloud) to 100 per road signs. We consider the use of DNN to compare the results. We will acquire our data set for road signs detection, taking care of the recommendations. We will fusion the visual and spatial data through GPS data and extrinsic parameters between LiDAR and camera. We will build feature vectors for descriptions and annotations. The feature vector will be a scoring system to measure the damage or condition of road signs.

Acknowledgments. Acknowledgments to CONACYT México and IPN México for the PhD study grants 584190. All authors acknowledge the financial support given by SIP-IPN (20195716 and 20201942).

References

1. Arcos-García, A., Álvarez García, J.A., Soria-Morillo, L.M.: Evaluation of deep neural networks for traffic sign detection systems. Neurocomputing **316**, 332–344 (2018)
2. Arcos-García, A., Soilán, M., Álvarez García, J.A., Riveiro, B.: Exploiting synergies of mobile mapping sensors and deep learning for traffic sign recognition systems. Expert Syst. Appl. **89**, 286–295 (2017)
3. Balado, J., Sousa, R., Díaz-Vilariñoc, L., Ariasa, P.: Transfer learning in urban object classification: online images to recognize point clouds. Autom. Construct. **11**, 103058 (2020)
4. Bruno, D.R., Sales, D.O., Amaro, J., Osório, F.S.: Analysis and fusion of 2D and 3D images applied for detection and recognition of traffic signs using a new method of features extraction in conjunction with deep learning. In: 2018 International Joint Conference on Neural Networks (IJCNN), pp. 1–8 (2018)
5. Buyval, A., Gabdullin, A., Lyubimov, M.: Road sign detection and localization based on camera and lidar data. In: International Conference on Machine Vision (2019)
6. Börcs, A., Nagy, B., Benedek, C.: Instant object detection in lidar point clouds. IEEE Geosci. Remote Sens. Lett. **14**(7), 992–996 (2017)
7. Cao, J., Song, C., Peng, S., Xiao, F., Song, S.: Improved traffic sign detection and recognition algorithm for intelligent vehicles. Sensors **19**(18), 4021 (2019)
8. Cheng, Yizong: Mean shift, mode seeking, and clustering. IEEE Trans. Pattern Anal. Mach. Intell. **17**(8), 790–799 (1995)
9. Cui, T., Ji, S., Shan, J., Gong, J., Liu, K.: Line-based registration of panoramic images and lidar point clouds for mobile mapping. Sensors **17**(1), 70 (2017)
10. Deng, Z., Zhou, L.: Detection and recognition of traffic planar objects using colorized laser scan and perspective distortion rectification. IEEE Trans. Intell. Transp. Syst. **19**, 1485–1495 (2018)
11. Fernández Alcantarilla, P.: Fast explicit diffusion for accelerated features in non-linear scale spaces. In: British Machine Vision Conference (BMVC), at Bristol, UK (09 2013). https://doi.org/10.5244/C.27.13
12. Fischler, M.A., Bolles, R.C.: Random sample consensus: a paradigm for model fitting with applications to image analysis and automated cartography. Graph. Image Process. **24**(6), 381–395 (1981)
13. García-Moreno, A., González-Barbosa, J., Hurtado-Ramos, J., Ornelas-Rodríguez, F., Ramírez-Pedraza, A.: Análisis de la sensibilidad en un modelo de calibración cámara-lidar. Rev. int. métodos numér. cálc. diseño ing. 32(4), 193–203 (2016)
14. Gargoum, S., El-Basyouny, K., Sabbagh, J., Froese, K.: Automated highway sign extraction using lidar data. J. Transp. Res. Boar **2643**(1), 1–8 (2017)

15. Geiger, A., Lenz, P., Urtasun, R.: Are we ready for autonomous driving? the kitti vision benchmark suite. In: Conference on Computer Vision and Pattern Recognition (CVPR) (2012)
16. Guan, H., Yan, W., Yu, Y., Zhong, L., Li, D.: Robust traffic-sign detection and classification using mobile lidar data with digital images. IEEE J. Sel. Top. Appl. Earth Obs. Remote Sens. **11**(5), 1715–1724 (2018)
17. Guo, J., Cheng, X., Chen, Q., Yang, Q.: Detection of occluded road signs on autonomous driving vehicles. In: 2019 IEEE International Conference on Multimedia and Expo (ICME) (2019)
18. Im, J.H., Im, S.H., Jee, G.I.: Extended line map-based precise vehicle localization using 3D lidar. Sensors **18**(10), 3179 (2018)
19. Karpathy, A.: Convolutional neural networks (cnns/convnets) (2018). http://cs231n.github.io/convolutional-networks/
20. Li, X., Du, S., Li, G., Li, H.: Integrate point-cloud segmentation with 3D lidar scan-matching for mobile robot localization and mapping. Sensors **20**(1), 237 (2020)
21. MacQueen, J.: Some methods for classification and analysis of multivariate observations. In: Proceedings of the Fifth Berkeley Symposium on Mathematical Statistics and Probability, Volume 1: Statistics, pp. 281–297. University of California Press, Berkeley, California (1967). https://projecteuclid.org/euclid.bsmsp/1200512992
22. Patil, A., Malla, S., Gang, H., Chen, Y.T.: The H3D dataset for full-surround 3D multi-object detection and tracking in crowded urban scenes. IEEE International Conference on Robotics and Automation (ICRA) (2019). arXiv:1903.01568 [cs.CV]
23. Powell, V.: Principal component analysis, explained visually. http://setosa.io/ev/principal-component-analysis/ (2018)
24. Riveiro, B., Díaz-Vilariño, L., Conde-Carnero, B., Soilán, M., Arias, P.: Automatic segmentation and shape-based classification of retro-reflective traffic signs from mobile lidar data. IEEE J. Sel. Top. Appl. Earth Obs. Remote Sens. **9**(1), 295–303 (2016)
25. Soilán, M., Riveiro, B., Martínez-Sánchez, J., Arias, P.: Traffic sign detection in MLS acquired point clouds for geometric and image-based semantic inventory. ISPRS J. Photogrammetry Remote Sens. **114**, 92–101 (2016)
26. Song, W., Zou, S., Tian, Y., Fong, S., Cho, K.: Classifying 3D objects in lidar point clouds with a back-propagation neural network. Hum. Centric Comput. Inf. Sci. **8**(29), 1–12 (2018)
27. Sánchez, N.: Máquinas de soporte vectorial y redes neuronales artificiales en la predicción del movimiento usd/cop spot intradiario. ODEON -(9), 113–172 (2015). https://doi.org/10.18601/17941113.n9.04
28. Wang, D., Wang, J., Scaioni, M., Si, Q.: Coarse-to-fine classification of road infrastructure elements from mobile point clouds using symmetric ensemble point network and euclidean cluster extraction. Sensors **20**(1), 225 (2020). https://doi.org/10.3390/s20010225
29. Wen, C., et al.: Spatial-related traffic sign inspection for inventory purposes using mobile laser scanning data. IEEE Trans. Intell. Transp. Syst. **17**(1), 27–37 (2016)
30. Wu, S., Wen, C., Luo, H., Chen, Y., Wang, C., Li, J.: Using mobile lidar point clouds for traffic sign detection and sign visibility estimation. IGARSS **2015**, 565–568 (2015)
31. Yang, B.: Developing a mobile mapping system for 3D GIS and smart city planning. Sustainability **11**(13), 3713 (2019)
32. You, C., Wen, C., Luo, H., Wang, C., Li, J.: Rapid traffic sign damage inspection in natural scenes using mobile laser scanning data. IGARSS **2017**, 6271–6274 (2017)

Person Authentication Based on Standard Deviation of EEG Signals and Bayesian Classifier

D. Farias-Castro and R. Salazar-Varas[✉]

Department of Computing, Electronics and Mechatronics,
Universidad de las Américas Puebla, 72810 San Andres Cholula, Puebla, Mexico
diego.fariasco@udlap.mx, rocio.salazarv@udlap.mx

Abstract. In the field of people authentication, the use of biometric identifiers has attracted the interest of the research community. In this sense, the brain activity pattern is an interesting candidate, since each individual has shown to possess a particular one. This document presents the results obtained by using the standard deviation of electroencephalographic signals as a feature and a naive Bayes classifier for the aforementioned authentication purpose. The proposed methodology is composed of two stages: in the first, a selection of the most suitable frequency bands is performed. In the second, the previously selected bands are used for authentication. In order to discover such promising frequency bands and to subsequently evaluate the performance of the whole methodology, the Data Set IIb from BCI competition IV was used. As it will be shown, the total success rate for all evaluated subjects was higher than 0.85 and, in most cases, higher than 0.95.

Keywords: Authentication · Brain activity patterns · Feature extraction · EEG classification

1 Introduction

The purpose of people authentication is to identify whether an individual is really who it claims to be. Such labor is necessary, in order to allow or deny the access to a restricted environment or resource. There are several modalities to carry out such process: 1) By the *knowledge* of a secret word, e.g. a password 2) By *possession* of a physical key e.g. a token, or 3) By the subject's unique *biometric* information, e.g. fingerprints.

The latter has been of great interest in recent years. This, mainly due to the fact that biometric information is part of the individual, a key condition that prevents theft, loss or forgetting. Within this authentication modality there exist several proposals, such as the recognition of venal and gait patterns, as well as electro-oculography signals [1]. It is important to mention that, in order to increase the reliability of authentication systems, a fusion of different kinds of biometrics can be used [2].

© Springer Nature Switzerland AG 2020
L. Martínez-Villaseñor et al. (Eds.): MICAI 2020, LNAI 12469, pp. 390–400, 2020.
https://doi.org/10.1007/978-3-030-60887-3_34

Going further to more personal biometric information, an emerging proposal is to use the brain activity pattern, which has been shown to be unique on each individual [3,4]. For this purpose, different approaches have been exposed, using both spontaneous and evoked brain activity [5–8]. This proposal, however, requires the consideration of electroencephalographic (EEG) signals stability over time. A study [9], carried out over a period of three years, shows that the EEG signals are affected by aging. Nonetheless, some strategies are provided to compensate it.

Although different approaches have been suggested, achieving successful results, there is still a lot to explore. In this sense, it is necessary to analyze the performance of different classifiers, as well as the most appropriate channels and features for this particular problem schemata [10,11]. This document proposes a methodology based on the standard deviation of EEG signals and a Bayesian classifier that aims to enrich the EEG-based authentication field with a simple and feasible paradigm for real-life applications. Said methodology is divided in two stages: in the first one, the more affordable cutoff frequency bands for preprocessing are selected. Meanwhile, in the second stage, the authentication is performed. The feature vector is built with the standard deviation of the EEG signals, filtered in the frequency bands that were previously selected, i.e. in the first stage. Afterwards, a classification is performed by means of a Bayesian classifier, demonstrating the system's capability of subject identification.

In order to acquire the most suitable frequency bands and to subsequently evaluate the performance of the whole aforementioned methodology, the Data Set IIb from BCI competition IV was used [12]. This data set is related to motor imagery tasks of nine subjects, whose recording was made using only three channels: C3, Cz and C4, following the international 10–20 convention. The results acquired from above procedures shown a total success rate higher than 0.85 for all of the subjects, while 0.95 was achieved in most cases.

This document is organized as follows: in Sect. 2 the theoretic concepts used in this work are presented. In Sect. 3 the proposed methodology is explained. The data set used to evaluate the aforementioned paradigm and the obtained results are shown in Sect. 4. Finally, the conclusions and future work are presented in Sect. 5.

2 Theoretical Framework

2.1 Bayesian Classifier

Considering M classes $\omega_1, \omega_2, \omega_3, ..., \omega_M$ and a feature vector x, which represents an unknown pattern, the classification problem consists of deciding to which class the x vector belongs to. As it is usually known, the Bayesian classifier distinguishes such feature vector's class x according to the probability of belonging to it [13].

Given $P(\omega_i)$, $i = 1, 2, 3, ...M$ the probability of occurrence of the i-class, and $p(x|\omega_i)$ the class conditional probability density function that describes the distribution of the feature vectors in each of the classes. The conditional probability $P(\omega_i|\boldsymbol{x})$ is defined by

$$P(\omega_i|\boldsymbol{x}) = \frac{p(\boldsymbol{x}|\omega_i)P(\omega_i)}{p(\boldsymbol{x})} \tag{1}$$

where $p(\boldsymbol{x})$ denotes the probability density function of the feature vector $\boldsymbol{x}$.

The Bayes classification rule can be expressed as:

$$\text{If } P(\omega_i|\boldsymbol{x}) > P(\omega_j|\boldsymbol{x}) \ i,j = 1,2,...M \,;i \neq j$$
$$\Rightarrow \boldsymbol{x} \text{ is classified as } \omega_i, \tag{2}$$

In case that the data follows a Gaussian distribution, $p(\boldsymbol{x}|\omega_i)$ is expressed as:

$$p(\boldsymbol{x}|\omega_i) = \frac{1}{2\pi^{l/2}|\Sigma_i|^{1/2}} exp(-\frac{1}{2}(\boldsymbol{x} - \mu_i)^T \Sigma_i^{-1}(\boldsymbol{x} - \mu_i)) \tag{3}$$

where l is the dimension of the feature vectors, μ_i is the mean vale of the feature vectors belonging to the i-class, and Σ_i represents the covariance matrix.

2.2 Evaluation

As it was mentioned earlier, an authentication system must verify that a person is actually who it claims to be. This, implies that the system must be capable of differentiating between a *client* and an *impostor*. To evaluate the performance of this process, two error types can be defined [2]:

– *False acceptance.* When an impostor is classified as a client and accepted.
– *False rejection.* When a client is classified as an impostor and rejected.

Based on these values, the global errors *False Acceptance Rate* (FAR) and *False Rejection Rate* (FRR) can be computed by

$$FAR = \frac{\text{Number of false acceptances}}{\text{Number of impostor accesses}} \tag{4}$$

$$FRR = \frac{\text{Number of false rejections}}{\text{Number of client accesses}} \tag{5}$$

Different operating points can be obtained based on the value of FAR and FRR [2]. Obviously, the desired point is when FRR is high, close to or equal to one, and FAR is low, close to or equal to zero.

In addition, the *Total Error Rate* (TER) can be defined as:

$$TER = \frac{\text{Number of FA} + \text{Number of FR}}{\text{Total number of accesses}} \tag{6}$$

Finally, the *Total Success Rate* (TER) is defined as the complement of TER

$$TSR = 1 - TER \tag{7}$$

3 Proposed Methodology

In this section, details concerning the proposed methodology are presented. As it was initially mentioned, such methodology consists of two stages: the selection of the most suitable frequency bands and the authentication stage.

First Stage: Frequency Bands Selection
In this stage the main objective was determining the appropriate cutoff frequencies for preprocessing. For this purpose, a Bayesian classifier was used, which was tested in all possible binary combinations between subjects, and in all possible cutoff frequencies for a single frequency band, i.e. specifying a low F_L and high F_H cutoff frequency value. The above, in order to determine the best cutoff frequencies for all pairs of subjects through a brute-force approximation. The filter that was used is inherently a bandpass, of fourth-order Butterworth type. The feature vector was built by computing the standard deviation on each available channel i.e. C3, Cz and C4. In order to refine the aforementioned classifications, the data went through an outlier removal process based on the Median Absolute Deviation (MAD), as suggested in [14].[1]

Once the most appropriate cutoff frequencies for each binary case were established, the final selection of the appropriate cutoff frequencies to be used in the preprocessing of the authentication stage was performed. This selection was based on a search of the frequency bands that contained the intersection of most of the selected frequency bands for each subject, although a subsequent trial and error refinement was made.[2]

Second Stage: Authentication
In this stage, the main goal was the actual person authentication, that is, to recognize if the subject who claims to be the client is really it or if, on the con-

---

[1] For example, on the first combination subjects' 1 and 2 data was classified by means of a Bayesian classifier. Such data was applied an outlier suppression process based on MAD and a single-band, fourth-order Butterworth filter i.e. only one low F_L and high F_H cutoff frequency was specified. All possible and coherent low and high values were tested in a brute-force mood i.e. ($F_L = 1$, $F_H = 2$), ($F_L = 1$, $F_H = 3$), ... ($F_L = 48$, $F_H = 49$). On all cases the average effectiveness of the classification was temporarily stored. Once all cutoff frequencies were analyzed, the best F_L and F_H values were extracted and permanently stored. Subsequently, another subject combination was tested i.e. subject 1 vs 3. This process was repeated until all subjects and all frequency bands were tested. Best frequency bands from the above procedures are shown in Table 2.

[2] For the sake of exemplification assume that the previous phase thrown 1–5 as an optimal frequency band for subjects' 1 and 2 first motor imagery task. Hence, frequencies 1 through 5 increased their number of appearances in the best frequencies record by one. Now, assume that the previous phase thrown 3–6 as an optimal frequency band for subject's 2 and 3 first motor imagery task. Now, frequencies 3 through 5 appeared twice in the best frequencies record, while 1, 2, and 6 did only once. The remaining 43 frequency bands didn't even appear. Therefore, a suitable frequency band would be 3–5 as it is the intersection of most selected frequency bands for each subject. The results of the aforementioned procedures constitute Figs. 2 and 3.

trary, it is an impostor. For this goal, two classes were built: client and generic class. The client class contains the information of the subject to authenticate and the generic class contains the training information of all other subjects.[3] Under these conditions, the EEG signals were filtered using the frequency bands selected in the previous stage. The feature vector contains the standard deviation of each channel in the different selected frequency bands and a Bayesian classifier was again used. Figure 1 shows a graphical overview of the aforementioned methodology.

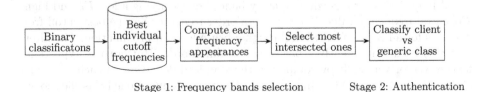

Stage 1: Frequency bands selection   Stage 2: Authentication

Fig. 1. Flow diagram of the proposed methodology

4   Experimental Results

In this section, the results obtained while using the proposed methodology are exposed. To start with such exposition, the data set used to evaluate the methodology is described. Subsequently, the technical details of the proposed paradigm are given. To conclude, a presentation of the obtained results is made in terms of accuracy, FRR, FAR, TER and TSR.

4.1   Data Set Description

The proposed methodology was evaluated using the Data set IIb from the international BCI competition IV. This data set is composed by EEG recordings of nine subjects. The EEG data was recorded while the subjects were performing two different motor imagery tasks: the left and right hand movement. Only three bipolar recordings are provided: C3, Cz and C4, the sampling frequency was 250 Hz.

Five sessions were carried out in different days. Two screening sessions without feedback were performed, each containing six runs with ten trials for each

---

[3] For the sake of exemplification assume that the client is set to be subject number one. Hence, subjects 2 through 9 will constitute the generic class. Now, assume that 50 is set to be the number of training experiments. In this scenario the training data for the client class will be the standard deviation and mean of 50 random experiments from subject 1. By its side, the generic class' training data will be composed of the standard deviation and mean of 400 experiments. This, results from randomly picking 50 experiments from all remaining subjects i.e. 2 through 9.

class, i.e. in total 60 trials per class per session. The imagination period lasted 4 s. Furthermore, three online sessions with feedback were carried out. In this case, each session consists of four runs, each containing 20 trials per class. The subject was asked to imagine the movement during 4.5 s. For more details refer to [12].

In this work, only the first three sessions were used: the first two sessions without feedback and the first feedback session. Therefore a total of 200 trials per class (60 × 2 + 80) are available for experimentation. However, trials containing artifacts were removed. It is important to mention that, from the 4 s that the subject imagined the movement, only the two intermediate seconds i.e. from 1.5 s through 3.5 s were used. This, in order to guarantee that the test subject was actually performing the required mental task.

In the next section, the results obtained under the described conditions are exposed.

4.2 Results

First Stage: Frequency Bands Selection

For this stage, a total of 36 binary combinations of subjects (S_i vs S_j, $i, j = 1, 2, ...9$, $i \neq j$) were obtained. The subject classification was performed for each imaginary movement. From the total available data from each subject, empirically, the 50% was used to train the classifier and the remaining 50% was used to evaluate it. To ensure that the data selected for training or evaluation did not influence the classification, such process was repeated 30 times. On each iteration, the training and evaluation data was randomly selected. Prior to classification, the removal of outliers was performed, using the Median Absolute Deviation (MAD)-based methodology suggested in [14]. The probability $P(\omega_i)$ was computed as N_i/N, where N is the total number of trials and N_i denotes the trials in the i-class. To guarantee that data followed a Gaussian distribution, the D'Agostino test was applied.

Tables 1 and 2 show the mean accuracy obtained for each subject combination for left and right movement, respectively. These accuracy levels were achieved after performing the brute-force evaluation of different cutoff frequencies and selecting the most suitable low F_L and high F_H cutoff frequency values. As it can be seen, in this approach of distinguishing between two subjects based on standard deviation, most cases resulted in an accuracy higher than 95%. Additionally, in several cases it was possible to differentiate between subjects with an accuracy of 100%.

The selected frequency bands for each combination, acquired through the brute-force approach, are shown in Tables 3 and 4 for left and right movement, respectively. Using this information, bar graphs shown in Figs. 2 and 3 were built. These show the times that each frequency appears contained in any frequency band from Tables 3 and 4. Said graphs were then used to explore which intersections of the frequency bands obtained in both movements are repeated the most.

Table 1. Accuracy for left movement (Average value)

	S2	S3	S4	S5	S6	S7	S8	S9
S1	99.71	100.00	98.83	98.98	100.00	99.91	90.71	98.29
S2		98.76	100.00	99.80	100.00	98.16	99.98	97.77
S3			99.42	99.95	94.59	100.00	99.11	98.08
S4				97.99	100.00	99.81	97.74	97.70
S5					99.48	100.00	99.54	98.52
S6						100.00	99.92	100.00
S7							98.16	82.14
S8								97.05

Table 2. Accuracy for right movement (Average value)

	S2	S3	S4	S5	S6	S7	S8	S9
S1	99.22	100.00	99.44	99.17	100.00	99.66	92.13	97.58
S2		98.41	98.76	99.36	100.00	98.96	99.88	99.41
S3			98.88	99.33	97.86	100.00	99.53	98.88
S4				94.61	99.95	99.92	96.64	96.62
S5					99.88	99.98	99.46	99.03
S6						100.00	99.98	100.00
S7							97.66	87.44
S8								98.73

This, served as a starting point for locating the most suitable frequencies, which were lately refined by automatized trial and error. From this analysis, the selected frequency bands that would be used in the authentication stage were obtained and are reported in Table 5.

Second Stage: Authentication
In the authentication stage, EEG signals were filtered in the frequency bands shown in Table 5. The feature vector was built by computing the standard deviation in these different ranges for each electrode, therefore, the dimension of the feature vector in this stage was 15 (3 electrodes × 5 frequency bands).

It is important to remember that, in this case, the discrimination was made between the client class and the generic class. As in the first stage, the classification was empirically performed using the 50% of the available data to train the classifier and the remaining data to evaluate it. Training and evaluation data was randomly selected in an iterative process. In this case, the removal of outliers wasn't performed. The results obtained for each movement are shown in Tables 6 and 7. As it can be seen, for all the cases, the TSR is higher than 0.85, and in several cases it is higher than 0.95. In a more detailed analysis, it is possible to

Table 3. Selected cutoff frequencies for each subject for left movement

	S2	S3	S4	S5	S6	S7	S8	S9
S1	23–34	3–9	33–47	20–48	4–9	36–48	34–40	39–48
S2		28–49	11–14	9–12	1–48	38–46	10–36	41–47
S3			32–41	15–49	23–30	4–7	31–49	35–48
S4				11–15	4–7	3–9	23–38	10–12
S5					15–49	4–9	18–24	1–7
S6						4–7	3–7	2–7
S7							27–49	33–49
S8								20–23

observe that, in most of the cases, the FRR is lower than 0.1. This, implies that, most of the times, the methodology correctly identifies the subject. In addition, the fact that the FAR is also low, guarantees that the methodology rejects who should be rejected.

Table 4. Selected cutoff frequencies for each subject for right movement

	S2	S3	S4	S5	S6	S7	S8	S9
S1	25–32	3–8	26–39	21–45	3–9	37–49	32–43	39–49
S2		31–42	11–12	10–12	4–49	37–49	11–36	2–7
S3			34–49	27–49	21–49	26–49	33–49	36–45
S4				18–21	3–6	3–10	23–28	17–20
S5					13–25	4–9	21–29	2–8
S6						5–9	4–7	3–6
S7							28–48	34–45
S8								18–22

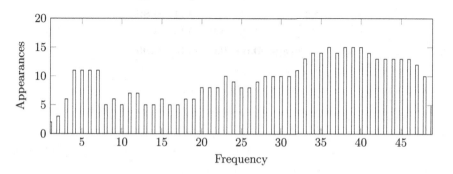

Fig. 2. Appearances of each frequency within any left-movement selected band

Table 5. Selected cutoff frequencies for authentication stage

Low	High
36	43
4	8
23	35
47	48
11	15

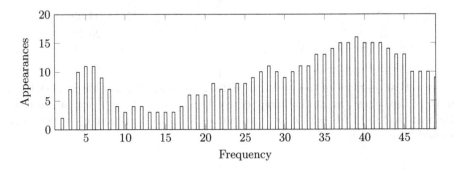

Fig. 3. Appearances of each frequency within any right-movement selected band

Table 6. FRR, FAR, TER and TSR for left movement

Subject	FRR	FAR	TER	TSR
S1	0.07	0.02	0.03	**0.97**
S2	0.04	0.05	0.05	**0.95**
S3	0.02	0.00	0.01	**0.99**
S4	0.07	0.06	0.05	**0.95**
S5	0.10	0.08	0.09	**0.91**
S6	0.01	0.01	0.01	**0.99**
S7	0.02	0.01	0.01	**0.99**
S8	0.10	0.12	0.11	**0.89**
S9	0.10	0.13	0.13	**0.87**
Average	**0.06**	**0.05**	**0.05**	**0.95**

Table 7. FRR, FAR, TER and TSR for right movement

Subject	FRR	FAR	TER	TSR
S1	0.05	0.03	0.04	**0.96**
S2	0.07	0.05	0.05	**0.95**
S3	0.03	0.01	0.01	**0.99**
S4	0.06	0.10	0.08	**0.92**
S5	0.11	0.12	0.12	**0.88**
S6	0.01	0.01	0.01	**0.99**
S7	0.01	0.01	0.01	**0.99**
S8	0.09	0.08	0.08	**0.92**
S9	0.10	0.14	0.13	**0.87**
Average	**0.06**	**0.06**	**0.06**	**0.94**

5   Conclusions and Future Work

Throughout this document a methodology based on standard deviation and Bayesian classifier was presented. The methodology is composed of two stages: frequency bands selection and authentication. The selection of the most suitable frequency bands was performed by taking into account the accuracy levels achieved when distinguishing between two subjects. This selection stage is crucial for the good performance in the authentication stage.

During authentication stage, two classes were built: the client class and the generic class. The EEG signals were filtered in the most affordable frequency bands and the feature vector was built by computing the standard deviation in this frequency bands. The classification was performed through a Bayesian classifier.

The data set used to evaluate the proposed methodology, as mentioned, proceeds from BCI competition IV and is related to motor imagery tasks. A total success rate higher than 0.95 in most of the cases, shows that is feasible to apply the proposed methodology in the problem of people authentication. In addition, the fact that both, FRR and FAR have a low value demonstrates the robust performance of proposed methodology. Finally, the fact that these results were achieved by using only three electrodes is very attractive when thinking about real applications.

As future work there are three interest points:

- Evaluate the performance of different features in both, frequency and time domain. This, in order to improve the presented methodology's performance.
- Determine a metric to identify the person without the need to built a generic class, minimizing the current computation time.
- Taking into account what was reported in [9], it is interesting to evaluate the performance of the proposed methodology using the data recorded in different days.

Acknowledgment. The authors would like to thank Universidad de las Américas Puebla for the economic support.

References

1. Jiang, R., Al-Maadeed, S., Bouridane, A., Crookes, D., Beghdadi, A.: Biometric Security and Privacy, 1st edn. Springer, Cham (2017). https://doi.org/10.1007/978-3-319-47301-7_9
2. Verlinde, P., Chollet, G., Acheroy, M.: Multi-modal identity verification using expert fusion. Inf. Fusion 1(1), 17–33 (2000)
3. Tavor, I., Parker Jones, O., Mars, R., Smith, S., Behrens, T., Jbabdi, S.: Task-free MRI predicts individual differences in brain activity during task performance. Science 352, 216–220 (2016)
4. Miller, M., Donovan, C.L., Horn, J., German, E., Sokol-Hessner, P., Wolford, G.: Unique and persistent individual patterns of brain activity across different memory retrieval tasks. NeuroImage 48, 625–635 (2009)
5. Marcel, S., Millán, J.D.R.: Person authentication using brainwaves (EEG) and maximum a posteriori model adaptation. IEEE Trans. Pattern Anal. Mach. Intell. 29(4), 743–752 (2007)
6. Das, R., Maiorana, E., Campisi, P.: EEG biometrics using visual stimuli: a longitudinal study. IEEE Sig. Process. Lett. 23(3), 341–345 (2016)
7. Khalifa, W., Salem, A., Roushdy, M., Revett, K.: A survey of EEG based user authentication schemes. In: 2012 8th International Conference on Informatics and Systems (INFOS) (2012). BIO-55-BIO-60
8. Maiorana, E., La Rocca, D., Campisi, P.: On the permanence of EEG signals for biometric recognition. IEEE Trans. Inf. Forensics Secur. 11(1), 163–175 (2016)
9. Maiorana, E., Campisi, P.: Longitudinal evaluation of EEG-based biometric recognition. IEEE Trans. Inf. Forensics Secur. 13(5), 1123–1138 (2018)
10. Zeynali, M., Seyedarabi, H.: EEG-based single-channel authentication systems with optimum electrode placement for different mental activities. Biomed. J. 42(4), 261–267 (2019)
11. Jayarathne, I., Cohen, M., Amarakeerthi, S.: Survey of EEG-based biometric authentication. In: 2017 IEEE 8th International Conference on Awareness Science and Technology (iCAST), pp. 324–329 (2017)
12. Leeb, R., Brunner, C., Müller-Putz, G., Schlögl, A., Pfurtscheller, G.: Dataset IIb [Mat file]. Institute for Knowledge Discovery, & Graz University of Technology (2008)
13. Theodoridis, S., Koutroumbas, K.: Pattern Recognition, 4th edn. Academic Press Inc., Cambridge (2008)
14. Leys, C., Ley, C., Klein, O., Bernard, P., Licata, L.: Detecting outliers: do not use standard deviation around the mean, use absolute deviation around the median. J. Exp. Soc. Psychol. 49, 2–3 (2013)

Intelligent Applications and Robotics

Using the Gini Index for a Gaussian Mixture Model

Adriana Laura López-Lobato$^{(\boxtimes)}$ and Martha Lorena Avendaño-Garrido

Faculty of Mathematics, University of Veracruz, Xalapa, Veracruz, Mexico
adrilau17@gmail.com

Abstract. A Gaussian mixture model is a weighted sum of parametric Gaussian components. These parametric density functions are widely used in data mining and pattern recognition. In this work we propose an efficient method to model a density function as a Gaussian mixture through an iterative algorithm that allow us to estimate the parameters of the model for a given data set. For this purpose we use the Gini Index, a measure of the inequality degree between two probability distributions. The Gini Index is obtained by finding the solution of an optimization problem. Our model consists in minimizing the Gini Index between an empirical distribution and a parametric distribution that is a Gaussian mixture. We will show some simulated examples and real data examples, with two widely used datasets, to observe the efficiency and properties of our model.

Keywords: Gini index problem · Gaussian mixture model · Density estimation

1 Introduction

In this work we consider the problem of modeling the behavior of a known dataset $\{p_1, p_2, ..., p_M\}$ with $p_m \in \mathbb{R}^N$. We usually take some of the known density functions, like the normal density for the multivariate case given by

$$f(x|\mu, \Sigma) = \frac{1}{(2\pi)^{N/2}|\Sigma|^{1/2}} e^{(-\frac{1}{2}(x-\mu)^T \Sigma^{-1}(x-\mu))}, \tag{1}$$

where μ is the mean and Σ is the covariance matrix. However, there are cases in which the data cannot be represented with simple distributions, so we use linear combinations of simple components, as the Gaussian mixture model.

The Gaussian mixture model is a linear combination of K Gaussian components, see [2] and [10]. In the articles [6,8] and [12], we can see that the Gaussian mixture model can be used for image segmentation, speech recognition, language identification and statistical representation, among others.

The Gaussian mixture model considers the density function

$$\sum_{k=1}^{K} \phi_k f(x|\mu_k, \Sigma_k), \tag{2}$$

© Springer Nature Switzerland AG 2020
L. Martínez-Villaseñor et al. (Eds.): MICAI 2020, LNAI 12469, pp. 403–418, 2020.
https://doi.org/10.1007/978-3-030-60887-3_35

where $f(x|\mu_k, \Sigma_k)$ are Gaussian densities (1), with mean μ_k and covariance matrix Σ_k. The ϕ_k parameters are mixing coefficients that must comply with $0 \leq \phi_k \leq 1$ and $\sum_{k=1}^{K} \phi_k = 1$. Then, to solve this problem, we have to find the values for the parameters $\phi = (\phi_1, \phi_2, ..., \phi_K)$, $\mu = (\mu_1, \mu_2, ..., \mu_K)$ and $\Sigma = (\sigma_1, \sigma_2, ..., \Sigma_K)$.

In several texts as [4,7,14] and [16], the authors maximize the likelihood function

$$L = \sum_{m=1}^{M} \log \left(\sum_{k=1}^{K} \phi_k f(x_m|\mu_k, \Sigma_k) \right), \tag{3}$$

when $\{x_m\}_{m=1}^{M}$ is the data sample. There is no analytical solution to this problem, so they use an iterative numerical optimization technique, known as EM-algorithm.

Once we have the values for the parameters ϕ, μ and Σ, we can classificate the data using the Total Probability Law and the Bayes Theorem. We can obtain the conditional probability $P(g_k|x)$, for $k = 1, ..., K$, by the following expression

$$P(g_{k'}|x) = \frac{P(g_{k'})P(x|g_{k'})}{\sum_{k=1}^{K} P(g_k)P(x|g_k)}, \text{ for } k' = 1, 2, ..., K. \tag{4}$$

$P(g_k|x)$ tell us the probability that given a data point x, it belongs to the parametric distribution g_k, when $P(x|g_k)$ is the probability that x comes from the parametric distribution k and $P(g_k) = \phi_k$ is the probability of the parametric distribution k. Once we obtain the probabilities $P(g_k|x)$, with $k = 1, ..., K$, we determine that the point x belongs to the parametric distribution with greater probabilistic value $P(g_k|x)$.

In this work, we propose to efficiently estimate the parameters for a Gaussian mixture model using a distance, known as Gini Index, between two probability distributions, the empirical distribution of the analyzed data and the density function of a Gaussian mixture.

We will give a brief introduction to the Gini Index in Sect. 2. In Sect. 3 we will show the proposed procedure to estimate the parameters of a Gaussian mixture model through the Gini Index problem. In Sect. 4 we will perform experiments with simulated data and real data, to compare the numerical results obtained by the EM-algorithm, the K-means method and the algorithm proposed in this work. We will mention conclusions and future work in Sect. 5.

2   The Gini Index

The Gini Index is a measure of inequality level between two probability distributions. It is applied in several fields of study like engineering, ecology, transport and in income distribution as an indicator of social and economic inequality. See [5,9] and [13].

For the Gini Index problem (GI), in its discrete form, we consider a discrete random variable X, two probability measures ν_1 and ν_2 in X and a distance function in $X \times X$, $d : X \times X \to \mathbb{R}$.

The *GI* problem consists in

$$\text{minimize: } \sum_{x \in X} \sum_{y \in X} d(x, y) \, \pi(x, y) \tag{5}$$

$$\text{subject to: } \pi \in \Pi(\nu_1, \nu_2),$$

where $\Pi(\nu_1, \nu_2)$ is the set of joint probability distributions in $X \times X$, whose marginals in the first and second components are the probability distributions ν_1 and ν_2, respectively.

In [11] and [15] has been shown that this problem always has a solution, it is a distance between the probability distributions ν_1 and ν_2 and it can be very expensive to find. In addition, the solution $\pi^* \in \Pi(\nu_1, \nu_2)$ is a probability measure and the optimal value of this problem defines the Gini Index between the probability measures ν_1 and ν_2, denoted by $GI(\nu_1, \nu_2)$, that is,

$$GI(\nu_1, \nu_2) = \sum_{x \in X} \sum_{y \in X} d(x, y) \, \pi^*(x, y). \tag{6}$$

In this work, we propose to efficiently estimate the parameters for a Gaussian mixture model that minimizes the Gini Index to an empirical distribution. We make this proposal based on the theory of the minimum dissimilarity estimators and the estimators of minimum distance of Kantorovich given in paper [1]. In this work, the distribution ν_1 is known, commonly associated with an empirical distribution, an the distribution ν_2 is a parametric distribution that must be estimated in such a way that the distance between ν_1 and ν_2, given by the Gini Index, is minimum. In this case, we propose the parametric distribution ν_2 as a Gaussian mixture model. Modeling multimodal data through a mixture of Gaussians distributions makes intuitive sense, due that the most used distribution in modeling unimodal data is the Gaussian distribution, as we can see with the Central Limit Theorem, see [3].

3   Parameters Estimation Minimizing the Gini Index

Suppose that we have a P dataset with M elements in dimension N, denoted by

$$P = \left\{ p_m = \left(p_1^{(m)}, p_2^{(m)}, ..., p_N^{(m)} \right) \right\}_{m=1}^{M}. \tag{7}$$

If we consider the P set as a data frame, we have the arrangement given in Table 1. We define data frame column sets as $C_n = \left\{ p_n^{(m)} \right\}_{m=1}^{M}$, for $n = 1, ..., N$.

Now, we must define the base elements to establish the Gini Index problem: the random variable $\boldsymbol{X}$ with its probability distributions ν_1 and ν_2.

Table 1. Data frame of P

	C_1	C_2	$\cdots$	C_N
p_1	$p_1^{(1)}$	$p_2^{(1)}$	$\cdots$	$p_N^{(1)}$
p_2	$p_1^{(2)}$	$p_2^{(2)}$	$\cdots$	$p_N^{(2)}$
$\vdots$	$\vdots$	$\vdots$	$\ddots$	$\vdots$
p_M	$p_1^{(M)}$	$p_2^{(M)}$	$\cdots$	$p_N^{(M)}$

3.1 Random Variable X

To define the random variable $\boldsymbol{X}$, we use a representative histogram of each data frame column C_n, with $n = 1, ..., N$. For each representative histogram, we obtain the intervals center points, denoted as X_n, for $n = 1, ..., N$. This X_n variables are represented as dots in Fig. 1. We will denote as N_n the number of elements in X_n, for $n = 1, ..., N$, then

$$X_n = \left\{ x_1^{(n)}, x_2^{(n)}, ..., x_{N_n}^{(n)} \right\}, \text{ with } n = 1, ..., N. \tag{8}$$

We define the random variable $\boldsymbol{X}$ for the Gini Index problem as $\boldsymbol{X} = X_1 \times X_2 \times ... \times X_N$. This random variable has $\boldsymbol{N} = N_1 \cdot N_2 \cdot ... \cdot N_N$ elements. So

$$\boldsymbol{X} = \left\{ \left(x_{j_1}^{(1)}, x_{j_2}^{(2)}, ..., x_{j_N}^{(N)} \right) \middle| x_{j_n}^{(n)} \in X_n, \ n = 1, 2, ..., N, \ j_n = 1, 2, ..., N_n \right\}. \tag{9}$$

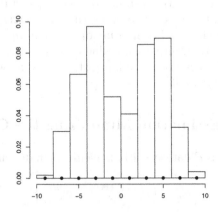

Fig. 1. Representative histogram and central points set X_n for column C_n.

3.2 Empirical Distribution ν_1

We consider the information regarding the density given by the representative histograms to define the empirical distribution ν_1 in the random variable $\boldsymbol{X}$ defined in (9).

Consider the column C_n, its representative histogram and its corresponding random variable $X_n = \left\{ x_1^{(n)}, x_2^{(n)}, ..., x_{N_n}^{(n)} \right\}$, for a fixed $n \in \{1, 2, ..., N\}$.

Denote as $d_{j_n}^{(n)}$ the density of $x_{j_n}^{(n)} \in X_n$, for $j_n = 1, ..., N_n$, see Fig. 2. To define a probability distribution γ_n in X_n, we consider the intervals endpoints for the corresponding histogram C_n, shown as squares in Fig. 2. We can obtain the width of the intervals with this endpoints, denoting them as A_n, due the columns in the histogram have the same width. Thus, we define the γ_n distribution in X_n as

$$\gamma_n \left(x_{j_n}^{(n)} \right) = A_n \cdot d_{j_n}^{(n)}, \text{ for } j_n = 1, ..., N_n. \tag{10}$$

So, we obtain the probability distributions γ_1, γ_2, ..., γ_N, for each of the columns C_1, C_2, ..., C_N, respectively.

To define the empirical distribution ν_1 in X, we consider the multiplication of the empirical distributions for each of the columns

$$\nu_1(x) = \gamma_1 \left(x_{j_1}^{(1)} \right) \cdot \gamma_2 \left(x_{j_2}^{(2)} \right) \cdot ... \cdot \gamma_N \left(x_{j_N}^{(N)} \right), \text{ for } x = \left(x_{j_1}^{(1)}, x_{j_2}^{(2)}, ..., x_{j_N}^{(N)} \right) \in X. \tag{11}$$

It is important to remember that for the i-th coordinate of x, $x_{j_i}^{(i)}$, j_i can take values from 1 to N_i, i.e. the value of each subscript j_n depends on the superscript n, corresponding to the x coordinate (see expression (9)).

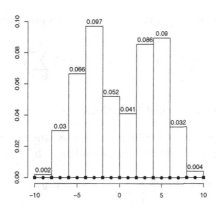

Fig. 2. X_n densities and extreme points of the histogram intervals in column C_n.

3.3 Parametric Distribution ν_2

We consider a Gaussian mixture model for the parametric distribution ν_2 in X, that is a density function with form

$$\nu_2(x) = \sum_{k=1}^{K} \phi_k f(x | \mu_k, \Sigma_k). \tag{12}$$

The f function denotes an independent multivariate normal distribution of dimension N, that is

$$f(x|\mu_k, \Sigma_k) = \frac{1}{(2\pi)^{N/2}|\Sigma_k|^{1/2}} \cdot e^{\left(-\frac{1}{2}(x-\mu_k)^T \Sigma_k^{-1}(x-\mu_k)\right)}, \tag{13}$$

where the mean μ_k is a real vector $(\mu_{k1}, \mu_{k2}, \ldots, \mu_{kN})^T$ and the covariance matrix is a real diagonal positive definite matrix of dimension $\mathbf{N \times N}$, i.e. $\Sigma_k = diag(\sigma_{k1}^2, \sigma_{k2}^2, \ldots, \sigma_{kN}^2)$. We know that, by the Σ_k form, the f function is

$$f(x|\mu_k, \Sigma_k) = \prod_{n=1}^{N} g(x_n|\mu_{kn}, \sigma_{kn}^2), \text{ for } x = (x_1, x_2, \ldots, x_N) \in \mathbb{R}^N, \tag{14}$$

where g is the univariate normal density function

$$g(s|\mu, \sigma) = \frac{1}{\sqrt{2\pi\sigma^2}} e^{-\frac{(s-\mu)^2}{2\sigma^2}}. \tag{15}$$

So, for the random variable X given in (9), we have

$$\nu_2(x) = \sum_{k=1}^{K} \phi_k \cdot \prod_{n=1}^{N} g(x_{j_n}^{(n)}|\mu_{kn}, \sigma_{kn}^2), \text{ for } x = \left(x_{j_1}^{(1)}, x_{j_2}^{(2)}, \ldots, x_{j_N}^{(N)}\right) \in X. \tag{16}$$

The ϕ_k parameters are the mixture proportions and must comply with

$$0 \le \phi_k \le 1, \text{ for } k = 1, \ldots, K, \text{ and } \sum_{k=1}^{K} \phi_k = 1. \tag{17}$$

3.4 Gini Index Problem and Solution

With the establishment of the random variable X and the probability distributions ν_1 and ν_2 we define the Gini Index problem, which consists of

$$\text{Minimize: } \sum_{x \in X} \sum_{y \in X} d(x, y)\pi(x, y)$$

$$\text{subject to: } \sum_{y \in X} \pi(x, y) = \nu_1(x), \forall x \in X$$

$$\sum_{x \in X} \pi(x, y) = \nu_2(y), \forall y \in X \tag{18}$$

$$\pi(x, y) \ge 0, \forall x, y \in X$$

$$\sum_{x \in X} \sum_{y \in X} \pi(x, y) = 1$$

considering d as the Euclidean distance in X, ν_1 as (11) and ν_2 as (16). We are looking for the proportions of the mixture $\phi = (\phi_1, \ldots, \phi_K)$, which must comply with (17), the means $\boldsymbol{\mu} = (\mu_1, \ldots, \mu_K)$ and the covariance matrices $\boldsymbol{\Sigma} = (\Sigma_1, \ldots, \Sigma_K)$.

If the solution exists, we can solve this problem using the Lagrange multiplier method. The Lagrangian for this problem is

$$\mathcal{L}(\phi, \boldsymbol{\mu}, \boldsymbol{\Sigma}) = \sum_{x \in X} \sum_{y \in X} d(x, y)\pi(x, y) - \sum_{x \in X} \lambda_x \left[\sum_{y \in X} \pi(x, y) - \nu_1(x) \right] \quad (19)$$

$$- \sum_{y \in X} \gamma_y \left[\sum_{x \in X} \pi(x, y) - \nu_2(y) \right] - \alpha \left[\sum_{x \in X} \sum_{y \in X} \pi(x, y) - 1 \right] - \beta \left[\sum_{k=1}^{K} \phi_k - 1 \right].$$

We denote as $\pi_{xy} = \pi(x, y)$ and $d_{xy} = d(x, y)$. If we derive the Lagrangian $\mathcal{L}$ with respect to π_{st}, for s and t fixed in X, we can simplify the equation (19).

If we derive this simplified expression with respect to μ_{tr}, with fixed $1 \leq t \leq K$ and fixed $1 \leq r \leq N$, and equals to 0, we have

$$\mu_{tr} = \frac{\sum_{j_r=1}^{N_r} y_{j_r}^{(r)} \exp\left(-\frac{(y_{j_r}^{(r)} - \mu_{tr})^2}{2\sigma_{tr}^2}\right)}{\sum_{j_r=1}^{N_r} \exp\left(-\frac{(y_{j_r}^{(r)} - \mu_{tr})^2}{2\sigma_{tr}^2}\right)}. \quad (20)$$

If we derive the simplified expression with respect to σ_{tr}, with fixed $1 \leq t \leq K$, $1 \leq r \leq N$, we have

$$\sigma_{tr}^2 = \frac{\sum_{j_r=1}^{N_r} \left(y_{j_r}^{(r)} - \mu_{tr}\right)^2 \exp\left(-\frac{(y_{j_r}^{(r)} - \mu_{tr})^2}{2\sigma_{tr}^2}\right)}{\sum_{j_r=1}^{N_r} \exp\left(-\frac{(y_{j_r}^{(r)} - \mu_{tr})^2}{2\sigma_{tr}^2}\right)}. \quad (21)$$

Expressions (20) and (21) can be taken iteratively. This estimate is made with respect to the P dataset, so the iterative expressions of the **IG-algorithm** are:

$$\mu_{tr} = \frac{\sum_{m=1}^{M} p_r^{(m)} \exp\left(-\frac{(p_r^{(m)} - \mu_{tr})^2}{2\sigma_{tr}^2}\right)}{\sum_{m=1}^{M} \exp\left(-\frac{(p_r^{(m)} - \mu_{tr})^2}{2\sigma_{tr}^2}\right)}, \quad \text{for } 1 \leq t \leq K \text{ and } 1 \leq r \leq N, \quad (22)$$

$$\sigma_{tr}^2 = \frac{\sum_{m=1}^{M} \left(p_r^{(m)} - \mu_{tr}\right)^2 \exp\left(-\frac{(p_r^{(m)} - \mu_{tr})^2}{2\sigma_{tr}^2}\right)}{\sum_{m=1}^{M} \exp\left(-\frac{(p_r^{(m)} - \mu_{tr})^2}{2\sigma_{tr}^2}\right)}, \quad \text{for } 1 \leq t \leq K \text{ and } 1 \leq r \leq N.$$

$$(23)$$

It is important to emphasize that to obtain the expressions (22) and (23) we made the assumption that the components of the Gaussian mixture ν_2 were independent multivariate normal distributions, and for this reason we only looked for the values of the covariance matrix that is a positive definite diagonal matrix.

4   Numerical Results

In this section we will compare the GI-algorithm with the EM-algorithm and the K-means method. We made this comparison due that the EM-algorithm and the K-means method are efficient heuristics that quickly converge to a local optimum. Both algorithms use the data group centers to model the data with the difference that K-means consider groups of comparable spatial extent, while the EM-algorithm allows different shapes in the groups. Because of these specifications the K-means method quickly converge to a local optimum and the EM-algorithm has a high level of efficiency, see [2].

We consider experiments with simulated data and real data. In the experiments with simulated data we consider data from 3 Gaussians. For experiments with real data we consider two databases: "Iris Data Set" and "Seeds Data Set", found in UCI Machine Learning Repository[1].

4.1   Simulated Data

In this section we will show the performed experiments with synthetic data. We perform the following process 100 times:

- We generate a P set with $M = 3000$ training data specifying the characteristics of the configuration, the same for 100 iterations.
  - Data proportion: the data amount can be *equal* or *different*.
  - Data intersection: Gaussians can be *separated* or *intersected*.
- We obtain a class vector that establishes from which Gaussian each data comes from, since we generated them.
- We adjusted each of the three models to the P dataset, assuming that there are 2, 3 and 4 groups.
- We obtain a data classification with each model and the percentage of classification success.
- We record the obtained success percentages through the three models and, for the univariate case, the Gini Index values given by the IG algorithm.

Once we obtain the results of the 100 iterations, we calculate an average percentage of success for each model and the average values of the Gini Index in the univariate case. These values are recorded in the tables in the following sections.

In each table we show in **bold** the best average success rate and in *italics* the second best average success rate. For the univariate case, the lowest Gini Index value is underlining. Let us remember that a small value for the Gini Index suggests that the empirical distribution is closer to the estimated theoretical model.

---

[1] https://archive.ics.uci.edu/ml/index.php.

Univariate Case

Configuration 1. (See Fig. 3(a)) 3000 data generated with equal proportions of 3 separate univariate Gaussians. The used parameters are: 1000 data with $\mu_1 = 4$, $\sigma_1 = 1$, 1000 data with $\mu_2 = 18$, $\sigma_2 = 2$ and 1000 data with $\mu_3 = 33$, $\sigma_3 = 2$.

The results are shown in Table 2. The best percentage of average success for the 3 models are those that consider the search for 3 Gaussians, because the data was generated from 3 Gaussians. With the IG algorithm we obtain a 100% percentage of average success. We obtain a 100% percentage when we are searching for 4 Gaussians because, even when it finds 4 Gaussians, one of them finds it with a proportion $\phi_k = 0$. For the value of the Gini Index, it can be seen that the minimum value is obtained for 3 and 4 Gaussians.

Table 2. Results of configuration 1, univariate case.

Adjusted Gaussians	IG algorithm	EM algorithm	K-means
2	66.66667 IG: 5.92747	66.36667	33.33333
3	**100** IG: 2.788711	*99.96667*	83.93333
4	**100** IG: 2.788711	89.23333	66.66666

Configuration 2. (See Fig. 3(b)) 3000 data generated with equal proportions from 3 univariate Gaussians, 2 intersected, with the following parameters: 1000 data with $\mu_1 = -3, \sigma_1 = 1$, 1000 data with $\mu_2 = 10, \sigma_2 = 2$ and 1000 data with $\mu_3 = 16, \sigma_3 = 1$. The obtained results are found in Table 3.

Table 3. Results of configuration 2, univariate case.

Adjusted Gaussians	IG algorithm	EM algorithm	K-means
2	66.66667 IG: 2.350686	66.66667	33.33333
3	**98.13333** IG: 1.963381	97.799667	96.36667
4	*97.8* IG: 1.963519	83.16667	84.63333

The models obtain a better average percentage of success for 3 Gaussians and the IG algorithm has the highest value. In this case, for some of the iterations, the IG algorithm did find 4 clusters, so the average success rate was lower than that of 3 Gaussians, however this is reflected in the Gini Index value.

In this configuration, we obtain a better percentage with our model than with the other ones when we adjust 4 Gaussians due the same reason than the previous configuration. In fact, if we adjust a higher quantity of Gaussians than the real number of it, our model, in automatic, calculated the parameters of the additional Gaussians with proportion ϕ_k near or equal to 0.

Configuration 3. (See Fig. 3(c)) 3000 data generated with different proportions from 3 intersected univariate Gaussians, with the following parameters: 1500 data with $\mu_1 = -5$, $\sigma_1 = 2$, 1000 data with $\mu_2 = 2$, $\sigma_2 = 1$, 500 data with $\mu_3 = 8$, $\sigma_3 = 2$. In Table 4 we show the results for this configuration.

Table 4. Results of configuration 3, univariate case.

Adjusted Gaussians	IG algorithm	EM algorithm	K-means
2	81.79333 IG: 1.662882	80.967	49.967
3	**97.635** IG: <u>0.91244</u>	*97.568*	75.75867
4	**97.635** IG: <u>0.91244</u>	85.70867	66.66667

In configuration 3 we obtain better percentage of average success with the IG algorithm. We obtain the same averages for 4 Gaussians, which means that the model found the same means and deviations values for 3 and 4 Gaussians, with a mixing proportion $\phi_k = 0$ for the 4 Gaussians case.

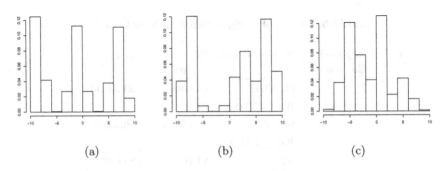

(a)   (b)   (c)

Fig. 3. Configuration examples for the univariate case

Bivariate Case. For the data generation in the plane, we consider independent bivariate normal distributions, that is, Gaussians whose covariance matrix is diagonal, and non-independent bivariate normal distributions, that is, Gaussians whose covariance matrix is a positive definite non-diagonal matrix, this in order to verify how efficient the models are when we use databases that might not meet the independence condition used by the IG algorithm.

Configuration 1. (See Fig. 4(a)) 3000 data generated with equal proportions of 3 separate bivariate Gaussians, with the parameters: 1000 data with $\mu_1 = (0, 16)$, $\Sigma_1 = \begin{pmatrix} 1 & 1 \\ 1 & 2 \end{pmatrix}$, 1000 data with $\mu_2 = (10, 6)$, $\Sigma_2 = \begin{pmatrix} 2 & 0 \\ 0 & 1 \end{pmatrix}$, 1000 data with $\mu_3 = (21, -5)$, $\Sigma_3 = \begin{pmatrix} 2 & 0 \\ 0 & 2 \end{pmatrix}$.

In the Table 5 you can see the obtained results. We have the best percentage of average success with the IG algorithm for 3 and 4 Gaussians and the K-means method for 3 Gaussians. In addition, we obtain a higher average with the EM algorithm when we consider 4 Gaussians, which would mean that this algorithm did not correctly locate the means of the analyzed data for three Gaussians and made a bad classification.

Table 5. Results of configuration 1, bivariate case.

Adjusted Gaussians	IG algorithm	EM algorithm	K-means
2	66.66667	66.66667	66.66667
3	100	50.16667	100
4	100	83.43333	83.6

Configuration 2. (See Fig. 4(b)) 3000 data generated with *different proportions* from 3 *separate bivariate* Gaussians, with the following parameters: 1500 data with $\mu_1 = (2, -1)$, $\Sigma_1 = \begin{pmatrix} 1 & 0 \\ 0 & 2 \end{pmatrix}$, 1000 data with $\mu_2 = (11, 19)$, $\Sigma_2 = \begin{pmatrix} 2 & 0 \\ 0 & 1 \end{pmatrix}$, 500 data with $\mu_3 = (20, 9)$, $\Sigma_3 = \begin{pmatrix} 2 & 0 \\ 0 & 2 \end{pmatrix}$.

In Table 6 are shown the results for this configuration. In this case, we obtain high averages for two adjusted Gaussians since the three models find the classes with 1500 and 1000 data. We obtain a 100% averages with the EM algorithm and the IG algorithm.

Configuration 3. (See Fig. 4(c)) 3000 data generated with different proportions of 3 bivariate Gaussians, 2 intersected, with the following parameters: 1500 data with $\mu_1 = (2,2)$, $\Sigma_1 = \begin{pmatrix} 2 & 0 \\ 0 & 1 \end{pmatrix}$, 1000 data with $\mu_2 = (7,-1)$, $\Sigma_2 = \begin{pmatrix} 1 & -1 \\ -1 & 2 \end{pmatrix}$, 500 data with $\mu_3 = (11,5)$, $\Sigma_3 = \begin{pmatrix} 1 & 0 \\ 0 & 2 \end{pmatrix}$.

Table 6. Results of configuration 2, bivariate case.

Adjusted Gaussians	IG algorithm	EM algorithm	K-means
2	83.33333	83.33333	83.33333
3	100	100	58.342
4	100	86.78867	52.28367

The results are in Table 7. Here, we obtain the best average with the K-means method for 3 Gaussians. However, we obtain values not so far apart with the additional advantage that we have the considered standard deviations values in the data generation.

Also, in the same way than previous examples, when we adjust one more group than the original number of groups, the proposed model has less fails than the other ones due that we obtain a proportion ϕ_k equals to zero for one of the calculated groups.

Table 7. Results of configuration 3, bivariate case.

Adjusted Gaussians	IG algorithm	EM algorithm	K-means
2	82.057667	81.633	82.141
3	99.41333	99.33733	**99.56733**
4	99.40267	83.80633	77.10833

Configuration 4. (See Fig. 4(d)) 3000 data generated with *different proportions* of 3 intersected bivariate Gaussians, with the following parameters: 1500 data with $\mu_1 = (4,5)$, $\Sigma_1 = \begin{pmatrix} 2 & -1 \\ -1 & 2 \end{pmatrix}$, 1000 data with $\mu_2 = (7,0)$, $\Sigma_2 = \begin{pmatrix} 2 & 1 \\ 1 & 2 \end{pmatrix}$, 500 data with $\mu_3 = (13,1)$, $\Sigma_3 = \begin{pmatrix} 2 & -1 \\ -1 & 1 \end{pmatrix}$.

The results for this configuration are shown in Table 8. We obtain the best percentage of average success with the IG algorithm when we consider 3 clusters.

Table 8. Results of configuration 4, bivariate case.

Adjusted Gaussians	IG algorithm	EM algorithm	K-means
2	81.01367	80.56667	60.93333
3	**95.96667**	93.80067	81.96667
4	*95.9*	78.835	79.30333

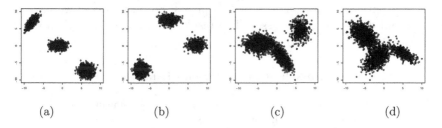

(a)   (b)   (c)   (d)

Fig. 4. Configuration examples for the bivariate case

Remarks

- In most cases we get better results with the IG algorithm than with the other two models.
- The difference between the average for 3 and 4 clusters of the IG algorithm is very small, because when we consider 4 clusters, this model finds the parameters for the 3 Gaussians and the parameters of a fourth Gaussian but with proportion ϕ_k equal or close to 0.
- The Gini Index values for the univariate case are lower when 3 clusters are considered, so it could be said that the Gini Index value is lower when we consider the actual number of data classes.

4.2 Real Data

To carry out classification experiments with real data, we consider "Iris Data Set" and "Seeds Data Set" of UCI Machine Learning Repository. We form a general P dataset for each of these data sets and then we use the GI-algorithm, the EM-algorithm and the K-means method to classification in the following way:

- We split the data set by classes. Then we randomly divide each class into training set and test set, with 70% and 30% of elements, respectively.
- We generate a general training set and a general test set by joining the training sets and test sets of the considered classes.
- We adjust the general training set with the EM-algorithm, the GI-algorithm and the K-means method, to obtain the values of the sought means and standard deviations.

– With the found parameters, we make a classification of the general test set, obtaining a percentage of success for each of the analyzed algorithms.

We carry out this process 100 times obtaining percentages of success on each occasion.

Iris Data Set. In this data set the authors examine 3 different varieties of Iris plant: Iris Setosa, Iris Versicolour and Iris Virginica, 50 instances each, considering 4 physical characteristics of this plants. So, in this case we have a 4-dimensional dataset with 3 classes. The results obtained for this database are shown in Table 9.

Table 9. Results for the Iris Data Set

Adjusted Gaussians	IG algorithm	EM algorithm	K-means
2	64.44444	50.22222	66.66667
3	**82.83333**	*79.44444*	79.22222
4	78.11111	63.88889	65.44444

We obtain the best percentage of success for the 3 models when we consider 3 clusters, because the database has 3 differentiated classes. We obtain a better percentage of average success with the IG algorithm for 3 Gaussians.

Seeds Data Set. In this data set, the authors examine 3 different varieties of wheat seeds: Kama, Rosa and Canadian, 70 instances each, considering 7 geometrical parameters of wheat grains, then we have a 7-dimensional dataset with 3 classes.

By adjusting this dataset through the GI algorithm, the EM algorithm, and the K-means method, we obtain the average percentage of success show in the Table 10.

Table 10. Results for the Seeds Data Set

Adjusted Gaussians	IG algorithm	EM algorithm	K-means
2	**65.06349**	63.44444	**65.06349**
3	*64.95238*	52.38095	63.44444
4	58.76191	42.93651	57.06349

In this case, with the 3 models we obtain a better percentage of average success when we consider 2 adjusted Gaussians, which would suggest that 2 of the classes are not well differentiated, although this cannot be corroborated due the dimension of the data. Even with these characteristics, we obtain the best averages with the IG algorithm.

5 Conclusions and Future Work

Thanks to the experiments carried out in this work, we can say that with the proposed model we obtain favorable results, because our model seeks to minimize the Gini Index between the empirical distribution and the proposed parametric distribution.

It is important to mention that the GI-algorithm performs fewer iterations than the EM-algorithm to find the sought parameters and consequently the average time it takes to find them is smaller. This leads us to declare that the proposed model helps us to efficiently estimate the parameters of a Gaussian mixture model through the Gini Index problem.

It should be noted that when we analyze the obtained results, we can see that the values for the covariance matrices with the IG algorithm are smaller than those obtained with the EM algorithm, which graphically translates into sharper or steeper curves. We believe that this characteristic is of great importance to carry out the pertinent classifications, because although the groups to be classified are intersected, the small value of the variances helps that the points to be classified correctly.

As a future work, we want to study the theoretical properties and the convergence of the IG-algorithm and search for applications with real data.

References

1. Bassetti, F., Bodini, A., Regazzini, E.: On minimum Kantorovich distance estimators. Stat. Probab. Lett. **76**(12), 1298–1302 (2006)
2. Bishop, C.M.: Pattern Recognition and Machine Learning. Springer, New York (2006)
3. Capinski, M., Kopp, P.: Measure, Integral and Probability. Springer Undergraduate Mathematics Series. Springer, London (2013). https://books.google.com.mx/books?id=5d6PBAAAQBAJ
4. Dempster, A.P., Laird, N.M., Rubin, D.B.: Maximum likelihood from incomplete data via the EM algorithm. J. Roy. Stat. Soc.: Ser. B (Methodol.) **39**(1), 1–22 (1977)
5. Giorgi, G.M., Gigliarano, C.: The Gini concentration index: a review of the inference literature. J. Econ. Surv. **31**(4), 1130–1148 (2017)
6. Greenspan, H., Ruf, A., Goldberger, J.: Constrained Gaussian mixture model framework for automatic segmentation of MR brain images. IEEE Trans. Med. Imaging **25**(9), 1233–1245 (2006)
7. Meng, X.L., Rubin, D.B.: On the global and component wise rates of convergence of the EM algorithm. Linear Algebra Appl. **199**, 413–425 (1994)
8. Povey, D., et al.: The subspace Gaussian mixture model–a structured model for speech recognition. Comput. Speech Lang. **25**(2), 404–439 (2011)

9. Rachev, S.T., Klebanov, L.B., Stoyanov, S.V., Fabozzi, F.: The Methods of Distances in the Theory of Probability and Statistics. Springer, New York (2013). https://doi.org/10.1007/978-1-4614-4869-3

10. Reynolds, D.A.: Gaussian mixture models. In: Encyclopedia of Biometrics, vol. 741 (2009)

11. Rubner, Y., Tomasi, C., Guibas, L.J.: The earth mover's distance as a metric for image retrieval. Int. J. Comput. Vis. **40**(2), 99–121 (2000). https://doi.org/10.1023/A:1026543900054

12. Torres-Carrasquillo, P.A., Reynolds, D.A., Deller, J.R.: Language identification using gaussian mixture model tokenization. In: 2002 IEEE International Conference on Acoustics, Speech, and Signal Processing, vol. 1, p. I-757. IEEE (2002)

13. Ultsch, A., Lötsch, J.: A data science based standardized Gini index as a Lorenz dominance preserving measure of the inequality of distributions. PloS One **12**(8), e0181572 (2017)

14. Vaida, F.: Parameter convergence for EM and MM algorithms. Statistica Sinica **15**, 831–840 (2005)

15. Villani, C.: Topics in Optimal Transportation. American Mathematical Society, Providence (2003)

16. Xu, L., Jordan, M.I.: On convergence properties of the EM algorithm for Gaussian mixtures. Neural Comput. **8**(1), 129–151 (1996)

MASDES-DWMV: Model for Dynamic Ensemble Selection Based on Multiagent System and Dynamic Weighted Majority Voting

Arnoldo Uber Jr.[1]([⊠])(iD), Ricardo Azambuja Silveira[1](iD),
Paulo Jose de Freitas Filho[1](iD), Julio Cezar Uzinski[2](iD),
and Reinaldo Augusto da Costa Bianchi[3](iD)

[1] Department of Informatics and Statistics, Federal University of Santa Catarina,
Florianópolis, Brazil
arnoldo.u.jr@gmail.com, ricardo.silveira@ufsc.br, pjffbr@gmail.com
[2] State University of Mato Grosso Faculty of Exact and Technological Sciences,
Cuiabá, Brazil
uzinski.julio@unemat.br
[3] Electrical Engineering Department, Centro Universitario da FEI,
São Bernardo do Campo, Brazil
rbianchi@fei.edu.br

Abstract. One of the main challenges of machine learning algorithms is to maximize the result and generalization. Thus, the committee machines, i.e., the combination of more than one learning machine (an approach also called in the literature by ensemble), together with agent theory, become a promising alternative in this challenge. In this sense, this research proposes a dynamic selection model of machine committees for classification problems based on the multi-agent system and the weighted majority voting dynamic. This model has an agent-based architecture, with its roles and behaviors modeled and described throughout the life cycle of the multiagent system. The steps of generalization, selection, combination, and decision are performed through the behavior and interaction of agents with the environment, in the exercise of their respective roles. In order to validate the model, experiments were performed on 20 datasets from four repositories and compared with seven state-of-the-art dynamic committee selection models. At the end, the results of these experiments are presented, compared and analyzed, in which the proposed model obtained considerable gains in relation to the other models.

Keywords: Machine committee · Ensemble · Multiagent system · Dynamic weighted majority vote

1 Introduction

Usually, most people who need to make a decision, seek the opinion of other people or experts about the problem. This behavior has the objective of obtaining

© Springer Nature Switzerland AG 2020
L. Martínez-Villaseñor et al. (Eds.): MICAI 2020, LNAI 12469, pp. 419–434, 2020.
https://doi.org/10.1007/978-3-030-60887-3_36

more points of view on the subject, allowing in this way to make a more assertive decision. This way of solving problems is also present in situations where there is voting; in this case, opinions are returned by votes.

In the Machine Learning area, the idea of committee formation had its first proposals made in Ablow's, Kaylor, and Nilsson's 1965 research [20]. In the first works, there was a proposal to combine different classifiers for the final improvement of a classification system. However, the area was only intensively developed later, with computational improvements, through the work of [12] and [21], who had improvements in the performance of Artificial Neural Networks (RNA) from the use of machine committees. See [24] and [23] for details.

In supervised classification problems, data instances are represented as pairs (x, y_i), where x represents an instance described in the form of an attribute value vector and y_i, are class names, where y_i belongs to a set of classes $Y = y_1, y_2, y_3, ... y_n$. The learning algorithm specifies a classifier whose output must be the prediction of a class y_z, with $z \in 1,2,3, ..., n$, the given instance.

Machine committees, as the name implies, are the union of more than one learning machine in generating a solution to a problem. In addition to having the main objective of maximizing the generalization capacity, they have two other motivational factors: the availability of computational resources (to generate the union of several techniques in the solution of a problem) and the demonstration made by [25] which demonstrates that there are no generic models of machine learning that, on average, perform better than any other model for any class of problems [23,24].

The composition of a machine committee can take into account all components, only one of them, or select a subset of components from the set of available components. The strategy of considering all also carries with it the components with low aptitude. The idea of considering only one component leads to algorithm matching rates for the individual component. Therefore, selecting a subset of components and matching them appropriately leads to higher hit rates, as mentioned by [1,5,24] and [10]. Dynamic selection of machine committees is a multiple classifier approach that selects a subset of classifiers best suited to classify a query pattern. Therefore, the dynamic selection of machine committees (which will be referred to in this article by the ensemble) is an excellent strategy used to maximize results in solving complex decision problems.

This paper proposes a dynamic ensemble selection based on multiagent system (MAS) and the dynamic weighted majority voting (DWMV), which was called MASDES-DWMV. It is also the purpose of this article to apply the proposed model in different case studies of the literature and present the results obtained, and in the end, compare with other existing models. Dynamic selection of ensemble consists of choosing a subset (possibly unitary) of classifiers, maximizing their combination, and thereby predict the class of a particular instance of the problem, according to [5]. In these stages, many decision-making is performed, and each stage can be improved. Therefore, agent theory is proposed in this research as an alternative to assist in the autonomous decision making of the agents and the coordination of the ensemble, besides providing mechanisms to scale and distribute processing in the learning of the instances of the classifiers.

The proposed multi-agent system (MAS) for dynamic ensemble selection (DES) seeks to maximize the results of the main stages of ensemble formation: generalization, selection, and combination (decision). In the generalization, it aims to maximize the individual results of the classifiers, seeking to maximize their competence through parameterization and training. Also, it introduces the concept change (Drifting Concepts), in which classifiers are added or removed depending on their competence. See [3,7,16] and [1] for details. As for selection and combination, it assists in the definition of the classifier or subset of classifiers by proposing a weight-adjusted majority voting method based on the WAVE method [14], that is, the Dynamic Weighted Majority Voting method (DWMV), which extends the WAVE algorithm, making the necessary improvements to DES.

This article is organized into sections. In the next section, a theoretical revision of the basic concepts of machine committees, agent theory, and DES is presented, also presenting a retrospect and state of the art. Section 3 presents the MASDES-DWMV model, describes the architecture, agent roles, and life cycle, detailing the algorithms used in agent behavior. Section 4 describes the experiments and presents the methodology and results, and Sect. 5 describes the conclusions.

2   Theoretical Background

The use of Ensembles as an alternative to increasing the efficiency and accuracy in pattern recognition [18] is mentioned in the literature in several applications [1,10,26]. The work found involves two different approaches: the combination of classifiers and the dynamic selection of classifiers [9]. In the combination of classifiers, all available classifiers are used and their outputs combined to form the final decision. This approach presents two main problems: in the first problem, classifiers are assumed to make independent, and possibly different, errors, which is a difficult situation to find in real pattern recognition applications [9]. The second problem is that not every available classifier is a specialist for the test pattern. Therefore only some classifiers can predict the correct classification for a given set of data. The dynamic selection precisely seeks to estimate the competence level of the classifier for each query separately. Only the most competent classifier about the input pattern are selected to compose Ensemble. Figure 1 shows the basic scheme of dynamic selection.

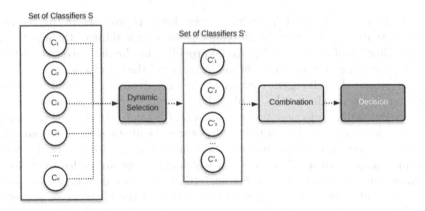

Fig. 1. Dynamic selection operation scheme

The selection comprises three ways of being performed: static, classifier dynamics and ensemble dynamics. Figure 2 exemplifies the three forms of selection.

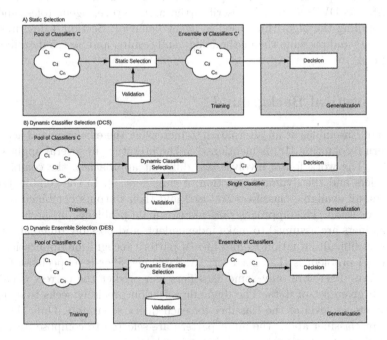

Fig. 2. Selection methods source adapted from [10]

In Fig. 2, item (a), the static selection is presented, which is made the selection of the classifiers based on the training data while the dynamic selection (b) is based on the selection in the test data. The difference between dynamic selection of classifiers and ensemble (c) is that the former selects only one classifier and in the second, dynamic ensemble selection (DES), a set of classifiers [10].

In dynamic selection, sorting a new input pattern usually involves three steps:

1. Define the region of competence: area around the place of consultation;
2. Define the selection criterion: method of calculating the competence of the classifier, for example: accuracy and complexity;
3. Define the approach or selection mechanism: whether a classifier (DSC) or a set of classifiers (DES) will be used.

A review of the most relevant work on dynamic selection is presented in Table 1 according to [1,3,10] and [8].

Table 1. Comparison between DES models

Reference	Year	Model	Region competence	Selection criteria	Approach
Sabourin	1993	Classifier Rank (DCS-Rank)	K-NN	Ranking	DCS
Woods	1997	Overall Local Accuracy (OLA)	K-NN	Accuracy	DCS
Woods	1997	Local class accuracy (LCA)	K-NN	Accuracy	DCS
Giacinto	1999	A Priori	K-NN	Probabilistic	DCS
Giacinto	1999	A Posteriori	K-NN	Probabilistic	DCS
Giacinto et al.	2001	Multiple Classifier Behavior (MCB)	K-NN	Behavior	DCS
P.C. Smits	2002	Modified Local Accuracy (MLA)	K-NN	Accuracy	DCS
Soares et al.	2006	DES-Clustering	Clustering	Accuracy & Diversity	DES
Soares et al.	2006	DES-KNN	K-NN	Accuracy & Diversity	DES
Ko et al.	2008	K-Nearest Oracles Eliminate (KNORA-E)	K-NN	Oracle	DES
Ko et al.	2008	K-Nearest Oracles Union (KNORA-U)	K-NN	Oracle	DES
Woloszynski et al.	2011	Randomized Reference Classifier (RRC)	Potential function	Probabilistic	DES
Woloszynski et al.	2012	Kullback-Leibler (DES-KL)	Potential function	Probabilistic	DES
Woloszynski et al.	2012	DES Performance (DES-P)	Potential function	Probabilistic	DES
Cavalin et al.	2013	K-Nearest Output Profiles (KNOP)	K-NN	Behavior	DES
Cruz et al.	2015	META-DES	K-NN	Meta-Learning	DES
Cruz et al.	2016	META-DES.Oracle	K-NN	Meta-Learning	DES
Brun et al.	2016	Dynamic Selection On Complexity (DSOC)	K-NN	Accuracy & Complexity	DCS
Almeida	2016	DYNSE	K-NN	Behavior	DES
Albuquerque	2018	DESDD	Potential function	Accuracy & Diversity	DES

As shown in Table 1, in the region of competence, many works use the K-NN (K Nearest Neighbors) technique, using clustered methods (K-Means) using the decision of the classifiers or via competence map defined through a function potential. The selection criterion varies according to the algorithm; the first works directly use the accuracy or a ranking of the best classifiers, followed by probabilistic functions, in which the grouping of several criteria are used, which are examples of meta-learning and behavior. Finally, the approach is also defined, that is, the criteria for selecting a set of classifiers (DES) or only one (DCS).

The MASDES-DWMV model about the region of competence continually analyzes the performance of the classifiers based on the concept of context change, that is, if the set of training data used leads to a specific performance during the period. Training and later, the data set used in the execution, leads to another performance, the classifier about penalties and bonuses by adjusting the weights and parameters used in the execution of the model. This adds and removes classifiers as needed and maintains an in-stance history to align all classifiers with the same criteria. At this point, the extension to the WAVE method was crucial for performance improvement as it enabled the removal of low-efficiency classifiers and added new ones, based on a model performance history, with possibilities to improve the classifier set, preserving the instance history. These actions are not present in the original model, as well as the improvements in the execution of the models made by the agents, seeking the best set of execution parameters of the classification model. As for the selection criterion, it uses accuracy and complexity to define the weights of the classifiers to be selected to form the set of classifiers used in the model decision, allowing a DCS and DES approach (configured through parameters in the model execution). Also, all steps are performed in a distributed and autonomous manner in agents, an approach also used by [13] and [6]. At runtime, it may include or exclude classifiers, distribute the execution of algorithms in several environments.

3   Model MASDES-DWMV

The MASDES-DWMV model presents a MAS-based Ensemble dynamic selection architecture in which the combination step is performed through a WAVE method extension proposed by [14], called DWMV (Dynamic Weight Majority Voting). Therefore the MASDES-DWMV model describes an agent architecture, a model of interaction between these agents with their roles and the respective life cycle of the MAS so that it can fulfill the steps of generalization, dynamic selection and decision. The scheme of the MASDES-DWMV model is shown in Fig. 3.

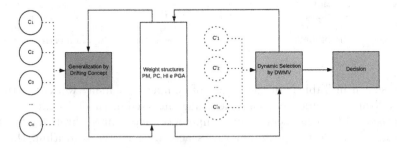

Fig. 3. MASDES-DWMV model layout

The Fig. 3 shows the classifiers linked to the generalization. The generalization has a behavior based on change of concept, adding and removing components through the analysis of the weights updated by dynamic selection. The dynamic selection updates the PM, PC, PGA and HI weights (which are the weight of the models, weights of the aComponents, global weight of hits and the history of the instances, respectively) and selects the component or a set of components to compose the decision of model. Between the generalization and dynamic selection stages, there is a set of weights that unifies the algorithms used by the agents' behaviors.

3.1 Architecture

The proposed agent architecture defines the role of five agents: Monitor, Generalization, Component, Dynamic Selection, and Decision. In this model, these agents are called aMonitor, aGeneralization, aComponent, aDynamicSelection, aDecision respectively. Figure 4 presents a diagram of the proposed agents in the model and their interaction.

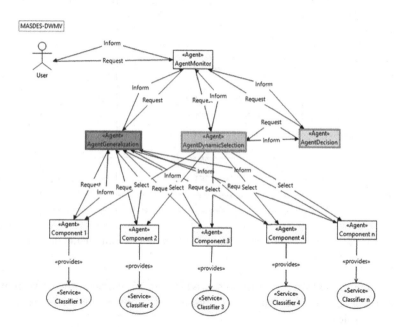

Fig. 4. MASDES-DWMV agent model

The Fig. 4 presents a macro model of agents in interaction with the user, the interrelationship between agents through protocols and, in the case of Agent-Components, their direct relationship with the classification models providing the classification service. In this Fig. 4, only a few request and response messages (Request and Inform) were cited, as a way of exemplifying the communication between the agents.

The communication protocol between agents, however, has a very diverse set of messages that are exchanged between agents, using Request and Inform. In addition to these, some other messages are exchanged with the environment by the agents. For example, in the case of AgentMonitor, to check existing agents, it checks the Directory Facilitator (DF), which agents are available and which provide a certain service, and then select the agent for the MASDES-DWMV structure. Another example is shown in Fig. 5.

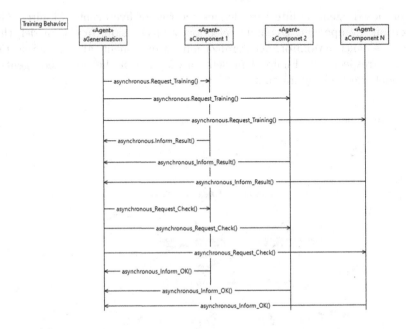

Fig. 5. Protocol between aGeneralization and aComponents agents

The communication protocol presented in Fig. 5 is carried out between aGeneralization agents and aComponents present in the environment, as described in the agents' roles.

3.2   Agent Roles

The roles of the agents that make up the MASDES-DWMN are related to the actions and decisions that the agent performs during its life cycle, that is, its behavior, and are described as:

- AgentMonitor (referred to as aMonitor): This agent is intended to enable user interaction with SMA, and also to create and monitor the base agent structure. The interaction can occur for several reasons, such as: informing a set of parameters, requesting results presenting a problem, modifying the composition of the set of components, etc.;

- AgentGeneralization (referred to as aGeneralization): responsible for forming and maintaining the set of aComponents. To form the set of components, you can execute algorithms known in the literature as Bagging [4], Boosting [22], etc., or use customized training by the user. As for maintenance, it has some behaviors modeled in relation to the components, such as: training aComponents, adding aComponent and removing aComponent;
- Component (referred to as aComponent): this agent encapsulates the classifier and provides interaction behaviors with it, such as: instantiation, parameterization, execution and improvement (depending on the results, aComponent can, or through aGeneralization demand, execute the classifier again other parameters, seeking to maximize the classifier's performance). It is worth remembering that, depending on the classifier model, these interactions may result in different implementations;
- AgentDynamicSelection (referred to as aDynamic Selection): it interacts with aComponents and aGeneralization and performs the selection of the set of aComponents through the DWMV algorithm, to define the aComponents that will be part of the decision of the agent aDecision;
- AgentDecision (referred to as aDecision): checks with the aDynamicSelection the selected set of aComponents and executes the DWMV algorithm, ultimately generating the model decision.

3.3   Life Cycle

The life cycle of the MASDES-DWMV model is based on cooperative-competitive behavior. The agents aMonitor, aGeneralization, aDynamicSelection and aDecision cooperate with each other in the search for the best result. AComponents agents compete among themselves for permanence in the execution of the model.

It starts with the instantiation of aMonitor by the user, who also defines the classification problem through the data set, how to use this data set (division of the data in training and testing with cross-validation methods: holdout, k-fold, etc.), the limits on the number of classifiers, aComponents model and training parameters, in addition to the generation method of the aComponents set (mentioned in the aGeneralization behavior).

AMonitor receives the information and instantiates the other agents of the model: aGeneralization, aDynamicSelection and aDecision, informing the initial definitions, therefore it is responsible for verifying that all the necessary agents for the SMA are present, and if they are not, perform the creation of the same. AMonitor also manages the weight structures used in the model, which are: PM, PC and HI, which are respectively: model weights, aComponents weights and instance history. AGeneralization, when created, runs an initialization algorithm described in Fig. 6.

```
aGeneralization.Initialization ({mC, nMi}, C, dsT)
    mC = aComponents model
    nMi = number of components of the initial model
    C = maximum number of components
    dsT = structure with the total test dataset
    PM = structure with model weights
    PC = structure with aComponents weights
    HI = structure with instance history

    1.  Define the initial training data set, observing the previously chosen
        form: Bagging, Boosting or Customized;
    2.  Check the types of aComponents and create instances of these agents,
        defining the initial parameters. By definition, at least two instances must
        be created.
    3.  Assign identical weights to components and PC models {c..C} = 1; PM
        {m ... M} = 1;
    4.  Ask aComponents to initialize their models.
```

Fig. 6. AGeneralization initialization behavior algorithm

The AComponents, when initialized, create instances of the models that encapsulate and apply the initial settings and parameters informed by aGeneralization. Subsequently, they inform the initial training dataset, but do not execute it, returning to aGeneralization that they are ready to start.

AGeneralization then proceeds to perform the training behavior, described in Fig. 7. As they are in parallel processes, they occur independently and autonomously, each aComponent has its perception of the environment and different parameters, and are trained with the same data set provided by aGeneralization, that is, the training instance.

```
aGeneralization.Training (dstn)
    dstn = structure with the Test dataset
    γ = Constant for updating the weight of the aComponent with error

    1. Asks aComponents to run their models with training data;
    2. Requests that aComponets execute a parameter improvement algorithm;
    3. If you can improve the model parameters, perform step 1 again.
```

Fig. 7. AGeneralization training behavior algorithm

The competitive form of the agents appears in the sense of the search for the best individual results, with the objective of remaining in the set of aComponents available for selection. Having executed the training algorithm in all aComponents, aGeneralization asks aDynamicSelection to choose the classifiers that will compose the ensemble, so aDynamicSelection executes the DynamicSelectionByDWMV algorithm described in Fig. 8.

```
aDynamicSelection.DynamicSelectionByDWMV ()

  DWMV ()
     1.  Reads weights from the executions of the AComponents.
     2.  Updates HI with the current instance.
     3.  Updates the weights of aComponents using the P * formula of the
         WAVE algorithm (Kim, 2011) presented below:

         ?  P ^ * = X?   ^ '(J_nk-X) (J_kk-I_k)

     4.  The convergence of the use of this formula is proven in Theorem 1
         detailed by Kim (2011), in which:
     5.  X = matrix n x k (n instances and k classifiers) containing 1s and 0s for
         successes and errors, respectively;
     6.  X '= matrix transposed from X
     7.  Jij = matrix i x j consisting of 1s for dimension i and j.
     8.  I k = identity matrix k x k;
     9.  After calculating the P *, the PC weight structure is updated.

  1.  Asks aMonitor for PM, PC and HI weight structures;
  2.  Updates the weights of aComponents using the DWMV method and
      returns to aMonitor;
  3.  Select the set of aComponents that will be part of the final decision
      through the filter of aComponents that have weights greater than the
      defined parameter, or if none meets the criterion, then q aComponents with
      greater weight are chosen, in which q is a parameter initially defined in
      model.
  4.  Asks aDecision to return the final decision on the input data set.
```

Fig. 8. aDynamicSelection selection behavior algorithm

For each training run, the weights of the aComponents are adjusted in the PC structure, the weights of the models in the PM structure and the training instances are stored in the HI structure. ADecision then executes the decision algorithm presented in Fig. 9.

```
aDecision. Decision ()

  1.  Asks aMonitor for the weight structures of aComponents PC;
  2.  Executes weighted majority voting algorithm generating the final
      decision regarding the classification;
  3.  Informs aGeneralization of the return;
  4.  Updates weight structure of PM models.
  5.  If there was an error, it penalizes by decreasing the Global Weight of
      Corrections (PGA) and increases the counter of the class with the
      highest number of errors.
```

Fig. 9. A decision behavior algorithm

The AGeneralization retrieves aDecision and updates the SMA according to the algorithm described in Fig. 10.

Generalization. UpdateSMA ()

1. Check if aComponents are below the desired minimum efficiency limit,
   that is, observe the PC weights, if any, remove it from the set of
   available aComponents and update PM weights;
2. Check if the PGA is lower than the minimum hit limit, if it is, check
   which model is more efficient for the highest error class through the
   PM, and add a new aComponent. After training in the instances already
   stored in the HI structure, using the algorithm in Table 3.

Fig. 10. Algorithm for updating SMA through aGeneralization

After the aComponents training stage is finished, the MASDES-DWMV model is ready for execution. At that time, aMonitor is notified by aGeneralization that the training has been completed and the model is now in execution mode. Thereafter the addition and removal of aComponents is suspended.

4   Experiments

A total of 20 datasets were used in comparative experiments, of these, five come from the Machine Learning Repository (UCI) repository [11], three from the STATLOG project [15], eight from Knowledge Extraction based on Evolutionary Learning [2] and four from the Ludmila Kuncheva Repository Collection of real medical data [17].

The experiments were carried out following the same methodology mentioned in the survey prepared by [10], to compare the results of the models analyzed by the author, using the same criteria. This methodology uses 20 replications for each dataset. With each replication, the datasets are randomly divided as follows: 50% for training, 25% for dynamic selection, and 25% for testing. Divisions are made according to the probabilities of each class.

Table 2 presents the results obtained through the DES-RRC, META-DES, META-DES.O, DES-KL, DES-P, and KNORA-U techniques mentioned in [10] survey complemented by the results obtained by MASDES-DWMV model.

In Table 2, the 20 datasets are listed, and the hit percentage and standard deviation for each model tested against the datasets are described. The result with the highest percentage of accuracy among the models is highlighted in bold. In the Position column the classification of the MASDES-DWMV model is mentioned about the others for each dataset. The last row of Table 3 shows the average of the models in all datasets and the first place quantities obtained by the models. The graph shown in Fig. 11 shows the models' performances about the datasets.

Table 2. Mean and standard deviation of dynamic selection techniques

Dataset	Position	MASDES-DWMV	DES-RRC	META-DES	META-DES.O	DES-KL	DES-P	KNORA-U	
Adult	1	88,63	1,78 87,16	1,53 87,15	2,43 87,74	2,04 86,06	2,90 87,10	2,76 80,21	2,26
Banana	1	95,67	1,12 86,56	1,76 91,78	2,68 94,54	1,16 85,58	2,40 93,61	1,80 92,40	2,87
Breast Cancer (WDBC)	3	97,32	0,81 96,94	0,61 97,40	1,07 96,71	0,86 97,13	0,59 96,78	0,78 97,41	1,02
Ecoli	1	81,62	3,28 80,66	3,58 77,25	3,52 81,57	3,47 79,95	4,30 79,83	4,26 77,12	3,41
Liver Disorders	2	71,56	3,76 68,01	4,14 70,08	3,49 72,02	4,72 67,11	5,62 67,46	3,84 61,29	3,76
German Credit	1	76,98	1,88 75,83	2,36 75,55	1,31 76,58	1,99 73,88	2,15 74,72	3,50 73,16	1,80
Glass	3	66,19	2,89 66,04	4,23 66,87	2,99 66,46	4,22 63,32	4,27 63,13	5,30 62,05	2,88
Heart	2	84,89	3,78 83,99	3,64 84,80	3,36 86,44	3,38 82,83	4,13 83,27	3,68 84,15	4,05
Laryngeal1	5	86,11	3,85 87,21	3,79 79,67	3,78 87,42	2,98 86,54	4,25 86,35	3,82 82,38	4,45
Laryngeal3	4	94,78	2,23 97,61	0,06 96,78	0,87 73,67	0,75 72,67	2,37 73,34	2,81 96,71	1,89
Magic Gamma Telescope	1	86,54	1,73 86,20	1,84 84,35	3,27 86,02	2,20 83,56	1,22 83,54	1,34 82,99	1,25
Mammographic	3	84,96	1,78 85,00	1,32 84,82	1,55 80,72	2,56 84,12	2,26 84,98	1,86 82,91	2,27
Monk2	1	95,42	1,58 80,98	2,58 83,24	2,19 94,95	1,88 80,85	2,68 79,93	2,57 77,88	4,25
Phoneme	1	86,84	2,19 74,65	1,55 80,35	2,58 85,05	1,08 77,13	1,32 81,64	0,53 79,94	3,33
Pima	3	78,71	2,58 77,64	2,73 79,03	2,24 77,53	2,24 77,97	2,64 76,87	1,87 78,84	2,18
Sonar	5	78,94	1,87 80,77	5,09 80,55	5,39 81,63	3,90 78,15	6,28 79,49	6,66 77,34	1,94
Thyroid	4	95,98	1,83 72,74	1,87 72,65	2,17 96,99	2,14 97,04	1,09 96,98	1,12 72,36	1,25
Vehicle	1	83,59	1,84 83,34	1,81 72,75	1,70 82,87	1,65 82,99	1,74 82,85	1,97 83,12	1,70
Weaning	3	81,02	3,68 78,96	3,29 79,00	3,78 81,73	3,14 78,74	4,44 78,08	4,07 82,02	3,65
Wine	4	98,64	2,13 98,77	1,57 99,25	1,11 99,52	1,11 98,26	2,45 98,48	2,30 98,03	1,62
Average/First Place		85,72	8 82,45	1 82,17	3 84,51	5 81,69	1 82,42	0 81,12	2

Adapted from [10]

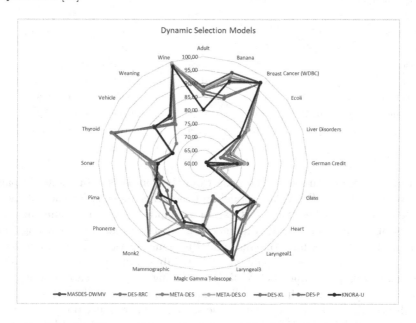

Fig. 11. The performances of the models about the datasets

A statistical analysis was performed to prove that the results obtained by the MASDES-DWMV model are statistically different from the other results of the other compared models, therefore, through the Welch t-test, the results were compared in MASDES-DWMV pairs with the results of the other models, for the 20 datasets, and the result is presented in Table 3.

Table 3. Statistical analysis by welch t-test

Welch t-test	DES-RRC	META-DES	META-DES.O	DES-KL	DES-P	KNORA-U
Adult	1,00	0,98	0,95	1,00	0,98	1,00
Banana	1,00	1,00	1,00	1,00	1,00	1,00
Breast Cancer (WDBC)	0,95	0,40	0,99	0,80	0,98	0,38
Ecoli	0,81	1,00	0,52	0,91	0,93	1,00
Liver Disorders	1,00	0,90	0,37	1,00	1,00	1,00
German Credit	0,95	1,00	0,74	1,00	0,99	1,00
Glass	0,55	0,23	0,41	0,99	0,98	1,00
Heart	0,78	0,53	0,90	0,95	0,91	0,72
Laryngeal1	0,18	1,00	0,12	0,37	0,42	1,00
Laryngeal3	0,01	0,05	1,00	1,00	1,00	0,00
Magic Gamma Telescope	0,72	0,99	0,79	1,00	1,00	1,00
Mammographic	0,47	0,60	1,00	0,90	0,49	1,00
Monk2	1,00	1,00	0,80	1,00	1,00	1,00
Phoneme	1,00	1,00	1,00	1,00	1,00	1,00
Pima	0,89	0,34	0,93	0,81	0,99	0,43
Sonar	0,07	0,11	0,05	0,70	0,36	0,99
Thyroid	1,00	1,00	0,06	0,02	0,02	1,00
Vehicle	0,67	1,00	0,90	0,85	0,89	0,80
Weaning	0,97	0,95	0,26	0,96	0,99	0,20
Wine	0,41	0,13	0,06	0,70	0,59	0,84

Table 3 presents the value resulting from the comparison of results in pairs, between the MASDES-DWMV model and the other models, for each of the 20 datasets. Based on these values, it can be concluded with 95% confidence that the algorithms are statistically different in all bold comparisons, reaching 100% of the tests for the Adult, Banana, and Phoneme datasets. The remaining 16 datasets had at least one test that was 95% reliable, and only one dataset, Wine, all comparisons had values below 95%. Analyzing the Wine case, it was found that it is the dataset with the smallest number of instances of the test dataset, and also the dataset whose tests have the smallest performance difference among the largest, 99.52% of META-DES.O, and the smallest, 98.03% of KNORA-U.

The prototype developed for model validation uses the Python 3.7 programming language in the Spyder 3.3 editor. Both belong to the Anaconda 3 distribution, which highlights the use of SciPy, scikit-learn, NumPy, Mlxtend, Pandas, matplotlib. It also uses the Python Agent Development framework (PADE) framework [19], which implements the FIPA standard.

5   Conclusion

In this paper, we present a dynamic ensemble selection model based on multiagent systems and dynamic weighted majority voting. The algorithms described are used in the behavior of the agents in which there is an interface between the generalization and dynamic selection steps. In the generalization stage, the classifiers are executed, and the weights are made available, and in the dynamic selection stage, the weights are updated, returning to the generalization stage to be used for permanence analysis or exclusion of the classifier.

The experimental results obtained were based on 20 classification datasets, from four different repositories (UCI, KELL, STATLOG and LKC) and compared between seven state-of-the-art models of SDE. The results of the experiments demonstrate that the proposed model obtained the best result in 8/20 analyzed datasets, that is, 40%, also obtaining the highest overall average of success, 85.72%. However, new experiments must be made using other models of classifiers, and in more use cases, thus being able to prove the results.

One of the characteristics noted in the proposed MASDES-DWMV model is that it performs better in problems with more instances of training and execution among the datasets selected in this research, which indicates that there is improvement in relation to the number of instances of the problem and the behaviors of inclusion and exclusion of classifiers and further analysis based on the DWMV algorithm.

New research can be done, in future works, changing the model for regression problems, thus being able to also use the datasets present in the literature for these classes of problems, and thus, validate the MASDES-DWMV model for this purpose.

Another suggestion for future work would be to use models based on Deep Learning as classification algorithms and make a comparison with MASDES-DWMV and other models

References

1. Albuquerque, R.A.S.: Seleção dinâmica de comitês de classificadores baseada em diversidade e acurácia para detecção de mudança de conceitos. Master's thesis, Universidade Federal de Santa Catarina (2018)
2. Alcalá-Fdez, J.E.A.: Keel - knowledge extraction based on evolutionary learning. http://www.keel.es. Accessed 06 May 2019
3. Almeida, P.R.L.D., Oliveira, L.S., BRITTO, A.D.S., Sabourin, R.: Handling concept drifts using a dynamic selection of classifiers. In: IEEE International Conference on Tools with Artificial Intelligence, pp. 989–995 (2016)
4. Breiman, L.: Bagging predictors. Mach. Learn. **24**(2), 123–140 (1996)
5. Britto, A.S., Sabourin, R., Oliveira, L.E.S.: Dynamic selection of classifiers - a comprehensive review. Pattern Recogn. **47**(11), 3665–3680 (2014)
6. Calderón, J., López-Ortega, O., Castro-Espinoza, F.A.: A multi-agent ensemble of classifiers. In: Sidorov, G., Galicia-Haro, S.N. (eds.) MICAI 2015. LNCS (LNAI), vol. 9413, pp. 499–508. Springer, Cham (2015). https://doi.org/10.1007/978-3-319-27060-9_41

7. Chen, H., MA, S., Jiang, K.: Detecting and adapting to drifting concepts. In: Proceedings of 9th International Conference on Fuzzy Systems and Knowledge Discovery, FSKD 2012, pp. 775–779 (2012)

8. Cruz, R.M.O., et al.: DESlib: a dynamic ensemble selection library in python. https://deslib.readthedocs.io/en/latest. Accessed 06 May 2019

9. Cruz, R.M.O., et al.: Meta-des: a dynamic ensemble selection framework using meta-learning. Pattern Recogn. **48**(5), 1925–1935 (2014)

10. Cruz, R.M.O., et al.: Dynamic classifier selection: recent advances and perspectives. Inf. Fusion **41**(41), 195–216 (2018)

11. Dua, D., Graff, C.: UCI machine learning repository. http://archive.ics.uci.edu/ml. Accessed 30 July 2020

12. Hansen, L., Salamon, P.: Neural network ensembles. IEEE Trans. Pattern Anal. Mach. Intell. **12**(10), 993–1001 (1990)

13. Helmy, T., Al-Harthi, M.M., Faheem, M.T.: Adaptive ensemble and hybrid models for classification of bioinformatics datasets. Trans. Fuzzy Neural Netw. Bioinform. **3**, 20–29 (2012)

14. Kim, H., Kim, H., Moon, H., Ahn, H.: A weight-adjusted voting algorithm for ensembles of classifiers. J. Korean Stat. Soc. **40**(4), 437–449 (2011)

15. King, R.D., Feng, C., Sutherland, A.: Statlog: comparison of classification algorithms on large real-world problem. Appl. Artif. Intell. **9**(5), 289–333 (1995)

16. Kolter, J.Z., Maloof, M.A.: Dynamic weighted majority: an ensemble method for drifting concepts. J. Mach. Learn. Res. **8**, 2755–2790 (2007)

17. kuncheva, L.: Ludmila kuncheva - real medical data sets. http://pages.bangor.ac.uk/~mas00a/activities/realdata.htm. Accessed 06 May 2019

18. Kuncheva, L.I.: Combining Pattern Classifiers: Methods and Algorithms. Wiley-Interscience, Hoboken (2004)

19. Melo, L.: Python agent development framework. https://pade.readthedocs.io/pt-BR/latest. Accessed 30 July 2020

20. Nilsson, N.J.: Learning Machines. McGraw-Hill, New York (1965)

21. Perrone, M.P., Cooper, L.: When networks disagree: ensemble methods for hybrid neural networks. In: How We Learn; How We Remember: Toward An Understanding of Brain and Neural Systems: Selected Papers of Leon N Cooper, pp. 342–358. World Scientific (1995)

22. Schapire, R.E.: The strength of weak learnability. Mach. Learn. **5**, 197–227 (1990). https://doi.org/10.1007/bf00116037

23. Uber Junior, A., de Freitas Filho, P.J., Silveira, R.A., Mueloschat, J.: iEnsemble2: committee machine model-based on heuristically-accelerated multiagent reinforcement learning. In: Barolli, L., Javaid, N., Ikeda, M., Takizawa, M. (eds.) CISIS 2018. AISC, vol. 772, pp. 363–374. Springer, Cham (2019). https://doi.org/10.1007/978-3-319-93659-8_32

24. Junior, A.U., de Freitas Filho, P.J., Azambuja Silveira, R., Costa e Lima, M.D., Reitz, R.W.: iEnsemble: a framework for committee machine based on multiagent systems with reinforcement learning. In: Pichardo-Lagunas, O., Miranda-Jiménez, S. (eds.) MICAI 2016. LNCS (LNAI), vol. 10062, pp. 65–80. Springer, Cham (2017). https://doi.org/10.1007/978-3-319-62428-0_6

25. Wolpert, D.: The lack of a priori distinctions between learning algorithms. Neural Comput. **8**(7), 1341–1390 (1996)

26. Wozniak, M., Graña, M., Corchado, E.: A survey of multiple classifier systems as hybrid systems. Inf. Fusion **16**, 3–17 (2014). https://doi.org/10.1016/j.inffus.2013.04.006

Prediction of Social Ties Based on Bluetooth Proximity Time Series Data

José C. Carrasco-Jiménez⬤, Ramon F. Brena$^{(\boxtimes)}$⬤, and Sigfrido Iglesias

Tecnologico de Monterrey, 64849 Monterrey, Mexico
A00810665@itesm.mx, {ramon.brena,sigfrido}@tec.mx

Abstract. Personal ties in most social networks are explicitly declared by its participants, like on Facebook and LinkedIn. Nonetheless, the accuracy of self-declared relationships has been contested. Empirical studies show that behavioral data is much more accurate than self-reported data, as it relies on objective evidence of social link formation. Evidence collected from online interactions have been widely used to elicit social linkage, while physical interactions are rarely exploited to uncover underlying social structure. In this paper, we use proximity data taken from the Bluetooth detection of mobile devices and show that from the analysis of physical proximity, social relationships can be accurately inferred considering time and order of encounters. We also show that moments of time in which there was no proximity are as relevant for social network elicitation as moments of physical proximity. The purpose of this research work is to infer social ties based solely on behavioral proximity data; our experiments substantiate the claim that we are able to infer social structure with high accuracy exploiting proximity clues only.

Keywords: Social networks · Proximity · Pattern recognition · Bluetooth · Context

1 Introduction

In recent times, as the COVID-19 pandemic spread over the world, contagion tracking became utterly important, so researchers and the technology giants rushed to develop tracking apps for mobile phones [3]. These apps use Bluetooth detection in order to infer proximity, due to the short-range nature of Bluetooth communication, compared to cellular and even WiFi networks. Bluetooth has been used as proximity "proxy" for more than a decade, for purposes other than contagion tracking. In this paper we focus on the problem of eliciting social ties using only Bluetooth proximity information.

The term "Social Network" became popular in recent years, mostly associated to online services such as Facebook, LinkedIn, etc. [1] In a broader sense, social networks are as old as human civilization: it is just the social tissue of relations like who is a friend, enemy, family, leader of each other. Needless to say, an online social network is only an inaccurate subset of the actual human network: self-reported data is often unreliable, and in some cases even at odds with reality;

© Springer Nature Switzerland AG 2020
L. Martínez-Villaseñor et al. (Eds.): MICAI 2020, LNAI 12469, pp. 435–446, 2020.
https://doi.org/10.1007/978-3-030-60887-3_37

it is well known that in some organizations there are people who are frequently asked for opinions, who are not the official advisors [6], and it is of great value to those organizations to know to which degree the actual interactions correspond to the formal organization chart.

Social relationships are often thought to be influenced by physical proximity and the amount of time that two individuals allocate for social interactions [8]. Stronger relationships such as those shared by family members or close friends are regularly associated to prolonged times of collocation that involve physical proximity, while weak relationships are thought to reduce the amount of physical interaction. In other words, the type of social relationship that links two individuals is described by the amount of time allocated for personal encounters. However, the absence of personal relationship does not necessarily imply absence of physical proximity, and the other way around is also true: there are situations in which two users co-locate without having a true relationship (for instance, if they are both waiting for a bus). Therefore, we have to analyze patterns of proximity that uncover the presence/absence of relationship.

In this paper we propose a method to infer social networks from Bluetooth detection of proximity, and analyze the relationship between physical proximity and social linkage. For doing so, we identify the social encoding extracted from physical proximity that allow us to infer social linkage among individuals.

Our findings indicate that we are able to identify whether or not two individuals are related, and that temporal patterns in proximity are enough to infer social linkage. Our analysis also indicates that identifying multiple types of relationships using only proximity clues still remains a very difficult problem for which we did not find a satisfactory solution. Although the use of multiple kinds of information might be devised, physical proximity and the temporal aspect of it play a crucial role in identifying social relationships, but in order to infer a specific type of social relationship proximity clues alone appear to be insufficient.

The aim of this work is to approximate the structure of the underlying social network using only proximity related physical clues, and to investigate the impact of physical clues to identify social relationships. The purpose of this work does not include justification for certain types of relationships, nor investigation on finding the elements that rule certain types of relationships. We are aware of the different dimensions that rule how social links are formed [6,8,10], but our objective has been intentionally restricted to measure how much precision can be reached by exploiting physical clues only. Other studies described in Sect. 2 examine the impact of virtual clues or a mixed set of sources of information from both, the physical and virtual world, but our work is limited to physical clues only, as no other research work has studied in detail how physical clues serve as social indicators of relationship.

The paper is organized as follows. Section 2 presents some related work. In Sect. 3 we give an overview of our approach. Then, in Sect. 4 we examine the steps included in our approach, and the experimental results are presented in Sect. 5. Finally, conclusions are presented in Sect. 6.

2   Related Work

The main focus of this work is to understand whether or not we can infer the social relationships based on physical proximity. Recent works in the field of social computing can involve online communication, physical proximity, or both.

Huang et al. [10] analyze the extent to which distance shapes the structure of interactions. They investigate the role of proximity in establishing online relations. The authors suggest that offline proximity is an important factor to bring people together in virtual worlds. Other works [14], study the likelihood of creating friendship relationships when people are geographically closer (e.g. living in the same city); they complement location data collected from user check-ins from a social media website called Gowalla with the number of common friends shared by two individuals. Their results conclude that individuals who are within geographical proximity are more likely to establish physical interactions. Other works support these conclusions. For example, Chin et al. conclude that physical proximity plays an important role in making new friends and that encounters with longer duration will result in an increased probability of a person adding another as a friend [2].

Other works study the impact of physical proximity, not related to virtual links, but in the creation of real world social links. They generalize the fact that individuals who share common physical places are more likely to become friends. Some studies results [19] point towards a strong association between social linking and physical proximity by exploiting WiFi positioning technology to detect proximity. Their approach extracts information including duration of encounters and frequency, that are correlated with the strength of social links. Another work [15] claims that Bluetooth detection does not correspond to social interactions in some cases. As a consequence, the authors make use of Bluetooth signal strength, which is used to add or remove links in a social network based on a signal strength threshold to avoid false positives when registering encounters. In our work, we diminish the impact of false encounters by considering temporal patterns of encounters as opposed to single Bluetooth encounters.

Some researchers [2,14] conclude that virtual linkage is strongly correlated to physical proximity. Other works conclude that people who are socially linked, tend to live geographically close and as a consequence tend to interact more often [15,19]. They talk about the strong relationship between physical proximity and social linkage. But if this is the case, we should be able to infer the type of relationship given proximity information between a pair of individuals. This is where our contribution falls since, to our knowledge, no other work has executed an analysis that studies relationships at the personal level just from proximity data.

As we can see from the mentioned related works, it has been firmly established that proximity does correlates with human relationship and networking, but other authors do not intend to make an accurate prediction of human networking just using proximity information, and this is exactly what we do in this paper.

3   Methodology and Approach

For our research, we needed an information source (dataset) containing both Bluetooth detection and also social ties information, such as friendship. We found such a dataset at the well known Social Evolution dataset collected by the MIT in 2008 [11]. This dataset contains many data categories that could be useful for other research works, but we avoid here, such as phone calls log, because, as we have mentioned above, our purpose is to examine how well can we predict social ties using only the Bluetooth proximity information and nothing else. We are also omitting the WiFi hot-spot information, which is related to location and thus to proximity, because it is less precise than Bluetooth detection (connection to the same WiFi hot-spot defines a distance range of around 20 m between two cellphones, while Bluetooth detection means around 3 m between the two devices).

The ground-truth for the social ties in this dataset (that is, if two given people have a friendship link or not) is not directly given by the dataset, but can be inferred –with some caveats– as we will see later.

So the steps included in our methodology are as follows (depicted in Fig. 1):

1. The first step, *preprocessing*, is the time series dataset construction and refinement. This step is not limited to information gathering, and problems like Bluetooth encounters asymmetry, as well as friendship asymmetry, are solved at this point.
2. We visualized the encounters (as given by Bluetooth detection) as a sort of heat-map that we called *Social proximity maps*. Though this step was not in principle necessary for the next steps, it gave us very useful insights for choosing the features we needed to extract from the dataset, which are the next step.
3. A number of *social encodings* are features related to social connections can be derived from the dataset. Social encodings capture different aspects of social relationships such as frequency of encounters, among others, described in Sect. 4.3.
4. Social encodings serve as the basis for *training a social ties predictor* from Bluetooth detection recordings, which is the next phase. Social ties prediction is done with a supervised Machine Learning approach that learns the different social encodings and infers the underlying social structure.
5. Prediction performance evaluation is the last step, where we use typical Machine Learning performance indicators such as accuracy, precision, recall, F1 score, etc.

Details about each step in our approach will be explained in the following sections.

4   Time Series Dataset Construction

The data used in this research work is part of the Social Evolution data set collected from October 2008 to May 2009 [11]. It consists of *physical proximity*

BT Data

Fig. 1. Main steps in our approach

records, collected through Bluetooth-enabled devices, along with *location*, which was approximated using WiFi sensors, and *call-logs* of 84 subjects. The ground-truth information about the users' relationships is also part of the dataset, which additionally includes health habits, nutritional information and other items that were not used in this study. The types of relationship reported in the dataset include *no-relationship, close friends, political discussant, at least two common activities per week, shared all tagged Facebook photos, shared blog/twitter activities*. In order to track the evolution of social relationships throughout time, the surveys were conducted in time intervals of one month, but in this paper we do not analyze the dynamic aspect of social ties, and to avoid changes in the social network that could contaminate our results, we took just 33 days, ranging from October to November 2008. The range of 33 we took counted for 8,065 sampling points, with vectors of Bluetooth detection for each of the subjects of the study.

Everything in the original dataset which was not directly related to Bluetooth detection or social ties ground-truth was discarded (like the phone calls logs), as it does not serve the purpose of this work. We are not going to give here all the details of the information included in the MIT dataset and rather refer the interested reader to their reference [11].

4.1 Data Challenges

The dataset considered in this project shows several *asymmetries*: *asymmetry in reported social relationships* and *asymmetry in the number of encounters recorded by the Bluetooth-enabled devices*. Asymmetry in reported social relationships occurs when two individuals report different types of relationship between each other. This happens frequently, since friendship is subjective and depends on the perspective of each individual [5]. Asymmetry observed in the encounters occurs when an encounter between two individuals is observed by only one user. These two kinds of asymmetries are often ignored in other research works.

In the dataset we are considering, only 2282 links out of the 3486 possible links are symmetric, yielding about 65% of symmetry. This degree of asymmetry could not be ignored, as it could damage the prediction performance of the classifiers we present in Sect. 5. A solution to the problem of asymmetry restricted to the hierarchical relationships is given by Freeman [7], but this method does not apply to non-hierarchical relationships. So, we adopt a simple approach to deal with asymmetry by choosing the triangle of the adjacency matrix that is the most balanced in terms of instances that represent the different types of relationships.

Encounter asymmetry -which is the fact that one handset detects the other's signal, but not the other way around- is also a common issue that might be due to noise caused by hardware failure, environmental noise, or even the mobile phones' capabilities. A total of 2136 out of 3486 possible links (the combinations of 84 users taken pairwise) are symmetric in terms of encounters, representing about 61.3% of symmetry. We assume that false positives are not possible, thus we consider an encounter if at least one Bluetooth device detected the other, which is the "or" combination of the Bluetooth detection for each pair of users.

Yet another data challenge in our dataset refers to an unbalanced distribution of relationships reported among the participants of the study. In particular, social links of type "no-relationship" represent about 70% of the total number of social relationships reported. Further, "close-friends" category was more than 20 times smaller than any other kind of relationship. The distribution of relationship classes is remarkably unbalanced and measures needed to be taken to avoid this fact to damage prediction performance. In order to reduce the impact of the unbalanced distribution of relationships, we used down-sampling by randomly sampling from the data set to end up with classes having similar frequencies, at least from the point of view of "no relationship" compared to any relationship.

4.2   Social Proximity Maps Visualization

A visualization of encounters between pairs of individuals, that we generated as a sort of Bluetooth detection heat-maps maps, was very useful to identify patterns of encounters. For example, Fig. 2a shows the encounter patterns between individuals with ID 51 and 34 from the dataset described in Sect. 4. Red dots represent instants of time in which no encounters were recorded, white dots show times in which only one user recorded the other, and green dots represent times in which there was a mutual encounter. Figure 2b shows a pair of individuals linked by a type of relationships called "Socialize Twice per Week", and we can see that the number of encounters differs from close friends, shown in Fig. 2a. On the other hand, users with ID 44 and 5, who reported "Political Discussant" relationship, reported mutual encounters for the most part, but show significantly different patterns from people who have different type of relationship. Lastly, users with ID 62 and 8 exhibit even less encounters, in addition to asymmetry in the number of mutual encounter.

4.3   Social Encodings

We will call "social encodings" the features extracted from social interactions. But how exactly to define social encodings (which is a form of *feature engineering* [20]) so that they are useful for social network elicitation, is not an easy problem. Further, to find out relevant features from the raw Bluetooth detection data (that is, a matrix with 8,065 sampling points with 3,486 pairs of subjects) is a daunting task. We have tried a number of social encodings, as we explain in the following.

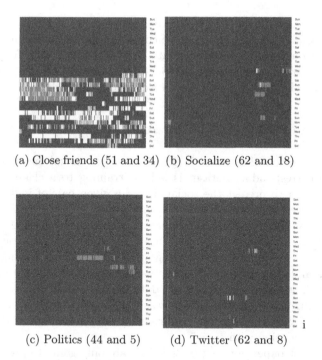

(a) Close friends (51 and 34)   (b) Socialize (62 and 18)

(c) Politics (44 and 5)   (d) Twitter (62 and 8)

Fig. 2. Proximity maps for different relationships (user numbers in parenthesis). (Color figure online)

The social encodings we tried are the following:

1. A "brute force" approach would be to take each of the 8,065 data points as one feature in a multidimensional space. Let us call *full time series* the brute force raw encounter time series data; it considers a matrix with 3,486 rows and each row with 8,065 columns, where cells have a 1 for an encounter (Bluetooth detection) and 0 otherwise. One interesting advantage of the full time series is that it contains information both about encounters and lack of encounters between two individuals; in most of the literature we found, the entire focus of their efforts lies in discovering patterns of encounters, leaving aside the information that can be extracted from *lack of encounters*.

2. Another social encoding we tried with good results (see below) was "encounter + time", which considers time frames and number of encounters observed within them [5,17], which is of course a radical dimensionality reduction compared to the full time series. This social encoding includes 18 features: total number of encounters, number of weekday encounters, number of weekend encounters, number of encounters during morning (6:00–11:00), lunch time (12:00–14:00), afternoon (15:00–19:00), evening (20:00–23:00) and early morning (00:00–5:00) for both weekdays and weekend days.

3. Other social encodings we tried, less successfully, include the frequency of encounters, total duration (total number of encounters recorded by Bluetooth

devices), average duration of encounters between two individuals during the period of study, cumulative weekly encounter patterns (overlapping the four weeks into one), and weekly encounter patterns along with frequency and duration approach. In the present paper we are only going to present the results from the first two ones, which were more successful.

5 Experimental Results and Analysis

Following the process illustrated in Fig. 1, once the social encodings are calculated, the corresponding dataset is fed for training to a classifier, which is expected later on to predict the social ties of a given pair of individuals. The classifiers we used include: *Support Vector Machines* (SVM) [12], *K-Nearest-Neighbor* (KNN) [9], *Penalized Discriminant Analysis* (PDA) and *Linear Discriminant Analysis* (LDA) [9], as they are the most commonly used ones and we did not had a specific reason for using other ones instead. We ruled out neural networks and deep learning, as they usually need data in the order of the millions, which was not the case for our datasets.

The performance of these classifiers was evaluated in terms of accuracy, precision, sensitivity and specificity. Performance estimation was validated using K-fold cross validation with $k = 5$.

In the present paper, for length's sake, we are only going to present the two social encodings which gave the best performance.

First, we examine the results of classifiers trained with the full time series.

Due to computational limitations (we would need a supercomputer for processing the full time series), we reduced the number of features (corresponding to timestamps in the time series) to less than one tenth of the original 8,065 timestamps, but keeping most of the variance explanation (75% of it). We used Principal Component Analysis (PCA) to reduce the impact of times that exhibit low variability. We set a cumulative sum of variance threshold of 75%, that will allow us to preserve the principal components that describe most of the variance from the original system of variables. Once the principal components are extracted, we retain the original variables that contribute substantially to them (that is, have coefficients greater than a threshold) in the principal components extracted in the previous step. In this way, we were able to reduce the dimensionality from 8,065 to 790 variables (times). Other forms of feature selection can also be used [9], but as we see in the results below, the one we chose was good enough.

Of course, to find out that some times in the sampled month are more significant than others for the detection of social ties is itself an interesting result, which needs to be explained (see Sect. 6).

Table 1 shows a high predictive power to infer relation vs non-relation (the binary case) with comparable accuracy for most classification methods, except SVM with sigmoid kernel, which achieved about 47% compared to 99% accuracy for the other statistical methods. Most of the classifiers perform similarly, indicating that the sequences of patterns that consider both encounters and non-encounters are good predictors to identify whether or not two individuals are related (friend vs non-friend). But the same table indicates a low predictive power to identify more specific types of relationships.

Table 1. Full time series performance results.

	Multiclass				Binary			
	Acc.	Prec.	Sen.	Spe.	Acc.	Prec.	Sen.	Spe.
KNN(k = 5)	0.251	0.306	0.251	0.924	0.982	0.982	0.982	0.990
KNN(k = 10)	0.242	0.324	0.243	0.906	0.995	0.995	0.995	0.999
KNN(k = 20)	0.239	0.222	0.239	0.885	0.999	0.999	0.999	1.000
SVM(rad.)	0.191	0.189	0.191	0.767	0.999	0.999	0.999	0.999
SVM(sig.)	0.194	0.195	0.195	0.747	0.472	0.470	0.472	0.497
PDA	0.316	0.195	0.316	0.845	0.999	0.999	0.999	0.999

Now we are going to examine the results of the encounters + time encoding, where we have 18 variables that split time frames of encounters as follows: total number of encounters, number of weekday encounters, number of weekend encounters, number of encounters during morning (6:00–11:00), lunch time (12:00–14:00), afternoon (15:00–19:00), evening (20:00–23:00) and early morning (00:00–5:00) for both weekdays and weekend days.

As it is shown in Table 2, these features have the ability to extract information about social linkage, bound to the presence/absence of relationship. Classification results vary widely among different classifiers, with SVM and PDA providing the best classification results in the binary case; though there is some loss compared to the full time series encoding, it is not too much considering that the number of features goes down from 790 to 18. The multi-class case gives similar results to other analyses, which result in poor performance. We may conclude that our social encoding is good at inferring presence/absence of relationship, but lacks the ability to infer complex relationships, as was the case of the brute force encoding.

Table 3 shows a comparison between the social encoding with the best performance found in this research work, and state-of-the-art methods. As it is shown, the social encoding of *Full Time Series* approach yielded the best performance in our analysis followed by *Encounters + Time*, thus, we compare the performance achieved to that of state-of-the-art methods that aim to accomplish the same goal, that is, infer social networks from objective data.

Although the analysis shows better performance for the method we propose, we notice that the different approaches have different data sources. For example, some works exploit non-physical clues such as voice calls, email, on-line data, SMS, and surveys [5] and other works that use physical clues such as proximity and location [5,17,18], either combine more than one source of information, or use different data sets with different number of individuals in specific settings

Table 2. Encounter + time results.

	Multiclass				Binary			
	Acc.	Prec.	Sen.	Spe.	Acc.	Prec.	Sen.	Spe.
KNN(k = 5)	0.327	0.319	0.327	0.914	0.842	0.843	0.842	0.858
KNN(k = 10)	0.311	0.295	0.311	0.923	0.846	0.847	0.846	0.872
KNN(k = 20)	0.297	0.268	0.297	0.936	0.859	0.862	0.859	0.899
SVM(rad.)	0.235	0.254	0.235	0.716	0.973	0.973	0.973	0.981
SVM(sig.)	0.220	0.249	0.220	0.759	0.942	0.942	0.942	0.932
PDA	0.232	0.244	0.232	0.708	0.971	0.971	0.971	0.976
LDA	0.217	0.225	0.217	0.672	0.485	0.485	0.485	0.434

(e.g. work, school, trips). So, we need to be cautious when we interpret our results in comparison with other works. Concluding which approach works best is subject to a number of circumstances as it was mentioned, but we can conclude that our approach shows good performance and that we were able to reduce the number of sources of information required to infer social relationships to just Bluetooth detection, which was the goal of this research.

Table 3. Comparison with state-of-the-art works.

Reference	Data source(s)	Accuracy
Ours	Bluetooth (Full time series)	99%
Ours	Bluetooth (Encounters + time)	97%
[5]	Bluetooth, Cell Towers, Communication Events	95%
[4]	Communication, Demographic, SMS, Voice, Bluetooth, WiFi	95%
[17]	Location (Gowalla check-ins)	93.92%
[18]	Call logs, Bluetooth, Location	83.1%
[16]	Voice calls, Email, On-line Data, Surveys	82.4%
[13]	SMS and Voice calls	68.6% avg.

6   Conclusions

We have analyzed the implications of several social encodings to predict the presence/absence and the kind of relationship available in the *Social Evolution* dataset. We may say that presence/absence of relationship can be accurately predicted by *Full Time Series* indicating proximity at each time, with very good performance for most of the classifiers considered. The general consistency of results among classifiers suggests that the ability of the predictors proposed to distinguish among classes is not linked to a particular classification procedure. We also found that the commonly used *Encounters + Time* set of features provides good predictive power but linked to specific classifiers. Every other set of predictors considered in this work yielded prediction results of low or no value. The absence of personal relationship does not necessarily imply absence of physical proximity, highlighting the challenges and thus the value of this study.

The main contribution of this work relies on exhibiting the power of physical proximity clues in predicting social linkage. We have successfully shown that Bluetooth-based proximity clues by themselves account sufficiently for the presence/absence of personal relationships. In doing so, we have considered a number of procedures to validate results and a wide range of features related to physical proximity clues, some of them not studied in previous works.

The specific type of relationship between two individuals is not predictable with spatial-temporal features as considered in this work and might well not be explainable restricting attention to this kind of data.

Further research might include the explanation of why some time periods are substantially more important for the relationship prediction than others, as was pointed out when we made a dimensionality reduction, from 8,065 time points to just 790 relevant ones. We also want to explore the possibility that extending the number of days of sampling with larger databases could better approximate the relationship between two individuals or try deep learning approaches.

Acknowledgements. Authors gratefully acknowledge the support from the Mexican National Council for Science and Technology (CONACYT), in the form of a PhD scholarship for the main author.

References

1. Carrington, P.J., Scott, J., Wasserman, S.: Models and Methods in Social Network Analysis, vol. 28. Cambridge University Press, Cambridge (2005)
2. Chin, A., Xu, B., Wang, H., Wang, X.: Linking people through physical proximity in a conference. In: Proceedings of the 3rd International Workshop on Modeling Social Media, MSM 2012, pp. 13–20. ACM, New York (2012). https://doi.org/10.1145/2310057.2310061, https://doi.org/10.1145/2310057.2310061
3. Cho, H., Ippolito, D., Yu, Y.W.: Contact tracing mobile apps for COVID-19: privacy considerations and related trade-offs. arXiv preprint arXiv:2003.11511 (2020)
4. Dong, W., Lepri, B., Pentland, A.S.: Modeling the co-evolution of behaviors and social relationships using mobile phone data. In: Proceedings of the 10th International Conference on Mobile and Ubiquitous Multimedia, MUM 2011, pp. 134–143. ACM, New York (2011). https://doi.org/10.1145/2107596.2107613
5. Eagle, N., Pentland, A.S., Lazer, D.: Inferring friendship network structure by using mobile phone data. Proc. Nat. Acad. Sci. **106**(36), 15274–15278 (2009). https://doi.org/10.1073/pnas.0900282106. http://www.pnas.org/content/106/36/15274.abstract
6. Feld, S.L.: The focused organization of social ties. Am. J. Sociol. **86**(5), 1015–1035 (1981)
7. Freeman, L.C.: Uncovering organizational hierarchies. Comput. Math. Organ. Theor. **3**(1), 5–18 (1997). https://doi.org/10.1023/A:1009690520577
8. Granovetter, M.: The strength of weak ties. Am. J. Sociol. **6**(78), 1360–1380 (1973)

9. Hastie, T., Tibshirani, R., Friedman, J.: The Elements of Statistical Learning: Data Mining, Inference, and Prediction, corrected edn, Springer, New york, August 2003. http://www.worldcat.org/isbn/0387952845, https://doi.org/10.1007/978-0-387-84858-7

10. Huang, Y., Shen, C., Contractor, N.S.: Distance matters: exploring proximity and homophily in virtual world networks. Decis. Support Syst. **55**(4), 969–977 (2013). https://doi.org/10.1016/j.dss.2013.01.006. http://www.sciencedirect.com/science/article/pii/S0167923613000158 1. Social Media Research and Applications 2. Theory and Applications of Social Networks

11. Madan, A., Moturu, S.T., Lazer, D., Pentland, A.: Social sensing: obesity, unhealthy eating and exercise in face-to-face networks. Wirel. Health **2010**, 104–110 (2010)

12. Meyer, D., Dimitriadou, E., Hornik, K., Weingessel, A., Leisch, F.: E1071: misc functions of the department of statistics (e1071), TU Wien, R package version1.6-4 (2014). http://CRAN.R-project.org/package=e1071

13. Mirisaee, S., Noorzadeh, S., Sami, A., Sameni, R.: Mining friendship from cell-phone switch data. In: 2010 3rd International Conference on Human-Centric Computing (HumanCom), pp. 1–5 August 2010. https://doi.org/10.1109/HUMANCOM.2010.5563332

14. Nguyen, T., Chen, M., Szymanski, B.: Analyzing the proximity and interactions of friends in communities in gowalla. In: 2013 IEEE 13th International Conference on Data Mining Workshops (ICDMW), pp. 1036–1044, December 2013. https://doi.org/10.1109/ICDMW.2013.60

15. Sekara, V., Lehmann, S.: The strength of friendship ties in proximity sensor data. PLoS ONE **9**(7), e100915 (2014). https://doi.org/10.1371/journal.pone.0100915

16. Sun, D., Lau, W.C.: Social relationship classification based on interaction data from smartphones. In: 2013 IEEE International Conference on Pervasive Computing and Communications Workshops (PERCOM Workshops), pp. 205–210, March 2013. https://doi.org/10.1109/PerComW.2013.6529482

17. Tan, R., Gu, J., Chen, P., Zhong, Z.: Link prediction using protected location history. In: 2013 5th International Conference on Computational and Information Sciences (ICCIS), pp. 795–798, June 2013. https://doi.org/10.1109/ICCIS.2013.213

18. Tang, W., Zhuang, H., Tang, J.: Learning to infer social ties in large networks. In: Gunopulos, D., Hofmann, T., Malerba, D., Vazirgiannis, M. (eds.) ECML PKDD 2011. LNCS (LNAI), vol. 6913, pp. 381–397. Springer, Heidelberg (2011). https://doi.org/10.1007/978-3-642-23808-6_25

19. Xu, B., Chin, A., Wang, H., Wang, H., Zhang, L.: Social linking and physical proximity in a mobile location-based service. In: Proceedings of the 1st International Workshop on Mobile Location-based Service, MLBS 2011, pp. 99–108. ACM, New York (2011). https://doi.org/10.1145/2025876.2025895

20. Zheng, A., Casari, A.: Feature Engineering for Machine Learning: Principles and Techniques for Data Scientists. O'Reilly Media Inc, Sebastopol (2018)

Energy Saving by Using Internet of Things Paradigm and Machine Learning

Josimar Reyes-Campos[1]([⊠]), Giner Alor-Hernández[1], Isaac Machorro-Cano[1],
José Luis Sánchez-Cervantes[2], Hilarión Muñoz-Contreras[1],
and José Oscar Olmedo-Aguirre[3]

[1] Tecnológico Nacional de México/I. T. Orizaba, Av. Oriente 9, 852. Col. Emiliano Zapata,
94320 Orizaba, Veracruz, Mexico
josi.reyescampos@gmail.com, {galor,hmnoz}@ito-depi.edu.mx,
imachorro@gmail.com
[2] CONACYT - Instituto Tecnológico de Orizaba, Av. Oriente 9 no. 852. Col. Emiliano Zapata,
94320 Orizaba, Veracruz, Mexico
jsanchezc@ito-depi.edu.mx
[3] Department of Electrical Engineering, CINVESTAV-IPN, Av. Instituto Politécnico Nacional,
2,508 Col. San Pedro Zacatenco, Delegación Gustavo A. Madero, 07360 Mexico City, Mexico
oolmedo@cinvestav.mx

Abstract. Nowadays, energy consumption is acquiring growing attention for the economic and environmental implications in our society due to the growing number of electronic home devices. From this perspective, the Internet of Things (IoT) and Machine Learning have emerged as technologies that allow monitoring and controlling devices installed in houses to detect behavioral patterns that identify feasible scenarios of energy saving. For this reason, intelligent configuration approaches for home automation are of utmost importance. This paper proposes a mobile application (called IntelihOgarT) that optimizes energy consumption through Machine Learning and IoT, while improving, at the same time, comfort at home. The proposed application makes use of Machine Learning algorithm C4.5, which automatically takes decisions based on attributes of a training data set. Furthermore, the case study presented validates the effectiveness of the mobile application, where efficient use of energy at home is a primary concern.

Keywords: Energy saving · Home automation · Internet of Things · Machine Learning

1 Introduction

IoT consists of interconnected physical devices and software components. These connected things or objects exchange information to provide a service to the end-user [1]. Currently, there are large numbers of devices for daily use, such as smartphones, sensors, actuators, smart televisions, cameras, among others. IoT is increasingly used for the realization of intelligent home environments. IoT largely improves the solutions inspired in the service-oriented principles to enhance the quality of service and security

© Springer Nature Switzerland AG 2020
L. Martínez-Villaseñor et al. (Eds.): MICAI 2020, LNAI 12469, pp. 447–458, 2020.
https://doi.org/10.1007/978-3-030-60887-3_38

of the modern lifestyle of home residents [2]. Domotic comprises a set of methods and technologies designed for home automation, including security, energy management, welfare and communications schemes. Energy saving and comfort are essential in a home automation system that becomes highly benefited when combined with IoT leading to an improvement in the well-being of users [3]. Additionally, energy efficiency is a topic of enormous interest among the scientific community and society in general, since in recent years a serious concern has emerged to alleviate the negative impact of energy consumption in the environment [4].

IoT applications to Domotics face important challenges such as the lack of platforms that allow suitable communication among devices, the availability of security protocols that protect users' private data along with automatic configuration tools through smart environments that analyze behavioral patterns of users [5]. In [6], Machine Learning is defined as the study of methods for constructing and improving software systems by analyzing examples of their desired behavior rather than by directly programming them. Consequently, Machine Learning methods are appropriate in application settings where people are unable to provide precise specifications for the desired system behavior, but where examples of this behavior are available [6].

In the literature appear reported research efforts focused on domotic and smart homes following the IoT paradigm regarding energy efficiency. However, such efforts do not consider intelligent customization of energy use and comfort for the benefit of the user. This paper presents a mobile application (called IntelihOgarT) that optimizes energy consumption through Machine Learning and IoT, allowing, among other things, to improve user comfort at home. The proposed application makes use of Machine Learning algorithm C4.5, which allow predicting one or more discrete variables, based on attributes of a data set. Besides, a case study is presented to validate the effectiveness of the mobile application, where energy consumption was monitored.

This work is structured as follows: Sect. 2 presents a review of related research works. Section 3 describes the architecture and functionality of IntelihOgarT. Section 4 presents a case study where IntelihOgarT's was validated. Finally, Sect. 5 presents some conclusions based on the results obtained and the future work.

2   Related Work

We currently live in an information age and the introduction and acceptance of new communication paradigms such as IoT, have allowed the improvement of residential automation schemes, which leads to the evolution of the interaction between users, their devices and their functionalities. In this section, several works are discussed to settle the context under which our project was developed. Filho et al. [7] proposed STORm (Smart Solution for Decision Making in a Residential Environment), a solution that combines fog computing and computational intelligence. The solution was able to recover, treat, disseminate, detect and control the information generated by the sensors installed in the residential scenario to apply it to the decision-making process. Silva et al. [8] proposed a multi-criteria model and a framework to analyze the most appropriate solution for the design of an IoT system. When tested the framework, the results presented showed that the proposed methodology helped in the selection of an IoT system, considering criteria such as energy consumption, implementation time, difficulty of use, cost, among other attributes.

Malina et al. [5] presented a security framework for the Message Queuing Transport Telemetry (MQTT) protocol that allow to improve the security and privacy services of the Internet of Things. Additionally, Castro-Antonio et al. [9] presented an approach based on Robotics Operation Systems (ROS) that integrated different types of services into an Intelligent House Services System (IHSS). The central idea of the system was to provide services in smart homes without the need to have recognized the entire house through a collection of sensors and cameras. Kasnesis et al. [10] proposed an integrated platform that allows dynamic injections of automation rules based on semantic Web technologies within a collective intelligent environment. Saba et al. [11] presented a contribution to the modeling and simulation of multi-agent systems for a residence powered by a hybrid renewable energy system, whose objective was to reduce energy consumption.

Frontoni et al. [12] developed a framework to allow the rapid development of complex hardware and software systems, the integration of new device classes into existing systems and the control and centralization of information. Besides, Chacon-Troya et al. [13] explained the design of an intelligent residence application for the control and monitoring of electricity quotas. A hybrid application that combines Web and native technology was developed to estimate the costs of the devices of the residence. Buono et al. [14] presented the use of a cheap and easy-to-apply Non-Intrusive Load Monitoring System (NILM) that shows, in mobile devices, historical and real-time energy consumption and sends alerts if it is about to occur an energy overload. Lanfor and Perez [15] implemented a security system that uses video streaming to monitor the environment.

Weixian et al. [16] presented a self-learning home management system that integrated a series of subsystems in charge of classifying and learning from data generated in smart homes through the use of computer and Machine Learning with the aim of strengthening an energy alert system, as well as providing users price forecasting. In [17], Elkhorchani and Grayaa proposed the architecture of a home energy management system that employs wireless communication, renewable energy principles and a power shedding algorithm. By implementing this system, they achieved the reduction of CO_2 emissions as well as the reduction of energy rates. Matsui [18] proposed a system that displays information according to the user's indoor comfort preferences in order to suggest ways to reduce electricity consumption. Fensel et al. [19] presented OpenFridge, an IoT semantic platform that enable information and data analytics useful for building services on top of typical home appliance data. Likewise, Baker et al. [20] developed E2C2, a multi-cloud IoT service composition algorithm. This algorithm search and integrates the least possible number of IoT services to fulfil user requirements, to create an energy-aware composition plan. Additionally, it is important to mention that in this context, previous works (in which studies of optimization methods that make use of nature-inspired metaheuristic algorithms have been developed) should be considered in order to be used to obtain important contributions in the field of energy demand predictions with the objective of promoting energy saving [21–24].

As shown in this section, the related works expose issues related to the configuration, security, energy saving and intercommunication of home automation systems. However, only some of them are focused on improving comfort according to the analysis of resident behavior. After reviewing these works, it was concluded that there are no intelligent configuration schemes that maintain a balance between energy savings and comfort through the analysis of the user's interaction with domotic devices. Taking this into account, IntelihOgarT is a solution, implemented through a mobile application, capable of analyzing the historical interaction of residents with their devices. Useful behavioral patterns are recognized from this analysis that makes possible to automate home configurations and also to reduce energy consumption. For these reasons, IntelihOgarT is an effective tool that extends and improves the functions of a home automation system. The following section describes the architecture and functionality of IntelihOgarT.

3 IntelihOgarT: Architecture and Functionality

The need for an intelligent configuration for domotic devices that makes use of computational learning arises from the lack of an infrastructure that automates repetitive tasks. In particular that is missing is an automatic configuration scheme that analyzes the historical data of users so that it can accurately predict the actions that would take the user if he/she had manual control of the system. To attend this need, IntelihOgarT, a mobile application capable of applying classification algorithms to the data used in a home automation system was developed.

3.1 Architecture

The design of the IntelihOgarT architecture is built upon a home automation system, which includes the necessary domotic technology for the control and monitoring of the connected devices. The home automation system is composed of sensors and actuators that act as face-to-face agents in the facilities where the intelligent system is implemented, collecting real-time information from the surrounding environment and controlling the actuating devices. In turn, the home automation system comprises a set of Web services that allows obtaining the data coming from the sensors, and performing the control functions of the actuators.

The IntelihOgarT architecture includes a module for the acquisition and analysis of information that, through the use of Web services, obtains the data from the sensors, validates them, and ensures their integrity. After ensuring the quality of the incoming data, they are redirected to a data mining submodule that analyzes its content and determines its usefulness, preserving the historical records of the data produced that is used for the modules of automatic configuration and presentation of historical and real-time data. Besides, a GUI provides the means to the user to access the mobile application from which the system information can be consulted, the domotic devices can be controlled and the automatic configuration scheme can be managed. The IntelihOgarT's architecture consists of five modules: 1) Presentation of acquired information, 2) Presentation of historical information, 3) Automatic configuration, 4) Manual control, and 5) Energy-saving recommendations. The IntelihOgarT's architecture is shown in Fig. 1, where the modules and their relationship are depicted.

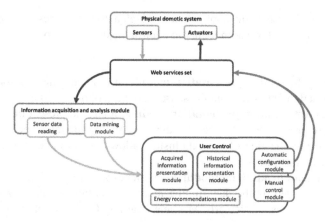

Fig. 1. IntelihOgarT's architecture.

On the other hand, the main functions of the architectural elements of the proposed solution are described below.

- **Physical domotic system:** includes the necessary communication technology for controlling and monitoring domotic devices. Upon this domotic system the intelligent system is implemented.

  - **Sensors:** are devices used to detect changes in data values coming from relevant environmental variables (temperature, light, people, among others) for the home automation system.
  - **Actuators:** are devices that exert changes into the environment after receiving the respective commands to do so.

- **Web services set:** contain methods to perform actions through a communication interface between the physical domotic system and the mobile application. They allow for collecting data from the sensors and for preparing and sending instructions to the actuators.
- **Information acquisition and analysis module:** stores, validates and filters the information obtained from physical devices through the invocation of web services.

  - **Sensor data processing:** consists of receiving data from the sensors which are validated to ensure their consistency and integrity and of finally redirecting them to the data mining sub-module.
  - **Data mining module:** analyzes the information content of data, determines its usefulness, and stores historical records of the collected data. These records are grouped into three types of energy consumption categories (low, medium, high) through the use of a K means approach in order to obtain the data for the presentation of historical and real-time information to the user, and for directly feeding the automatic configuration module.

- **User control:** represents the application to which a user has access, and from which he/she can consult information, either to control the devices or to let the system take decisions for him.

  - **Acquired information presentation module:** it is a real-time monitor to present the information obtained from the sensors connected to the system.
  - **Historical information presentation module:** it aims to present the information previously processed and stored by the data mining module. The ordering and structured visualization of the data history allows users to keep track of the use they give to the system.
  - **Energy recommendations module:** This module is responsible for processing the data provided by the data mining module. Additionally, making use of prediction algorithms, particularly Machine Learning C4.5 algorithm that generates rules based on decision trees on input variables, the module provides energy-saving recommendations considering the usage history of the domotic devices. However, the user retains the final decision if such recommendations are accepted or not. The C4.5 algorithm was selected for its ability to avoid overtraining or over classing data. It addresses such issues through a one-step pruning process that can work with discrete and continuous data, in addition to handling incomplete data problems [25].
  - **Automatic configuration module:** in this module the users can establish comfort settings according to their preferences, considering the energy saving recommendations.
  - **Manual control module:** is responsible for providing the user with a friendly interface from which is possible to issue direct orders to the actuators connected to the system. This module does not use prediction models.

3.2 Energy-Saving Recommendation Module

The energy-saving recommendations module of IntelihOgarT uses the Machine Learning algorithm C4.5 to generate the most suitable rules based on decision making trees derived from the usage history of the devices. Once generated the rules, they are applied to automatically configure the devices with no human intervention, for example, when turning the lights on after entering a dark room. The application of classification algorithms is very broad, as they can be used in areas ranging from economic and financial frameworks to health and safety services, where they have even been highlighted as tools that facilitate the diagnosis of diseases such as cancer [26] or tumor detection [27]. Likewise, C4.5 has established itself as a reliable algorithm and whose performance surpasses other algorithms, such as CART and Random Forest [28]. In order to discover energy consumption patterns and generate the automatic configuration with the rules is necessary to obtain, store and analyze the data related to the actions that the user executes in the IntelihOgarT daily. When a usage history is available for each device, the predictive algorithm C4.5 uses the data to perform the configuration automatically. The algorithm starts processing large sets of sensors readings that belong to known settings. The sensor readings, described by a combination of numerical and nominal properties, are examined to discover behavioral patterns, allowing the device settings to be reliably discriminated. These behavioral patterns are then expressed as models, in the form of decision trees or sets of conditional rules, which are used to classify new readings [29] and to improve the device settings. It is worth mentioning that the more the IntelihOgarT is operated by the user, the more and better usage data will be obtained, allowing the automatic configuration module to be trained more precisely that leads to more accurate automatic configurations.

Figure 2 shows the manual configuration interfaces of the application (created for the Android platform) for a room and an individual device. The frequent use by the user allows IntelihOgarT to obtain data that serve for the analysis and subsequent prediction of automatic configurations for each device.

Fig. 2. Interface for configuring a room.

Figure 3 shows the interface from which the entire house can be set in intelligent mode according to the user's behavioral patterns (for example, the hours the bedroom lights are turned on or off, the temperature of the air conditioning at night, among other settings). The automatic configuration module uses the C4.5 Machine Learning algorithm to analyze data from home automation devices and establish usage patterns. These patterns function as training values to predict future configurations without the user needing to interact with IntelihOgarT.

Fig. 3. Interface for menu and house profiles.

4 Case Study: Monitoring Energy Consumption in Residential Home

In this section, a case study is presented to validate IntelihOgarT's and consolidate it as a tool that reduces energy consumption and comfort in a house. The case study scenario is described as follows:

- A home automation system integrated into a house must be monitored to obtain data from the devices connected to it and to ensure that the system can provide efficient energy consumption and comfort to the residents. The system uses four types of sensors: temperature, presence, artificial and natural light.

Figure 4 presents a visual representation of the scenario. As shown in the figure, the real-time information of the house is obtained by the installed sensors. Every reading coming from a sensor is accessed by IntelihOgarT through Web services to analyze information and to automate the configuration settings of the connected domotic devices. The user can interact with IntelihOgarT through an application installed on his/her mobile device. To ensure good experience in such interaction, IntelihOgarT deploys an interface showing notifications and information received from the connected devices. The application also has the option to set manual control of them.

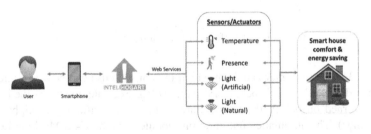

Fig. 4. Smart house monitoring scenario.

Four houses (located in Mexico) were selected to validate the case study. Each house was inhabited by three residents and had installed home automation devices to measure energy consumption. The characteristics of the houses were predefined to make fair comparisons among similar environments. Table 1 presents the common features of all the houses.

It is important to mention that to compare the energy consumption of the houses with and without the use of the IntelihOgarT´s intelligent configuration module, the quotas received over two periods (of two-months each one) were taken into account.

- Users were asked to interact normally with their home automation devices during the first bimonthly period (mid-November 2019 to mid-January 2020), controlling them manually through our mobile application. This period was used to obtain specific data on the energy consumption of the house, as well as to obtain the values necessary to train the automatic configuration module. It is important to highlight that at the end

Table 1. House characteristics and devices.

Room	Devices
Living room	Television, stereo, three lights, one air conditioner
Dining room	Three lights
Kitchen	One refrigerator, one blender, one microwave, one stove, two lights
Laundry	One washer, two lights
Bathroom	Two lights, one mini fan
Bedroom 1	One television, two lights, two lamps, one air conditioner
Bedroom 2	One television, two lights, two lamps

of this period it was possible to realize that users tend to leave appliances, like the lights or the television, turned on in their rooms when they were no longer present. The usage patterns of the devices obtained during the first period were combined with the energy saving patterns that allow the smart home to be automated to improve the comfort conditions of users without raising energy consumption rates.

- During the second bimonthly period (mid-January to mid-March 2020), residents were instructed to make use of IntelihOgarT's intelligent configuration module and let the system control the domotic devices as autonomously as possible. Energy consumption information was also collected, and at the end of the period the amounts established in the energy consumption receipts were compared (issued by CFE - Mexican electricity company), as well as the data obtained from the sensors during each period.
- The comparison made between the first and second monitoring period is presented in Table 2, which shows that the energy consumption decreased 12% on average.

Table 2. kWh consumption comparison

	first bimonthly period (avg)	Second bimonthly period (avg)
kWh	549	485

It was confirmed that energy consumption is related to user behavior. However, applying energy saving rules adjusted to the personalized home comfort scheme allows users to improve their comfort experience without affecting their economy. It should be remarked that, even though IntelihOgarT has an intelligent control module, the user is the one who decides whether to activate it or take manual control of the devices connected to the system. Finally, it is important to mention that there are factors that affect energy consumption at home, such as geographical location, the season of the year, the age or health status of the inhabitants, among others.

5 Conclusion and Future Work

People have always sought to improve the comfort of their homes. However, nowadays, they also seek to optimize energy consumption without affecting their comfort. Thanks to technological paradigms such as IoT and Machine Learning, today is possible to develop domotic systems that integrate energy saving recommendations in the house. Year after year, residents include in their homes larger amounts of devices that already have IoT technology, which improves the ability to obtain the energy consumption per day. On the other hand, with Machine Learning is possible to analyze all the information obtained through IoT to define device patterns of user behavior to establish house schemes that adjust to particular preferences.

This paper proposed IntelihOgarT, a mobile application that performs recommendations about the reduction of energy consumption through Machine Learning and IoT, allowing to improve comfort at the same time. IntelihOgarT used Machine Learning algorithm C4.5 to find a decision tree for the automatic configuration of the devices that matches the user's preferences. Finally, a case study was analyzed to validate the effectiveness of the mobile application. However, the effectiveness of the application is affected by factors such as the user's habits of energy consumption, the season of the year, or the geographical location.

As future work, we are considering incorporating into IntelihOgarT different configuration scenarios of domotic control, such as security scenarios and comfort for people with disabilities. These scenarios will allow establishing the rules needed by the automatic configuration module for correct real-time decision making. In this way, the user can select the rules that best suit his/her needs, depending on his/her current situation. Improving the availability of the mobile application for other platforms, extending the number of devices that can be installed. In order to validate user satisfaction of the platform regarding usability, improvement of energy saving and comfort at home, a User-Centered Evaluation is intended to be carried out by applying the User-Centric Evaluation Framework for Recommender Systems. In addition, to consolidate the effectiveness of IntelihOgarT, a larger evaluation is planned with at least 10 houses, considering at least two different house types during longer energy monitoring periods (8 months to a year). Finally, the addition of an invocation module for external service providers (such as product or security suppliers) is also considered.

Acknowledgments. This work was supported by Tecnológico Nacional de México (TecNM) and sponsored by the National Council of Science and Technology (CONACYT), the Secretariat of Public Education (SEP).

References

1. Krishna, A., Le Pallec, M., Mateescu, R., Noirie, L., Salaun, G.: IoT composer: composition and deployment of IoT applications. In: Proceedings - 2019 IEEE/ACM 41st International Conference on Software Engineering: Companion, ICSE-Companion 2019. pp. 19–22. Institute of Electrical and Electronics Engineers Inc. (2019). https://doi.org/10.1109/ICSE-Companion.2019.00028

2. Kaldeli, E., Warriach, E.U., Lazovik, A., Aiello, M.: Coordinating the web of services for a smart home. ACM Trans. Web. **7**, 1–40 (2013). https://doi.org/10.1145/2460383.2460389
3. Filho, G.P.R., Villas, L.A., Gonçalves, V.P., Pessin, G., Loureiro, A.A.F., Ueyama, J.: Energy-efficient smart home systems: Infrastructure and decision-making process. Internet of Things **5**, 153–167 (2019). https://doi.org/10.1016/j.iot.2018.12.004
4. Thema, J., et al.: The multiple benefits of the 2030 EU energy efficiency potential. Energies **12**, 2798 (2019). https://doi.org/10.3390/en12142798
5. Malina, L., Srivastava, G., Dzurenda, P., Hajny, J., Fujdiak, R.: A secure publish/subscribe protocol for internet of things. In: ACM International Conference Proceeding Series, pp. 1–10. Association for Computing Machinery, New York, USA (2019). https://doi.org/10.1145/3339252.3340503
6. Reilly, E.D., Ralston, A., Hemmendinger, D.: Encyclopedia of computer science. Nature Pub. Group (2000)
7. Filho, G.P.R., Mano, L.Y., Valejo, A.D.B., Villas, L.A., Ueyama, J.: A low-cost smart home automation to enhance decision-making based on fog computing and computational intelligence. IEEE Lat. Am. Trans. **16**, 186–191 (2018). https://doi.org/10.1109/TLA.2018.8291472
8. Silva, E.M., Agostinho, C., Jardim-Goncalves, R.: A multi-criteria decision model for the selection of a more suitable Internet-of-Things device. In: 2017 International Conference on Engineering, Technology and Innovation: Engineering, Technology and Innovation Management Beyond 2020: New Challenges, New Approaches, ICE/ITMC 2017 - Proceedings. pp. 1268–1276. Institute of Electrical and Electronics Engineers Inc. (2018). https://doi.org/10.1109/ICE.2017.8280026
9. Castro-Antonio, M.K., Carmona-Arroyo, G., Herrera-Luna, I., Marin-Hernandez, A., Rios-Figueroa, H. V., Rechy-Ramirez, E.J.: An approach based on a robotics operation system for the implementation of integrated intelligent house services system. In: CONIELECOMP 2019 - 2019 International Conference on Electronics, Communications and Computers, pp. 182–186. Institute of Electrical and Electronics Engineers Inc. (2019). https://doi.org/10.1109/CONIELECOMP.2019.8673166
10. Kasnesis, P., Patrikakis, C.Z., Venieris, I.S.: Collective domotic intelligence through dynamic injection of semantic rules. In: IEEE International Conference on Communications, pp. 592–597. Institute of Electrical and Electronics Engineers Inc. (2015). https://doi.org/10.1109/ICC.2015.7248386
11. Saba, D., Degha, H.E., Berbaoui, B., Laallam, F.Z., Maouedj, R.: Contribution to the modeling and simulation of multiagent systems for energy saving in the habitat. In: Proceedings of the 2017 International Conference on Mathematics and Information Technology, ICMIT 2017. pp. 204–208. Institute of Electrical and Electronics Engineers Inc. (2017). https://doi.org/10.1109/MATHIT.2017.8259718
12. Frontoni, E., Liciotti, D., Paolanti, M., Pollini, R., Zingaretti, P.: Design of an interoperable framework with domotic sensors network integration. In: IEEE International Conference on Consumer Electronics - Berlin, ICCE-Berlin. pp. 49–50. IEEE Computer Society (2017). https://doi.org/10.1109/ICCE-Berlin.2017.8210586
13. Chacón-Troya, D.P., González, O.O., Campoverde, P.C.: Domotic application for the monitoring and control of residential electrical loads. In: 2017 IEEE 37th Central America and Panama Convention, CONCAPAN 2017, pp. 1–6. Institute of Electrical and Electronics Engineers Inc. (2018). https://doi.org/10.1109/CONCAPAN.2017.8278471

14. Buono, P., Balducci, F., Cassano, F., Piccinno, A.: EnergyAware: a non-intrusive load monitoring system to improve the domestic energy consumption awareness. In: EnSEmble 2019 - Proceedings of the 2nd ACM SIGSOFT International Workshop on Ensemble-Based Software Engineering for Modern Computing Platforms, co-located with ESEC/FSE 2019, pp. 1–8. Association for Computing Machinery, Inc, New York, USA (2019). https://doi.org/10.1145/3340436.3342726

15. Lanfor, O.G.F., Perez, J.F.P.: Implementación de un sistema de seguridad independiente y automatización de una residencia por medio del internet de las cosas. In: 2017 IEEE Central America and Panama Student Conference, CONESCAPAN 2017. pp. 1–5. Institute of Electrical and Electronics Engineers Inc. (2018). https://doi.org/10.1109/CONESCAPAN.2017.8277600

16. Li, W., Logenthiran, T., Phan, V.T., Woo, W.L.: Implemented IoT-based self-learning home management system (SHMS) for Singapore. IEEE Internet Things J. **5**, 2212–2219 (2018). https://doi.org/10.1109/JIOT.2018.2828144

17. Elkhorchani, H., Grayaa, K.: Novel home energy management system using wireless communication technologies for carbon emission reduction within a smart grid. J. Clean. Prod. **135**, 950–962 (2016). https://doi.org/10.1016/j.jclepro.2016.06.179

18. Matsui, K.: An information provision system as a function of HEMS to promote energy conservation and maintain indoor comfort. In: Energy Procedia. pp. 3213–3218. Elsevier Ltd (2017). https://doi.org/10.1016/j.egypro.2017.03.705

19. Fensel, A., Tomic, D.K., Koller, A.: Contributing to appliances' energy efficiency with Internet of Things, smart data and user engagement. Futur. Gener. Comput. Syst. **76**, 329–338 (2017). https://doi.org/10.1016/j.future.2016.11.026

20. Baker, T., Asim, M., Tawfik, H., Aldawsari, B., Buyya, R.: An energy-aware service composition algorithm for multiple cloud-based IoT applications. J. Netw. Comput. Appl. **89**, 96–108 (2017). https://doi.org/10.1016/j.jnca.2017.03.008

21. Ganesan, T., Vasant, P., Elamvazuthi, I.: Advances in metaheuristics: applications in engineering systems. CRC Press (2016). https://doi.org/10.1201/9781315297651

22. Vasant, P., Kose, U., Watada, J.: Metaheuristic techniques in enhancing the efficiency and performance of thermo-electric cooling devices. Energies **10**, 1703 (2017). https://doi.org/10.3390/en10111703

23. Zelinka, I., Tomaszek, L., Vasant, P., Dao, T.T., Hoang, D.V.: A novel approach on evolutionary dynamics analysis – a progress report. J. Comput. Sci. **25**, 437–445 (2018). https://doi.org/10.1016/j.jocs.2017.08.010

24. Vasant, P., Marmolejo, J.A., Litvinchev, I., Aguilar, R.R.: Nature-inspired meta-heuristics approaches for charging plug-in hybrid electric vehicle. Wireless Netw. **26**(7), 4753–4766 (2019). https://doi.org/10.1007/s11276-019-01993-w

25. Saha, S.: What is the C4.5 algorithm and how does it work? - Towards Data Science. https://towardsdatascience.com/what-is-the-c4-5-algorithm-and-how-does-it-work-2b971a9e7db0. Accessed 03 Apr 2020

26. Pattanapairoj, S., et al.: Improve discrimination power of serum markers for diagnosis of cholangiocarcinoma using data mining-based approach. Clin. Biochem. **48**, 668–673 (2015). https://doi.org/10.1016/j.clinbiochem.2015.03.022

27. Mutaz, A., Abdalla, M., Dress, S., Zaki, N.: Detection of masses in digital mammogram using second order statistics and artificial neural network. Int. J. Comput. Sci. Inf. Technol. **3**, (2011). https://doi.org/10.5121/ijcsit.2011.3312

28. Kureshi, N., Abidi, S.S.R., Blouin, C.: A predictive model for personalized therapeutic interventions in non-small cell lung cancer. IEEE J. Biomed. Heal. Informatics. **20**, 424–431 (2016). https://doi.org/10.1109/JBHI.2014.2377517

29. Quinlan, J.R. (John R.: C4.5 : programs for machine learning. Morgan Kaufmann Publishers (1993)

Vision-Based Autonomous Navigation
with Evolutionary Learning

Ernesto Moya-Albor[1]([✉])(iD), Hiram Ponce[1](iD), Jorge Brieva[1](iD),
Sandra L. Coronel[2](iD), and Rodrigo Chávez-Domínguez[1]

[1] Facultad de Ingeniería, Universidad Panamericana, Augusto Rodin 498, 03920
Ciudad de México, México
{emoya,hponce,jbrieva,0204555}@up.edu.mx
[2] Departamento de Ingeniería, Instituto Politécnico Nacional, UPIITA,
Av. IPN No. 2580. Col. La Laguna Ticomán, 07340 Ciudad de México, Mexico
sgomezc@ipn.mx

Abstract. In this paper, we propose a vision-based autonomous robotics navigation system, it uses a bio-inspired optical flow approach using the Hermite transform and a fuzzy logic controller, the input membership functions were tuned applying a distributed evolutionary learning based on social wound treatment inspired in the *Megaponera analis* ant. The proposed method was implemented in a virtual robotics system using the V-REP software and in communication con MATLAB. The results show that the optimization of the input fuzzy membership functions improves the navigation behavior against an empirical tuning of them.

Keywords: Optical flow · Hermite transform · V-rep · Evolutionary robotics · Metaheuristic optimization · Vision-based control navigation

1 Introduction

Robotics navigation and obstacle avoidance problems have been addressed using deterministic and non-deterministic algorithms [1], where the robot must be able to navigate taking decisions in unknown or partially known, dynamic environments. For this, the robot takes observations through its perception system that includes different sensors such as odometry, visual sensors, sonar system, global positioning system (GPS) unit, and light detection and ranging system (LIDAR). Nowadays, the visual sensor-based approaches have been popularized due to the machine learning advances that use visual information, for example in [2], the authors presented an autonomous obstacle avoidance control using a perceptive-based optical flow method. In [3], it is presented a vision-based obstacle avoidance algorithm for micro aerial vehicles using the optical flow in 3-D textured environments, where they calculate the horizontal optical flow within a window as input a proportional-derivative controller. In [4], was presented a visual path tracking lateral control method using only one webcam, into a fuzzy logic controller, the proposal was tested in virtual and real scenes.

© Springer Nature Switzerland AG 2020
L. Martínez-Villaseñor et al. (Eds.): MICAI 2020, LNAI 12469, pp. 459–471, 2020.
https://doi.org/10.1007/978-3-030-60887-3_39

In general, the environments where the robot navigates are dynamic, that is to say, there are static and dynamic objects that can collide with it. To address this kind of scenarios, the robotic navigation system must adopt some strategy, where two main strategies have been proposed, deliberative and reactive or a mix of both. Thereby, the purest deliberative systems generate their trajectory based on prior environment knowledge (planed navigation) [5]. Whereas that in the reactive systems, the robot does not plan the trajectory a priori, instead, it navigates reacting to the environment in real-time.

In the robotics navigation reactive systems, soft computing methods have been applied to show that intelligent approaches allow autonomous navigating and avoidance obstacles in dynamic environments [1]. In [6], the authors shown that bio-inspired intelligent algorithms are a good option to intelligence robotics navigation and autonomy. In [7], it is presented an autonomous control using a hybrid software design and a fuzzy evolutionary artificial neural network.

In [8], it is presented a navigation system for autonomous robotic vehicles using fuzzy control techniques and the application of computational vision, where the proposed system was tested in poorly structured environment. On the other hand, evolutionary robotics [9] considers that the robots must be adapted to the conditions of the environment, that is, they must evolve. In this regards, in [9], a distributed evolutionary learning method, based on a social metaheuristic wound treatment optimization (WTO) proposal, for mobile robot navigation was presented. In [10], it is presented an educational robotics platform, it uses ant colony optimization, and the robot uses stereoscopic vision.

In this paper, we present a vision-based method for autonomous obstacle avoidance. Our proposal uses the bio-inspired optical flow method of [2] to represent the relevant visual features in an image, in such a way, that the optical flow highlights the obstacles in the scene, and a fuzzy logic controller takes the decisions to avoidance of the objects. The input fuzzy membership functions were tuned using a WTO method as was reported in [9], which is inspired on the social wound treatment in an ant species namely *Megaponera analis* located in the sub-Saharan Africa. To test the proposed vision-based autonomous navigation method, we used a virtual robotics environment, allowing us testing, tuning, and improving the proposed method.

Rest of the paper is organized as follow: in Sect. 3, we present the virtual robotics system. Section 3 presents the bio-inspired optical flow method used. In Sect. 4, we show the proposed vision-based autonomous navigation method with evolutionary learning. Experiment results of control to obstacle avoidance are given in Sect. 5. Finally, the conclusions are given in Sect. 6.

2   Virtual Robotics System Implementation

To test and evaluate the proposed autonomous navigation method, we used V-REP to define virtual scenarios, and MATLAB to implement the control system. V-REP is a general-purpose robotic simulator, which incorporates an integrated development environment including sensors, mechanisms, scenarios, furniture, and whole robot systems, where the last ones can be modeled and simulated allowing fast prototyping, controller testing, and product presentation. Besides, MATLAB a programming platform used to analyze data, develop algorithms and create models, allowing to implement complex mathematical models.

We used V-REP to simulate real environments conditions and the physics of the robots, which allow us to test the proposed controller to avoid objects, accelerating the design and tuning phases. In this sense, we used the Pioneer P3-DX model as testing robot, which includes two independent wheels that control the motion of the robot through the velocity of them, also, we place a vision sensor on the top of the robot to acquire RGB images used in the vision-based control. Figure 1 shows the settings of the simulated robot, where the vision sensor is highlighted in a dotted yellow circle.

On the other hand, MATLAB was used to communicate the fuzzy controller with the virtual robot in V-REP. For simulation, we set up the robot tested in different environments, the simulation rate was configured to 200 ms running in a Dell Inspiron 7000 Intel Core i7 up to 1.8 GHz, with 16 GB in RAM and an Intel UHD Graphics 620 graphics card.

3 Bio-inspired Optical Flow Estimation Method

Several methods have been proposed to calculate the optical flow in an image sequence, the great majority of the methods are based on the differential method proposed by Horn and Schunck [11]. In this work, we used the bio-inspired optical flow method using the Hermite transform proposed by Moya et al. [2]. It uses the Hermite Cartesian coefficients [12] to define the intensity constraint found in Horn and Schunck approach, furthermore, it uses the Steered Hermite coefficients [12] to overcomes the intensity changes and a new term sensitive to orientated structures such as edges and corners. Equation (1) shows the optical flow used in the proposed vision-based autonomous navigation method.

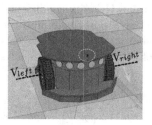

Fig. 1. Virtual robot showing its sensors and actuators. (Color figure online)

$$E_{HH} = \int_V \int_V \left[\left[uL_{01}(X) + vL_{10}(X) + L_{00_t}(X) \right] \right.$$

$$+ \gamma \sum_{i=0}^{N} \left[u\mathcal{L}_{n,\theta(n+1)}(X) + v\mathcal{L}_{n,\theta(m+1)}(X) + \mathcal{L}_{n,\theta_t}(X) \right] \right]^2 \tag{1}$$

$$+ \alpha^2 \left[\left(\frac{\partial u}{\partial x} \right)^2 + \left(\frac{\partial u}{\partial y} \right)^2 + \left(\frac{\partial v}{\partial x} \right)^2 + \left(\frac{\partial v}{\partial y} \right)^2 + \beta \left[c(X) \left(u^2 + v^2 \right) \right] \right] dX,$$

where $L(X)$ is a digital image with $X = (x, y)^\top$ representing the spatial location of each pixel, t is the temporal variable, and $(u, v)^\top$ are the horizontal and vertical displacement components.

The energy functional of Eq. (1) uses the Hermite coefficients to represent the data term, where $\mathcal{L}_{n,\theta,(k+1)}(X)$ are the Steered Hermite coefficients with an order increment in the k direction, terms of the form Z_{k_t} represent temporal partial derivatives of the Cartesian Hermite coefficients $(L_{00_t}(X), \mathcal{L}_{n,\theta_t}(X))$. The tuning parameters of the proposed functional are defined by N as the maximum order of the Hermite expansion, α and γ as the smoothed parameter and intensity invariant weight, respectively, and $C(X)$ as a term sensitive to oriented structures which is controlled by the weight parameter β.

The Cartesian and the Steered Hermite coefficients are obtained from the Hermite transform [12], a special case of polynomial transform inspired in the human vision system (HVS), it performs an image decomposition applying local processing and using a Gaussian derivative model of the receptive fields [12,13]. The Cartesian coefficients are obtained by convolving an image $L(X)$ with the separable Hermite filter functions $D_{m,n-m}(X) = D_m(x)D_{n-m}(y)$ as is shown in Eq. (2):

$$L_{m,n-m}(X) = \int_{-\infty}^{\infty} L(X)D_{m,n-m}(X)dX, \tag{2}$$

where $L_{m,n-m}$ are the Cartesian Hermite coefficients of decomposition order m and $n - m$, in x and y direction respectively, and the filter functions are determined by selecting a Gaussian analysis window, with variance σ^2 and unitary energy for $v(X)^2$, which define a family of associated orthogonal polynomials. Equation (3) shows the Hermite filter D_k of order k for $k = 0, 1, 2, \ldots, \infty$:

$$D_k(x) = \frac{(-1)^k}{\sqrt{2^k k!}} \frac{1}{\sigma\sqrt{\pi}} H_k \left(\frac{x}{\sigma} \right) \exp \left(-\frac{x^2}{\sigma^2} \right) \tag{3}$$

and $H_n \left(\frac{x}{\sigma} \right)$ represents the generalized Hermite polynomial defined by Eq. (4) [12]:

$$H_n \left(\frac{x}{\sigma} \right) = (-1)^n \exp \left(-\frac{x^2}{\sigma^2} \right) \frac{d^n}{dx^n} \exp \left(-\frac{x^2}{\sigma^2} \right). \tag{4}$$

On the other hand, the Steered Hermite transform is a compact representation of the Hermite coefficients, the Steered Hermite coefficients $(\mathcal{L}_{n,\theta}(x, y))$

are obtained by rotating the Cartesian coefficients towards an estimated local orientation as is shown in Eq. (5):

$$\mathcal{L}_{n,\theta}(X) = \sum_{k=0}^{n} \Big(L_{k,n-k}(X)\Big)\Big(g_{k,n-k}(\theta)\Big), \tag{5}$$

where $g_{m,n-m}(\theta)$ are the angular functions, of order n, that expresses the directional selectivity of the filter for the angle θ (e.g., the gradient direction), as it is showed in Eq. (6):

$$g_{m,n-m}(\theta) = \sqrt{\binom{n}{m}} \Big(\cos^m(\theta)\Big)\Big(\sin^{n-m}(\theta)\Big). \tag{6}$$

Finally, $C(X)$ is used to reward the optical flow estimation on highly oriented structures such as edges and corners. In [2], the authors proposed two approaches to define $C(X)$, in the first approach, the first-order steered Hermite coefficients was used to find the borders of the obstacles, while in the second method a perceptive mask, using the Hermite transform, was proposed to identify the relevant perceptive regions. As was reported, both approaches gave good results in collisions and trajectories trace, however, in this work the Steered Hermite transform approach was selected so as not to make more operations as required in the perceptive mask approach. Equation (7) shows the reward term where $\epsilon \ll 1$.

$$C_{SHT}(X) = \Big(\big[\mathcal{L}_{1,\theta}(X, t+1) - \mathcal{L}_{1,\theta}(X)\big]^2 + \epsilon\Big)^{1/2}. \tag{7}$$

4   Vision-Based Obstacle Avoidance Control with Evolutionary Learning

In [2], the authors proposed an obstacle avoidance fuzzy controller using the optical flow estimation of Eq. (1) and including an oriented reward term (Eq. 7). In this work, we redefine the fuzzy sets and the fuzzy rules to improve the obstacle avoidance control.

In Fig. 2, we show the input and output fuzzy membership functions of the proposed controller. It consists of three inputs with two membership functions and two outputs with four membership functions.

The input fuzzy membership functions correspond to the optic flow density acquired for the vision sensor, Fig. 3 shows an example of the optical flow estimated using the energy functional of Eq. (1), where the visual field was divided into three horizontal regions equally spaced, all with the height of the image. Following, the average of the optical flow magnitude ($\sqrt{(u^2 + v^2)}$) of each section was calculated and normalized respecting to the maximum average, thereby, we obtain the optical flow densities $flow_{left}$, $flow_{center}$ and $flow_{right}$. In the proposed fuzzy inference system, each one the densities is fuzzified through the input membership functions *thin* (T) and *dense* (D), where the range of the flow densities varies from zero as minimum value (zero dense) up to 1 as maximum

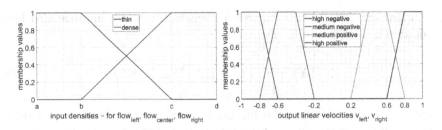

Fig. 2. The input membership functions represent the optical flow density, and the output membership functions are the velocities of the wheels of the robot.

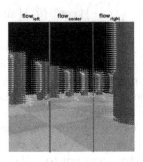

Fig. 3. Example of the optical flow densities obtained.

value (very dense). As it is observed in Fig. 3, the input membership functions are entirely defined for the parameters a, b, c, d, i.e., the lower limit, the initial and final transitions, and the upper limit of the trapezoidal functions, respectively. These parameters are obtained by a distributed metaheuristic optimization method as is shown in Sect. 4.1.

Regarding the output fuzzy membership functions, they were defined to control the velocities and direction of rotation of the robot wheels by means of the output variables v_{left} and v_{right}, the fuzzy controller has two output membership functions associated to each linear velocity of the wheels with partitions *high negative* (HN), *medium negative* (MN), *medium positive* (MP) and *high positive* (HP). The output values for both velocities varies from -1 (higher negative velocity) to $+1$ (higher positive velocity).

Finally, the set of fuzzy rules, as knowledge base, is show in Table 1. In this work, we used a Mamdani-type inference process as fuzzy engine.

4.1   Distributed Metaheuristic Optimization Method

To completely define the input membership functions of Fig. 2, it is necessary to assign values to the parameters (a, b, c, d), it can be performed empirically or through an optimization process. In this regard, this work proposes to tune the input membership functions using a distributed metaheuristic optimization method namely wound treatment optimization (WTO) [9]. This method

Table 1. Fuzzy rules implemented in the fuzzy controller.

$flow_{left}$	$flow_{center}$	$flow_{right}$	v_{left}	v_{right}
T	T	T	HP	HP
T	T	D	MP	HP
T	D	T	HP	MP
T	D	D	MP	HP
D	T	T	HP	MP
D	T	D	HP	HP
D	D	T	HP	MP
D	D	D	MN	HN

is inspired on the social wound treatment in an ant species namely *Megaponera analis* located in the sub-Saharan Africa. These ants hunt termites, and many of them are injured during rides. After hunting, healthy ants carry out the injured ones, and then they are treated. This animal behavior inspires WTO.

For this work, consider a decentralized system with $\mathcal{N}$ homogeneous robots, representing the ants, that communicate among them. The goal of this decentralized system is to find the optimal parameters (*params*) that setup the fuzzy controller of each robot in order to learn the task of avoiding obstacles and move freely around the environment. Currently, $params = \{a, b, c, d\}$ refers to the parameters that define the input membership functions (Fig. 2).

Then, each robot performs as an ant in the WTO method. At each iteration, the robot $i \in \mathcal{N}$ ($robot_i$) performs an attempt of T steps and then it obtains the optical flow densities $(flow_{left}, flow_{center}, flow_{right})_i^\top$ of its visual field, which give information about distance and motion between the closest obstacles and itself. After the attempt completion, the robot is able to compute a fitness function value f as defined in Eq. (8); where, WS is the free workspace of the robot, $penalty > 0$ is a large integer number that penalizes the performance if the robot is outside WS or bumped into an obstacle, $|v_i| = |0.5(v_{left} + v_{right})|$ is the robot velocity, and t is the step time in the attempt.

$$f(robot_i) = \begin{cases} \sum_{t=0}^{T} d_i(t) + \frac{1}{|v_i(t)|} & robot_i \in WS \\ penalty & robot_i \notin WS \end{cases} \tag{8}$$

Each time the robot completes an attempt, it restart the initial step time t to make a new one. Also, it communicates its own $f(robot_i)$ value and its values $params_i$ to the other robots. Once all the robots received the f values, each robot ranked itself as *healthy*, *average* or *heavy*-injured robot. If $robot_i$ is *average*, then it updates its own parameters $params_i$ using the parameters of a random robot considered *healthy* and the best robot so far, as shown in Eq. (9), where x_i is the current $params_i$ (candidate solution), g_{best} is the best set of parameters found so far by the robots, $x_{healthy}$ is randomly picked up from a subset of healthy individuals (the best ranked), h_{rate} is called the help rate, and r^1 and r^2 are random values in the range $[0, 1]$. If $robot_i$ is *heavy*, the parameters $params_i$ are replaced with the parameters of a random robot considered *healthy*. After the global algorithm stops, most of the robots will perform well avoiding obstacles and moving freely, and the best parameters will broadcast to all the robots. For a detailed description of the WTO used method, see [9].

$$x_{i,j}(t+1) = h_{rate} \times x_{i,j}(t) + r^1_{i,j} \times (x_{healthy,j}(t) - x_{i,j}(t))$$
$$+ r^2_{i,j} \times (g_{best,j}(t) - x_{i,j}(t)). \tag{9}$$

5   Testing and Results

In this section, we show the tests performed to evaluate the proposed vision-based obstacle avoidance control. First, we present the testing scenarios using the V-REP environment. Second, we show the results of the parameter optimization using the WTO method. Finally, we compared the performance of the obstacle avoidance control using both the optimized and not optimized parameters.

5.1   Virtual V-REP Scenarios for Testing

In Fig. 4a, we show the V-REP scenario where the distributed WTO method was performed, for this, five robots $\mathcal{N} = 5$ were placed to found the best input membership parameters, this number was selected arbitrarily, trying to have a balance between the computational resources used and allowing to simulate the social wound treatment. Here, each robot was implemented with a vision-based obstacle avoidance control as was defined in Sect. 4.

To test the vision-based autonomous navigation control we define three different scenarios in the V-REP environment, which are shown in Fig. 4. The complexity of these scenarios starts from the simplest to the more complex: a wall of cylinders at different distances (Fig. 4b), a curve path limited by cylinders (Fig. 4c), and a maze-type path formed by textured surface cylinders (Fig. 4d).

5.2   Parameter Optimization Using the WTO Method

We place five robots in the V-REP scenario of Fig. 4a, for this, we manually defined the meta-parameters of the WTO method as follows: $\mathcal{T} = 10,000$, $penalty = 1,000,000$, $p_{healthy,} = 0.2$, $p_{heavy} = 0.5$, $p_{alive} = 0.1$, $h_{rate} = 0.7$, $B_L = 0$ and $B_U = 1$.

After of $10,000$ injury function evaluations the simulation was stopped and we obtain the best parameters. In this sense, we obtained the best injury function evaluation $f_{best} = 47.2732$ and the best solution (best ant) $params_{best} = \{0, 0.4921, 0.7910, 1\}$. In Fig. 5(a), we show the evolution of the injury function evaluation (f), and Fig. 5(b) shows the input fuzzy membership functions with the optimized parameters after the applied the WTO method.

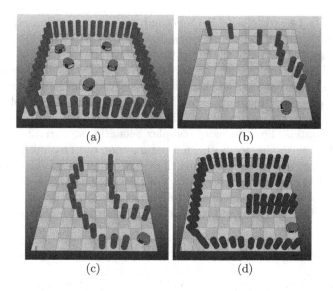

Fig. 4. (a) Scenario for testing the WOT method. (b-d) V-REP scenarios to test the vision-based autonomous navigation control.

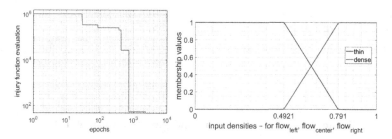

Fig. 5. (a) Evolution of the injury function evaluation. (b) Optimized input membership functions of the fuzzy controller.

5.3 Performance Evaluation of the Vision-Based Obstacle Avoidance Control

Finally, we present the performance of the vision-based obstacle avoidance control. For this, two metrics were used to evaluate the performance of the control: the collisions number and the execution time, also, we used the trajectory traced by the robot to visually evaluate the performance of both controllers, the optimized and not optimized. These metrics, together with the visual trajectories, were evaluated on three scenarios of Fig. 4, where three attempts were carried out per scenario.

First, we evaluate the proposed method using a set of not optimized parameters, which was defined empirically by dividing the range of flow density ($[0, 1]$)

into equal parts, i.e., $\{0, 0.25, 0.75, 1\}$. Second, we used the optimized parameters shown in Fig. 5(b), with values $params_{best} = \{0, 0.4921, 0.7910, 1\}$.

To compare the performance of both sets of parameters (not optimized and optimized), Tables 2 and 3 shows the number of collisions and the execution time for the three scenarios and the three attempts. From the results, for scenarios 1 and 2, we observed that both sets of parameters present zero collisions, but the optimized set of parameters (WTO) shows the minor time of execution to finish the scenarios. For the more complex scenario, scenario 3, we observed that the robot, with the not optimized parameters, did not finish the scenario for any attempt, and therefore, both the collision number and the execution time, presents an infinity number (inf). On the other hand, the robot with the optimized parameters finished successfully scenario 3 in all attempts. It should be noted that in attempt 3, only two collisions occurred to find the exit.

Table 2. Number of collisions and execution time (not optimized parameters).

Attempt	Scene1		Scene2		Scene3	
	Collisions	Time (s)	Collisions	Time (s)	Collisions	Time (s)
1	0	112	0	283	Inf	Inf
2	0	100	0	262	Inf	Inf
3	0	104	0	237	Inf	Inf
Average	0	105.33	0	260.67	Inf	Inf

Table 3. Number of collisions and execution time (WTO).

Attempt	Scene1		Scene2		Scene3	
	Collisions	Time (s)	Collisions	Time (s)	Collisions	Time (s)
1	0	81	0	100	29	264
2	0	70	0	90	57	180
3	0	70	0	82	2	262
Average	0	**73.67**	0	**90.67**	**29.33**	**235.33**

With regard to visual trajectories, Fig. 6 shows the best results for the trajectories traced by the robot using the not optimized and optimized parameters (WTO). Here, each row corresponds to one scenario, and the first column shows the results for the not optimized parameters and the second and third column (scenario 3) for the optimized parameters. For scenario 3, we present two different trajectories taken for the robot with the optimized parameters. We can observe that in scenarios 1 and 2, the results using the optimized parameters present a more smooth trajectory comparing with the not optimized parameters, nevertheless, for scenario 3, in addition to successfully finishing the scenario, the robot using the optimized parameters found two different exits from the route, whereas the robot using the not optimized parameters was not able to finish the route in any attempt.

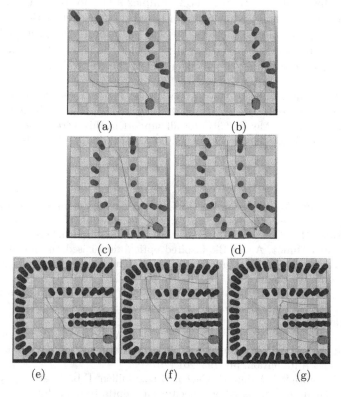

Fig. 6. Examples of the trajectories traced by the robot using the two sets of parameters. (a, c, e) not optimized parameters. (b, d, f, g) optimized parameters (WTO).

6   Conclusions

We present a vision-based autonomous robotics navigation approach, which uses a monocular camera instead of other approaches that use a vision-stereo strategy. Besides, we adopted a bio-inspired optical flow strategy into a fuzzy logic controller to objects avoidance. The optical flow method rewards the motion estimation at high oriented structures. On the other hand, the input fuzzy membership functions were tuned applying a distributed evolutionary learning method based on social wound treatment, inspired in the *Megaponera analis* ant. The results show that the tuned robotics navigation system by WTO overcomes to that with the empirical parameters, allowing to robot obstacle avoidance and to finish successfully more complex scenarios.

As future work, we are considering to analyze the performance of the optimization of parameters using the WTO approach while varying the number of robots and the effect of the meta-parameters. Also, we will test the proposed vision-based navigation method in more complex scenarios (e.g., mazes). On the other hand, we will incorporate a vision-based distance estimation to improve the navigation performance. Besides, we will incorporate the proposed method in a physical robot to demonstrate as the transfer the knowledge to real robots, allows decreasing the tuning time-consuming of the controller.

Acknowledgements. Ernesto Moya-Albor, Hiram Ponce, Jorge Brieva and Rodrigo Chávez-Domínguez would like to thank the Facultad de Ingeniería of Universidad Panamericana (Campus Mexico City) for all support in this work. Sandra L. Coronel thanks to Instituto Politécnico Nacional (UPIITA) for the support in this work.

References

1. Pandey, A., Pandey, S., Parhi, D.: Mobile robot navigation and obstacle avoidance techniques: a review. Int. Rob. Auto. J. **2**(3), 00022 (2017)
2. Moya-Albor, E., Coronel, S.L., Ponce, H., Brieva, J., Chávez-Domínguez, R., Guadarrama-Muñoz, A. E.: Bio-inspired optical flow-based autonomous obstacle avoidance control. In: 2019 International Conference on Mechatronics, Electronics and Automotive Engineering (ICMEAE), pp. 18–23 (2019)
3. Cho, G., Kim, J., Oh, H.: Vision-based obstacle avoidance strategies for MAVs using optical flows in 3-D textured environments. Sensors **19**(11), 2523 (2019). Switzerland
4. Zhu, L., Wang, W., Yang, W., Pan, Z., Chen, A.: Visual path tracking control for park scene. In: Proceedings of the 3rd International Conference on Robotics, Control and Automation, pp. 195–201 (2018)
5. López, J., Sanchez-Vilariño, P., Cacho, M.D., Guillén, E.L.: Obstacle avoidance in dynamic environments based on velocity space optimization. Robot. Auton. Syst. **131**, 103569 (2020)
6. Ni, J., Wu, L., Fan, X., Yang, S.X.: Bioinspired intelligent algorithm and its applications for mobile robot control: a survey. Comput. Intell. Neurosci. **2016**, 1 (2016)
7. Tedder, M., et al.: An affordable modular mobile robotic platform with fuzzy logic control and evolutionary artificial neural networks. J. Rob. Syst. **21**(8), 419–428 (2004)
8. Hernandez, B., Vitor, G., Moreno, R., Ferreira, J.: Fuzzy control for navigation of a mobile robot using real time computational vision. J. Eng. Appl. Sci. **13**(14), 5665–5673 (2018)
9. Ponce, H., Moya-Albor, E., Martínez-Villaseñor, L., Brieva, J.: Distributed evolutionary learning control for mobile robot navigation based on virtual and physical agents. Simul. Model. Pract. Theory **102**, 102058 (2019)
10. Montiel-Ross, O., Sepúlveda, R., Castillo, O., Melin, P.: Ant colony test center for planning autonomous mobile robot navigation. Comput. Appl. Eng. Educ. **21**(2), 214–229 (2013)
11. Horn, B.K.P., Schunck, B.G.: Determining optical flow. Artif. Intell. **17**, 185–203 (1981)

12. Martens, J.-B.: The Hermite transform-theory. IEEE Trans. Acoust. Speech Signal Process. **38**(9), 1595–1606 (1990)
13. Young, R.A.: The Gaussian derivative theory of spatial vision: analysis of cortical cell receptive field line-weighting profiles. Technical Report GMR-4920, General Motors Research Laboratories, Detroit, Mich, USA (1985)

Mobile Robotic Navigation System With Improved Autonomy Under Diverse Scenarios

Elizabeth López-Lozada$^{(\boxtimes)}$ ⓘ, Elsa Rubio-Espino ⓘ,
Juan-Humberto Sossa-Azuela ⓘ, and Víctor H. Ponce-Ponce ⓘ

Instituto Politécnico Nacional, Centro de Investigación en Computación,
Av. Juan de Dios Bátiz Esq. Miguel Othón de Mendizábal S/N,
Nueva Industrial Vallejo, 07738 Gustavo A. Madero, Mexico City, Mexico
elizabeth.l.lozada@gmail.com, {erubio,hsossa,vponce}@cic.ipn.mx

Abstract. Mobile robots integrate a combination of physical robotic elements for locomotion and artificial intelligence algorithms to move and explore the environment. They have the ability to react and make decisions based on the perception they receive from the environment to fulfill the assigned navigation tasks. A crucial issue in mobile robots is to address the energy consumption in the robot design strategy for prolonged autonomous operation. Therefore, the battery charge level is an input variable that is commonly monitored and evaluated at all times, in this type of robots, in order to influence the decision-making with the least user intervention, during the navigation phase. Hence, the robot is capable to complete its tasks successfully. To achieve this, a navigation approach based on a fuzzy Q-Learning architecture for decision-making in combination with a module of artificial potential fields for path planning is introduced. The exhibited behavior of a six-legged robot obtained under this approach, demonstrates the robot's ability of moving from a starting point to a destination point, considering the need to go to the charging station or to remain static, if necessary.

Keywords: Mobile robot · Navigation · Path planning · Fuzzy Q-learning · Reinforcement learning · Fuzzy inference system · Artificial potential fields

1 Introduction

Navigation tasks performed by mobile robots involve any action that allows the robot to move from its current position to a destination point [5]. To accomplish these tasks, the navigation approach is commonly based on the use of artificial intelligence (AI) for obstacle avoidance maneuvers. Therefore, it improves the path planning and reduces power consumption, with the resulting increase in the robot's autonomy level. Autonomous mobile robots are an exciting technology getting a great deal of attention due to their adoption in many areas such as

© Springer Nature Switzerland AG 2020
L. Martínez-Villaseñor et al. (Eds.): MICAI 2020, LNAI 12469, pp. 472–485, 2020.
https://doi.org/10.1007/978-3-030-60887-3_40

space exploration, search and rescue tasks, inspection and maintenance operations, in agricultural, domestic, security, and defense tasks, among many others [2]. To deal with the complex decision making problems faced during navigation obstacle avoidance, a variety of AI approaches are employed, such as fuzzy inference systems (FIS) [8,12,19], potential fields methods [4,10,18], neural networks [6,14], and reinforcement learning [1,3,11,17].

For navigation tasks, different methods have been applied, where path planning and obstacle avoidance are sub-tasks that are solved during navigation, under the assumption that the environment corresponds to static scenarios that have obstacles and destinations that do not change over time. At one hand, from a reactive point of view, at [8] they propose to use a rule-based method to detect obstacles with ultrasonic sensors to control the movements of a differential robot while moving in an environment with a target and static obstacles. While in [10], a dynamic path planning scheme is introduced where a modified version of the artificial potential field method is used, demonstrating its performance on a group of robots, in this case, the scenario implies three different targets; one for each robot. A function is added to consider the other robots in the group as obstacles. On the other hand, the authors on [3] propose to use the Fuzzy Q-Learning method to make a six-legged robot learn to move in an environment with obstacles, where they implement a function so that the learning rate and the discount factor change during each time step. Their proposed system has 27 rules and allows the robot to learn to move to the left or right or to move forward. In all these cases, authors only consider the information coming from the distance sensors. However, they do not consider the electrical discharge of the battery that occurs during robot operation. In [16], they determine that the robot makes a path planning where there is a starting point, five nodes that the robot has to visit, and last node corresponds to a service station where the robot can replenish its battery. In the same direction, in [13] they propose an algorithm for an electric vehicle to generate an optimal strategy to go to a charging station, using a reinforced deep learning algorithm, considering the traffic conditions, the cost of the charge, and the time spent at the charging station. Hence, the algorithm may decide to decrease the number of visits to the charging station.

In this work, a system that allows a mobile robot to have the ability to make decisions autonomously is proposed, based on the continuous measurement of its battery charge level, to decide which actions to execute to reach its destination or a battery charging station. The main parts of the system consist of a module for decision making, utilizing a Fuzzy Q-Learning architecture and a module for path planning, employing the artificial potential field method. These modules are implemented in a six-legged robot increasing its stability and functional autonomy margin as it is capable of going to the charging station to refuel energy if necessary or waiting in a certain place for a new trajectory order.

The paper is organized as follows. Section 2 describes the basis of fuzzy Q-Learning and artificial potential field methods, Sect. 3 describes the hardware used in this work, Sect. 4 describes the architecture proposed for the navigation system, Sect. 5 shows simulations conducted and the experimental results obtained during the development of this work. Finally, in Sect. 7, conclusions are presented.

2   Background

This section describes the methods used for mobile robot navigation. First, Fuzzy Q-Learning method used for decision-making is exposed, then the path planning method is described.

2.1   Fuzzy Q-Learning

Fuzzy Q-Learning (FQL) method can be view as an extension of Fuzzy Inference Systems (FIS), where fuzzy rules define the state of a learning agent. When the rule entries are converted into fuzzy values, the system enters in a rule which is the agent current state; each rule has a numerical value associated, called the strength of the rule, represented by α_i, where i is the number of the rule, this value helps to define the degree that the agent is in a state. This allows the agent to choose an action from the set of actions A, the jth possible action in rule i is called $a[i, j]$ and its corresponding q-value $q[i, j]$. With this, the FIS is constructed in the following way:

If x **is** S_i **then** $a[i, 1]$ with $q[i, 1]$ **or** ... **or** $a[i, J]$ with $q[i, J]$

At the end, the learning agent must find the best solution for each rule, i.e., the action with the best q-value. This value is on a table that contains $i \times j$ q-values, the dimensions of the table correspond to the number of rules per number of actions.

Actions are selected with an exploration-exploration policy, which is based on the quality of a state-action pair, corresponding to a numerical q-value. For this work the exploration-exploitation probability is given by $\varepsilon = \frac{10}{10+T}$, where T corresponds to the step number. To calculate the inferred action and the q-value we use $a(x)$ and $Q(x, a)$ functions in (1), where x is the state or rule, a is the inferred action, i is the current state of agent, i^o is the inferred action index, α_i is the strength of the rule and N is a positive number, $N \in \mathbb{N}^+$, which is the total number of the rules.

$$a(x) = \frac{\sum_{i=1}^{N} \alpha_i \times a(i, i^o)}{\sum_{i=1}^{N} \alpha_i(x)}, \quad Q(x, a) = \frac{\sum_{i=1}^{N} \alpha_i \times q(i, i^o)}{\sum_{i=1}^{N} \alpha_i(x)} \tag{1}$$

The state-action value is obtained with function $V(x, a)$ in (2), where i^* corresponds to optimal action index, i.e., the action index with the highest q-value and x is the current state.

$$V(x, a) = \frac{\sum_{i=1}^{N} \alpha_i(x) \times q(i, i^*)}{\sum_{i=1}^{N} \alpha_i(x)} \tag{2}$$

The derivative of the Q-function is defined as in Eq. (3), where r corresponds to the reward and γ is the discount factor:

$$\Delta Q = r + \gamma V(x, a) - Q(x, a) \tag{3}$$

An eligibility value is also obtained, which is used during q-value updating, with Eq. (4) where i is the current state, j is the selected action, i^o is the inferred action, γ is the discount factor $0 \le \gamma \le 1$, λ is the decay parameter whose value is between 0 and 1, i is the number of the rule, and j is the selected action.

$$e[i, j] = \begin{cases} \lambda \gamma e[i, j] + \frac{\alpha_i(x)}{\sum_{i=1}^{N} \alpha_i(x)} & \text{if } j = i^o \\ \lambda \gamma e[i, j] & \text{other case} \end{cases} \tag{4}$$

Finally, Eq. (5) updates q-value, where ϵ is a small number, $\epsilon \in (0, 1]$, which affect the learning rate and $e[i, j]$ is the eligibility value.

$$\Delta q[i, i] = \epsilon \times \Delta Q \times e[i, j] \tag{5}$$

2.2   The Artificial Potential Field Method

Khatib [7] introduced the artificial potential field method for robots in 1986. This method is based on the use of attractive and repulsive forces to reach a goal and avoid obstacles. These forces can be calculated using the functions in (6) and (7) for attractive and repulsive forces respectively.

$$F_{attr}(q, g) = -\xi p(q, g) \tag{6}$$

where ξ is the attractive factor, $p(q, g)$ is the euclidean distance between robot and goal, q is robot position and g is goal position.

$$F_{rep}(q) = \begin{cases} \eta(\frac{1}{p(q_i, q_{o_j})} - \frac{1}{p_{o_j}}) \frac{p^2(q_i, q_{o_j})}{p(q_i, q_{o_j})}, & \text{if} \quad d < p_{o_j} \\ \eta(-\frac{1}{p_{o_j}}) & \text{if} \quad d = p_{o_j} \\ 0, & \text{if} \quad \text{other case} \end{cases} \tag{7}$$

where η is the repulsive factor, $p(q_i, q_{o_j})$ is the distance between robot and the obstacle j, p_{o_j} is obstacle radius threshold and d is the distance between robot and obstacle. Finally, the resultant force is the sum of attractive and repulsive forces.

3   Hardware

The King Spider robot platform from ROBOTIS [9,15], was chosen as the robot platform for this work, which consists of 18 servo Dynamixel, AX-12A, which are connected in a network to establish communication with them through a serial port and the processing unit.

A Raspberry Pi board is used as an embedded computer with the Rasbian Buster operating system. For obstacle detection, the ultrasonic sensor HC-SR04 is used while for battery level measurement the ADAFRUIT energy monitor INA260 is used. The robot is equipped with an 11.1 V battery, with battery capacity of 4250 mAh.

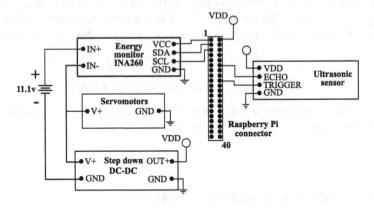

Fig. 1. Electrical connections

4   Navigation Proposed

In this section, the navigation methods proposed for the King Spider robot are explained. First, FQL architecture developed for decision-making, and then the path planning method is described. Figure 2 shows a flow diagram with the proposed method using an action selection with FQL to reach the robot destination.

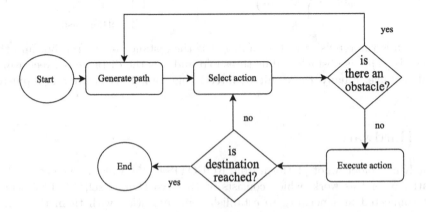

Fig. 2. Navigation flow diagram

4.1   Decision-Making Module

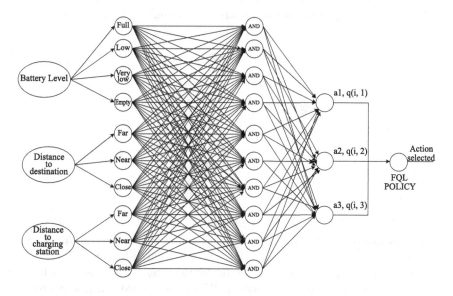

Fig. 3. FQL architecture for decision making

Figure 3, shows the proposed architecture for the decision-making module, using FQL. Crisp inputs correspond to battery level, distance to the destination, and distance to the charging station. For the first input, four fuzzy sets are defined labeled as full, low, very low, and empty battery level; the second and third inputs are applied to three the fuzzy sets labeled as far, near, and very close. There are a total of 36 fuzzy rules defined and three possible actions that are associated with a numerical q-value, the system selects an action according to the exploring-exploitation policy described in Sect. 2.1.

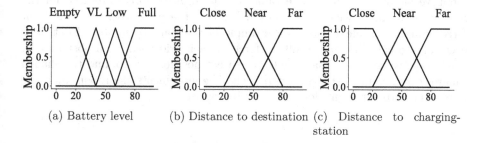

(a) Battery level   (b) Distance to destination (c) Distance to charging-
                                                        station

Fig. 4. Membership functions

Figure 4 shows the membership functions used for each entry. The Fig. 4a on the left corresponds to the battery level membership function, Fig. 4b corresponds to the distance to the destination, and Fig. 4c corresponds to the distance membership function to the battery charging station.

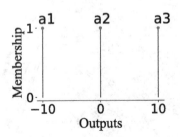

Fig. 5. Output functions

The possible outputs are defined by singleton type membership functions in Fig. 5, a's correspond to the actions that can be executed by robot where *"go to the destination"* is a1 with index 0, *"go to the charging station"* is a2 with index 1 and *"remain static"* is a3 with index 2. While the reward function is given by the expression (8), the discount factor and the learning rate are equal to 0.5 and 0.01 respectively.

$$r(t) = \begin{cases} +10 \text{ if robot is close to destination} \\ +5 \text{ if robot is close to the charging station} \\ -20 \text{ if robot is far of destinations and battery level is very low} \\ 1 \text{ other case} \end{cases} \tag{8}$$

4.2 Path Planning Module

Path planning module uses the artificial potential fields method using the equations presented in Sect. 2.2 where the attractive factor is equal to 2.3, while the repulsive factor set in 61.5. The values of the resulting forces are used for the path generation, in which the robot will follow the path where the resultant force is higher. Table 1 shows an example of the route generated. First, it is necessary to calculate the resultant force in each grid space, the obstacles, in black, take values lower than −100, the target position is zero, and the shaded squares correspond to the generated path. The movements the robot can make were limited to left, right, forwards and backwards.

Table 1. Example of path with 20 obstacles using artificial potential fields

-21.7	-19.3	-16.9	-14.6	-12.2	-9.9	-130.6	-5.4	-126.4	-125.4
-21.6	-19.2	-16.8	-14.4	-12.0	-9.6	-7.2	-4.8	-2.4	0
-21.7	-19.3	-16.9	-14.6	-12.2	-132.9	-7.6	-128.4	-3.4	-2.4
-22.1	-19.8	-17.5	-15.2	-135.9	-10.7	-8.7	-6.8	-128.4	-4.8
-145.8	-20.5	-18.3	-16.1	-13.9	-21.0	-133.2	-8.7	-7.6	-7.2
-23.6	-144.5	-19.3	-17.3	-15.4	-136.6	-135.0	-10.7	-132.9	-9.6
-24.7	-22.6	-143.6	-141.7	-16.9	-15.4	-13.9	-12.9	-12.2	-135.0
-25.9	-147.0	-22.1	-20.4	-18.7	-17.3	-16.1	-138.2	-14.6	-14.14
-27.4	-25.5	-23.8	-145.1	-20.6	-19.3	-141.3	-17.5	-16.9	-16.8
-28.9	-27.2	-25.5	-24.0	-22.6	-21.5	-20.5	-19.8	-19.3	-19.2

5   Simulations and Experimental Results

This section shows the obtained results from the navigation tests, conducted with the King Spider robot. The following figures depict system behavior during simulations and tests. In every case, the scenarios have a grid of dimensions 10 × 10, and the learning agent, which represents the robot, always starts in the left below corner at the (0, 0) coordinates. The learning agent is free to move forwards, downwards, to the right or to the left, depending on the resultant forces. Based on the navigation flow diagram, see Fig. 2, a path generation example is shown next. The trajectory execution starts with a two path generation as is depicted in Fig. 6, where Fig. 6a is the path to the destination and Fig. 6b is the path to the charging battery station. In both figures, the rectangles represent the obstacles, the star is the destination point, the pentagon is the charging station and the line, connected with dots, is the entire generated path by the potential field module.

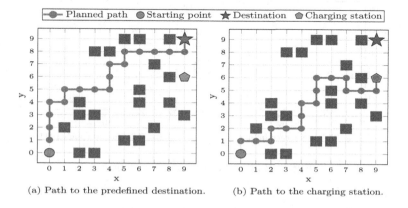

(a) Path to the predefined destination.   (b) Path to the charging station.

Fig. 6. Generated paths to destinations

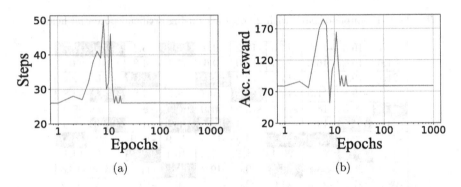

Fig. 7. System behavior during training

To ensure that the agent learns to select the right actions that lead to the completion of its task, the system was trained until the learning agent reaches its destination during 1000 epochs. The system learning behavior during training is shown in Fig. 7a with a graph of the number of steps versus the number of epochs is shown. The y axis means the number of the steps the learning agent requires to reach the destination point during every epoch, represented on the x axis, and in Fig. 7b shows the accumulated reward behavior in each epoch. With these two graphs, it can be observed that during the early training epochs, the behavior is unstable, i.e., the worst case took 50 steps to reach the destination point. Furthermore, in about 20 epochs, the system started to reach stability, as the necessary number of steps as well as the accumulated reward to get to the destination point remained constant. At the end of the training, it took 26 steps for the task completion. That is to say, the system learned to select adequate actions that lead to the full task completion in 26 time steps.

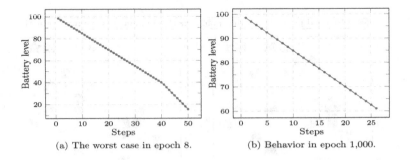

Fig. 8. Battery behavior during training

Additionally, in Fig. 8 is possible to see the resulting battery behavior during training, in Fig. 8a the worst-case battery charge duration is depicted with a remaining charge level less than 20% of the total charge at the end of the simulation. In Fig. 8b, the battery remaining charge level finished within 60%. This last case corresponds when the agent finishes its task in 26 steps. On the other side, Fig. 9 shows the actions selected and the states where the system fell during the training in the last training epoch.

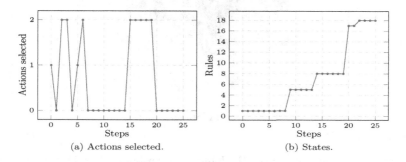

(a) Actions selected.   (b) States.

Fig. 9. Actions and states behavior during training

At the end of the simulations, learning agent took a path depending on the actions selected by the decision making module. Figure 10 shows two paths selected, Fig. 10a during the worst-case and Fig. 10b during the last training epoch. It can be seen than in Fig. 10a, the agent was undecided to go the charging station or to the destination point, and for several steps it exchanged between the possible destinations without really advancing to any of them until, at the end, it decided to proceed to the destination point. In contrast, figure Fig. 10b shows that the agent chose a direct route to the destination and it did not have such an intermittent selection between the possible destinations.

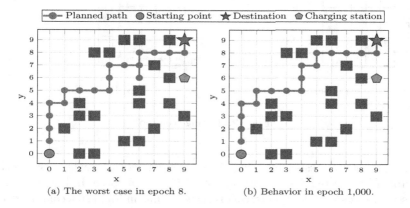

(a) The worst case in epoch 8.   (b) Behavior in epoch 1,000.

Fig. 10. Path followed during training

Fig. 11. King Spider robot

For the experimental testing, the King Spider robot, in Fig. 11, was employed
with the main electronics components depicted in Fig. 1, where the energy monitor INA260 is the essential component to determine the actions selected by the
system, due that it provides the system with the battery charge level measurements. For the implementation, the robot was provided with two files containing
the destination and charging station positions, two scripts for sensors measurements and another file with the Q-table obtained during the training phase.
Sensor scripts execution was implemented on the program used for simulations.
Every file was embedded and executed on the Raspberry Pi board.

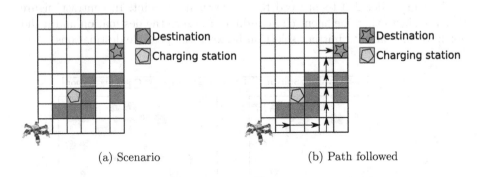

(a) Scenario   (b) Path followed

Fig. 12. Testing scenario

The robot was tested by using three different scenarios where the working
environment selected for the testings is a 2.1 × 2.8 m space. In Fig. 12a, one of
the scenarios used for system testing is shown, while in Fig. 12b a computed path
followed by the robot is shown. During testing, an erratic behavior was observed
when the battery charge was below 50%, since in some cases the servomotors were
not properly energized, consequently torque was insufficient to move the robot
correctly. However, when the system started at full battery charge, the robot

managed to accomplish its full trajectory to the destination point, without any issues. Finally, Fig. 13 shows a set of scenes with with the displacement of the robot on the scenario of the figure Fig. 12.

Fig. 13. Robot behavior in experimental testing scenario

6 Discussions

This work presented a navigation system for mobile robots based on an FQL architecture, that allows a robot to take decisions autonomously based on its battery charge level while it moves from a starting point to a destination point. By using FQL, the system learns through trial-error with a reinforcement learning paradigm, where the reward function definition took an important role in system learning process. There were tested different functions until function (8) was chosen. Among all the advantages that were found when selecting this architecture, the possibility that the expert can define the number of states that the system can fall in is distinguished as well as the number of rules by which the FIS is composed. If a classical Q-learning is used, the number of states could be extended to the interval of measurements of the battery voltage, i.e., 100 states for a integer percentage voltage level perception, or a even more states, if a millivolt scale is afforded by the charge measurement sensor device. While among the disadvantages, the system could not necessarily choose the shortest path at all times. This behaviour was observed during the simulations as in some steps the learning agent selected to remain in a stand-by mode while determining which destination to go. However, with the acquired learning, it finally managed to select the right actions that allowed it to complete its task.

7 Conclusions

The navigation approach proposed in this work, based on a fuzzy Q-Learning architecture in combination with artificial potential fields allows a robot to learn and execute actions autonomously while moving to its destination point, according to the monitored battery level charge. The obtained results are satisfactory under the given tests scenarios. Furthermore, this system can be expanded to accomplish a wide number of mobile robotic complex tasks, through learning according to the interactions with new environments and reward functions. During the conducted experiments with the robot, some technical issues were faced, concerning mainly with the materials used to fabricate the limbs as well as with the floor texture, provoking that the robot limbs did not have a correct grip with the ground and sometimes sliding. Consequently, this generated some movement errors. Besides, for these experiments the movements of the robot were limited to moving forward, backward, left, or right. In the future, it would be sought to have a control with more degrees of freedom for movement, in addition to operating on less structured terrains. Among other improvements that must be considered in the hardware design is the use of a more number of sensors which can be extended to the robot legs to pick information that allows avoiding some undesired situations, i.e., when the legs touch a wall; a situation that the current system proposal does not detect.

Acknowledgments. We appreciate the support to develop this project by the Instituto Politécnico Nacional (IPN) and Secretaría de Investigación y Posgrado (SIP-IPN) under the projects SIP20180943, SIP20190007, SIP20195835, SIP20200630, SIP20201397 and SIP20200569, also to Consejo Nacional de Ciencia y Tecnología (CONACYT-México).

References

1. Arvind, C.S., Senthilnath, J.: Autonomous vehicle for obstacle detection and avoidance using reinforcement learning. In: Das, K.N., Bansal, J.C., Deep, K., Nagar, A.K., Pathipooranam, P., Naidu, R.C. (eds.) Soft Computing for Problem Solving. AISC, vol. 1048, pp. 55–66. Springer, Singapore (2020). https://doi.org/10.1007/978-981-15-0035-0_5

2. Zhao, Y.-X., Hao, R.-X.: Navigation and navigation algorithms. In: Yang, X.-S., Zhao, Y.-X. (eds.) Nature-Inspired Computation in Navigation and Routing Problems. STNC, pp. 19–56. Springer, Singapore (2020). https://doi.org/10.1007/978-981-15-1842-3_2

3. Hong, J., Tang, K., Chen, C.: Obstacle avoidance of hexapod robots using fuzzy q-learning. In: 2017 IEEE Symposium Series on Computational Intelligence (SSCI), pp. 1–6 (2017). https://doi.org/10.1109/SSCI.2017.8280907

4. Rostami, S.M.H., Sangaiah, A.K., Wang, J., Liu, X.: Obstacle avoidance of mobile robots using modified artificial potential field algorithm. EURASIP J. Wirel. Commun. Netw. **2019**(1), 1–19 (2019). https://doi.org/10.1186/s13638-019-1396-2

5. Huskić, G., Buck, S., Zell, A.: GeRoNa: generic robot navigation. J. Intell. Robot. Syst. **95**(2), 419–442 (2018). https://doi.org/10.1007/s10846-018-0951-0

6. Jalali, S.M.J., Hedjam, R., Khosravi, A., Heidari, A.A., Mirjalili, S., Nahavandi, S.: Autonomous robot navigation using moth-flame-based neuroevolution. In: Mirjalili, S., Faris, H., Aljarah, I. (eds.) Evolutionary Machine Learning Techniques. AIS, pp. 67–83. Springer, Singapore (2020). https://doi.org/10.1007/978-981-32-9990-0_5

7. Khatib, O.: Real-time obstacle avoidance for manipulators and mobile robots. In: Proceedings. 1985 IEEE International Conference on Robotics and Automation, vol. 2, pp. 500–505 (1985). https://doi.org/10.1109/ROBOT.1985.1087247

8. Kumar, A., Guha, A., Pandey, D.A.: Dynamic motion planning for autonomous wheeled robot using minimum fuzzy rule based controller with avoidance of moving obstacles. Int. J. Innov. Technol. Explor. Eng. (IJITEE) 9(1), 4192–4198 (2019). https://doi.org/10.35940/ijitee.A6114.119119

9. López-Lozada, E.: Navegación y evasión de obstáculos con un robot móvil. Centro de Investigación en Computación IPN (2020)

10. Matoui, F., Boussaid, B., Metoui, B., Frej, G., Abdelkrim, M.: Path planning of a group of robots with potential field approach: decentralized architecture. IFAC-PapersOnLine 50(1), 11473–11478 (2017). https://doi.org/10.1016/j.ifacol.2017. 08.1822. https://www.sciencedirect.com/science/article/pii/S2405896317324448, 20th IFAC World Congress

11. Pambudi, A.D., Agustinah, T., Effendi, R.: Reinforcement point and fuzzy input design of fuzzy q-learning for mobile robot navigation system. In: 2019 International Conference of Artificial Intelligence and Information Technology (ICAIIT), pp. 186–191 (2019). https://doi.org/10.1109/ICAIIT.2019.8834601

12. Park, J.W., Kwak, H.J., Kang, Y.C., Kim, D.W.: Advanced fuzzy potential field method for mobile robot obstacle avoidance. Comput. Intell. Neurosci. 2016, 1–13 (2016). https://doi.org/10.1155/2016/6047906

13. Qian, T., Shao, C., Wang, X., Shahidehpour, M.: Deep reinforcement learning for EV charging navigation by coordinating smart grid and intelligent transportation system. IEEE Trans. Smart Grid 11(2), 1714–1723 (2020). https://doi.org/ 10.1109/TSG.2019.2942593

14. Reis, D.H.D., Welfer, D., Cuadros, M.A.D.S.L., Gamarra, D.F.T.: Mobile robot navigation using an object recognition software with RGBD images and the yolo algorithm. Appl. Artif. Intell. 33(14), 1290–1305 (2019)

15. ROBOTIS: ROBOTIS e-manual (2019). https://emanual.robotis.com/docs/en/ edu/bioloid/premium/

16. Shidujaman, M., Samani, H., Raayatpanah, M.A., Mi, H., Premachandra, C.: Towards deploying the wireless charging robots in smart environments. In: 2018 International Conference on System Science and Engineering (ICSSE), pp. 1–6 (2018). https://doi.org/10.1109/ICSSE.2018.8520063

17. Shuhuan, W., Xueheng, H., Zhen, L., Keung, L.H., Fuchun, S., Bin, F.: NAO robot obstacle avoidance based on fuzzy q-learning. Ind. Robot: Int. J. Robot. Res. Appl. 2019 (2019). https://doi.org/10.1108/IR-01-2019-0002

18. Singh, N.H., Devi, S.S., Thongam, K.: Modified artificial potential field approaches for mobile robot navigation in unknown environments. In: Das, K.N., Bansal, J.C., Deep, K., Nagar, A.K., Pathipooranam, P., Naidu, R.C. (eds.) Soft Computing for Problem Solving. AISC, vol. 1048, pp. 319–328. Springer, Singapore (2020). https://doi.org/10.1007/978-981-15-0035-0_25

19. Subbash, P., Chong, K.T.: Adaptive network fuzzy inference system based navigation controller for mobile robot. Front. Inf. Technol. Electron. Eng. 20(2), 141–151 (2019). https://doi.org/10.1631/FITEE.1700206

Author Index

Printed in the United States
By Bookmasters